Stray
Voltage

An Association of the U.S. Army Book

Stray Voltage

War in the Information Age

Wayne Michael Hall

NAVAL INSTITUTE PRESS
ANNAPOLIS, MARYLAND

Naval Institute Press
291 Wood Road
Annapolis, MD 21402

Library of Congress Cataloging-in-Publication Data
Hall, Wayne M., 1946-
Stray voltage : war in the information age / Wayne M. Hall.
p. cm.
Includes bibliographical references and index.
ISBN 1-59114-350-0 (alk. paper)
1. United States—Defenses. 2. Information warfare.
3. Asymmetric warfare. 4. Military art and science—United States—History—
21st century. I. Title.
UA23 .H34 2003
355.02—dc21

                                        2002012419

Printed in the United States of America on acid-free paper ∞
10 09 08 07 06 05 04 03   9 8 7 6 5 4 3 2
First printing

Contents

Preface

Imagine the year is 2015 and the war against terrorism is raging around the globe. While conventional military operations have gone exceptionally well in the fight, asymmetric warfare, which the terrorists have chosen as their "strategy du jour," has not gone as well for U.S. security forces and U.S. allies. The terrorists are using an asymmetric strategy of knowledge war (the competition for data, information, and knowledge with which to make better decisions faster than the adversary) to attack and defeat the United States. The U.S. national command authority is recommitting its military forces overseas to destroy terrorist enclaves in the jungles of Southeast Asia. Our adversaries have unleashed a wide range of synchronized asymmetric strategies and tools, from low-tech car bombs to sophisticated cyber, chemical, and biological attacks. Interestingly, the terrorists orchestrate their activities in a synchronized way, subjecting the citizens of the United States to waves of violence and terror across the country. They coordinate and synchronize events into waves of terror by using modern telephony, the Internet, computers, and off-the-shelf security software. Their macro goal is to create second- and third-order effects that cross the large economic, social, informational, political, financial, and military systems that make up our country's system of systems. Hackers and terrorist bombers attack U.S. civil aviation and challenge the Federal Aviation

Administration's control of U.S. airspace. Commercial air travel stops, un-leashing immense second- and third-order effects on the economy and on the social and psychological lives of citizens around the globe. Hackers and kinetically oriented terrorists attack Wall Street communications, servers, databases, and the knowledge workers who run those systems with dia-bolical cunning and ferocity. The stock market shuts down and the result-ing financial insecurity contributes to the disturbed psyche of the popula-tion. Domestic and foreign hackers launch cyberattacks against key infrastructure sites including electric stations in large population centers that house troop staging areas and sea and aerial ports of embarkation. The lights are out in critical locations and some activities are in disarray. Televi-sion and web sites are full of manipulative images, subliminal messages, cyber deceptions all designed to sow the seeds of confusion, despair, and fear. People are afraid and confused. A loosely knit group of foreign and do-mestic terrorists use Internet and wireless communications to detonate multiple explosives in key locations in population centers, compounding the confusion and chaos. The coordinated explosions create casualties and damage. Citizens begin to doubt the ability of law enforcement agencies to cope with the violence. The terrorists have unleashed biological weapons and nerve gas in large metropolitan areas. Their goal is to create terror in the minds of citizens, overwhelm the medical infrastructure, and, most im-portant, cause people to question the capability of the country's govern-ment to gain control of a chaotic situation.

In the military realm the terrorists use incapacitants to disrupt the flow of troops to staging bases and target areas near terrorist enclaves. They use incapacitants to affect knowledge workers operating the software and hardware controlling the movement of people and supplies. Hackers dis-rupt the computerized management of personnel and supplies flowing from bases, ports, and forts. Terrorists harass, kidnap, and frighten families of soldiers, sailors, airmen, and marines either deploying or preparing to deploy from their home stations. To disrupt the flow of military power they use suicide bombings, assassinations, chemical weapons, and cyber-attacks. They murder Coast Guard officers in charge of seaports of em-barkation and the key knowledge workers who control automated port activities. They blow up railroad tracks and bridges, which slows the flow of equipment to ports of embarkation. Compounding the chaos, terrorists conduct synchronized chemical attacks on military aerial ports of em-barkation, causing the greatest chaos and impediments to a previously un-challenged arm of the U.S. military—the synchronized, rapid deployment of military power to overseas locations. Exacerbating the many deployment problems the cyberterrorists disrupt communications and send false e-mail messages into the Internet to influence the morale of deploying troops and the decision-making abilities of their officers.

Once on the ground in their specific target areas, U.S. soldiers find themselves purposefully drawn into urban areas, which negates their typical advantages in maneuver, mobility, fire control, and equipment used for C4ISR (command, control, communications, computers, intelligence, surveillance, and reconnaissance). Commanders on the ground face asymmetric foes wise and sophisticated enough to distort data and information causing them to question its veracity. Hesitancy creeps in to command-and-control processes; the ability to make decisions becomes increasingly ineffective and slow. Terrorists operating in urban areas capitalize on the growing hesitancy of U.S. commanders to make decisions. Risk becomes accentuated by the stifling effects of complex constraints and rules of engagement (ROE). As a tactical strategy the country's asymmetric foes create large numbers of casualties and escalate collateral damage. The fighting takes a strange turn of events when loosely organized groups of terrorists, conventional soldiers, criminal elements, and drug gangs cooperate, coordinate, and synchronize their efforts through the Internet.

In urban target areas the asymmetric foes scatter in loosely connected, nonlinear organizations in buildings, under the ground, and in cultural icons such as churches. They make themselves difficult to locate, and their intent is difficult to discern. Hierarchically organized forces, such as those belonging to the United States, have trouble coping with this foe's alacrity in decision making, resiliency, security, self-synchronization, co-evolution, and ability to adapt. Asymmetric foes attack friendly force command-and-control sites and precision weaponry enhancers such as satellite ground stations at key locations around the world using kinetic and information attacks. Once again they coordinate their activities over the Internet using commercial satellite communications. U.S. opponents purposefully create widely scattered spasms of violence, ensuring that the world's media sends snapshots and video clips to populations in the United States and other countries to manipulate public opinion and continue the fear. Intensely emotional images of extreme violence, heavy casualties, panic in the U.S. populace, widespread collateral damage, all synchronized with waves of violence and biological and chemical terror in the homeland, quickly sway public opinion against the deployment and involvement of U.S. troops. When faced with violence and fear at home, the populace has trouble understanding the relationship between the need to attack terrorist enclaves in Southeast Asia and the political, social, and economic conditions in the U.S. proper. Immense pressure builds on politicians to return the economy and the society to its normal status. The immediacy phenomenon, a creature of the Internet age, surfaces in the U.S. press and on CNN; asymmetric foes quickly capitalize on it. People become distraught over the stock market decline and job losses, their inability to travel, the unavailability of necessities (thanks to assaults on the infrastructure), and a general lack of law and order in some major population areas. Because of the immense second- and third-order effects cascading from the U.S. homeland to other countries, coalition strength and resolve become increasingly fragile. Moreover, deployed soldiers in Kosovo feel paralyzed by the complicated ROE, by constraints imposed by member governments of the crumbling coalition, and by a confounding inability of commanders to make fast decisions.[1]

Does this scene sound far-fetched? Possibly, but it is entirely plausible. It is only a matter of time before our country's foes become smart enough to create the actions leading to the effects that are described above. Our foes of the future will, by necessity, engage in asymmetric warfare—the strategy, tactics, and tools a weaker adversary uses to offset the superiority of a foe by attacking the stronger force's vulnerabilities, using both direct and indirect approaches to hamper vital functions or locations for the explicit purpose of seeking and exploiting advantages. We are in a "Hundred Years' War" against formidable, adaptive, and creative opponents. The struggle involves a zero-sum triumph of will—there will be no compromise from either side until one side wins and the other loses. Our previously superior military might will be only partially effective because our adversaries will have adopted asymmetric strategies, tactics, and tools against our conventional military forces. The new war will be a struggle devoid of sanctuaries—the homeland will be either as important as or more important than areas overseas. It will be a struggle in which social, political, military, economic, financial, and informational systems are woven together to the point than none can experience perturbation without affecting the others. It will be a struggle for control of the intangibles that constitute will: fear, morale, and surprise. This war will involve both kinetic elements—such as bullets, bombs, and attrition—and nonkinetic elements—such as perception management, deception, knowledge, decision making, and activities in cyberspace. The war will be a struggle for the survival of a way of life and conflicting cultural and religious perspectives.

Think for a moment about the essence of the challenges in national security that this scenario presents. In national security, a great transformation is occurring—a transformation from conflict involving narrow, military aspects of kinetic warfare to a broader, full-world system competition that is knowledge war. At least part of this transformation comes with the changing strategies and tactics of our adversaries. The country's future adversaries will not engage our military forces in direct, force-on-force conventional military confrontations. They would lose in such a contest, and they know it. Besides, there are other ways of competing with the United States under a condition of improved odds of success. Evidence exists in open source, unclassified literature that potential adversaries have started thinking about alternative ways of competing with the United States. Chinese military officers have written that knowledge war and information operations clearly provide a way that weaker forces can compete with stronger forces in the future.[2] That is, a way must be found to engage in knowledge war in which each side seeks valuable information and knowledge to enable its leaders to make faster and better decisions than the adversary. Out of superior decision making come numerous advantages such as intellectual, knowledge, technical, positional, and action advantages. Such advantages provide one side the potential to dominate the other in tempo, initiative, and momentum.

Knowledge war deals with how people use information and knowledge in an intellectual sense—it is about superior thinking and planning. To be certain, information technology is important, but it is only a tool to assist in the quest for what is really important—superior thinking and planning that lead to superior decision making. One of the principal tools for waging knowledge war is information operations (IO). IO consist of tools that are used to affect the information, knowledge, thinking, planning, and decision making of one's opponents. Information operations protect friendly decision-making activities and attack, slow down, or deceive the adversary's.

Information operations can be difficult to understand because they involve invisible, unquantifiable, squishy things like perceptions, decision making, and disturbances in the collective psyches of citizenry, along with the more noticeable assaults on computers and databases. Information operations use invisible, indirect approaches to assault sensitive, important things; with IO, it is difficult to pin blame. Information operations are tools, among many, of foes using asymmetric strategies. Information operations involve activities that enhance one's own decision making while negatively affecting a competitor's decisions. IO are the perfect tool with which to wage knowledge war. Is the country prepared? The Department of Defense, though making many meaningful and significant efforts to change, is not prepared to the extent needed to engage in a knowledge war against a smart, adaptive, asymmetric adversary highly skilled in information operations. Many people believe it will take a major radiation event, a chemical or biological event, a Task Force Smith type of event, or a "digital Pearl Harbor" to cause the country and the Department of Defense to initiate needed changes in how we train and educate our forces, how we develop leaders, how we plan, organize, and conduct research and development, and how we procure materiel. Others disagree, believing that our defense apparatus will change without the impetus of a major calamity. Their reasoning is that the United States has men and women who can lead people and organizations to change sufficiently to meet the challenges of knowledge war. The country cannot wait for a digital or electronic catastrophe to spur them into taking action. Quite simply, the outcomes are potentially too devastating to allow such a calamity to occur. None of us want another July 1950 in Korea, when our country committed a feeble, ill-prepared, ill-trained, and under-resourced fighting force to conduct an ill-defined mission. Asymmetric warfare threats are far more insidious and potentially more devastating to our national sense of security than the security problems we faced in 1950. Our national leaders need to be active, creative, and innovative to recognize a rapidly changing environment and to take steps to prepare for asymmetric assaults that imperil our futures and place our way of life at great risk.

In the fight our greatest assets lie in our ever-expanding minds and the technology that helps those minds think and plan. Any advantage that we gain with our intellects and technology, however, will be short-lived. Our terrorist adversaries think, learn, and use technology to help execute their nefarious deeds. Thus, we must be aggressive without arrogance. We cannot be complacent. We must always seek to learn, change, and figure out better ways to think and to leverage technology to help us triumph against the terrorist. We must be proactive instead of reactive—too much is at stake to allow our adversaries to make the first move or initiate the next strike. Anticipating and acting to deny terrorist activities before they occur in an open society, ever mindful of our cherished civil liberties, is going to be difficult. Yet with good thinking and effective use of technology we can "be there waiting when they come."

No doubt our future adversaries will choose knowledge war as an effective way to compete with us. They know how the U.S. defense establishment depends on valuable knowledge to enhance its decision making and its precision weaponry. They know the U.S. defense establishment primarily focuses on the past and continues to be tied to the basics of kinetic warfare. They know that the U.S. defense establishment is slow to change. They know that our knowledge workers (analysts) sometimes engage in mirror imaging that distorts reality by reflecting their backgrounds, intellects, and values on to the perceptions and activities they believe inherent to a foe, even when those perceptions and thoughts are dramatically different than theirs. They know that the world's politics and economies interweave and inextricably relate, which opens windows of opportunity for exploitation. They recognize the paradoxical position of the United States when it comes to combating terrorism yet maintaining the civil liberties and freedoms the country was founded upon. The terrorists will use this knowledge—and, even more sinister, their understanding of relationships —to their advantage. They will manipulate our country's openness, its legal codes, the peculiarities of its bureaucracies, and the rules of engagement that cascade from the philosophical and societal inclinations of the United States in a relentless pursuit of victory.

The country must do a better job in preparing for knowledge war. We must start now to prepare for the day when our foes are a formidable force across the full range of asymmetric warfare, and when they engage in sophisticated knowledge war in particular. Quite simply, training and education, organization, leadership, and materiel development are too complex and take too long to develop and perfect for us to wait until an opponent, one that uses mature concepts and sophisticated activities like knowledge war, appears.

This book provides thoughts and insights into improving and empowering people to enter the murky, competitive realm of knowledge war and win every time. And, while it may encourage further deep thinking on the defense capabilities that our country will need in the future, it most assuredly does not provide the only answer. Our country must accept the premises that a knowledge war will occur, that it will be dangerous, and that it will be difficult to prepare for (beyond the typical three-to-five-year timeframe). We must start now to develop sufficient capabilities to meet the "new" foe head-on and win in every engagement.

This book is about the future; it is not a research work. Instead, the thoughts it presents focus on theory, projection, and hypothetical situations that could very well have meaning for defending the United States in the twenty-first century. Some of the thoughts presented are merely conjecture since, quite simply, evidence, history, and facts are in short supply when dealing with many of the aspects of asymmetric warfare, knowledge war, and information operations. Moreover, it is not a discourse on the specifics of technology; technology is subordinated here to the workings of the human mind and is described simply as a tool to help people think and make decisions. Please reflect on the ideas presented here, think about the implications for future global conflicts, then consider for yourself the steps that must be taken to prepare ourselves for the engagements we will face in the coming years.

Acknowledgments

I began writing this book on one of my many trips across the Pacific Ocean during the time I was stationed in the Republic of Korea (from 1994 to 1998). I had many solitary hours to consider the complexities of life, the here-and-now implications of modern communications, information, and knowledge, the state of current and future conflict, and the steps that the security establishment must take to prepare for a dangerous future. But the crucible of Korea and the absolute importance of data, information, and knowledge focused my thinking into the concept of "knowledge war." During my first two years in Korea I led a brigade of soldiers that collected data and information and then turned that information into usable knowledge. I also was responsible for safeguarding their lives and my organization. During my last two years in Korea I was the J2 (intelligence officer) for all the U.S. forces in Korea. I gave a lot of thought to the North Koreans and how they might attack us if war broke out. Being responsible for the intelligence operations of the entire peninsula opened my intellectual vistas and caused me to think about various aspects of conflict: the element of surprise, wargaming, numerous intangibles, and the absolute importance of mental perceptions. I was surrounded by the best and the brightest people in the military during those four years, and my interactions with them produced many extraordinarily creative thoughts and actions. I am in-debted to them and our many intense and lengthy interactions, for their talent caused me to think about the many complex subjects this book com-prises.

Writing this book has been a labor of love. My first thanks must go to the countless soldiers, sailors, airmen, and marines with whom I had the privilege of serving as an active duty army officer. During my career I have been blessed in two ways. First, the people under my command were su-perb individuals. I learned a great deal from them and enjoyed watching the magic they created. Second, I had the great fortune to work for many fine senior officers who patiently mentored me over the years. I owe each of them for the intellectual, moral, and leadership foundations they at-tempted to instill in me.

My experiences in the Army's School of Advanced Military Studies (SAMS) were a defining year for me. I was surrounded by twenty-three great army officers and we all shared with and challenged each other on a daily basis. Our superb mentors provided constant challenges, directions, and questions to ponder. The stimulation and guidance I received in the SAMS program served me every day of my career from that point forward, and I used what I learned in SAMS every day, particularly when serving as a general officer, and even today as a private consultant.

I want to thank Lt. Col. Roger Cirillo, U.S. Army (Ret.). Roger encour-aged me to write this book when the task seemed impossibly beyond either my desires or capabilities. He believed that my ideas needed to be heard. He stuck with me during many emotional phone calls and always encouraged me to continue on. I am in his debt.

I thank my sister, Sherry, for her continuing and unwavering support during the writing of this book. Although not a military person, Sherry took the time to read an earlier version of the book and vigorously encour-aged me to proceed. She has a spirit that is always upbeat, and her outlook has been a true inspiration during my times of emotional and spiritual anx-iety.

While I owe much to the people I have worked for throughout my mil-itary career and after I retired, there are a few people I would like to men-tion specifically. My former brigade commander, great friend, and mentor, Lt. Gen. Johnny J. Johnson (Ret.), has always been a role model, father fig-ure, and teacher to me. I shall always be in his debt for encouraging me to think creatively and showing me how senior people lead. Brig. Gen. Frank Akers (Ret.) brought me to the 82nd Airborne Division, mentored me in the ways of the division, and encouraged me to think about the future and the present. Frank also was instrumental in bringing back to life the creativity that I had lost owing to a severe personal tragedy. After my retirement from the military Frank hired me to work for him and encouraged me to think broadly and creatively, and to share my thoughts with others. His patience, counsel, and trust brought me back to life, in an intellectual sense. My senior leaders during my time serving as a brigade commander in Korea—Lt. Gen. Rich Timmons (Ret.) and Gen. Gary Luck (Ret.)—established a great command climate that allowed me, indeed encouraged me, to experiment with different ways of accomplishing intelligence missions and to try some creative ways of solving problems, underwriting my failures when the creative ideas didn't always work or meet our expectations. After my retirement General Luck brought me into the future conflict consulting business. He worked hard to present my thoughts to many senior people charged with experimenting with transforming the military of the United States. I am in his debt for being such a good friend, a wise mentor, and an inspirational leader to me and countless others.

I also owe much to my great friends Mike and Beverly Tanksley. Along with being cherished friends since our Leavenworth days, Mike and Beverly have constantly and consistently encouraged me to pursue my thoughts and to articulate them in writing. They have stood by me in periods of happiness and sorrow. I don't know what I would have done without their friendship and support over the years.

I thank my wife, Sandy, for her encouragement, for her thoughts, for her care for me, and for the countless hours she spent editing my work. Sandy and I spent many hours talking about many of these complex subjects, and her penetrating intellect helped me think through and shed light upon the complexities that resulted. I could not have finished this project without her able, patient, and loving assistance.

Stray
Voltage

Since the dawn of time men have competed with
each other—with clubs, crossbows, or cannon, dollars,
ballots, and trading stamps. Much of mankind, of course,
abhors competition, and these remain the acted upon,
not the actors. Anyone who says there will be no
competition in the future simply does not understand
the nature of man . . . men must compete.

T. R. Fehrenbach, *This Kind of War*

The Setting 1

Events of 11 September 2001 tell us much about the nature of future conflict. That one terrible day has highlighted the belief that the United States
faces tough, ruthless, adaptive adversaries who hate America and all it
stands for. These adversaries will sacrifice anything to further their cause
and to win, including their lives. They will compete ferociously with the
United States for triumph of their way of life and to reduce the pervasiveness of U.S. prestige, influence, and presence. Nonetheless, methods for
future conflict in all realms of national security will be different than in
the past.

In the future global security environment the struggle for usable, valuable information and knowledge dwarfs other efforts to coerce and compel
opponents by engaging them using kinetic means (bullets, bombs, and missiles), thereby destroying their people, equipment, or cities in the traditional physical way. Instead, we are rapidly approaching the age in which
struggles for winning, for achieving advantages, and for imposing our will
on others directly center on the control and use of knowledge. Knowledge
has the potential to empower people to make good decisions. The trick is to
make better decisions faster than one's adversary is able to do so—and
knowledge gives people that ability, given they and their organizations can
make good, fast decisions themselves. Our nation's military has put forth its

vision for the future in "Joint Vision 2020." The visionary document describes the future environment, adversaries, role of technology, and how U.S. military power will retain its advantages. The document clearly makes the case for knowledge-based operations as being decisive in any future battle space.

Our country's future adversaries will have goals and objectives quite different from those of people living in a Western democracy. The list of reasons for these differences is long and includes distinctions among people such as culture and cultural icons, religion, societal beliefs, and thinking and planning. People living in a well-established democracy often have a difficult time understanding why their adversaries hate them or, even more important, the rationale for the hatred.

The future adversaries I speak of possess capabilities and a formidability that demands our begrudging respect. They have lots of money, and they arm themselves with the latest information technology and high-tech weaponry. Increasingly these threats gather information and knowledge from the Internet and the media and learn from the mistakes of others. After the events of 11 September it should be self-evident to all people that our asymmetric foes readily learn, calculate, and change. They aggregate and migrate from a few individuals and loose groups into learning, adaptive organizations. Their hatred, at once profound and resilient, revolves around ideological differences, jealousy, competitive advantage, and what they consider a long-term struggle for cultural and religious domination. Their hatred seethes, and is manifested in actions antithetical to the interests of the United States—then boils into acts of terror and extreme violence against innocent civilians.

Knowledge War

"Knowledge war" will be the preeminent form of future conflict in the twenty-first century. Knowledge war can be defined as an intense competition for valuable information and knowledge that both sides need for making better decisions faster than their adversary. The goal in this type of conflict is to seek, find, and sustain decision dominance, which leads to an overall advantage in decision making and results in a triumph of will by one side or the other. (For our purposes here "will" is defined as the resolution, sacrifice, and perseverance of individuals and groups of people to win in a competitive struggle.) Decisions influence the all-important will. Without question will, the "center of gravity" for the twenty-first century, and the concomitant struggle for triumph of will most likely will occur in the homeland of the United States. This struggle for triumph of will, though, rests not with U.S. military forces and traditional military targets. Instead, the struggle for triumph of will rests within the minds of the populace writ large that comprises the body politic of the United States.

While knowledge war applies to all levels of war, the thoughts presented here will be focused primarily on operational and strategic levels of conflict. Thus, while it may be an interesting topic, the tactical sensor-to-shooter part of an action cycle will not be assessed. Instead, the focal point will be on leader decision cycles that turn knowledge into actions that create effects (outcomes). Consequently, the strange, man-machine symbiotic acts of decision making in cyberspace will be considered later.

In the world described here two sets of propositions underpin most intellectual reasoning. A first proposition is that, because of its influence on decision making and results, knowledge also forms the intellectual foundation for strategies guiding the conduct of activities in competitive struggles. Knowledge-based strategies have at their core knowledge, which is "familiarity, awareness, or comprehension acquired by experience or study."[1] The core concept of a knowledge-based strategy primarily involves seeking, finding, creating, and using knowledge to make effective decisions— and, secondarily, enabling people to improve their thinking capabilities. Knowledge-based operations derive from knowledge-based strategies. (Knowledge-based operations range from nonkinetic actions and activities that are guided by knowledge of a foe's decision cycle and thinking process, to the knowledge that guides munitions unleashed to create precise kinetic effects or outcomes.) A second proposition pertaining here is that a trail of logic exists that can explain how knowledge originates or becomes discovered; that is, data becomes information through the manipulation of machines and knowledge workers turn information into knowledge through thought, experience, intuition, and creativity. Knowledge also leads to understanding, which occurs when decision makers combine several pieces of related knowledge into an intelligible collage. With understanding comes the potential to make effective decisions.

Victory in future conflicts will go to the side whose leaders make the best use of knowledge to make the most effective and, in some cases, quickest decisions. Decisions lead to actions, which in turn create effects (outcomes) that influence the abstract but critically important concept known as will. While technology can help establish conditions for winning a struggle, those who make the best use of their collective intellects to create the *most relevant* knowledge faster than their adversaries will triumph. The stakes are high in these cerebrally oriented contests, since surely outcomes will affect the way citizens of the United States live and experience life.

Information operations (IO) will be one of the principal tools of asymmetric adversaries who engage in a knowledge war with the United States. IO involve the tangible and intangible activities that affect an adversary's decision making, combined with information technologies that support decision-making processes. IO also affect the thinking and planning that support decisions throughout all levels of command. When one considers the reasons for this shift—from conventional, kinetically oriented conflict to knowledge war—the rationale always seems to center on the invisible hand of the information revolution. The forces of the information revolution are complex and affect thinking, planning, decision making, and materiel development. The changes demanded by the information revolution are significant and touch virtually every aspect of the defense establishment. Thus, in a collective security sense all individuals and organizations dealing with future conflict have to accept the mantra that change is on the horizon of the United States security apparatus, and anyone connected to it must constantly respond innovatively and creatively, regardless of either the difficulty or the complexity of the situation.

While it is true that the United States does not face a true conventional peer, it nevertheless does face increasingly formidable adversaries armed with extremely cunning intellects and the latest technology that money can buy. These adversaries use the speed and pervasiveness of the Internet, an uncanny sense about the "soft spots" of the United States, and an increased base of knowledge and intellectual power that levels the playing field. The United States security apparatus must prepare to meet and defeat this new threat in any area of competition, whether on physical terrain or in cyberspace, or in the minds of the U.S. populace, in the minds of its trading partners and allies, and, of course, in the minds of its adversaries.

The age in which we reside is much different and potentially much more dangerous than anything human beings have experienced in the past. This danger has many faces, including the face of a conventional conflict. Therefore I am not arguing that conventional conflict is anachronistic. But an asymmetric conflict also looms. America's armed forces must prepare for two types of conflict: physical, kinetic, conventional, force-on-force; and the shadowy, nuance-laden, sometimes digital and largely invisible asymmetric warfare.

These two types of conflict differ dramatically. The first, kinetically oriented warfare, involves dropping bombs and shooting bullets and artillery rounds. It requires traditional troop movements, maneuvers, attrition, command and control, and the occupation and control of physical terrain. Its principal focus lies in using kinetic energy to compel and defeat an adversary through death, destruction, and the psychological and physical effects arising from attrition. The mind-set of people waging kinetic war orients on killing people and breaking things, and thereby causes an enemy to submit once a particular threshold of pain arrives. The principal theories for waging this type of conflict can be found in the works of Carl von Clausewitz, the German philosopher of war. Kinetic warfare soldiers are humans moving at the speed of the fastest available purveyor of transportation.

In contrast, asymmetric warfare involves more emphasis on the non-kinetic aspect of a conflict. It may include the use of kinetic mechanisms to create the actions that lead to effects that influence behavior and, ultimately, influence will. But, instead of focusing on attrition as the principal means to compel, asymmetric warfare orients more toward an opponent's knowledge, decision-making processes, and perceptions. The principal theoretician for this type of war was Sun Tzu, the Chinese philosopher of war. For example, when waging a knowledge war, victory may come with only minimal casualties. This way of waging future conflict includes both visible and invisible realms. Its terrain lies in the minds of human beings, in the souls of terrorists and their human targets, inside fiber-optic cables, inside databases, in computer software code, and along radio and satellite frequencies. Asymmetric warfare soldiers may be terrorists who sacrifice anything and everything to further their cause. We know they are increasingly clever in their planning, targeting, and execution. Eventually our adversaries will have another ally: "cyberbots," the sophisticated software programs that operate in cyberspace, that learn with experience, and that perform complex tasks such as intelligence collection, communications, attacks on computer servers, and deception, all at the speed of light. These cyberbots will move and maneuver at machine speed.

Asymmetric warfare deals with the maneuver of knowledge and the manipulation of psyches (individual and aggregate); it concentrates on effective decision making for success. Opponents' use of asymmetric warfare and the strategies of a knowledge war can and will impede a country's ability to engage in a kinetic, conventional, force-on-force conflict. The ascension of asymmetric warfare and the likelihood of a knowledge war suggest the need for the adaptation of a strategy that combines kinetic energy with direct and indirect attacks or manipulations on an adversary's psyche, knowledge, decision cycle, and perceptions.

It should come as no surprise that the demands of preparing for a kinetic, conventional war do not necessarily prepare combatants for planning and waging an asymmetric war, particularly knowledge war. The requirements for thinking, training and education, organization development, materiel acquisition, and reliable leadership are dramatically different when preparing for asymmetric warfare than when preparing for force-on-force operations. Asymmetric warfare is intellectually more challenging and its activities are characterized by seeking, finding, and understanding relationships; it involves a nuance and subtlety that is often invisible and difficult to measure (since it resides in the minds of an adversary). In addition, in "cyberwar" (a tool of asymmetric strategy) the actions of attacking, surveilling, reconnoitering, and protecting most often are executed by cyberbots moving and acting invisibly at the speed of light, watched over by human controllers thousands of miles from the scene of the digital fray. The philosophy of preparing for the worst case and then working lesser problem sets simply does not work when the worst case is a kinetically oriented conventional conflict and the lesser problem set is an asymmetric one. With the passage of time our asymmetric foes will become smarter, more adaptive, and technically capable. They will learn to use decentralized decision making and they will synchronize activities over the Internet and modern telephony to create massive perturbations in decision making and in the will of the collective mind of the U.S. populace. Thus we have to prepare for a very long struggle against a capable, determined foe. Preparing for this scenario is much different than preparing for its antithesis.

Before proceeding further, a quick review of what has changed and what has remained the same in the security environment seems appropriate. First, the American people have always faced formidable enemies. Yet, while dangerous, these enemies were sometimes slow on the uptake, they experienced difficulty in figuring out America's motives and intentions, or they could not figure out how to compete with us with any sustained degree of success. As an example, think about the Germans in World War I and World War II. Who could forget the memorable comment by Hermann Goering that Americans are "only good at making razor blades"? Or consider the Soviet Union's experience during the latter half of the twentieth century. Soviet leaders had great difficulty understanding the American mind-set forged by years of free choice, a capitalist market philosophy, democratic institutions, belief in the human spirit, and freedom of religion.

Until 1989 the principal threat to the security of the United States was the Union of Soviet Socialist Republics. In 1989, however, this threat turned from a roar to a whisper with the demise of the Soviet Union and its massive conventional warfare power. Since then no other country having the capability to compete in a conventional sense with the military prowess of the United States has emerged. Yet conflict and war did not disappear when the Soviet Union did. Unfortunately, the clash of ideals among nations will continue and conflict and war will always exist. Given that, how do nations or organizations intend to compete with the United States if they cannot compete in a traditional, kinetically oriented conventional way?

The answer, of course, lies in asymmetric warfare and its strategy of knowledge war. Smart, formidable adversaries will compete with America by using a wide variety of asymmetric strategies, tactics, and tools. They will focus their efforts on striking at U.S. vulnerabilities, and they will most often use an indirect approach. Although other forms of it may exist, in the context here asymmetric warfare means "seeking, finding, and using offsets against a stronger power and striking at vulnerabilities with ways and means difficult to counter."

Our future adversaries will follow the thinking that underpins asymmetric warfare and knowledge war, and will achieve their goals using psychological assessments, information operations, coordinated terror, nonlethal weapons, weapons of mass destruction, weapons of mass effect, and assassination. Their aims lie in affecting the decision-making processes of individuals, of aggregates of individuals, and especially of political leaders of the United States and other countries. The ultimate goal is to affect the will of the American populace. The country's future adversaries will spare neither expense nor numbers of lives to win, nor will they have qualms about killing large numbers of innocent Americans abroad or within the continental United States to further their aims.

A second point to consider in relation to the current security environment is that the human mind is wonderful; it is the quintessential representation of God's gift to mankind. People have the ability to create, adjust, learn, and adapt, regardless of the speed and pervasiveness of change going on around them. Yet within these same minds are the seeds of self-destruction, seeds that can easily and quickly move from a state of dormancy to one of growth and activism. The catalysts that could cause these seeds of destruction to grow include intellectual atrophy, remaining wedded to the status quo, and possessing a reductionist, closed system or singular view of an increasingly complex and interrelated world.

Third, the Internet, modern communications, and advances in computing have created an explosive, unifying force—a truly global earth. Digits, and thus information and knowledge, move at the speed of light through a growing network enshrouding and uniting the world. This is a wonderful development politically, economically, and socially. Yet the phenomenon has at the same time increased the power of terrorists by allowing them to instantly exchange information and coordinate activities around the world. For these foes distance has become irrelevant and time has become meaningless. In contrast, time and distance continue to be extremely relevant to the United States because of the need to move large numbers of people and things quickly over vast distances.

Finally, as interrelationships and connectivity continue to expand in size and grow in importance, problems occurring in one part of the world (social, ecological, political, economic, financial, military, and so forth) will increasingly influence events taking place in other parts of the world—causing tremendous frustration and consternation to people who experience difficulty in perceiving the "shroud" of interrelationships cloaking the world and the second- and third-order effects that influence the interaction of these systems. Indeed, as we progress in the new millennium the catalysts energizing events and relationships that lead to effects (outcomes) will become more abstract, less observable, and, in many cases, completely invisible. Thus, recognizing relationships and developing the intellectual capability to combine them into thinking practices will become increasingly important. One of these practices comes as a result of the thinking skill called synthesis, which is a "fusion of separate elements or substances to form a whole."[2] The rigor needed for this level of intellectual synthesizing skill is a clarion call for the development and improvement of people in national security, and serves as a unifying theme throughout this book.

All thinking people understand how difficult it is to accept change—particularly rapid change—and how difficult it is to use change for creating an advantage. Most people have a strong innate desire to retain the status quo and will resist any change that threatens their self-esteem, their knowledge or expertise, or their position. I do not condemn people for resisting change, since change is always difficult to accept. But change must be harnessed to be able to use it as a shaping force.

To accept and cope with rapid change people should consider viewing the future in a visionary way. Such vision presents a future full of potential and possibility, and relies on the notion that people can shape the future by searching for and using relationships—and accepting, indeed instigating, change using good thinking and a positive outlook.

The reader will be well served to keep in mind several points: First, the world is interconnected, with social, political, financial, economic, ecological, informational, and military systems all inextricably intertwined. These interrelationships, which I call the "world tapestry of systems," interrelate and change rapidly from even the mildest of perturbations. A tapestry shows a picture, icon, or scene that someone has developed by weaving together numerous threads into one whole. The front or viewing side of the tapestry is often graceful beauty. If the tapestry is flipped over, however, there is no picture—just many knots and threads that appear disjointed. In reality, though, the disjointed back of the tapestry is inextricably related to the graceful and coherent visual depiction on the front. The relationships between and among the seemingly disparate threads and knots on the back of the tapestry are both difficult to discern and sensitive to each other. When a disturbance in one thread occurs, the rest of the tapestry changes and contorts. So it is with discerning the obvious and seemingly disparate relationships (that is, the cause and effect that occur) within and among the world's system of systems. A disturbance in the financial system causes change in the intertwined social, political, military, and economic systems. In addition, the density, sensitivity to variables, and complexity of texture of these relationships increase with the passage of time. This tapestry of intertwining systems is reality; this reality constantly vibrates, dramatically changing the world's body politic. Interestingly, because of the tightly connected relationships comprising the world's functional systems, second- and third-order effects can occur as a result of a perturbation at a far-off location. The attack on the World Trade Center caused massive perturbations and effects in the social, political, economical, and financial systems around the globe. Ominously, America's future adversaries will surely see the tapestry's threads as "a way" to attack the whole of the United States and its partners.

Second, the human race has the capability to learn and change; this capability is a gift at birth. People can improve their capability to adapt to change by improving their minds. With such improvement people think about the world differently. Heretofore some people believed in the notion that technology would eventually solve all the world's problems. This outlook led to a distinctive arrogance and yielded notions such as "truth" and "one right way" (their way). Such people enmeshed themselves in a closed rather than an open system of thinking, and believed they could predict the future because they "knew the truth" or "knew the right way." This rigid view of truth presents vulnerability to an asymmetric foe. A rigid view of truth causes one to discount that which is believed to be not true when it is possible that "not true" could be "true." Many people of the twenty-first century have discovered, however, that truths are difficult to find, they change frequently, and there appears to be a decreasing number of absolute truths. People of this intellectual ilk also have discovered that often there are many ways to accomplish tasks, and innovative and creative ways to accomplish tasks are indeed valuable. Moreover, these human beings have come to accept the aphorism that great insights into the essence of life come to those who learn *how* to think rather than concentrating entirely on *what* to think.

The phenomenon of concentrating entirely on *what* to think is particularly erroneous in an era of rapid change. When people concentrate on what to think they often find themselves trapped by riddles and paradoxes; extricating themselves from the mental trap they have set results only in more confusion and more dogmatic thinking. Instead of relying on dogmatic thinking, America's security professionals need to engage in creative thinking for developing solutions to problems and challenges posed by the rapid changes and complexity of problems inherent to conflict in the modern world.

Third, people of the world have never before experienced the impact of technology like that the near future holds. Previously the world's masses did not have access to the data, information, and knowledge that has been produced with the explosive growth of personal computers, cellular telephones, personal digital assistants (PDAs), and the Internet. Change was experienced at the speed of the fastest moving medium technology could support—by word of mouth, handwriting, or the telegraph. Now and in the future changes will affect all of us literally at the speed of light, and, if historical trends hold true, the pace of change will only accelerate with the passage of time.

Competition is a root cause of all conflict, regardless of the form a conflict takes. But future conflict will be more dangerous than in the past because of its influence on vast numbers of individuals, on societies around the world, on politics and economics, and on the world's ecology. The capability to use weapons of mass effect, for example, lies in the arsenals of asymmetric opponents and could very well affect all people in the future. Future conflict also will be more complex. The principal culprit in this growing complexity lies within the notion that conflict will be highly relational, invisible, and will involve chaos and complexity theories. To understand the role of relationships, the invisible nature of conflict, and chaos and complexity theories demands a growing dependence on wits instead of physical might in conflict resolution.

To gain a better appreciation of the nature of future competition, American citizens and security professionals will have to think about how to compete and protect themselves from opponents who are vastly different, more capable, and better armed with technology than opponents of the past. They also will have to understand the constantly changing impact of information technology on the world writ large and specifically on national defense. These subjects are important to our country's well-being, and therefore we must deal with them in a mature and thoughtful way. We cannot leave important matters involving knowledge war to people wedded to the past or resistant to change. Put simply, national security implications of the twenty-first century are rapidly materializing, they appear highly relational, and they have potentially immense outcomes. The thoughtful and creative people in the national security apparatus must engage seriously in seeking solutions to these issues now.

Leaders from all areas and at all organizational levels must understand this future environment, they must put forth a vision, they must lead people toward that vision, and they must attempt to shape the future in a direction that is consistent with the democratic underpinnings of the country but sensitive to nature of the complex environment in which we reside. Those responsible for the nation's security must commit to shaping the future in an information-rich environment, one in which all people, no matter where they are located or what their wealth allows, have open opportunity to access and learn from the media and the Internet.

This book will attempt to help all of us think just a little differently about the nature of the competitive environment in which our country is inextricably intertwined, and suggest actions that must be taken to protect the people of the United States and other freedom-loving countries of the world. Asymmetric warfare and the knowledge war are important subjects when thinking about future conflict, and they will garner lengthy discussion and deep thought. Information superiority will enable U.S. forces to triumph in any situation. But people need to understand that information superiority does not come automatically—it will be repeatedly contested—and it is only a condition that permits advantages to form. People also need to understand just exactly what information superiority is, what it enables, and why it is important. No subject represents the complexity of the age like information operations. IO are critical for the country's security because asymmetric foes intend to use the tools of IO to wage a knowledge war in return. People must understand how a capable and smart asymmetric foe thinks and plans to influence his or her decisions and will. All of these issues are discussed. In addition, the book provides ideas for coping with very sophisticated deception. Some of these ideas include improving thinking and planning, improving how leaders think about and use deception, and how people can learn to recognize when an intelligent and skilled opponent is employing deception as a tool for attacking decision making. As a different twist on future conflict, the book puts forth the proposition that asymmetric advantages come with adroitness in managing and using knowledge to make better decisions faster than the adversary. Knowledge management, as well as the individual human mind and cyberspace, constitute new and very different "battlefields" of the future. The book also presents some ideas about technologies that would improve collection, automation, communications, and analysis and synthesis. (The discussion of intelligence analysis and collection includes an initial inquiry into the nature of intelligence operations in cyberspace.)

In addition, the book introduces the notion of knowledge advantage centers (KAC). Knowledge advantage centers provide a way to focus hundreds of minds in a vast collaborative network that connects state, local, Department of Defense (DOD), and national government agencies and, eventually, private corporations and coalition partners as well. These entities are solidly connected in the thickening web of security in wake of the 11 September 2001 terror attacks. It follows that all elements need to share data, information, and knowledge to enable people to make better decisions faster than the adversary. The book puts forth the proposition that the defense establishment needs an asymmetric opposing force (OPFOR) to train soldiers, sailors, airmen, and marines, and the need for a joint information operations proving ground (JIOPG) for training leaders, for experimenting with new organizations, for researching and developing new automation tools to support offensive and defensive information operations, and for testing and evaluating IO weapons (lethal, nonlethal, and knowledge). Along this line of thought, the country's military establishment and state and local governments need an Internet replicator to simulate the environment of cyberspace in which the country defends itself and attacks its antagonists in future conflicts. Finally the book assesses the characteristics of "cyberwarriors." Cyberwarriors will ascend in importance, as upon their minds and shoulders will fall the absolute requirement for matching wits with adversary handlers controlling attacking cyberbots at lightning speed along the avenues of approach and lines of communication of the future—that is, cyberspace. If the country develops a core of cyberwarriors as discussed, it will enter any future conflict possessing highly skilled cyberwarriors fully capable of defeating any adversary waging knowledge war.

> Those skilled in war subdue the enemy's army without battle. They capture his cities without assaulting them and overthrow his state without protracted operations.
>
> Sun Tzu, *The Art of War*

2   Knowledge War

Knowledge war will descend in full fury upon the United States in the early part of the twenty-first century. Knowledge war involves a struggle between adversaries over accessing and using valuable information and knowledge to gain advantages in decision making.

To return to the definition provided earlier, "knowledge war" is the struggle for valuable information and knowledge that both sides need for making better decisions faster than their adversaries. Knowledge leads to understanding, which occurs when decision makers combine knowledge into a collage. With such understanding comes the potential to make effective decisions. Decisions lead to actions that lead to effects—some that are behaviors, others that influence will (both individual and aggregate). Valuable, relevant knowledge is important enough that increasingly people will be willing to fight over it.

Production of Knowledge

Producing knowledge is expensive. It is expensive to buy, update, and maintain the information machines that collect, process, store, retrieve, manipulate, visualize, and report data. It is expensive to transform data

into usable information, and to train knowledge workers capable of turning that information into knowledge. Moreover, thanks to the rapidity of
change that currently affects the world, it is expensive to help knowledge
workers achieve a desired level of knowledge proficiency, to maintain that
proficiency, and to develop new expertise continuously. Thus, any organization competing with a capable adversary must acknowledge the need to
orient its processes within a framework of knowledge-based strategies and
operations. In this type of operational environment three essential elements interact constantly: knowledge workers, technology, and organizations. The implications of knowledge-based operations for each are indeed
profound.

Knowledge Workers

In this age of knowledge-based operations, knowledge workers will constitute the most important element of conflict (until computers surpass the
human in intellectual capability). Knowledge workers are analysts, knowledge experts, academics, planners, and operators—indeed, the concept of a
knowledge worker is more far-reaching than the simple term "analyst" to
which we have grown accustomed. Suffice it to say, our leaders and organizations must develop, nurture, and improve the intellects of their
knowledge workers with the same degree of commitment and energy as
they developed and maintained their mechanical engines of war in the
past.

Technology

Knowledge workers must have information technology (IT) machines to
help them find relevant information, and then turn it into knowledge.
Moreover, this same machinery must help both knowledge workers and
the decision makers they support to monitor the actions leading to effects
envisioned in the decisions. Every conceivable means of IT must be used
to help make effective decisions—decision-making aids such as visualization, collaborative software, communications conduits—to transmit guidance and orders and create actions, and to monitor the effectiveness of
those actions. This equipment becomes obsolete quickly, mandates precise
and lengthy training, and requires intense manpower to manage and
maintain.

Organizations

Military structures have been hierarchical for hundreds of years. This hierarchy has existed for good reasons and will probably be around far beyond the time any one of us has on this earth. Yet to engage in knowledge-based operations people must share information and knowledge, and decisions must occur quickly at lower levels, owing to the fact that today's foe's organization is decentralized, it makes decisions quickly, and its individuals are smart, adaptive, and technologically capable. The environment and asymmetric adversaries we face suggest the need for a decentralized, flat organizational structure and horizontal sharing of information. Current military structure is firmly hierarchical. How will a flat, collaborative, decentralized organization coexist within a hierarchical, traditional military one? How do we overcome this contradiction? The answer is quite simple: today's military organizations must engender mentally agile leaders and flexible organizational structures that can quickly slip back and forth between the poles of flat and hierarchical.

Knowledge-Based Strategies

Knowledge-based strategies are the result of thinking through how our actions will influence an adversary's behavior, decision making, and, ultimately, that adversary's will. Effects don't occur by accident—they come as the result of actions. Taking actions only make sense if in the end those actions alter someone's behavior, decision making, or will. To know which effects we want, which actions lead to those effects, which effects have cascading or second- and third-order effects, and to be able to measure the effectiveness of those effects demand that we bring to bear every bit of our knowledge to every situation. Knowledge of the adversary is critical in designing actions leading to the effects we desire. Effects are the outcomes leading to the influencing of behavior, the making of decisions that enable advantages to arise, and the imposition of our will on our adversary. Interestingly, the adversary will be doing the same thing. We can surmise, then, that decision cycles, perceptions, and organizational and individual behaviors—all integral elements of good decision making—are particularly important in devising knowledge-based strategies and operations. We must make a clear-headed assessment of actions from several additional points of view: such as from how effective our actions are and how the adversary views his own effects-focused campaign. We must also think about how the adversary views our view of our effects-focused campaign. In the most difficult aspect of adversarial thinking we must think about how the adversary views our view of how the adversary views the success of his own effects-focused campaign. (This relational thinking between friend and foe is important for understanding Red [adversary] and Blue [friend] wargaming described later.) Thus measuring both tangible and intangible actions that lead to effects and being able to evaluate overall effects and quickly adjusting our plans will be critically important in waging successful knowledge-based operations.

Knowledge War

Knowledge war has two aims: attacking the knowledge that the adversary uses to make decisions, and protecting our knowledge workers and information systems from assault. We must be able to attack our adversary's perceptions, his knowledge workers' way of thinking, and the information systems and infrastructure comprising the machinery that supports his decision-making processes and his ability to turn data into information and information into knowledge. We must protect our knowledge workers, decision makers, and the information systems and infrastructure that support them from assaults by a determined and capable foe using both tangible and intangible attacks, including cyberspace. For the purpose of discussion here, cyberspace can be defined as the boundless territory where human and machine activities occur, where time exists, and where computers, computer networks, and communications exist and exert influence, all together blending into a whole. We must recognize that many of the struggles in a knowledge war will be invisible and some of these struggles will occur at the speed of light.

Roles and Missions

We have to determine which mental activities people do well and which ones they don't do well. At the same time we must determine which activities machines do well and which ones they don't do well. Then, of course, we must efficiently meld the two—human and machine—and drive them to work together in a complementary way. Once humans are doing what they do best and machines are doing what they do best, a truly symbiotic relationship will allow us to do well in any knowledge-based strategy or operation that the future may hold.

Knowledge war will grow in influence and importance. Eventually it will be the preeminent way of engagement in future conflict. Decisions will serve as the focal point of contests between asymmetric foes. Most likely the stronger foe (the United States, for example) will espouse the use of knowledge as the quintessential element of strategy and operations and will leverage its technology as a distinct strength. The weaker foe will seek to offset this superiority by attacking the sources of that knowledge and leveraging the knowledge and technology that are available over the Internet and in open society. The winner of this struggle will be the side that is best able to make decisions that lead to meaningful actions, effects, and feedback on the effects. Victory will go to the side whose people are the smartest about their opponent and the side that makes the best use of technology and derived knowledge to affect the other's knowledge and technology. Decision making in the twenty-first century will be critically important in determining the outcomes of twenty-first-century conflict.

Knowledge war will revolutionize the way leaders and subordinates responsible for our nation's security think about, plan for, and wage war. Knowledge war will occur on a continuum. In knowledge war deterrence will be as important as the compel and defeat stages of conflict. Knowledge war will even affect the transition phase of conflict. Quite simply, the age in which atoms and kinetic energy dominated conflict is quickly coming to a close. Knowledge war will expand to influence the social fabric of all societies.

Knowledge war is an asymmetric strategy that involves the use of indirect approaches and avoids direct, force-on-force confrontations with a vastly superior conventional force. One of the principal tools that asymmetric adversaries will use to put their strategies into motion and wage knowledge war lies in information operations. A working definition of IO is "the activities and actions that occur in physical space and cyberspace to create effects (outcomes) in the thinking, planning, and decisions of adversary leaders and information systems, while protecting 'friendly' decision-making processes and supporting machinery." The goal of IO is to enable friendly decision makers to make better decisions faster than adversaries, thus creating exploitable advantages.

Information operations are integrated in a variety of offensive and defensive strategies. A useful way to view and understand IO is to divide the many operations into offensive and defensive while still recognizing that offense and defense are part of the same whole. Generally, offensive IO include electronic warfare, computer network attack (CNA), deception, physical destruction, and psychological operations. Defensive IO generally include counter-deception, computer network defense (CND), operational security (OPSEC), counterintelligence, and counter-propaganda. The concepts and tools of IO by design affect thinking, planning, and decision making, as well as the information technology that supports such processes. To maximize their effectiveness the diverse aspects of IO must operate as a whole, both in their operation and through organizationally coherent roles, missions, and responsibilities.

Information operations won't deal solely with governments, economies, and societies. No, information operations of the future will involve assaults—mental and physical, direct and indirect—on human minds. In particular, asymmetric adversaries of the future will specifically target the minds and thought processes of human beings who add value to information and turn it into knowledge; they will target the minds of leaders who use knowledge to gain insight and understanding that are germane to making fast, effective decisions. Opponents in knowledge war will steal, influence, or manipulate the information, knowledge, or understanding that knowledge workers or leaders have in their minds because, in any final analysis of who wins or loses in a conflict, valuable knowledge will be the most important variable, indeed the most important asset (tangible or intangible) that organizations or individuals possess.

Information operations involve struggles over the access to and denial of valuable information and knowledge. Information and knowledge will be highly valued, to the point that people will actively seek them for the potential advantage in decision making they bring. Nevertheless, it takes the art of leadership and command to turn that potential into reality through actions that create effects.

What kind of information will be needed? In a general sense people need four types of information: scanning, problem-solving, learning, and advantaging. Scanning information is the massive amount of information we regularly sort through, both consciously and subconsciously, searching for something of value. Scanning information helps people build conceptual frameworks and intellectual databases. People usually discard most scanning information, but some of it becomes useful and allows them to turn information into knowledge. Scanning information is not sufficiently valuable to consider it knowledge, since it is often narrow, outdated, and relevant to life on a micro level only.

In every walk of life people need problem-solving information and knowledge to make rational decisions about problems or challenges they face and tasks they want to accomplish. This problem-solving ability improves as people find and use information of value and turn it into knowledge. To find the information and knowledge that will help them solve problems, however, people must first learn to define their information needs. A vast sea of information—neutral, passive, and available to anyone astute enough to seek, access, manipulate, and use it for making good decisions—exists, much more than individuals can ever absorb. Thus the race for achieving advantage in the future involves the use of automation to search for, find, sort, and present relevant information in a useful form for analysts (knowledge workers) to apply their intellects, experience, and intuition, and develop that information into knowledge.

Information and knowledge are fundamental to learning, the lifeblood of analysts, knowledge workers, and leaders who make decisions. People ponder and reflect upon the information and knowledge they gather. They apply it to their inner worlds and often put it into practice in their particular walks of life. People have unlimited potential to learn and expand their knowledge bases thanks to the availability of information and knowledge. Of course, it goes without saying that people must first be inquisitive and intellectually ambitious to seek and use information and knowledge to learn.

Last, and most important for the purpose of this discussion, people seek information and knowledge to find and sustain advantage—intellectual, decision making, knowledge, technical, positional, and in their actions. Advantage allows people to create the conditions leading to the desired effects they envision. Each of these advantages is useful in its own right, but when combined they yield a truly powerful force.

If we can agree that information and knowledge are important for making fast, effective decisions, why is it so difficult to define information needs? A partial answer is that people generally don't know how they process information and think. And why is that? Because people usually don't spend much time pondering the abstract art of thinking, particularly their own. Yet once people can intelligently define the information and knowledge they need to fuel their thoughts and decisions, they are then able to provide very precise descriptions of the information and knowledge they need for making particular decisions. This precise information leads to better and faster decisions.

In a perfect world everyone would know and understand how they think. No time would be lost sorting through all the information available, and the right decisions would come fast. Different ways of thinking require different amounts and types of information. Inductive thinkers, for example, need lots of information as they build their thoughts in an aggregated way and form conclusions after processing a large amount of information. Generally these thinkers don't have a preconceived notion and arrive at what they consider to be a truth or an answer only after they have enough information to form a conclusion. Deductive thinkers, on the other hand, generally start with a view, vision, hypothesis, or conclusion they want to prove and select only the information that supports that conclusion.

Analytical thinkers shred information into constituent elements to gain increased knowledge about each element. Typically analytical thinkers need access to a great deal of information, which through continuing analysis leads them to heuristic or discovery-based understanding about particulars and increasing levels of complexity; that is, they discover gaps in information and seek answers as they continue to analyze. Often analytic thinkers are reductionist at heart and find contentment when they believe the constituent element cannot be further broken apart. Typically they don't synthesize information well. Holistic thinkers, on the other hand, employ the thinking process of synthesis and actively seek relationships, combinations, and understanding about how and why things interact and relate. Holistic thinkers gather pieces and form conclusions about trends. Holistic thinkers generally don't need a lot of information—precise and select information helps convince them that the conclusion(s) coming to mind quickly and easily is, in fact, the right one.

Once people understand *how* they think, the second step of defining the information and knowledge they require to enhance their thinking and decision making becomes much easier. When armed with an understanding of what they need, setting out to get it is much easier (a process that holds true for our opponents and for us). Therefore it only makes sense for people to understand how their adversaries think and what information and knowledge they seek and trust, because then manipulation of that information and knowledge becomes increasingly possible. We must anticipate how our adversaries believe we think, too, as the adversary attempts to influence our thought processes just as surely as we attempt to influence theirs.

To enter and win in knowledge war, adversaries must know their opponents' decision cycles or must know at least enough information to enter into and affect the activities that take place in the cycle. Unfortunately, not much is usually known about the specifics of decision cycles. If this is the case, people can model a generic cycle that could look something like the process shown in figure 1.

People making decisions in adversarial situations go through a similar process. But, admittedly, differences exist in how people think, how they decide, and in whom and what they trust. Each of these decision-making elements should receive much attention from intelligence-gathering efforts. The effort in understanding adversaries' decision cycles and how they view ours takes high-grade knowledge and is time-consuming.

Differences also exist in the hardware and software components of competitors' decision cycles, as well as in the data, information, and knowledge they seek. Their decision processes, thought processes, and perceptions may look and sound different than our own. "Cyberplanners" must consider these differences and consult experts knowledgeable of them to help them think their way through associated issues and avoid falling into the trap of "mirror imaging" (replicating or projecting our ways of doing things—our perceptions, thought processes, and value sets—onto our adversary).

One of the most important implications for people involved in knowledge war lies in the absolute requirement for expending resources to find and identify the intricacies of an opponent's decision-making apparatus. We must understand the decision logic that resides ever so subtly in information system databases and that resides in the code of other decision-supporting software. In addition, considerable effort must go into ascertaining the individual thinking skills and styles of an adversary's leadership structure, and the key people supporting that person or persons. Thinking styles and processes vary from one individual to the next, depending on intellectual capabilities, personality preferences, and favored ways of processing, absorbing, storing, and using information and knowledge. But the environment, goals, objectives, military capabilities, and perceived vulnerabilities of an opponent are the shaping forces and constraints and they help to influence at least some aspects of an adversary's thinking.

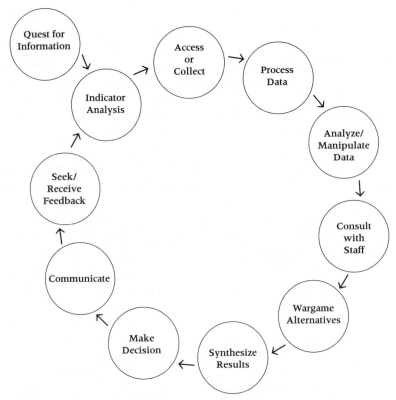

Figure 1. Decision cycle

While acknowledging the many differences among the thought processes of people as well as the machines that support those processes, people from all cultures making decisions in a competitive environment generally seek relevant, valuable information and knowledge to help them make effective decisions. It follows that the staff or support personnel of any organization in such an environment must go through some type of sifting, sorting, fusing, visualizing, analyzing, synthesizing, evaluating, and correlating to transform data into information and information into knowledge. With knowledge (especially knowledge relevant to the problem or issue at hand) people gain the necessary understanding for making decisions, initiating actions to create effects, and comprehending feedback on those actions or effects.

At the end of a decision cycle or process, decision makers must decide either to take action or not to take action, then clearly communicate to people with the capability to create effects through actions. This decision-communiqué should provide guidance for initiating feedback mechanisms for monitoring the progress of the decision. Interestingly, feedback mechanisms are critically important for continuing, changing, or halting the action or activity once it has started. Feedback mechanisms among military forces lie, for the most part, in intelligence-collection systems. Unfortunately, because of the trust invested in collection systems that provide feedback to decision makers, they may prove to be the most susceptible to manipulation by an adversary. It seems only natural that adversaries will expend intelligence-collection resources to monitor an opposing force's attempts to gain feedback to confirm or deny their hypotheses and the veracity of their decisions. Such efforts, though, constitute a distinct stress on scarce intelligence-gathering assets.

People need knowledge of several aspects of an opponent's decision cycle before engaging in an intense competition for decision advantage. First, they need to know how the opponent uses technically specific automation to help make decisions; that is, they need almost perfect and detailed knowledge about the hardware, software, information assurance tools, communication paths (including bandwidth), collaborative environment, database management, and knowledge management of an opponent's decision processes. Second, they need to know how the opponent seeks, accesses, absorbs, and shares information and knowledge. Along this line of thought, opponents need to gain insight into how their adversary intellectually processes and uses collected information and knowledge. Third, people need to know how their opponents communicate information. Once a decision is made, a decision maker's decision apparatus will disseminate the decision and direct actions. If the decision becomes action, the paths or conduits of connectivity between the decision maker and those turning the decision into action will be vulnerable to intrusion and manipulation. Fourth, people need to have a clear understanding about how an opponent perceives the world and how he or she thinks. This knowledge cannot be simply a mirror image of a preferred way of viewing the world or the opponent. Fifth, people need to know not only an opponent's collection and feedback mechanisms, but also the strengths and weaknesses of the opponent's collection and feedback mechanisms *in relation to their own*. Moreover, people must evaluate their own susceptibility to deception, their sensitivity to nuance, and the potential effects of perturbations of these two sets of feedback mechanisms. A vulnerability analysis from the perspective of friend or enemy is important and must include assessing one's vulnerability to acting cyclically, communication paths, degree of accuracy required for action to occur, and timeliness of information and knowledge, all of which lead to belief in veracity. Sixth, people need to know potential source(s) of the data, information, and knowledge that contribute to a state of plausibility, and hence believability, of the information and knowledge used in making decisions. Opponents can insert false data, deny access, slow down the flow and retrieval of the data, or insert deceptive images into the source of information and knowledge.

One interesting and complex outcome cyberplanners seek is to slow down an adversary's decision cycle by causing that adversary to disbelieve the veracity of information he has and to feel compelled to seek confirmatory evidence before deciding and acting. Without belief in the veracity of information and knowledge, decision makers can very easily suffer a degree of decision paralysis. Because of environmental controls or political constraints (such as the desire for a limited number of casualties, limited collateral damage, or a low tolerance for risk), military decision makers often desire a great deal of information and knowledge before deciding and acting, which of course slows down the decision cycle—while an opponent's decision cycle could very well be accelerating, particularly if it is not faced with similar constraints. Therefore any decision-making system must somehow incorporate mechanisms that guarantee the veracity of the information used.

Needless to say, achieving this degree of knowing and understanding rarely comes easily. Nevertheless, we must seek it, as the quest for such knowledge and understanding of an opponent's decision-making apparatus and decision cycle, and the potential advantages that come forth from them, constitute the essence of knowledge war. In this type of conflict both sides will surely be aware of the importance of the veracity of information and knowledge and their direct linkage to deception or a slowdown of decision cycles. Both sides will expend valuable resources to initiate such action and determine if the opponent is attempting the same.

Two important implications involving decision cycles emanate from this line of thinking. People must determine how to affect their opponent's decision cycle when and how they desire (for aiding and abetting their own plans). In addition they must anticipate their opponent's efforts to affect their decision cycle. They must also use collection assets in the physical world and in cyberspace to warn of such efforts, and they must decide what to do should their predictions prove true. In all cases people must protect their precious support systems and actual decision-making information and knowledge, for they certainly constitute the information "centers of gravity" in the Information Age.

In knowledge war people collect information, turn that information into usable knowledge for making decisions, form conclusions, protect against surprise from an opponent, and determine the adversary's information and knowledge requirements and possible intent. With increasingly sophisticated and capable software and constantly advancing processing power, entrée into virtual reality to obtain information about these variables becomes increasingly possible.

We must also be able to deny our adversaries the information and knowledge they seek for making their own decisions. An adversary waging a knowledge war has information requirements, too. In fact, both sides in a competitive struggle need valuable information—but this value is relative to the effectiveness of decisions people make, the viability of actions they take, the effectiveness of outcomes the actions produce, the advantage that ensues, or the speed of making decisions. If both sides in a contest have access to the same information at the same time, the value of that information generally dissipates. Here the advantage goes to the person with the highest intellectual prowess who can, in the absence of valuable information, make the best, fastest decision from the information at hand.

Consider an illustration of the idea: Two countries are competing against each other for technological supremacy that can lead to achieving advantages in military communications in the twenty-first century. One country (call it country one) has a spy working in another country's (country two's) ministry of economics who reports to his own leader on the knowledge, plans, and intentions of country two. Country one's leader can either use the information and openly win, or hold back and not destroy the credibility and access of his spy. The knowledge of the plans and intentions of the opponent gives that leader a powerful advantage, even if he doesn't take any action: he knows what is in his opponent's mind, what relationships country two seeks to understand, which things are important, which decision country two seeks, and the strategies country two intends to use to accomplish its goals. This knowledge also enables country one's leader to know how country two seeks and receives feedback on decisions, which clears the way for manipulating input to distort or shape country two's decision-making processes. Eventually, when the time is right and the stakes are sufficiently high, country one will use the valuable, spy-produced insider information and knowledge and turn it into action to defeat country two in some significant conflict. Clearly, as this situation demonstrates, information and knowledge are the means to power, not power in themselves.

A second example involves the value of obtained information. During World War II the British broke the code of Germany's Enigma encoding machine, which gave the Allies access to reliable decision-making information that was coming from the German high command. The Allies had gained a tremendous information advantage, even though they didn't always act on it because they didn't want to compromise the valuable source of it. The Allies were able to gain insight into Hitler's thinking, plans, and intentions. They used the information only when and where the potential outcome was the greatest and they could provide an easily explainable way for the information to have been accessed, thereby protecting their valuable source. The Allies' secret knowledge of Germany's high-level decision-making processes provided priceless insight into Germany's strategies, feedback mechanisms, and long-term goals, and gave the Allies a clear decision advantage when and where they needed it most.

Possessing secret access into the thinking and decision-making processes of an opponent provides one side a significant advantage. Generally, the knowledge that people use to make decisions in a highly intense competition has its greatest value relative to potential outcomes—protagonists will go to great lengths to protect their sources of knowledge if they know the knowledge is valuable. Without such protection, any compromise to the system could yield significant advantages to the other side.

Today, the national security apparatus of the United States faces a new kind of spy—a digital spy programmed to enter databases surreptitiously and "eavesdrop" on packets of digits as they swirl down fiber-optic cables or traverse space via radio and satellite frequencies. Significant growth in efforts to protect valuable sources of information from these invaders will have to occur. Though the task will be difficult, it is important to determine if surreptitious attempts to steal valuable information and knowledge are the work of an organized, determined foe or simply the work of an amateur.

Owing to the growing importance of information and knowledge in decision making, the notion of an "information center of gravity" has emerged. An information center of gravity is a confluence of streams of communication, collection, automation, thinking, planning, and decision making. This confluence, whether it is found in physical or virtual space, is so important that its demise or manipulation can seriously jeopardize the success of a mission or task.

It will be a rare situation in which only one information center of gravity exists. Instead, as a general rule many related centers of gravity will usually exist. These centers of gravity unfold into other, higher-order centers of gravity and enfold lower-order centers in amazingly complex relationships of information and knowledge collection, analysis, and production. Multiple centers of gravity connect horizontally, vertically, and diagonally. Some are easy to see—such as an important computer or communications node—while others are subtler—such as a generator that provides backup power to a decision-making node or a fiber-optic switch that channels all communications to another location. Still others are places (physical and virtual) where people gather and manipulate data, transform it into information and knowledge through collaboration and visualization, and make decisions. In a security sense and owing to the hierarchical way leaders in state and local emergency operations centers (EOCs) make decisions, key sites such as the Navy's combat information centers (CICs), the Army's tactical operations centers (TOCs), and the command centers of regional combatant commanders constitute information centers of gravity. These sites, which often combine both physical and virtual locations and activities, include the machines of information that help people turn information into knowledge and help leaders make effective decisions. By definition these places will be the focal point of many attacks. In addition, they will be, out of necessity, focal points for physical and cyberdefense, too.

Another consideration involves the complicated, invisible dynamic mental exchange that occurs between two sides in a conflict. One side can't simply study and then assault an opponent's information centers of gravity without also keeping in mind that opponent's perception of "friendly information" centers of gravity. In our dynamic and dualistic world, analysts and decision makers must assess how an opponent protects its centers of gravity and then contemplate how that opponent assesses its own protection. In this respect, our opponent's perceptions, not our own, are of primary importance. Similarly, the wise person judges how an opponent assesses the friendly mechanisms he himself uses for protecting his information centers of gravity, and how to perceive the opponent's perception of the measures in place to protect those centers.

The winner in the struggle to affect information centers of gravity accrues dual advantages in decision making—it's a zero-sum game involving the enhancement of one's own strategies while adversely affecting the opponent's. In a variable sensitivity sense, the complex interaction that contributes to the ability to make fast, effective decisions is critical to winning in knowledge war. If one is able to assault an adversary's centers of gravity and succeeds in gaining access to what the centers hold, the outcomes will undoubtedly be catastrophic for the other. Since all thinking people recognize the importance of communications and automation—constantly collaborating with other human beings, and seeking, finding, accessing, and using information and knowledge—it would be difficult to argue that centers of gravity don't exist. It follows that both being able to anticipate how an opponent intends to destroy or affect friendly information centers of gravity and then incorporating protective mechanisms that will deny them success will be critically important.

People can protect information centers of gravity, that is, their information and knowledge, from even the most determined assault, through anticipation, some sort of warning system, concentric rings of protection, a reliable web of information gathering and dissemination, and a flexible plan for either deceiving or counterattacking. What does each of these ideas imply?

First, any defensive system must anticipate all relevant possibilities. Our adversaries will be smart and they will learn through the Internet and from a wide and rich variety of learning resources. We know the importance of information and knowledge, and how that information and knowledge are put to use. Further, we know how important information centers of gravity will be to both sides in future conflicts. It follows that we must anticipate that our adversaries will attempt to influence our information centers of gravity, just as we attempt to influence theirs.

Part of this anticipation must involve thinking through how an opponent would attempt to steal relevant, valuable information and knowledge, why that opponent might desire that specific information and knowledge, and how these attempts might be masked. Anticipation helps people prepare physically and mentally for assaults yet to come. Anticipation acknowledges, indeed accepts as a given, that a struggle for valuable information and knowledge is at hand. People must be ready for the intensity and vagaries of the struggle, and acknowledge that part of the struggle will involve clandestine attempts to steal valuable information and knowledge. Clearly the value of information and knowledge increases if one side possesses advantage-giving information and knowledge and the other side does not know about that possession.

Second, a warning system must inform people that an adversary is attempting to penetrate or affect systems involving decision making within the information centers of gravity. In this regard, giving warning will be a physical function if the adversary attempts to use kinetic energy in the assault. Warning will be a digital function if the adversary attempts to use cyberspace as a route of maneuver for accessing information and knowledge. Arguably, warning is a significant desired state for operating in cyberspace. Programmers must develop smart software systems that will provide warning when IO assaults are pending. These cyberbots will serve as sentinels along with surveilling specific points within a digital system (such as switches) and reconnoitering paths along digital highways.

Third, the notion of using concentric rings of protection is simple: invisible and physical barriers and alarm systems that identify and thwart any type of intrusion must be developed and emplaced. The goals, of course, are to delay or deny access, to manipulate the attempted entry, or to gain access to the adversary's information and knowledge by denying entry into friendly information centers of gravity, as well as having the capability to observe what happens. Increasingly, opponents' assaults will be intense and determined. An opponent's assault could be an attempt to destroy the contents of a friendly center of gravity (such as data files), or destroy the function of a center of gravity (such as communicating decisions via a satellite terminal). An opponent's assault could be an attempt to steal data, information, or knowledge surreptitiously or manipulating data to cause us to formulate the wrong conclusions. An opponent could also try to inject a virus to disrupt or destroy files at the time most conducive to his plans. Or an opponent could attempt to manipulate or corrupt data to present false images. All those involved need, indeed demand, warning of attempts to steal valuable information and knowledge.

To protect against such determined, complex, and multifaceted assaults on our information centers of gravity, we must first imagine and then conceptually construct concentric rings of protection like the rings of a tree, with the tree's core being the valuable information and knowledge that a competitor desires. The competitor must be able to make its way through numerous physical and electronic barriers (the rings) to reach valuable information and knowledge. These barriers can include physical guards, electronic alarms that warn of assault, or firewalls, antivirus software, intrusion-detection systems, cyber sentinels, and actual electromagnetic sensors and barriers that protect switches, databases, and routers.

Another way to protect centers of gravity involves the use of webs of information and knowledge to defeat plans to attack information centers of gravity. Defenders build webs, links, and nodes of information and knowledge to create internal and external networks. When thinking about these webs of information, the most illustrative metaphor is the web of an ordinary spider in which the links (communication paths), cells (servers), and totality of the cells and links work as one (knowledge, knowledge workers, and decision makers).

In the cyberworld, attempts to steal, destroy, or manipulate information and knowledge occur at machine speed. Thus it will be critical to develop counteractions that detect, act, and deny at machine speed. On their own human beings have no hope of winning in this world—human thinking and responses are too slow. Man and machine must work symbiotically to deal with the speed and pervasiveness of future cyberbot assaults. To help gain insights about possible assaults and potential perpetrators, autonomous, intelligent agents or "cyberbots" will act as sentinels and deniers to the nefarious activities of opponent cyberbots.

Information links managed by smart cyberbots will provide early warning of impending threats. Information moves through the links and nodes and warns people to erect barriers, or to switch probes to "honeypots" (false servers deliberately set up to lure hackers and cyberbots and specifically designed to take a particular action to deny, delay, or degrade an opponent's attempts to destroy or gain access to information centers of gravity). With the arrival of such information and knowledge, people turn it into power—they decide to take action to thwart the assault (unless they actually want the assault to occur for the purpose of manipulating the situation to attain an advantage).

As a final possibility, people interested in defending their information centers of gravity need to develop a viable deception plan. Such a plan can deceive an adversary about what is known about his attempts to penetrate, or about the success the defender experienced in denying or manipulating the penetrators. This type of defensive deception confuses an adversary's knowledge, understanding of relationships, and decision making. (A later chapter is devoted to this important subject.) In the competition this country will face in the future we must seek to secure and use valuable information and knowledge before our adversaries do. They will attempt to obscure or hide the fact that they have accessed information. To have any hope of outwitting an intelligent, learning, and adapting adversary, people involved with information assurance must plan for and engage in cyber deception as a forethought in their planning, not as an afterthought.

Wargaming is the high point of intellectual encounter in knowledge war. Wargaming in this complex mental arena leads to valuable insights into an adversary's thought processes, planning steps, decisions, action execution, and feedback on the effects created by actions. Yet leaders waging knowledge war need to know that wargaming, though useful, will be a costly, time-consuming effort.

Why will wargaming be one of the most important aspects of planning for and waging a knowledge war? Because it allows one to put specificity into an act, then be able to react and counteract cycles of competition (the act–react–counteract cycle). With every action there will either be a single response or a range of responses, then a second response, and so forth. The goal is, of course, to anticipate how an adversary will respond to our actions or reactions and then deny, delay, or defeat such actions to achieve an advantage. Wargaming allows gamers to anticipate possibilities that an adversary might bring forth. Recall that possibilities are potentialities unbounded by constraints. Wargaming helps antagonists think about what an adversary *could* do rather than what that adversary has the specific capability *to* do, and it therefore helps people hedge against complete surprise. Wargaming encourages the formation of hypotheses about an adversary's act–react–counteract cycles. The smart gamer knows where to focus intelligence-gathering machines to get the right information for confirming or denying these hypotheses. Wargaming helps the gamer know where to take risks and how to cover the risks with information-gathering efforts. It helps adversaries understand how an opponent will attempt to gain feedback on the veracity of decisions and the effectiveness of actions designed to create certain outcomes. Wargaming in virtual reality, modeled accurately, enables the gamer to anticipate how friend and enemy cyberbots conduct activities in cyberspace and how to conduct intelligence-gathering operations in that domain. Wargaming helps gamers anticipate and understand the sensitivity of variables and how those variables can be manipulated from two points of view—our view and our adversary's view. Finally, wargaming against people who think and act like real adversaries helps hedge against the mental trap of projecting one's thought processes and value sets onto the adversary.

A fundamental aspect of anticipating an adversary's actions, reactions, and counteractions is the need for eight mental views or constructs (here Blue is friendly and Red is unfriendly):

Red and Blue Wargaming Construct

Red's view of self	Red's view of Blue	Red's view of Blue's view of self	Red's view of Blue's view of Red
Blue's view of self	Blue's view of Red	Blue's view of Red's view of self	Blue's view of Red's view of Blue

Red's view of self: Red's view of itself sometimes will be realistic, sometimes fanciful. The important thing is for country experts and other knowledge experts to take the time to help friendly (Blue) consider how an adversary views its own strengths, weaknesses, and centers of gravity.

Red's view of Blue. Red's view of Blue could very well be wrong, but it will drive the direction of Red's thoughts, and hence Red's actions.

Red's view of Blue's view of self. Red's view of how Blue views Blue could very well be a distortion of reality and simply be a mirror image of Red's own value set. This could be an advantage for Blue, just as it is for Red when Blue attempts to project Blue's values onto Red.

Red's view of Blue's view of Red. Finally, and probably the most important, is Red's view of Blue's view of Red. How an enemy thinks his enemy is thinking about him is critical to gaining decisional, intellectual, positional, and action advantages, and leads to insights into how his enemy will come at him and what he can do to detect and reduce the damage from such advances.

Blue's view of self. Any force must look introspectively at its own strengths and weaknesses.

Blue's view of Red. Identical to Red's sequence, Blue must understand Red's capabilities and limitations, and consider them in relation to Blue's view of itself and Red's view of itself.

Blue's view of Red's view of self. Blue must find the right knowledge experts to help assess Red's view of itself and what that means for possible actions, reactions, and counteractions.

Blue's view of Red's view of Blue. Again, a crucial step. Blue must have hypotheses and use information collection to confirm or deny their validity.

The goal in all wargaming is attaining advantage—in making decisions, in thinking and knowing, in technology, in maintaining position, or in taking actions. The construct may seem like intellectual gymnastics, and to a certain degree it is. Until computers become smart enough to represent an adversary's mind-set adequately, however, we must work our way through this kind of thought process to find "a way" to anticipate an adversary's act–react–counteract cycle.

Once we have mentally cleared these hurdles we then have three specific things to work with: a basis for formulating hypotheses about the how and why of potential terrorist attacks; a basis for identifying what the terrorists could need in the way of information and knowledge of us to perform the actions identified in the hypothesis; and a basis for detecting terrorist efforts to collect information and intelligence on us that he needs to conduct the attack.

An invisible competitive dynamic is at work here—a struggle or duel of sorts for achieving advantage, in particular an intellectual and knowledge advantage, over an adversary. Both sides in the competition know they face a formidable foe. Each also knows that the side that is able to develop the best knowledge and use it more quickly and ably than the other stands a superb chance of achieving intellectual advantage, perhaps even a decision advantage. Since every situation involves two sides (or more) in the quest to find the most accurate, timely, relevant, and specific information and knowledge, each opponent will attempt to acquire it first, manipulate it in his favor, deny success, and use it to gain advantage.

How, then, do we proceed to think through this murky, complex, and abstract field of intellectual competition? As a first step competitors must define their information needs since information leads to knowledge. They then must seek useful information aggressively, using the best sensors available, accurate human reporting, the fastest available search engines, and autonomous intelligent agents to find, sort, and gather the information and knowledge they need. Competitors deny information and knowledge to an adversary by first anticipating that adversary's information and knowledge requirements, and then denying it access or manipulating the information and knowledge the opponent happens to find.

What will information wargaming involve? What kind of intellectual expertise will it take to enter this competitive world and emerge a winner? For one thing, information wargaming will be extraordinarily complex and involve intense cerebral work and forays into virtual reality. In virtual reality both sides receive representations in a conceptual arena of competition such as in urban terrain, in cyberspace, or even on open terrain. To accurately represent how an opponent thinks and views the world in these varying contexts, very smart "avatars" will come into play. (Avatars as used here are complex software representations of human beings that are programmed to learn with experience and repetition; avatars are smart to begin with and are becoming smarter over time, thanks to genetic algorithms.) When avatars are first introduced into wargaming, the programs will have only minimal capabilities and humans will have to assist them. Yet technological advancements, self-learning in an ideal setting, and program development by software engineers, possibly even properly cleared U.S. government workers, from the country or culture in question will provide a way to start overcoming the scourge of mirror imaging. Since the avatar "learns" and becomes smarter with its encounters, eventually it will learn to beat the opponent in the act–react–counteract wargaming cycle. Eventually humans will be on the sidelines watching and learning, while smart avatars will compete in the wargaming environment.

The actual process of informational wargaming is difficult to design but fascinating to conduct. It is the essence of the struggle for gaining a decision advantage. It is a "mental" contest between and among entities that are surging, questing, and competing for advantages embedded in invisible bits and pieces of data and information scattered throughout the world in libraries, machines, random noise, chaos, and the unfathomable minds of human beings.

In wargaming's most basic form planners have to think through the situation from their own perspectives and then think through the situation from their opponent's perspective. Planners must identify goals, objectives, tasks, subtasks, actions, effects, outcomes, ways and means, centers of gravity, vulnerabilities, and strengths from both sides. Increasing levels of complexity come with depth of thinking. If the opponent is smart and sophisticated the complexity of mental wargaming will be immense. If the adversary is unsophisticated it will be easier to challenge. Cultural awareness and knowledge are, of course, the dominant variables in this interaction. If one side is intellectually and technically superior to the other, the inferior side will need a "miracle" to remain in the game. If the adversaries are intellectually and technically equal, the outcome of the contest is open to chance and the interaction of the variables mentioned above. Among intellectually equal opponents the game can, and most often will, go to the side possessing the best intuition, the most disciplined mental faculties, and the best thinking skills.

The great equalizer in uneven contests involving knowledge and decision advantage is the wise use of information and knowledge through the tools of information technology. Conversely, if the two sides in a competitive situation are evenly matched in intellect, the side possessing the best information technology and using it most effectively to think and turn information into knowledge (and therefore being able to make rapid, effective decisions) will win. Unfortunately, one side could easily "outsophisticate" itself against an opponent of lower intellectual or technical capability. Thus, it is essential to perform a thorough assessment of the intellectual and technical capabilities of the opponent either before or early on in the competition. As has been true throughout the ages, people need to know the enemy as well as they know themselves—and never underestimate the opponent.

Compared to conflicts of the past, the most significant difference now (and in the future) is the invisibility of the competition. In this sense the enemy's location or disposition will be irrelevant because in the cyberdomain "there" is no more, opposing forces can move at the speed of light, decisions will occur so quickly that the human mind cannot fathom the speed at which they occur, and the domain's "soldiers" (cyberbots) can be everywhere and nowhere.

In the twenty-first century the dominating struggle will be to deny, affect, or retard an adversary's capabilities to use knowledge for making fast, effective decisions. To a great extent the outcomes will often be intangible and reflected through nontraditional, yet-to-be-discovered means of intelligence collection (see chapter four, "Information Superiority," and chapter seven, "Knowledge Management"). In addition, the struggle will center on adversaries seeking to know and understand their opponent's way of thinking, their use of information and knowledge, their views of nuance, ambiguity, and hue, and how they possibly will react to surprise and fear. Winning the competition for valuable information and knowledge could be as simple as denying one side or the other access to the information and knowledge they value for a time just long enough to provide an edge in decision making.

The means for acquiring and processing information and knowledge are dangerously vulnerable to both crude and sophisticated means of overt and surreptitious entry, manipulation, distortion, and denial. Arguably, firewalls, intrusion-detection systems, antivirus software, mental discipline, and straightforward standards help mitigate risk to one's information system. Having these protections, though, should never lead to complacency or arrogance. For every action there is always a reaction, and thus for every measure of protection there is always a countermeasure. In addition, many other ways exist for affecting an adversary's decision making, such as manipulating the media or adeptly using psychological spin, perception management, or deception, to name a few. Thus it is only logical to go to extraordinary lengths to protect all information activities that contribute to turning information into knowledge and decision-making processes.

Many implications exist in the gathering of information, turning it into knowledge, and forming relationships to achieve advantages through the decide–act–effect cycle. Yet two of those implications seem to stand out in importance: the stresses on intelligence, surveillance, and reconnaissance (ISR), and the challenges of analysis and synthesis in a complex informational environment. Commanders in a knowledge-based organization must realize that knowledge, when coupled with good leadership and cutting-edge technology, provides a terrific advantage. But they must also realize that valuable data, information, and knowledge do not appear out of the ether. No, they must be sought, gathered, processed purposefully, and turned into information, and the brains of analysts and planners (knowledge workers) to produce knowledge must be used wisely. ISR provides a significant portion of the "seek and gather" part of the process. Bandwidth certainly contributes, as well as computers and software. But far into the future ISR will continue to be found in a constrained environment, and leaders will have to manage its use carefully.

A recognition of the constraints on ISR should be a part of any future exercise construct, just as surely as weapon systems are constrained. Constraints require prioritization and resolving conflicts among components and users of the ISR output—the data, information, and, with human intervention, knowledge. Constraints also involve using the resource of human intellectual power, since the production of knowledge is a costly endeavor that cannot be relegated to a staff function. In a knowledge-based organization the production of knowledge has to be a commander (decision maker) function. Compounding the problem, and because of an increasingly complex and varied operational environment, many more tugs and pulls on ISR will occur. Thus it behooves us to engage in some deep thinking about the role of ISR in future conflict scenarios. Undoubtedly ISR will have several roles. First, ISR will be the vehicle for finding the data and information that enable an assessment of the actions that lead to desired effects. Second, ISR will satisfy priority intelligence requirements and a commander's critical information requirements that contribute to situational development. Third, the surveillance component of ISR will show "a way" to mitigate risk and economize resources. Fourth, ISR will allow one side to track how an adversary's plan is unfolding by purpose-fully gaining data and information about that adversary's decision feed-back mechanisms, that is, *his* ISR. Fifth, time-sensitive targeting will become increasingly important, for example, attacking a mobile missile that carries chemical warheads. ISR will discover, track, and provide feedback on attacks of these mobile targets. Last, ISR will either confirm or deny possible adversary courses of action and the variables that surface during the act–react–counteract aspects of friendly-enemy wargaming. With this knowledge in hand planners, commanders, and homeland security decision makers will have to decide how to use ISR, what the priorities are, and how many optimum mixes and synergistic relationships can realistically exist.

A second implication—that the people who turn information into knowledge must be cared for as treasured resources—carries our discussion into the realm of intellectual development. Recall the importance of transforming data into knowledge and the role that knowledge workers play in the process. Because of this importance, the vast amounts of information demanding the attention of knowledge workers and the physical stresses of deep thinking through analysis and synthesis demand that leaders must exert care not to overburden their scarce intellectual resources. Thus, as one approach to solving the issue, more attention must be paid to the roles and missions involved in analyzing and synthesizing information and turning it into knowledge. Knowledge-based strategies, knowledge-based operations, and participating in knowledge war place enormous mental demands on organizational knowledge workers and even on knowledge experts outside the organization. Certainly heavy analysis occurs during planning. But heavy analysis also occurs during the execution of actions to determine the effectiveness of effects (effects assessment). Moreover, when decision makers adjust actions leading to an effect, even more analysis will then have to occur. In addition, if analytic efforts will be adjusting, changing, and useful, a combination of machines and analysts will have to put forth energy on a continuum, not aperiodically (to perform even mundane tasks such as updating databases). This level of energy and focus is a drain on human resources. Therefore, to effectively wage a knowledge war against a smart adversary, analysis and synthesis activities must be federated among levels of war (tactical, operational, and strategic) and coordinated over a collaborative network. A knowledge manager or "pit boss" will need to coordinate and synchronize such efforts. Moreover, analysts and operators must have access to a constantly updated knowledge expert database within the host staff, at lower level headquarters, at higher level headquarters, on a national level, and within academia, industry, and the like. Intellects and what they produce by way of knowledge are indeed critical to waging and winning a knowledge war. The minds of our best and brightest are treasured resources and must be trained, nurtured, synchronized, and orchestrated just like more traditional physical resources.

Knowledge war, in which opposing sides compete for valuable information and knowledge with which to make better and faster decisions, will be the preeminent way of executing activities in future conflicts. Information operations will be the principal tool used by the asymmetric foe that engages the United States in a knowledge war. Information operations influence an adversary's decisions because along the way they affect the information and knowledge coming to decision makers, as well as the information systems that support their decision processes. Decisions are critically important in a knowledge war because they are the key element in coalescing the will of individuals, of the general populace, of political leaders, and of the nation.

When competing for valuable information, opponents will always exist—if they don't the situation won't be competitive. Opponents will vary widely in physical attributes, in intellectual capabilities, in how they seek and find information and knowledge, in how well they use information technology, and in the effectiveness of their decision making. Therefore, before entering any particular knowledge war conflict, the wise person will assess an opponent's intellect, how he thinks, how cultural influences shape his perceptions, the information technology tools he possesses, how he uses information technology, and his capability to make rapid and effective decisions.

In knowledge war deception will figure prominently as a way to confuse or manipulate the thinking of an adversary. The goal is to slow or confuse decision making processes, thereby enabling decision advantage to accrue. Both sides in an information struggle will attempt to mask the identification of the information and knowledge they seek and receive, and mask what they know about the information and knowledge the adversary already possesses. In the quest to manipulate the situation to their advantage, wise people always anticipate deceptive or manipulative activities by their foes and collect information and knowledge to confirm or deny such activities. In knowledge war a mechanical transformation process changes data to information. It is, though, an intellectual process that takes place within knowledge workers' minds, who take information, add value, and create valuable knowledge. Knowledge leads to understanding and astute decision making. Besides being important in its own right, this decision making often results in actions that directly lead to effects that influence the will of people, organizations, politicians, and a country writ large. Thus, decision making is the key linkage among desired effects, actions, and will. As general access to information and knowledge improves, the struggle for advantage will grow in complexity and intensity. Situations will arise in which computers actually take over aspects of decision making because of the complexity and speed of needed decisions, particularly in cyberspace.

Information has great power embedded within it, waiting to spring forward into energy and action. But someone must make a decision for that information and knowledge to become power. Of course, quick-witted people with good information and knowledge will make good decisions. Fools having good knowledge but poor understanding (because they don't recognize the value of knowledge) will make foolish decisions. As a corollary, intellectually competent people armed with bad information and knowledge will have better odds in making good decisions than intellectually deficient people armed with good information and knowledge. Thus, it is an imperative of the age to develop with purpose the intellects of current and future decision makers to know and understand the mechanics of information technology and the art of thinking and planning sufficiently well to make decisions better and faster than an opponent.

During a struggle for decision advantage, there is an absolute need for thinking about the thinking of the adversary. Each side in such a struggle must understand the other's decision cycle. Each side must know and understand in detail the information systems its opponent uses to facilitate the development of knowledge, which then leads to effective decision making. This clear focus must include communications connectivity, hardware and software, functional items such as intrusion detection systems and firewalls, and the vast array of databases, switches, routers, and the like. Finally, each side in knowledge war must affect the adversary's decision making through both direct and indirect assaults on the information systems supporting the development of knowledge and decision making. These assaults also include anticipating and denying similar strategies, tactics, and actions by the adversary, as well as affecting both the inputs of information and the feedback mechanisms for ascertaining the effectiveness of the decisions once made. Deception is crucial to the processes that take place in knowledge war, having defensive and offensive elements as well as physical and intangible elements. Deception will occur in the physical world and in the cyberworld; it must be planned and planned for at the outset of an operation, because it is knowledge intensive and once an operation is underway it is too late to plan and execute.

A mental bridge must be built for moving from the art and science of war in a kinetic world (in which attrition reigns supreme) to war in a digital world (in which numerous intangibles, nuance, and a multitude of interrelated effects reign). Of course, some form of violence, death, and destruction will continue throughout the twenty-first century. That being said, the pendulum is swinging from the kinetic to an emphasis on more nonkinetic forms of conflict. Because information and knowledge are so critical for making good and timely decisions, the asymmetric tool of information operations will supplant the heavy reliance on traditional, kinetic, and atomistic perspectives of conflict in the twentieth century. Military services must prepare for future conflicts and the inevitable transformation of war having a more invisible, intangible, cerebral nature. People must consider and decide what information they need for making decisions faster and better than their antagonists. People will have to determine how they access and use information, and how they relate to the ends they seek—they must understand how to collect, process, and visualize information for it to be most useful to them. Increasingly, autonomous intelligent agents will provide the help that cyberstrategists need to think through decision branches representing alternatives and possibilities. In such processes, knowledge war strategists will seek identification of variables and potential second- and third-order effects. Once variables and possible effects become known, they are subject to manipulation for personal gain. The goal in all these machinations is, quite simply, to reduce one's own friction (unintended chance events) extant in any dynamic situation and to increase friction working on opponents.

Subtle and insubstantial, the expert leaves no trace;
divinely mysterious, he is indelible. Thus, he is the
master of his enemy's fate.

Sun Tzu, *The Art of War*

Asymmetric Warfare 3

The primary challenge facing the United States in the twenty-first century
will be to deal effectively with adversaries who are very different from
those of the past. New opponents will increasingly use asymmetric strategies, tactics, and tools. These asymmetric adversaries undoubtedly intend
to attack the country's will, and they see decision making as the principal
way to affect that will. They will design their actions to lead to effects (outcomes) that cut across social, political, military, informational, economic,
financial, and ecological systems. Their goal: to create the greatest second-
and third-order effects possible, effects that eventually cause the country's
will to implode. It will be a contest about will, and there will be a winner
and a loser. Our asymmetric foes of the future undoubtedly hope to recreate the massive second- and third-order effects that struck our country's
social, political, economic, financial, military, and information systems on
11 September.

Asymmetric warfare involves a strong force either using or threatening to use an advantage that the weaker opponent cannot respond to; it
also involves a weak force seeking offsets against the stronger force; and
it usually presents a social or political dilemma to the stronger force. In
almost any situation a stronger force flaunts its advantage openly and

powerfully—it combines direct and indirect tactics with the use of physical advantages such as tanks, wealth, technological prowess, and organizational strength. Although primarily a physical advantage, asymmetric activities in this sense provide a psychological advantage too. On the other hand, a weaker force uses an indirect approach to strike at a stronger force's vulnerabilities, and occasionally at its strengths. Weaker-force offsets can be physical, psychological, intellectual, or technological (a principal area for investigation here). It seeks to create effects by manipulating intangibles such as perceptions or by the use of deception and surprise. A weaker force also purposefully presents a stronger force with situations in which choice for counteraction is constrained by political or social influences. The weaker force seeks to establish the effects necessary for exploiting such dilemmas.

When an opponent engages the United States using asymmetrical strategies, tactics, and tools, it seeks advantages to affect U.S. decision making. Thus, as a new twist in thought, military planners must consider, indeed anticipate, the type(s) of advantage an asymmetric opponent will seek and the actions (and effects) they will take. It follows that strategists must always ask themselves, What kind of advantages do our opponents seek, why do they seek them, and what causes me to think this way?

Asymmetric warfare involves intangibles—achieving offsets through surprise, shock effects, and the ability to influence information sources to create aggregate effects in command and control systems, the decision making of leaders, and the will of the publics that support them. Our future adversaries will attempt to use actions (tangible and intangible) to create the greatest and most lasting second- and third-order effects that connect horizontally across the social, political, economic, financial, military, and information systems that make up the systems fabric of our country.

In our current situation, our asymmetric adversary—the terrorists—either are attempting to or will create actions that they hope will lead to effects that cause our national resolve to crumble. Effects determine the creation or seizure of advantage; they determine who wins and who loses. Often the best effects deal with intangible psychological elements arising from the inherent power of aggregation.

Effects are critically important in the struggle for decision dominance and supremacy of will. Thus, any smart asymmetric foe of the future will try to create conditions that facilitate the rise of second- and third-order effects that have a direct inroad to the decision making and will of the political and military decision makers of the United States. Often asymmetric adversaries will create kinetic actions leading to nonkinetic or intangible effects, such as creating terror and fear that serve to degrade or even stop rational individual and organizational thought.

In an environment of ever-increasing complexity, future military planners will have to acknowledge that our adversaries are attempting to create first-, second-, and third-order effects. Some of these effects will be played out in the military realm, but many others will be felt in the social, political, informational, financial, economic, and perhaps even religious arenas. Thus, along with anticipating the adversary's attempts to use action to create dramatic first-order effects, military and security planners must anticipate its efforts to create second- and third-order effects and conclude that thinking about these ancillary effects must take into account the adversary's point of view and our view of the adversary, and the perspectives of both adversary and friendly neutrals and allies. Indeed, this is cerebrally challenging work.

It will be a rare moment when a weaker opponent engages a stronger foe in force-on-force conflict. Instead, that opponent will come at its foe (us) indirectly, at unexpected times, and with unanticipated goals and will be able to create tangible and intangible effects that lead to the soft underbelly of decision making and eventually to will. This will occur with great stealth, cunning, and, in some cases, significant violence.

When attempting to use asymmetric warfare strategy, tactics, and tools, the dominant theoretician whose thoughts guide our thinking is the venerable sage of war, Sun Tzu. Sun Tzu wrote centuries ago about indirect approaches, the importance of information, the psychological effects of surprise and fear, the importance of taking the moral high ground, the importance of perceptions, the importance of intelligence, and the importance of deception.

Sun Tzu's world was a complex environment for that time, and his thought-provoking aphorisms remain subtle and hard to decipher to this day. He interpreted the art of war as possessing obvious and subtle relationships vulnerable to aggressive manipulation and human foibles that could be directly or indirectly exploited. Sun Tzu saw human weaknesses as "a way" to seek and attain conditions of advantage.

A few thoughts about aggregation theory provide some relevance to the discussion of what asymmetric opponents hope to accomplish. A great force comes with the power of aggregation. Those who learn to find relationships and combine and exploit them are aggregating. A certain degree of synergy (in which the whole is greater than the sum of its parts) comes through aggregating or connecting things, activities, or actions. When things are connected in tightly woven relationships, they move together whenever one is disturbed. The slightest perturbation or variance in one element can cause dramatic changes to the whole. Thus, we can surmise that in future struggles our asymmetric foes (as well as our own planners and decision makers) must engage in wargaming that enables one to discern which sensitive variables will cause the greatest perturbations to the adversary while anticipating and denying such variable manipulation aimed at us. (This thought also helps to explain the incredible power inherent within second- and third-order effects.)

An asymmetric opponent's use of tactical psychological operations (PSYOPS) illustrates aggregation theory. PSYOPS, in this instance, denote the ability to affect individuals and eventually the collective intellects of many people, which ultimately causes an effect in morale. Affecting morale influences the decision making of individual leaders and organizations. The struggle in such competitive activities lies in the invisible realm of manipulating collective-aggregate psyches, and the influence they have on fighting performance and hence on decisions. When perturbation disturbs the morale of fighting service people, it has the first-order effect of adversely influencing fighting capabilities and sapping will. A second-order effect is the cognitive dissonance that takes root *directly* in the minds of commanders, whose combat effectiveness and command and control systems and decision cycles are at risk. The third-order effect would be the indirect cognitive dissonance that occurs in the psyche of the leader. Dissonance piled on a mind already experiencing stress, fatigue, and fear will undoubtedly cause a corresponding decrease in ability to make fast and effective decisions.

To help explain the impact of asymmetric warfare in our country's current situation, several important ideas must be understood:

- First, in a military sense asymmetric warfare affects all levels of war—tactical, operational, and strategic.

- Second, no one, civilian or military, will be safe; no sanctuary will exist.

- Third, asymmetric attacks will focus on windows of opportunity in which activities compound in importance owing to the incredible power of event simultaneity. Events occurring simultaneously around the world, for example, have both a physical and psycho-

logical effect—through the use of the Internet, directions and execution instructions sent at the speed of light to create the right
actions can cause the right effects at the right time and at the
right places.

- Fourth, with the advent of global communications and computing systems, unique opportunities have been presented to opponents to synchronize activities, share information and knowledge,
  and make decisions instantly across space and time. Global communications have empowered individuals and small groups who
  constitute our future asymmetric foes.

- Fifth, physical distance and time will dissipate in importance to
  our foes. But distance and time will remain important for the
  United States because it will still have to move people and things
  over long distances. This situation presents asymmetric advantages to our country's future adversaries. National intelligence
  and law enforcement agencies, the U.S. military, state and local
  governments, and private sector individuals involved with homeland security must prepare for asymmetric attacks targeting the
  synchronization of people and equipment in all aspects of training, simulation, modeling, and wargaming.

- Sixth, in order to understand, use, and preclude opponents from
  attacking our vulnerabilities, we must change the way we think
  and plan. Learning to think better by understanding relationships
  through synthesizing information and knowledge is the principal
  point here. Our thinking processes must help us decide what information and knowledge we need to anticipate, tip and cue, and
  block and parry.

- Finally, information operations will be an important element in
  the tool kit of asymmetric opponents. The validity of this assertion lies in the dependence of the United States on information
  technology for seeking, finding, processing, and using information and eventually knowledge to make advantageous decisions.
  IO affect thinking, planning, and the data, information, and
  knowledge that people must have to make decisions.

The use and manipulation of information are critically important. Some of
our information requirements will center on seeking information about
what the terrorist is trying to find out. Some of our information require-

ments will center on how the terrorist thinks, plans, and perceives the world. Some of our information requirements will center on what is happening at the state, local, and national government levels, as well as pertinent information from locations throughout the world. Some of our information and knowledge requirements will come from experts who can provide educated opinions about our adversary's motives, targets, possible tools, and deceptive measures.

Terrorists have to obtain information and knowledge. Part of our readiness should be not only a physical anticipation of action but a mental anticipation of what information and knowledge the terrorist must possess to create the effects he seeks. There will be indicators in both the physical world and cyberspace that such collection activities are occurring. When armed with the results of such analysis and synthesis, our people must become proactive in deciding how to anticipate, deny, confuse-deceive, destroy, and so forth.

Terrorists, of course, also have to rely on an information-collection plan, but where will they find that information? The Internet has incredible amounts of useful and available information. Terrorists will conduct personal reconnaissance and surveillance of potential targets. Quite simply, however, they cannot obtain all they need from the Internet. They also will conduct human intelligence (HUMINT) operations of sorts, asking questions, and seeking guidance and advice from people who have the information and knowledge they want and need. They also will be forced to buy commercial imagery for their planning and execution. They will seek signals intelligence (SIGINT)-type information from cell phones and wireless computer communications. They will use cyberspace to gather information and knowledge about their targets, and they will do so surreptitiously.

What other things will terrorists require to wage their war against our country? For one thing, terrorists need communication devices to talk, coordinate, and orchestrate events, computers to support planning and rehearsals, and encryption tools to help protect what they are doing. Once they make their decisions and act, terrorists need feedback on the "effectiveness" of their attacks. The news media will help them immensely, but they also will require information and knowledge from the Internet and from personal interactions with people "in the know" to ascertain the true value of their actions and effects.

Terrorists will need to synchronize their activities (the obvious example being the attacks on the World Trade Center and the Pentagon). How did they plan? How did they synchronize? How did they rehearse? What were their backup plans? Why did they focus their attacks on the south side of the Pentagon? Orchestration and synchronization will become larger issues in the future. Because of globalization and the shrouds of relationships encompassing the world's tapestry of systems, terrorists will seek to orchestrate and synchronize in many locations, at the right time, to create the right effects that lead to acceptance of their will.

Terrorists have the distinct need to anticipate our responses and know the activities we will choose to thwart their actions. They have to have some kind of an act–react–counteract cycle for thinking through possibilities. The implication for us is, of course, to conduct a wargame from their perspective. This wargame should present an opportunity for us to anticipate and to "be there waiting" when they come forward from the shadows. It suggests, of course, a distinct need for our security knowledge workers and decision makers to have sophisticated virtual reality wargaming techniques at their disposal.

Information operations could very well be the centerpiece of future asymmetric warfare and conflict. Why is this so? Because the world is becoming increasingly dependent on information and knowledge. Also, webs of information encompassing the earth are becoming more plentiful, denser, and more interrelated. People are becoming better informed, smarter, and more connected, but also more dependent on information and attendant communications to move the information and manipulate the data and turn it into information and knowledge. Learning, adaptive foes cannot help but notice these vulnerabilities and will come at them with guile, cunning, and force.

What are some of the more salient ways that asymmetric adversaries differ from adversaries that U.S. armed forces have grown so comfortable with in the past? Our former foes had shape and form, which enabled people to template, and through the relationships among terrain, doctrinal templates, and organizational structure their future courses of action could be predicted. Our new foes defy strict, quantifiable definition, shape, and form, which renders such templating and the mental faculties needed for dealing with them hopelessly anachronistic. Our military forces need to rethink efforts to predict and concentrate more on relevant possibilities, forming hypotheses about those possibilities, and developing and looking for indicators that suggest the reality of those possibilities, confirming or denying such hypotheses with traditional and nontraditional information collection, and deciding to take action when appropriate. With this methodology we can anticipate terrorist actions and stop them.

Asymmetric activities are at once both physical and invisible, and their effects (outcomes) can be tangible and intangible. This characteristic, very different from the past, can cause confusion in our intellects and in the currently-in-use machinery of information collection. Quite simply, machinery that was designed to perform against hoards of tanks on the steppes of Russia will have a difficult time penetrating the cells and minds of terrorists.

The activities demanded by asymmetric foes require immense intellectual capabilities in the realm of analysis and synthesis. In the past, when the threat had shape and predictability and could be found and tracked, all that one needed was good, solid analysis. Now, as we chase shadows and anticipate the actions and effects of intellectually sophisticated, relational people and organizations, we will have to rely on thinking skills that lie in both analysis and synthesis. Unfortunately, neither our institutions of higher learning nor our military institutions of learning help people learn relational thinking.

Preparing to combat asymmetric foes requires an increasingly strong relationship between man and automated machines, as computers become smarter and are able to perform more of the activities and actions that cause effects. In the past, of course, neither friend nor enemy had computing capabilities. Now, both friend and enemy have tremendous capabilities in that arena, and a broadening symbiosis between man and machine. What we will see is a race to the future in which both sides seek to create advantages in cyberspace. Our country must prepare now for this challenge in our doctrine, in organizational design, training, and education, in materiel, leadership, facilities, and people.

Historically speaking, asymmetric warfare represents nothing new— weaker competitors always have sought an advantage through "offset" against a bigger, stronger foe. What is new lies in the increasing capabilities the information revolution presents for asymmetric adversaries to help create the offsets they desire. Examples that come to mind include the Internet, cellular phones, personal computers, personal digital assistants, powerful kinetic bombs, and individually placed weapons of mass destruction. Modern societies rely on these tools for thinking, planning, making decisions, and determining the efficacy of those decisions through modern information technology. Terrorists can use these same tools without fear, masked by the legitimate activities of millions of people. Moreover, the United States is unmistakably dependent on the tools of IT, a dependence that presents such a tantalizing target to future terrorist intentions. The movement of information around the world and the interconnectivity of previously isolated societies have heightened the importance of political constraints on the methods and limits of force in conflicts, such as rules of engagement. Ironically, strengths are also vulnerabilities. Moreover, the ascendancy of the digital age, the advent of the information revolution, the growing specter of terrorist capabilities in coordinating among themselves across time and distance, and the ability for terrorists to arm themselves with weapons of mass destruction all suggest a mortal danger to the world's well-being.

Several illustrations help in understanding the abstractions inherent to asymmetrical conflicts. People or groups who use asymmetrical strategies, tactics, and tools often use indirect approaches. An indirect approach relies on surprise and shock achieved through physical and mental actions designed to work against the individual and collective psyches of their foes. Such adversaries usually attack their opponent where least expected, with surprise, with ferocity, and rarely against the opponent's strengths. An indirect approach recognizes and uses both moral and physical domains of conflict. A specific example of an indirect approach would be to impede the flow of U.S. troops by manipulating the computer systems that control scheduling departures and arrivals of trains, ocean vessels, and aircraft at air or sea points of embarkation, thereby creating chaos in any attempt to synchronize people, materiel, and machines.

The moral domain of conflict, which lies at the heart of asymmetric warfare, involves perturbation in psyches and in individual and collective will, fear, moral resignation, fatigue, low morale, and surprise. The engine that provides strength to the entire domain is emotion. Effects engender change in decision making and will, and thus we can conclude that intangibles lie at the core of asymmetric conflict. It is extremely difficult to quantify, visualize, or measure these intangibles, or create quantifiable actions leading to desired effects, when dealing with the moral domain.

Use of the moral domain makes sense for our asymmetric enemies: why would anyone choose to compete with another's strengths, particularly their physical strength, if they could instead capitalize on their opponent's weaknesses, at minimal cost and risk? Why not choose the moral domain of conflict as the principal contextual area of competition and intangibles as the primary elements of the contextual environment to manipulate? Our foes of the "Hundred Years' War" we are engaged in will use the venue of the moral domain to wage their war with sound and fury.

A smart adversary will employ activities and weapons that the stronger opponent has no response to. Examples include using or threatening to use chemical or biological weapons, knowing that the target country could not provide a quid pro quo or reciprocal response. An intelligent opponent will think, quite simply, that even though the opponent has the capability (that is, the potential bound by constraints) to respond in equal or greater force, it nevertheless also would have too many constraints cutting across social, political, economic, financial, military, and ecological spheres to be viable. The interesting aspect of such a situation isn't about the truth or what one's perceptions would be. Instead, the important aspect of the situation is thinking about what the opponent's perception would be concerning whether or not the adversary (the United States) would or could find a useful quid pro quo to respond to or counter such asymmetric activities.

Terrorists foreign and domestic will attack, extort, and manipulate the families of U.S. service people while their loved ones are deployed away from home. Injecting false e-mail messages, morphing images, kidnapping children, and levying threats against families will have far-reaching and visible effects on the morale of U.S. service people. Morale affects operational capabilities, a degradation of operational capabilities affects decision making, and a deterioration in decision making affects will. Families will have to train to deal with such possibilities. Moreover, massive resources will be expended to protect this glaringly vulnerable Achilles' heel of the U.S. military.

Asymmetric foes could employ thousands of guerrillas in an opponent's rear area. The goal involves forcing a dilemma for the opponent's decision making: pull conventional forces from main battle areas to deal with enemy guerrillas, or allow them to cause great damage and harm to one's infrastructure, public psyches, and political perceptions. Such employment of guerrillas en masse and at the right time could cause public clamor so great that a government would have no choice but to weaken its front lines. Once the enemy sees such a weakening and momentary vulnerability, it will strike with full force to gain advantage of action, position, and decision making.

The asymmetric foes of the future will perform a sophisticated analysis of the stronger opponent's centers of gravity. They will seek to affect those centers of gravity indirectly but with significant impact, thereby affecting command and control (C2) to the point of decision-paralysis and creating conditions that make responses oscillate between impossible and inappropriate. Such an effort would be particularly powerful if the opponent timed such activities to occur at a moment of maximum vulnerability. An example of this idea would be to attack or manipulate a place (physical or cyberspace) in which a confluence of communications, collection, automation, thinking and planning, and decision making occurs and whose role in activities is so important that if lost or adversely affected would seriously jeopardize security. The United States has many vulnerable centers of gravity, ranging from military command posts and headquarters to state and local emergency operations centers, to incident commander operations centers.

Cyberspace is a perfect environment for asymmetric foes to strike. Adversaries will work indirectly, invisibly, and, from their perspective, undetected. They could, for example, plant hidden viruses on key command and control software if the software is developed in foreign countries. American vendors typically don't check software code for viruses and logic bombs, but instead check for functionality. Smart foes could design these viruses to work at the right time to cause disruption, employ deception, or render the center of gravity completely useless for a short time to accomplish effects fitting an IO campaign plan.

An asymmetric foe will leverage sophisticated psychological operations to affect decision making and will. It will attempt to manipulate the public psyche of the United States by using aggregation theory, which postulates the power of cumulative effects and phenomena inherent to energy contained in second- and third-order effects. It will work this lever by operating against the strengths of the United States, which are also its vulnerabilities: openness and widespread, instant access to information by its citizens.

Asymmetric competitors will attempt to manipulate the aggregate psyche of our public or communities of interest. They will use the power of images to provide information and knowledge that are, at once, both graphic and suggestive of cost and risk. Asymmetric foes will create situations in which they perceive that the aggregate psyches of the American populace won't cope with terror and heavy casualties. In this respect the opponent's goals lie in attempting to manipulate the age-old center of gravity historically recognized as "will." Perception management is crucial in this situation, particularly in anticipating and denying these efforts. From their perspective, asymmetric adversaries believe a violent, graphically communicated assault will cause sufficient angst in the American populace to cause political leaders to change policy. An example of this is the vivid image of the dead American soldier being dragged through the streets of Mogadishu, Somalia, in 1993. News reporters at the scene piped the image back to the States and other global media centers, immediately projecting it to millions of American homes and causing a significant perturbation in the aggregate psyche of the American populace and, more important, in the minds of the country's politicians. The result was that a single tactical event caused a change in the country's strategic national policy due to a change in politicians' perceptions about the will of the U.S. populace. Asymmetric adversaries of the future will attempt to recreate this kind of situation, as they believe the will of the country is not sufficiently strong to withstand large numbers of casualties visually brought into their homes in graphic detail. The country's strong response to the terrorist attacks of 11 September is but a first step in the first stage of the first phase of the "Hundred Years' War" in which we find ourselves.

Our future asymmetric competitors will be smart enough to attack a new age dichotomy—a strength (technology) is also a vulnerability. We must think through what is important to us and to our enemy, and we must anticipate his information and knowledge requirements to plan and execute such operations. We then have to establish nontraditional information-collection efforts among federal, state, and local law enforcement agencies, national agencies, and national intelligence services to tip and cue our planners and decision makers that such terrorist activity is occurring or about to occur. We must wargame how the adversary views himself and how he views friendly forces. We have to be on the lookout for deception in this intellectual drill. Only then will we be able to anticipate his activities and lie in wait for his approach, whether that approach is in a physical area or in cyberspace. The characteristics, strategies, tactics, and tools of forthcoming foes require us to wage a largely silent and invisible competition. Conventional, industrial-age approaches to information-age problems will be increasingly irrelevant and will often work to our disadvantage.

This different type of competitor represents several challenges to any organization involved with national defense. Leaders must write policies, and train and educate their people to operate against conventional, industrial-age as well as asymmetric adversaries. Thought processes—both organizational and individual—need adjusting. As such, people need to synthesize better and learn to use holistic planning. They need to think about possibilities, not just capabilities. In addition, we need to learn to wargame in our minds and with our machines to become more anticipatory in our thinking. In addition, nobody can wish away the specter of asymmetric strategies, tactics, and tools. Prudent leaders, therefore, must model realistic, asymmetric challenges in simulated wargames. In our training, whether civilian or military, people representing asymmetric adversaries have to think and act like our real foes, not like us. Such modeling requires the integration of tangible (attrition and movement) with intangible (morale, fear, surprise) effects. Asymmetric problems do not fit neatly into what people know as problem sets today—they are too different, too abstract, too relational, too cerebrally demanding, and often too invisible. Indeed, conventional foes will use increasingly asymmetric activities to advance their causes.

Regardless of how people view the future, asymmetric threats and activities are real, challenging, and dangerous. Asymmetric warfare represents a more significant challenge to the country than any threat that has heretofore surfaced in the country's history. How can the United States cope with such a threat? First, people have to change the way they think, plan, and view the world. Second, people in national security have to accept the concept of possibilities (potentialities not bound by constraints) and avoid falling into the well-camouflaged trap of following the past and concentrating only on capabilities (potentialities bound by constraints) and intent. Third, people must develop a way to avoid mirror imaging their own value sets, educational backgrounds, ethics, and religious values on opponents whose views are extremely different. Fourth, defense analysts and decision makers must consider the perceptions of people that asymmetric foes will target in the struggle over control of national will. Perception, so the saying goes, is truth in the eye of the beholder. It is the opponent's perception that is most important in this contest, not our own. Finally, a learning, adaptive threat could very well know and understand our systems, processes, and psyches almost as well as we do. If so, people responsible for U.S. national security have to recognize the ubiquity and potential availability of knowledge to all people. This availability of knowledge should cause defense analysts and decision makers to anticipate and expect asymmetric assaults at any time and in any place.

There is still another factor that can bring military action to a standstill: imperfect knowledge of the situation. The only situation a commander can know fully is his own; his opponent's he can know only from unreliable intelligence. His evaluation, therefore, may be mistaken and can lead him to suppose that the initiative lies with the enemy when in fact it remains with him. . . . Men are always more inclined to pitch their estimate of the enemy's strength too high than too low, such is human nature. Bearing this in mind, one must admit that partial ignorance of the situation is, generally speaking, a major factor in delaying the progress of military action.

Carl von Clausewitz, *On War*

4   Information Superiority

The goal of continuing the identification and improvement in technology and enhancing its effects on thinking, planning, and decision making should always be to ensure that the United States keeps its advantages over current and anticipated adversaries. These advantages, though, will fluctuate far into the future because our adversaries are becoming intellectually and technically smarter with the passage of time. Let us begin by gaining an understanding of the general ideas behind information superiority, then proceed to a forecast of what the U.S. military could need for meeting information superiority requirements to support military transformation.

Environment

Any discussion of a complex idea such as "information superiority" has to address its causal underpinnings. The four underpinnings to be considered here are information technology, nature of future conflicts, asymmetric warfare, and the knowledge war.

It is no secret that information technology has changed the world. People can instantly communicate from anywhere in the world through modern telephony and the Internet. Powerful yet inexpensive modern computing allows people to collect and manipulate information and knowledge in stunningly prodigious quantities at an amazingly fast pace. More important, people learn from the Internet by accessing libraries, entering chat rooms, surfing websites, and collaborating with other people to solve problems. Of course, people can use the knowledge they acquire in negative or antisocial ways, such as making bombs, hacking into systems, building biological and nuclear weapons, and exploiting the intricacies of societies. In a nefarious sense the Internet and personal computers have empowered small groups of people and individuals to coordinate and synchronize terrorism, criminal activities, and drug operations around the world at the speed of light. To many of our future adversaries, distance and time no longer constrain their activities, thanks to the Internet. But to the armed forces of the United States, which has to move people and equipment over long distances, time and distance are critical parts of any deployment strategy. These time and distance constraints are vulnerabilities any capable terrorist can attack.

The nature of future conflict doesn't bode well for the United States. Arguably, the United States has the best conventional armed forces in the world—no country now or in the short-term future can compete with the U.S. military in a conventional sense. But these conventional forces, with their force-on-force prowess, are experiencing an increasingly complex identity crisis: they will remain important for the sake of deterrence, and are a "must have" in the event of a conventional challenge, but only the very stupid opponent will choose to compete with them in a force-on-force situation similar to Desert Storm. That being said, people haven't stopped hating the United States or being envious of its strengths. Thus, opponents of the United States will undoubtedly choose to compete in dimensions of national power and the military sphere only when, where, and how it suits their purposes. They will undoubtedly focus their efforts on creating desired effects (outcomes) through both military and nonmilitary activities, actions, or, in some cases, combining the two.

Opponents of the United States will compete in economic, financial, political, informational, and social spheres of influence. They will seek offsets against the conventional power of the United States military and the multitude of advantages that such power brings. They will seek and use approaches that lead directly or indirectly to affecting the country's will. Employing all available weapons, these adversaries will neither fight nor compete fairly (from a Western point of view). In this type of competition, sanctuaries like the safe havens of past wars do not exist for either side. Often conflict will be in the invisible, digital cyberworld. Adding to the confusion inherent in a chaotic world, future adversaries likely will espouse objectives and goals very different from those of the United States—so different, in fact, that even the country's best analysts may fail to recognize them in advance.

Asymmetric adversaries understand the direct and obvious linkage between decisions and will. Clearly, future asymmetric foes aim to influence the supreme center of gravity—the will of the people of the United States—by affecting the decision making of its individual citizens, body politic, military, coalition partners, and political leadership.

With the presence of smart, adaptive, and learning adversaries, both sides in any conflict will quickly learn that the condition of information superiority is neither a given nor a lasting state of being. Instead, information superiority fluctuates, based on act–react–counteract cycles of interaction among smart, adaptive adversaries. Whoever controls this all-important condition will go far toward satisfying their objectives and making faster and better decisions than an opponent.

Knowledge war forms the underlying foundation of future conflict. Knowledge war is an abstract but very real struggle to gain valuable knowledge for the distinct purpose of making better and faster decisions than the adversary—and thereby gaining a decision advantage. Sometimes this competition for decision advantage involves only one person, but such a situation is the exception, not the rule. More commonly, adversaries try to affect the decisions of a group of people who constitute a decision-making apparatus. Decision advantage enables one side to take actions that are more beneficial than the actions of its opponent, thereby leading to other types of advantages, such as intellectual, knowledge, technology, positional, and action advantages. Decision advantage is arguably the most important advantage, because it sets the tempo and leads to gaining and sustaining initiative and momentum—all of which have been important for winning in military situations throughout history.

At a higher level of abstraction, decision making is the most important way to affect the will of opponents. From decisions come actions. From actions come effects. Effects lead to a manipulation of will. Thus, owing to this linkage, decision making is arguably the most critical aspect of future conflict. For as long as mankind has engaged in conflict, affecting will has been a preeminent goal. Even in its most abstract and intangible state, will is the energy, sacrifice, resolve, long-term perseverance, and collective desire of people to win or triumph over a foe. Affecting the decisions and decision-making apparatus of adversaries is the path toward influencing an individual's, an organization's, or an entire country's will. Will's influence is far-reaching, ranging from soldiers at the tactical level to aggregates of population and senior decision makers at the strategic level.

Making decisions that are decidedly better than an adversary's doesn't happen by accident. To the true artist of knowledge war, information superiority is a condition enabling the effects of superior decisions to come into being. Information superiority is only a temporary condition, though, since smart, adaptive, learning adversaries will quickly adjust and seek the condition's benefits. It follows that both sides will seriously and aggressively seek the benefits that derive from information superiority.

Leaders of our country's military establishment have publicly stated that information superiority is the key enabler in seeking, finding, and sustaining dominance over our future opponents. Along this line of thought, the Department of Defense's Joint Staff wrote a short, pithy document titled "Joint Vision 2020" to describe future environments, future threats, and probably roles and mission of the DOD's military forces and its technologies. It says: "These changes in the information environment make Information Superiority a key enabler of the transformation of the operational capabilities of the joint force and the evolution of joint command and control."[1] Curiously, this important document has a watered-down definition of information superiority: "the capability to collect, process, and disseminate an uninterrupted flow of information while exploiting or denying an adversary's ability to do the same."[2]

Yet when thinking about the definition of information superiority, one has to start asking questions such as, What exactly is information superiority? How does one know when information superiority is found, achieved, sustained, or lost? Is information superiority everywhere and all the time? Or is it localized and episodic, owing to the intellectual and technical capabilities of opponents? What does information superiority provide in the way of advantage? Anyone who reads "Joint Vision 2020" could easily be confused by the definition of information superiority provided there. Using imprecise wording the definition pertains to processes but doesn't tie to arguably the most important aspects of information superiority: the perpetual struggle for advantages in decisions, and gaining favorable situations in tempo, initiative, and momentum. A more meaningful and functional definition of information superiority is "the use of information technology and intellectual power to create conditions to make better and faster decisions than an adversary. These better, faster decisions provide advantages in tempo, initiative, and momentum against an enemy or opponent at a time and place of the commander's choosing, with the notion of creating conditions leading to the effects most conducive to rapid mission accomplishment and sustainment of advantage, at minimal cost." From this definition one can gather that information superiority is an elusive, changing condition. What, then, are the key elements of information superiority?

How people process information (visually, auditorily, and kinesthetically) and how they think and turn their thoughts into decisions and actions are the most important aspects of information superiority. But data, information, knowledge, understanding, and some decisions themselves are elements that provide the capabilities to process, think, and decide further. These elements do not stand alone—there must be software and hardware interfaces conjoining man and machine, coupled with the use of cognitive skills (of both machine and human) to make sense out of that information, to find patterns and relationships among aspects of seemingly different, but all the while highly related, elements of a macro-knowledge.

A purposeful transformation process of automated machines and people turns data into information. People have to seek and find the right data, then transform it into information through processing with the right computer software applications. Information becomes knowledge when skilled people add value to an issue or problem by way of their thoughts, experiences, intuition, and creativity. Knowledge, though powerful in and of itself, becomes extraordinary when combined with other knowledge to find patterns and relationships and to become available for use. This process involves the mental function of synthesis. Understanding, a precursor to making effective and timely decisions, comes from recognizing and exploiting obvious and disparate relationships among pieces of knowledge, then "weaving" these pieces into a meaningful whole that leads to decisions.

The word "advantage" suggests that one side or one person or one country has a favorable situation relative to an antagonist. The advantage in the context here lies in individual and collective brainpower commingled with technology, with information, and with knowledge. Information and knowledge, though, aren't static—they come to life when people use them to make decisions and act. In this context knowledge isn't power, it is the means to power. A leader has to make a decision to act or not to act using the knowledge at hand to enable what was once only knowledge to become power. The underlying lesson for all human beings is the importance of striving to do something with knowledge that creates a lasting advantage. It follows that what is truly important in any conflict is what people or machines do with information to turn it into knowledge, to manage that knowledge so people can use it, think, and plan, and to make decisions faster and better than the competitor.

When contemplating advantage, one dynamic comes into play quickly: while one side is attempting to find, use, and sustain advantage, the other side is not paralyzed. It is seeking ways to combat its disadvantage and find ways to regain the advantage. Achieving advantage could also be as simple as creating a disadvantage in a capability of an adversary. The smart participant in a conflict understands that advantage is short-lived, owing to the capability of the opponent to learn, adjust, adapt, and change. Thus, intense interaction fluctuates aperiodically—sometimes over space, sometimes over time, sometimes over both. Decisions, to be effective, involve taking advantageous actions, anticipating environmental factors and the actions and counteractions of the adversary, then adapting to the environment and protecting against those counteractions. It is neither an easy process nor a given state of being in any contest.

Several types of advantage seem pertinent to the subject under investigation. In addition to the age-old advantages of a fast tempo and gaining and maintaining momentum and initiative, adversaries should seek advantage in several domains: in decisions, in intellects (thinking), in knowledge (which includes knowledge management or how one uses knowledge), in technology, in gaining and maintaining position, and in taking action.

A first type of advantage, and probably the most important, lies in decision making. Finding and using a decision advantage provides a capability to seek and find other types of advantage. The worth of decisions is relative to the opponent's decision making; that is, the goal in a conflict is to make more timely (when relevant and appropriate) and better decisions than the opponent, then sustaining that advantage over time.

A second type of advantage lies in intellect: that is, one side has the distinct advantage over the other side in how well it thinks and plans. Intellectual advantage comes with 1) an ever-expanding ability to think (what to think and how to think); 2) training and education to improve thought processes; 3) technology to process, manipulate, see, hear, and feel emotion, and visualize information and knowledge; and 4) access to information and knowledge relevant to the contest at hand, whether through a website or in collaboration with knowledge experts. This advantage is perhaps the most difficult to sustain because of the growing capabilities of humans (opponents) to think, learn, innovate, create, change, and adapt, and to develop and use readily available information technology to find information and knowledge.

A third type of advantage, knowledge management, is the capability to use knowledge as a lever for gaining entrée into the mental domains of knowing, understanding, deciding, and acting; it is often crucial in a competition, particularly if opponents are evenly matched. Thus the way people organize, seek, process, analyze, synthesize, and protect information and knowledge becomes an advantage. Knowledge management is the purposeful and systematic retrieval, processing, organizing, analyzing, synthesizing, and sharing of data, information, and knowledge among knowledge workers, decision makers, and organizations.

A fourth type of advantage lies in technology. In essence, technology advantage helps people make better decisions than their adversaries and provides the means for achieving superior maneuverability in which positional advantage becomes possible. Technology advantage also provides the means to giving oneself better protection. It allows one adversary to kill and destroy with more precision than in the past. Moreover, technology helps people communicate faster, more clearly, and more effectively than opponents with inferior technology.

A fifth type of advantage lies in position. Positional advantage in a conventional military sense enables one side to seize the initiative, control the tempo of an operation, and seize and sustain momentum. Positional advantage can be viewed from two perspectives. In the traditional, holding the high ground or a key bridge is tantamount to outright victory, owing to the opponent's bad positioning and attendant decision-making disadvantage. For example, General Lee was in a position of positional inferiority at Gettysburg, since he had to attack a well-entrenched enemy that occupied high ground, while General Meade held a position of superiority because his forces occupied the high ground and could shift, using interior lines of communication. In the nontraditional perspective adversaries seek positional advantage in cyberspace. The soldiers of cyberspace—cyberbots—will get positional advantage along "avenues of approach" (read: fiber optics) and "key terrain" (read: switches, modems, routers, hard disk drives, and databases). Positional advantage could mean having the potential to emplace a logic bomb using hidden cyberbots in an adversary's database management system, ready to inject bad code into the database either on command or at a specified time. Positional advantage in cyberspace could mean positioning cyberbots to discern the opponent's reconnaissance and attacking cyberbots. Positional advantage in cyberspace could also mean placing an intelligent agent inside the firewalls and intrusion-detection systems of an adversary, thereby allowing for access to the decision planning and logic of an adversary without the opponent being aware of the penetration.

A sixth type of advantage lies in action. Action is the release of energy that causes effects. Each side in conflict causes latent energy to come forth through actions thereby creating effects that, in turn, create advantage.

Wise leaders are able to combine all these advantages to create desired effects. With wisdom military leaders gain insights into the adversary's actions, intentions, and goals. Moreover, leaders gain the same level of understanding about the friendly side, crucial for the thinking required for entering sophisticated net assessments. Over the course of military affairs of the past, only the military genius had wisdom. Now, with many people linked in collaborative networks, with the tremendous capabilities in information technology, and with the capability to learn over the Internet, the possibility of achieving wisdom is real and attainable for a wide variety of people and groups.

As these threads are pulled together into a pattern, people have to accomplish several things to create the condition of information superiority:

> Seek relative, timely, specific, and accurate data, information, and knowledge
>
> Initiate and sustain the transformation process (data to information to knowledge to understanding)
>
> Use knowledge and understanding to make a decision to act or not to act
>
> Understand that opponents are doing the same thing
>
> Send the decision to individuals and organizations to execute the act
>
> Seek feedback to provide measurement of the effectiveness of the decision, the actions that came from it, and the effects that the action(s) created
>
> Send intelligence collection to find information relative to the adversary's attempt to receive feedback on its decision

It follows that military planners and strategic thinkers need to acknowledge the temporary nature of advantage, owing to the fact that learning, adapting organisms constitute our opponents in future chaotic environments. As a further implication, the military establishment needs an automated information-collection system that reliably measures and displays the rapidly fluctuating conditions of information superiority. Additionally, since the best and latest information technology is for the most part available to anyone on the commercial market, opponents of the future often have information as good as the U.S. military's. It is safe to assume that over the next ten years the specificity, accuracy, timeliness, and relevancy of information and knowledge our adversaries will either produce or purchase will rival the best our military can find or create.

Thus, the U.S. military has to possess better ways to collect, process, manage, manipulate, fuse, integrate, visualize, hear, feel, collaborate, and actually use data, information, and knowledge in automated decision-making environments, and be able to present information and knowledge better than any potential foe. Leaders then have to use the art and science of decision making faster and better than their adversaries. It is only through a combination of technologies and intellectual power that the United States can hope to sustain an edge in the upcoming struggle for finding, achieving, and sustaining information superiority.

The Parts of Information Superiority

The new operating environment argues for changing the domains of conflict. Domains of conflict, defined as "spheres of activity, interest, or function,"[3] heretofore have included the air, ground, and sea, with the recent addition of space. To this time-honored list military leaders should add the cyber and cerebral domains. To help people understand these domains we need to explain their relationships to the physical world to the best of our ability.

Cyberdomain

The world increasingly depends on the cyberdomain for the functional operations of its social, political, economic, financial, military, and ecological systems. Through the "veins" of cyberspace flows the lifeblood of society—the zeroes and ones, the data, information, and knowledge that people need to exist. In this domain adversaries will struggle to gain information superiority and decision advantage. In the future, during a conflict of any conceivable sort, making decisions faster than an adversary will often depend on the effectiveness of activities in cyberspace.

The United States is growing increasingly dependent on the cyberworld for its health, well-being, leisure, money making, and work activities, to name just a few important areas. Because of our dependence on the cyberdomain, our opponents see cyberspace as both "a place" and "a way" to attack us to affect decision making, hence the will of the country. Because of its importance in everyday life and the magnitude of potential outcomes involving information it follows that this important area of influence and activity becomes coequal with the more traditional domains of conflict.

However, this domain is somewhat abstract, quite complex, and almost alien to traditional military ways of thinking and planning. Conflict in the cyberdomain will occur with no tangible or visible ground to hold. Cyberspace is everywhere and nowhere. Opponents in the cyberdomain will as often as not be cyberbots doing their human masters' bidding at machine speed. Thus it will become increasingly difficult for military people to "know" the enemy. Collecting data and information and turning it into knowledge will take place at the speed of light and will be invisible to the naked eye. Long-standing traditions of deception and espionage will become more abstract and sophisticated than the kinetic, physical world to which militaries are accustomed.

Cerebral Domain

Knowledge war involves thinking and planning, allowing knowledge workers to add value to information and turn it into knowledge. How humans think and view the world greatly influences how they accept the multitude of inputs into their minds, how they process information and seek relationships through synthesis, and how they apply context, experience, and intuition to produce knowledge that has value. Intellectual advantage for developing knowledge is crucial for making decisions sufficiently superior to create the condition of information superiority. Thus antagonists will try to improve their knowledge workers' minds and protect those minds from intrusion or manipulation. At the same time, opposing forces will attempt to manipulate the minds of the other's knowledge workers and decision makers to affect the quality of knowledge and decisions being produced. The cerebral domain arguably will become the most important domain and its value comes from the importance of the human mind in turning information into knowledge and developing knowledge that is at once more timely and more relevant to the issue at hand.

Additionally, knowledge war will involve traditional and nontraditional ways of conducting the pinnacle of cerebral activity: deception operations. To deceive one must have at least an approximate knowledge of how the opponent gathers information, places value on it, thinks and processes it intellectually, collaborates, makes decisions, and gathers feedback on how well the decisions are working. Confusing, slowing, or overloading an opponent's mind through deception is one of the advantages people will learn to seek.

Individual and collective perceptions are crucial in the cerebral domain. When thinking about perceptions people also must think about analysis, duality, impact of culture, psyches (individual and aggregate), relationships with decision making, relationships with information and knowledge, self-awareness (that is, organizational and individual introspection), manipulation, and management. Besides being fundamental for waging a knowledge war, perceptions and thought relate strongly to decision making and to will. Opposing sides will surely attempt to target the perceptions of opposing analysts (knowledge workers) who provide the knowledge that decision makers need. If efforts to alter perceptions succeed, favorable outcomes could very well be skewed toward the side cleverest in continuously manipulating its adversary's perceptions.

Challenges to Attaining
Information Superiority

To find and sustain the condition of information superiority the Department of Defense faces numerous formidable challenges. If the DOD does not address these challenges, which routinely and rapidly change in meaning and shape, the quest for superiority is doomed. The twenty-first-century warrior will be an aggressive change agent—new challenges brought about by advances and change elsewhere will dictate such an approach. In addition, the twenty-first-century warrior must be intellectually adaptive, technically astute, inquisitive, respectful of the capabilities of adversaries, and an ardent foe of the status quo. People who constantly argue for maintaining the status quo will stagnate the process of change and delay coherent responsiveness to an adversary's act–react–counteract cycles. Let us examine just a few of the challenges people face in national defense as they attempt to seek, find, and sustain information superiority.

Availability of Technology

Technology is readily available on the commercial market. For example, today anyone can purchase within twenty-four hours a half-meter resolution image of any place on earth. In the future the resolution will improve and the transaction will take just a few hours. In addition, anyone can buy GPS jammers, surveillance devices, and radio frequency jammers off the shelf. People with the right money can buy the technology they need to collect, process, analyze-synthesize, and communicate—all the essential ingredients for potentially making decisions better and faster than their foes. Moreover, the technology for making individually delivered weapons of mass effect, such as chemical and biological weapons, are easy to find, a situation which argues for a constantly searching and changing technological base for the defense of the country. Moreover, it demands that the country's intelligence system must focus on and judge the relative advantages that the United States possesses or that the country's adversaries possess—we must continually conduct a net assessment of technology. The country's defense forces can be neither complacent nor arrogant about any technological advantages it has, since others will surely catch up and surpass the United States if given the time and opportunity.

Technological Advancements

Technology advances and changes at an amazing pace. Closely guarded technological advances are a thing of the past and in many ways capitalism is driving the frenzy. People love to innovate and create for profit. In addition, the United States openly educates countless individuals from other countries in science and technology, some of whom do not possess our same goals and interests. Though some of these people will return to their home countries and develop technologies used to promote goals inimical to those of the United States, the phenomenon will not change because openness is one of the strengths of the United States. The country's defenders, though, must consider the education of our current or future adversaries when they anticipate technological advancements. Finally, some adversaries have more money to spend on technological advancements than the U.S. military controls (drug gangs come to mind). In the end, other organizations are or soon will be as technologically advanced as the U.S. military, thanks to education and money.

Effects-Based Operations

The Joint Forces Command (JFCOM), located in Norfolk, Virginia, is currently experimenting with the use of effects-based operations (EBO). Effects-based operations involve the analysis of a country's system fabric and the thoughts and perceptions of its people and leaders, then taking only the actions that create desired effects (outcomes). At the heart of effects-based operations lies the notion that many related elements of society contribute to a country's vigor and its strength of will. Recall the tapestry metaphor described earlier. The interrelated parts of a society's tapestry include the social, political, economic, financial, military, and ecological systems, at least. When we follow the logic of effects-based operations, decisions initiate actions and actions create effects that disturb the tapestry. Effects influence decision making, which in turn creates a pathway that can lead to individual, organizational, or societal will. EBO could be an excellent way to deal with adversarial nation-states of the future. The problem is that adversaries (including asymmetric threats) will choose to compete with the United States in the same way.

If adversaries are technologically astute, if they are guided by highly educated and trained planners, operators, decision makers, scientists, and technologists, and if they have access to the latest technologies, they could coherently and systematically attack the "threads" comprising the tapestry of the social systems of the United States. Our military planners must consider this possibility. They will need help from the scientific and technological communities, as the race to the future will be won by groups that can harness the technologies that best help them make decisions better and faster than their adversaries and protect the intertwining "threads" comprising their societal tapestry.

Paranoia, Decentralization, and Low-Tech Solutions

Opponents of the United States are becoming increasingly paranoid, and rightly so. Will this sense of paranoia cause them to cease their nefarious activities? Certainly not! Quite simply, they will adjust, change, counter, and deceive but they will definitely not cease and desist. When constantly worrying about the technological prowess of a country like the United States, adversaries assume that the U.S. security apparatus can see and hear them all the time and everywhere. Such worry drives their responses and actions. They will not, for example, transmit unencrypted cellular calls, except for the purpose of deception. They will not allow their computing servers to be unguarded, unless they want us to enter. They will not show themselves without camouflaging or hiding their identities. These opponents will not send e-mail messages without encrypting or hiding those messages in digital "water spots." In a related issue, the military activities of the country's future adversaries will increasingly decentralize, which means that they will be more difficult to locate with any degree of precision, particularly in urban areas. Moreover, even if U.S. forces take out a decentralized element during a conflict, it won't necessarily lead to or affect the core element or source of power.

Some foes will purposefully use low-tech solutions to compete with the United States. Low-tech mechanisms in some cases are very difficult for a high-tech country to cope with. As an example, think about the challenges the U.S. intelligence community faced as it grappled with the messengers our foes used in Somalia. The only good way to deal with low-tech adversaries is through human intelligence. HUMINT, though, must improve technically and intellectually to provide the accuracy, specificity, timeliness, and relevancy that operations in the twenty-first century will demand.

Counters, Countermeasures, and Counter-countermeasures

For every action the U.S. military takes there will be a reaction or counter. Because of the rapid pace of technological change and the way the United States depends on technology for its warfighting prowess, adversaries of the future will surely attempt to develop counters to our strengths, both tangible and intangible. Thus, to counter their counters, the United States will engage in countermeasures. The adversary will then respond with its own counter-countermeasures—and the spiral will go on and on. Thus the breadth, depth, and pace of technological change must advance unabated, fueled by the definite knowledge that the adversary is attempting to negate the power of our technology all the while attempting to improve its own to attain a technological advantage.

Cyberspace

Cyberspace will be a particularly challenging area for the U.S. military. To be certain, the military is waking up to the need for developing protection for its networks (computer network defense) and for developing ways to conduct attacks (computer network attack). That being said, cyberspace will continue to present challenges for the U.S. military from a variety of perspectives. First, cyberspace is invisible and therefore difficult to envision and impossible to hear or feel. If one can't envision, hear, or feel something, it is difficult to think deeply about it to the point that is sufficient to learn to use it fully. Second, conflict in cyberspace will involve cyberbots moving, attacking, and defending at machine speed. Human beings cannot hope to compete at this level. Humans need to combat these cyberbots with alerting sentinels, deceptive agents, and attacking agents and viruses. The battle space housing cyberoperations will be every bit as important as areas of operation on terrain of the past were. Third, U.S. armed forces will have to collect data and information, and turn it into knowledge in cyberspace from communications, networks, servers, databases, switches, routers, and the inner workings of servers and personal computers. This new arena is filled with challenges for the intelligence community. Fourth, cyberspace will be the battleground for much of effects-based operations. Cyberspace is the conduit for manipulating images, affecting financial centers, affecting the minds of politicians and decision makers, and conducting cyberdeception. Quite simply, this challenge is so significant that if the country fails to meet it intellectually or technologically, our adversaries could seriously jeopardize our national well-being.

Urban Areas

Foes of the future who desire to compete with the United States militarily will establish conditions that cause U.S. forces to engage in urban operations. Why? Because urban terrain, along with cyberspace, is the only place opponents can hope to accomplish their goals—in urban areas our opponents can cause casualties, cause extensive collateral damage, and play to the minds and emotions of Americans and coalition partners watching television or operating their computers. Moreover, competitors using asymmetric strategies can exploit the constraints that military and political exigencies here place on the U.S. armed forces.

In essence, the military needs flexibility to collect, process, analyze, synthesize, visualize, and communicate to meet the needs of these environments (open, urban, and cyber terrains) or institute separate yet related systems. In addition, opponents won't always operate as hierarchical organizations. Instead, they may be organizationally flat, spread throughout or under a city and only loosely connected. Moreover, owing to the potential of compromise of intent and personalities, they will change their structures and relocate frequently, thereby expanding the information challenges for analysts and commanders even more. For example, using the Internet some diverse organizations may come together to support one activity or protest and then disband (as occurred in Seattle during the spring of 2000 at the meetings of the World Trade Organization). Future foes operating in urban terrain will perform their activities inside buildings, underground, on the Internet, or in other areas of cyberspace. Intrabuilding and subterranean operations make sense owing to the inability of the United States to collect, process, fuse, and visualize data, information, and knowledge in urban terrain.

Intelligence collection, a complicated business, has improved over time; by the end of the cold war it worked very well. Yet perplexing shortfalls remain and have been carried over into the new operating environment that U.S. forces face. Collected information doesn't fuse well among the military services branches, and it is difficult to create conditions that could enhance synergy. Moreover, collectors in the cold war era concentrated on military communications, moving targets, and attrition, ranging from a simple headcount of destroyed tanks to understanding the damage inflicted by nuclear weapons. Unfortunately, this "counting" issue constitutes only part of the problem that commanders of the future will face. If a conventional force engages the United States in force-on-force combat situations, the holdover collection system will work well. If the situation changes much, the collection system will prove inadequate. That is, if our foes using asymmetric strategies engage our forces in urban terrain and cyberspace, existing and planned collectors will have a difficult time providing the information demanded by the situation.

Cover, Denial, and Deception

More and more future adversaries will go undercover and attempt to mask their activities and intentions from the "prying" eyes of the U.S. intelligence community in both the physical world and the cyberworld. Unfortunately, intelligence collection doesn't accomplish discerning activities and intentions in covered locations very well, such as underground or even within buildings. Nor has the intelligence community performed the "deep thinking" required to conduct operations in cyberspace. The U.S. intelligence community needs mechanisms for collecting underground or within buildings, using all senses and transmitting gathered data and information without the adversary knowing that the collection has occurred. It also needs collection and communications that operate in cyberspace (easier said than done), but it constitutes a growing concern in the country's efforts to seek, find, and sustain information superiority. In addition, adversaries will attempt to deceive our intelligence systems about their intentions, activities, and decisions. They will use age-old methods of deceiving (such as designing physical signatures, patterns, and activities) and traditional electronic deception (such as providing electronic signatures for collectors to detect). In addition, they will deceive in new ways by using elements of cyberspace in which they purposefully visit websites, for example, to show interest in a particular subject when in fact they are only deceiving.

Swarming and Miniaturization

Both combat activities and intelligence collection will experience the rising importance of swarming and miniaturization. Increasingly, asymmetric adversaries will use "swarms" of cheap, expendable mechanical devices to affect the capabilities of the United States to use a limited number of expensive, precision munitions in combat. Swarms can be big or small. Swarms self-adjust or adjust based on instructions from their controller. Swarms communicate among themselves and provide the means to adjust, depending on the situation or environment. Swarms of cheap missiles, for example, will attempt to confuse and defeat smart, precision missiles; swarms of cheap mines will seek to deny to friendly sea-based forces the unrestricted access to denied areas. Along with efforts to deny access, swarms such as these will attempt to make cost ratios exorbitant for forces of the United States. If a swarm of five hundred one-thousand-dollar missiles (or even cheaper dummy missiles) can cause significant error in one five-million-dollar missile, the cost ratio (500K vs. 5M) obviously favors the holder of the cheaper missile. Swarms of missiles and mines can also deceive military planners about main effort and intent. The miniaturization of implements on modern fields of strife will become increasingly important. For example, swarms of micro-sized multisensor intelligence collectors could flood areas to gain insight into the opposing force's activities. Organized forces will have only a limited ability to detect miniature swarms and, if detection occurs, countermeasures will be difficult to discern, particularly in a quickly changing technology-rich environment.

Information superiority creates the ability to seek, create, and possess advantages in decisions, intellects, knowledge, technology, position, and action. People need and therefore seek advantages to win competitive events—technology can help them achieve that advantage. Technology helps people search for, find, process, think about, synthesize, make, communicate, and receive and understand decisions, and then give and get feedback about the effectiveness of those decisions. As such, people must have access to information and knowledge to help them solve problems, make decisions superior to their opponent's, or achieve advantages they desire (in addition to the advantages that in and of themselves accrue when making superior decisions). Thus, technology helps people collect data and turn it into information. Knowledge workers add value to information, turning it into knowledge and forming the underpinning for successful understanding that decision makers need before they can make better and faster decisions than their opponents.

Conditions of superiority change rapidly, thanks to the presence of an adaptive, learning, and adjusting opponent. Both sides in a competitive endeavor will struggle to find the best information and knowledge, to protect what they find, and to use it to gain decision advantage when the timing is right. Thus, a continuously shifting state of being exists in the struggle for information superiority and advantages that surface once information superiority is established.

An Operational Venn Diagram for Winning the Knowledge War

Some things must gel for the condition of information superiority to come into being. Four elements of modern fields of strife are critical for enabling information superiority to arise: communications, automation, information collection, and analysis and synthesis. These elements work best when working in unison to allow the existence of a state of synergy between man and machine. The following thoughts are not inclusive and represent only a few of the future capabilities that the Department of Defense needs to harness in seeking, finding, and sustaining information superiority.

Communications

U.S. military forces will reduce their presence abroad over the next ten to fifteen years. Eventually very few military forces will be stationed overseas. But U.S. interests in foreign countries and regions will not subside. Thus, to protect its interests U.S. military forces must project military power rapidly when confronted with a conventional military situation. To summon its forces rapidly and use technology to the greatest possible extent to enhance knowledge and be able to make fast, effective decisions, the U.S. military needs the very best communications possible. The U.S. military standard must be an instantaneous, sustained, and flexible network of global communications. Support and access must be assured, from the highest to the lowest levels of military organizations. Communications provide the lifeblood of good decisions; that is, communications provide data, information, and knowledge. Communications also constitute the medium for turning knowledge into power. Military analysts and leaders have to realize that smart, asymmetric adversaries of the future also understand the importance of communications and will assault them kinetically and digitally, aiming for maximum disruption. Ten important characteristics of communications follow.

REACH

Operations in foreign lands will require a solid reach capability because of the austere infrastructure conditions usually encountered abroad. Reach communications will have to connect with various collaborative networks in the United States and around the world. Further reach connections will occur inside the continental United States, as forward-deployed forces come to depend on knowledge and technology experts who support them from virtual locations. Reach operations require an extensive network to support collaboration and visualization. The object is, of course, to communicate with any and every soldier, sailor, airman, marine, or first responder no matter where that individual is located, and to ensure that those communications are secure.

ASSURED COMMUNICATIONS

To avoid the necessity of dragging along a large infrastructure, military forces will come to rely on communications that connect to infrastructures located in the United States. No situation will cause more problems for troops deploying quickly to a foreign country than to be cut off from information emanating from the continental United States. Communications experts will have to build redundant means of communication to ensure that deployed military professionals on the ground are still connected, even if a primary or alternate method of communicating goes out. Moreover, those networks must be small, lightweight, and powerful (including having an adequate bandwidth).

SUPPORT TO ALL LEVELS OF COMMAND

It is obvious that different levels of command require different levels of communications capabilities. Heretofore only large headquarters required and received communications support sufficient to justify the cost of large bandwidth requirements and extensive security software. With the changing environment and a near-total dependence on information and knowledge for making effective decisions, leaders at all levels of command need access to communications pathways and bandwidth to accomplish missions that will vary from day to day and location to location. At the lowest level of operation, for example, Army and Marine squad leaders and individual soldiers will be involved in small unit military operations. They will need sophisticated communications armed with low probability of intercept and low probability of detection communications, imaging functions, and 3-D simulation for accomplishing tasks with minimum collateral damage and loss of life. First responders in homeland security operations have similar requirements. To satisfy these requirements is no small feat. Literally hundreds of disparate squads and civilian first responders could be conducting operations throughout a city, each requiring different sorts of data, information, and knowledge to flow over communications equipment. In such a case, communications receiver-transmitters (R/T) need to be very small, lightweight, and powerful. The R/T will in reality be a small computer or PDA that can present a clear, color graphic use interface (GUI) capable of receiving and sending imagery and receiving 3-D images of key targets (the insides of buildings and the like) in a secure, wireless environment.

SECURITY

Secure information enables commanders to have confidence in the veracity of the information they are using to make and sustain fast, effective decisions. The adversary, of course, wants to distort or fuzz the data or information in question to the point that decisions become slower while its
own become better and faster.

SWARMING

The use of swarming technology could be one way to provide communications. Swarming devices communicate among themselves and with
other entities. Miniature swarms use microelectromechanical systems
(MEMS) technology and could be the vehicle for providing the necessary
bandwidth on demand to the multitude of consumers the communications organizations will have to service. Swarms could move like loosely
connected entities (such as a cloud of dust) or like insects. Moreover,
swarms of communication entities could move to a location and perch on
high ground, treetops, building roofs, or even underground, with small
groups acting as communications relays for more stationary larger groups.
If communications falter, swarms could be used to automatically reestablish connections.

RESISTANCE TO JAMMING

Asymmetric foes will purchase cheap, off-the-shelf jammers to disrupt
friendly communications. The global positioning system (GPS), for example, is critically important to fighters on the ground, in the air, and on the
sea for figuring out their locations. GPS is also important for guidance of
precision-guided munitions. Asymmetric foes will surely attempt to fuzz
the accuracy of GPS, which would adversely affect one great advantage
that U.S. forces currently hold: precision strikes. Asymmetric opponents
will attempt to jam uplinks and downlinks to satellites. Eventually digital
communications will be jammed by cyberbots operating in cyberspace.
These cyberbots will attempt to influence the programming within packets or at opposing cyberbots by sending electrical signals that affect location and mission programming. A growing emphasis on disrupting the
communications coherency in cyberspace will develop, probably involving jamming among digital entities. Thus, counters to those jamming efforts will be a premium service and commodity.

WIRELESS COMMUNICATIONS

Wireless communications must be miniaturized and made even more lightweight and must solve demanding needs such as the integration of imagery, video, and streaming video without the trappings of a massive and vulnerable infrastructure, such as cellular phone towers. In the future wireless communications must be able to display single image, video, streaming video, and 3-D visualization. Wireless communications must improve to provide the direct input of communications into the human brain and eye. Wireless communications must provide clear data and information to small computers embedded within the skin or in the brain of a combatant.

LAYERS OF COMMUNICATION
(SATELLITES: HIGH AND LONG-EARTH ORBITERS)

Single sources of communications will not be sufficient because, quite simply, too many variables can affect throughput. When people deploy to foreign lands they depend on communications to provide the data, information, and knowledge they need for making effective decisions, and they cannot rely on a single source for these only to have it fail. A system of communications systems, in which communications devices are layered vertically and horizontally, all connecting with each other, might be a way to solve this problem. Satellites, unmanned aerial vehicles, unmanned aerial combat vehicles, balloons, and traditional aerial communications aircraft would work as a unified system to provide users, regardless of rank or location, the types of information and knowledge they need at the times and locations they need it. Thus, interface software would have to be interoperable. Moreover, it would be helpful if wireless data processing that connects with gateways leading to the Internet could occur on the communications aerial vehicle itself.

FIBER ENHANCEMENTS

Fiber-optic cable, the main vehicle for communicating in the digital world, continues to spread around the world regardless of economic conditions. With an ever-increasing demand for bandwidth, enhancing the capacity of fiber to carry digital communications will become increasingly important. Also, developing ways to conserve bandwidth while retaining the capability to send and receive essential communications will also be important.

(Compression and change detection come to mind as ways people currently conserve bandwidth.) The U.S. military will need fiber optics to engage in information superiority struggles, as the constant hunger for ever-increasing bandwidth grows and the country's adversaries use increasing amounts of commercial bandwidth available on the open markets. Our scientists must continue to find ways to improve the capacity of fiber.

INTERNET2

The Internet of today was never intended to provide services to millions of people around the world. Internet2, the next generation of the Internet, is being designed with superior features for sharing and moving information and knowledge. As *The Futurist* magazine says, "The key to Internet2 is vastly increased broadband capacity, permitting information to move 600 times faster than is possible using phone lines."[4] Armed adversaries and asymmetric forces around the world will take advantage of the dramatically improved Internet to further their aims of achieving information superiority. Cyberstrategists and the U.S. military must anticipate this enhanced way of communicating and use it to empower their individuals, organizations, and networks.

Automation

Automation dramatically assists in transforming data into information, and information into knowledge. Automation involves both hardware and software. Advances in both contribute to the dynamic nature of the struggle for information superiority, as each adversary constantly tries to gain at least a technological advantage by upgrading hardware and software thereby having the capability to "out-automate" the adversary. With such an advantage, the potential to achieve a decision advantage becomes more of a reality. Automation has many elements. The following thoughts are indicative of only a few of those that contribute to the struggle for information superiority.

VIRTUAL REALITY WARGAMING

Virtual reality wargaming will grow in importance for several reasons. First, wargamers will need a way to defeat the curse of mirror imaging that places their own value sets, ways of thinking, and cultures onto an opponent possessing different values and different ways of thinking. Second, the variables of a specific situation will be crucial. Act–react–counteract cycles of wargaming enable one side to consider how an adversary may react to an action and then which counteraction should be taken. Some variables will be more sensitive than others. Wargamers need to find the most sensitive variables and develop a plan to manipulate the opponent's while protecting their own. Third, wargamers need to develop a list of possibilities, working on the top priority possibilities first and the most unlikely possibilities later. Fourth, the wargamer needs these simulations to be able to develop a collection plan to seek, find, and obtain valuable data and information, and turn that information into knowledge. Moreover, the wargamer needs to understand how the opponent could very well be involved in information-collection operations, and in using collection assets to provide feedback on the worthiness of decisions and adjustments thereto. Such modeling in a simulated environment enables the wargamer to plan deception operations. Fifth, such a modeling and simulation environment enables human beings to plan, execute, and monitor operations in the invisible digital world in which cyberbots are the soldiers moving at the speed of light along the pathways of fiber-optic cables (the lines of communication in cyberspace), and the hard disk drives, servers, and databases that constitute key terrain.

COLLABORATION

Collaboration is the coalescing of individuals with varying degrees of knowledge expertise and intellectual capabilities, coming together either physically or virtually, to share their thoughts, creativity, and expert judgment for solving problems and contributing to effective decision making. Collaboration enables people to aggregate their intellects to solve problems. Quite literally hundreds or thousands of minds can come together in virtual space to solve problems and provide input to complex decision making regardless of their locations. Collaboration networks must be transparent and simple to use, involving only a few clicks with a mouse or perhaps a voice command to call up a web page. Users should be able to modify various aspects of the collaborative environment to meet their particular needs. Collaboration should involve multimedia and should grow to include all the senses. Users need to have sufficient resolution to view their co-collaborators' body language and sincerity.

TELEPRESENCE SOFTWARE

Telepresence software operating over Internet2 will allow the merging of videoconferencing and virtual reality, immersing the knowledge worker in multiple locations with multiple connections.[5] Telepresence will allow the integration of simple elements (data from collectors) to more complex ones (open source information, with media video clips, with audio, with text, with multilevel classified information and knowledge, with cyberbots, and with physical entities on the ground, in the air, or on the sea). Telepresence software also will support interactive virtual reality wargaming with players around the world.

Thanks to telepresence and Internet2, knowledge workers' ability to apply value to information will occur much more rapidly, theoretically enabling decision makers to make effective decisions. Commanders and other leaders, for example, will be able to get high pixel density approximations of reality on larger screens. Imagine being able to stick your arm into a screen so that it is represented on many other screens thousands of miles away, or being able to show people the database or switch that cyberbots will be probing as they look for valuable data and information. This is known as a digital sand table, one that can be transported over Internet2 and displayed around the world at extremely high speeds. The software possibilities for Internet2 are staggering. Anyone with advanced thinking about these possibilities possesses a distinct edge in the race for valuable knowledge.

DIGITAL MANEUVERING

Maneuvering in cyberspace will yield advantages in tempo, momentum, and initiative. Individual and groupings of cyberbots—swarms—will operate together to accomplish defensive and offensive missions. These cyberbot swarms, moving at the speed of light, will communicate among themselves and with their handlers. Leaders will be amassing swarms of cyberbots to gain advantage over their adversaries. Maneuver will occur at the speed of light, and many decisions will come from the cyberbots themselves. Undoubtedly not all cyberbots operating in swarms will be programmatically capable of making decisions. Instead, operative swarms will have both leaders and executors; leaders will be better protected because they have more capabilities and more-sophisticated resident source code.

Maneuver also will occur among Intranets operating virtual private networks (VPN). Leaders making decisions will amass intellects over the VPNs in their system of systems using collaborative tools and wide-bank networks (such as Internet2). They will shift emphasis at a moment's notice and "maneuver" digits, data, information, and knowledge to accomplish their goals and objectives and make decisions faster than their adversaries. Of course, their adaptive foes will be doing the same thing. Thus, we can conclude that the output from these machinations is the maneuver of knowledge and therefore superiority.

CHIPS EMBEDDED IN HUMAN BEINGS

Chips will become so small and unobtrusive that eventually they will be a part of the human anatomy. Chips operating at high speeds enhance the senses, enhance strength, and greatly improve thinking. Embedded chips will be particularly important for reconnaissance and surveillance operations and for HUMINT intelligence collection. Along with processing information and knowledge and storing it for future use, embedded chips will communicate with other embedded and external chips, passing data, information, and knowledge in a wireless network of man and machine interactivity. Embedded chips enable the human to interact with the machine to solve problems. This will be a complementary relationship in that humans will perform skills for which they are better suited than computers, and computers will perform tasks for which they are better suited than humans. Clusters of miniature chips will operate throughout the human body. One cluster will be more capable than others and will serve as the "control node"; other less-capable chips will enhance the senses, mental functioning, and physical capabilities.

KNOWLEDGE MANAGEMENT

How people and organizations manage data and information and how they then turn it into usable knowledge will be important in making decisions faster and better than the adversary. Knowledge management is a key element of knowledge war and it is critical in establishing and sustaining the condition of information superiority (see chapter seven for a full discussion of this topic).

INTELLIGENT AGENTS

Computer software in the form of intelligent agents must help human beings in several ways. First, agents help people understand complexity. Second, agents help people cope with the speed of action in cyberspace. People can neither think nor act at the speed of light, but intelligent agents can. Third, intelligent agents help people overcome shortfalls in synthesizing information and knowledge. Since much of the data, information, and knowledge people need for making fast, effective decisions flows through the "veins" of cyberspace, it follows that intelligent agents are crucial in the struggle for supremacy in decision making. How scientists develop the concept and actually program intelligent agents will be crucial for engendering activities most conducive to success in cyberspace.

RELATIONSHIP IDENTITY

Relationships, which add value to information and knowledge that have a value in their own right, will be critically important for gaining decision, intellectual, knowledge, positional, and action advantages. Typically, people perform analytic functions very well: they split ideas and concepts apart to fine levels of detail and granularity. Analysts come to know these details very well. Unfortunately, people neither habitually put the bits and pieces of information back into a whole nor search for relationships with other wholes. Obvious relationships are easy to spot, but disparate or widely scattered relationships are difficult to recognize, particularly if the relationships exist across categories. One role for software in the future, particularly that of autonomous intelligent agents, will be to search for, find, and present both the obvious and the hidden relationships that knowledge workers and decision makers need for creating the synergistic and combinatorial effects (outcomes) they seek.

CHARACTERISTICS OF AGENTS

Intelligent agents learn quickly. They present data, information, and knowledge in a format most readily acceptable to the way their human masters think and act. Eventually, in some aspects intelligent agents will become smarter than their human handlers. Intelligent agents form relationships between both obvious and disparate things—objects, data, information, and knowledge. Intelligent agents act quickly—at the speed of light in the cyberdomain. They adapt thoroughly to the requirements and preferences of their masters and provide the data, information, and knowledge needed for effective decision making. Intelligent agents antici-pate the future. Using experience and a solid sensing of the environment, missions, activities of rival cyberbots, and environmental constraints, they can both anticipate their master's future needs and predict those of the ad-versary as well—all of which can be programmed into the "brain" of the agent. Intelligent agents make rapid decisions and act. In cyberspace there isn't time for consultation between the intelligent agent and its human handlers. Thus, intelligent agents have to be programmed to decide and act at the speed of light.

Roles and Missions

Intelligent agents will perform many roles and missions in future military operations. Some will be offensive roles, leading these agents to roam cyberspace seeking data, information, and knowledge from adversaries. They will enter an adversary's databases and find the decision-making logic resident in the code of software being used for making decisions. The agents will collect data, information, and knowledge from their modems, switches, routers, and hubs, which channel digital flows—this is the cyberlandscape. Agents sometimes will be stationary, observing or surveilling locations or points on the terrain of cyberspace. Agents some-times will reconnoiter digital conduits or highways for activities such as the movement of particular digital packets. Agents sometimes will latch onto and "ride" packets and enter their structures to ascertain the data, information, and knowledge contained therein. Intelligent agents some-times will serve to help deceive opponent agents by routing them into honeypots and by acting as packets for opposing agents to ride and steal the valuable data, information, and knowledge. Intelligent agents some-times will be sleeper agents lurking in databases or along communica-tions conduits, waiting for a "wakeup call" from their handlers to execute cyberspeed operations. Intelligent agents sometimes will conduct attacks against opposing force information technology systems that support deci-sion making, including invisible strokes (such as the injection of viruses) and visible strokes (such as eating the silicon on processor chips by dis-pensing silicon-destroying chemicals or powering laser beams to burn key parts of chips).

Intelligent agents also can perform defensive roles. As such, intelligent agents act as sentinels that protect entrée into information technology networks supporting decision making. They serve to parry attempts by adversary intelligent agents (that is, cyberbots) to enter friendly decision-making systems. Intelligent agents sometimes will steer adversary agents into honeypots, thereby confusing them about the reality of the location or assisting in the conduct of cyberdeception. Intelligent agents sometimes will assist defenders by identifying the characteristics of the attacking agent and by locating the origin of the attack. Intelligent agents sometimes will assist the defender by surmising if the attack was local or related to other attacks. Intelligent agents sometimes will quickly learn the code of an attacking agent and ascertain its strengths and vulnerabilities for possible counterattack or manipulation.

Benign Roles

Intelligent agents will rapidly search for and sort data, information, and knowledge, thereby facilitating the potential for people to engage in rapid decision making. Intelligent agents sometimes will be involved in visually depicting information and knowledge for humans to gain better understanding of the complexity of the world, of cyberspace, or of the adversary's mind. The agent sometimes will package data, information, and knowledge to complement the way the individual thinks. Intelligent agents sometimes will seek, find, and explain obvious and disparate relationships, which leads to synthesis.

HARDWARE CHARACTERISTICS

Because of the absolute need for U.S. armed forces to move quickly and to retain its large, unwieldy infrastructures in the continental United States, technology must become more dependable, smaller, and lighter, all the while becoming more capable. Computing hardware must have large storage capacity used by extremely fast, intelligent software. Increasingly powerful processors direct and guide software, allowing millions of bits of data to process quickly, fusing data regardless of source, and integrating multiple sources of data, information, and knowledge almost instantly. Leaders making decisions need this type of support because their adversaries will certainly purchase the fastest and most capable processors available on the commercial market as well. Hardware will become so small and user friendly that it will merge with human beings' systems within the human body, leading to the true advent of man-machine symbiosis, in a hardware sense. What is still to be determined is when this technological advancement will occur and whether the implants will work using biological or signal processing. Implants will probably come into being within the next twenty years and become operational within twenty-five. The country needs to research the linking of hardware (embedded chips) and man to develop technology appropriate for what it is meant to accomplish or contribute. Human input to these implanted chips will likely be made by voice, touch, or electromagnetic sensing of changes, such as activity levels, fear, or variations in weather and humidity. Computer chips will become so smart that they will be able to anticipate what the human purveyor of the chip needs and when support is needed.

VISUALIZATION

People will depend more on visualization to help understand complexity quickly, a trait that will be enhanced by hardware implants in the human body. Streaming media and sounds will flow into the implanted chip, which will then communicate with brain- and eye-enhancing chips to form the relationships, identify follow-on needs, and display the visualization of complexity via a retinal enhancer or directly into the mind of the recipient. Visualization will be executed in the four traditional dimensions (time, depth, height, and width) and two nontraditional dimensions (cyberspace and cerebral). It will allow entrée into virtual reality wargaming regardless of location. Visualization will fuse data and information and display the result in a multimedia format. Visualization will allow the integration of data, information, and knowledge from all sources and will allow for the integration of numerous contributors. Visualization will give special operations soldiers a virtual entrée into target areas, including the nooks and crannies of cyberspace.

Collection

The information and intelligence collection challenge is much more difficult today than in the past because of the speed of the Internet and its treasure trove of data, information, and knowledge. Clearly, people engaged in knowledge war must have a dynamic collection operation that continuously seeks relevant, specific, accurate, and timely data, information, and knowledge. The following constitute just a few pertinent ideas about the collection challenges that the United States will face.

AGAINST TRADITIONAL
FORCE-ON-FORCE OPPONENTS

Current U.S. intelligence collection does well against conventional targets. Collection will continue to involve the traditional functions of determining the location and status of fixed and moving targets in all types of terrain. Conventional collection will aim at command and control for determining intent and identifying the decision-making activities of adversaries. It will also aim at adversary intelligence collection, which will be the means that opposing leaders employ to measure the effectiveness of their effects and decision making. Collection personnel and machinery must be able to find moving surface-to-air missiles systems (SAMS) and surface-to-surface missiles (SSM) (both of which are asymmetric counters to U.S. military strengths), within the constraints imposed in timeliness and accuracy to make fast decisions and minimize collateral damage and casualties.

AGAINST ANTI-ACCESS SYSTEMS

Collection networks will have to obtain information about weapons and information systems that provide anti-access capabilities to asymmetric foes, including the location and activities of swarms of devices designed for the purposes of intelligence collection, deception, or the physical destruction of friendly missiles, ships, and planes. Key to the effort to counter an asymmetric foe's anti-access capabilities will be the system of systems approach. That is, in an anti-access system of systems, our collectors will search for the foe's anti-access system command and control and communications, including the fiber-optic cables that connect the anti-access systems among themselves and in the military hierarchy. Friendly collectors will also search for the anti-access system's intelligence-collection systems, their various means of communications, and their missile launching mechanisms. Not to be forgotten in this broad intelligence-collection requirement is that any viable collection system will have to find anti-access missile and swarm "bed down" sites and storage facilities (the computer hardware and software that operate and fire the systems) and the people (knowledge workers) who operate and maintain the systems. Again, friendly collection devices will have to answer specificity, timeliness, accuracy, and relevancy criteria for success.

AGAINST HARDENED TARGETS

More and more of the country's future adversaries are moving underground because they believe that U.S. intelligence systems cannot see or hear underground. Thus, command posts will be deeply buried and will be centers for adversary decision making. It will be very difficult to know if the command post is active or inactive. Thus, in a general sense intelligence collection must enter these locations, surveil and reconnoiter without being discovered, and transmit collected data and information to external friendly fusion centers for quick use in friendly decision making. Intelligence collectors must determine if the command post is "hot" or merely in the caretaker mode. Intelligence collection must be capable and sensitive enough to take still photos and videos, and hear conversations. Intelligence collection against hard targets must collect against computer, communication, and collection systems housed within an underground command post. Collectors must operate through microelectromechanical systems. Sensors must see, hear, and smell. Sensors must perform onboard processing and possess long-lasting and extremely lightweight power capabilities. Sensors must not be vulnerable to detection. Sensors must collect against physical objects, against human activities, and against cyberbots and database logic in cyberspace. Sensors must use paths of communications that take advantage of existing adversary communications, for example, using fiber optics if the sensor cannot transmit because of some physical obstruction.

TO SUPPORT SPECIAL OPERATIONS

Given the types of people we will have as adversaries, it is a safe bet that they will become smarter and more technically capable with time. In a way that is similar to any mounted or dismounted force, special operations people need sensors to detect booby traps. They also need sensors that detect various types of nonlethal weapons such as superlubricants, sticky foam, and the like. Special operations personnel need to know, with as much precision as is possible, the location of countersurveillance devices such as infrared, thermal, and light-intensifying imaging equipment. For reconnaissance missions, special operations personnel need sensors that enhance sight and hearing from great distances. Special operations people need collection devices that micro-archive to explore miniscule objects (because of distance) in detail. These types of operations need miniature tracking and locating devices that operate with minimum power for the sake of preserving batteries and to help avoid detection. These types of operations need sensors that detect the location and composition of nuclear, chemical, and biological weapons. Special operations forces need devices that detect variation or interference in digital transmissions, visualization, or hearing devices. Special operations forces need sensors to discover the presence of remote detonations, to capture their thoughts and perceptions and transmit them to decision centers, and to find people and things in difficult terrain, such as urban areas, mountains, and triple canopy jungles.

IN URBAN TERRAIN

What collection capabilities do people operating in an urban environment need? First, U.S. forces need to improve the capability to operate against wireless, cellular phones, computers, and personal digital assistants. Second, U.S. forces need very fast access to human intelligence–derived information. HUMINT collectors need a way to digitize their thoughts, observations, and reflections, and to transmit to a fusion system where the data integrates with all sources of information collection into a dynamic, changing, situationally aware visualization. Technology must provide assurance to the point that the information's veracity is reliable. Commanders must believe in HUMINT as a dependable source rather than having to reject, delay, or confirm with other sources before accepting the information. Third, collection must get into the rooms where asymmetric foes plot their operations and whisper their instructions. Intent is critical; collection devices must gather spoken information and transmit it to analysis and synthesis centers. Language translation technology will be critically important for speed and accuracy. Fourth, collectors must be stealthy, miniature, and robotic. They must be capable of collecting, processing, and transmitting information to satellites or satellite surrogates so it can be gathered into fusion technology and immediately disseminated to people operating on the ground. Fifth, collection must sense the moods of crowds so leaders can anticipate potential crowd-related troubles. Sixth, collection must answer questions about the metrics associated with the effects of information operations.

IN CYBERSPACE

Collection efforts in cyberspace pose a particular challenge and are similar to collection operations in conventional settings from one perspective only: both seek valuable data, information, and knowledge. Cyberspace is invisible, and people need to visualize in virtual reality (wargame) what they cannot see or feel. The cyberbots in the virtual reality wargame will learn from the experiences of the wargame, from the analyst controlling the wargame, and from the adversary, and will then actually enter cyberspace to obtain specific, timely, relevant, and accurate data, information, and, when possible, knowledge. Principal collection devices will be highly specialized collection cyberbots—autonomous, intelligent agents programmed to collect data from hard disk drives, from random access memory, from random operating memory, and from communications pathways, switches, databases, and routers. Cyberbots will also collect data and information from digital telephone systems, digital television sets, and cellular phones. Cyberbots will enter databases; access and copy selected data, information, and knowledge; and forward the collected material to the controller. Cyberbots will track the activities of encrypted digital packets. In addition, cyberbots will track the activities and presence of opposing-force cyberbots. The activities of cyberbot intelligence collectors will be both active and passive; that is, some cyberbots will roam through cyberspace while others will concentrate on reconnoitering communications pathways or conducting surveillance on switches or routers. Some cyberbots will determine an adversary's digital defensive perimeter and relay to the programmer-analyst the types of firewalls, intrusion-detection systems, and encryption being used. Other cyberbots will attempt to locate the adversary's information centers of gravity where collection, communication, automation, thinking and planning, and decisions occur. Cyberbots will have the additional mission of searching for and alerting people to an adversary's attempts to engage in cyberdeception. Some cyberbot collectors will be "sleepers" that exist everywhere and nowhere—coming to life on command to perform their tasks with a small electrical pulse.

IN COUNTERTERRORISM OPERATIONS

Terrorists provide particularly difficult collection challenges. They operate in small groups whose members are difficult to bribe or influence into providing information. Increasingly they are concerned about not divulging secrets, insider information, patterns, or procedures to the intelligence-collection mechanisms of their opponents. In fact, often they resort to no- or low-tech communications and coordination solutions to foil the highly evolved efforts of their adversaries' intelligence agencies. If they engage in traditional, kinetically oriented operations, our collection systems will have a decent chance of finding and tracking them. If they operate as individuals and organizational cells in cities, or if they use off-the-shelf encryption and security devices for their cell phones and computers, our legacy collection systems—even the new Global Hawk unmanned aerial vehicle system—will have a difficult time providing meaningful information.

Terrorists will not decide, act, and create their effects in a vacuum. They need to act to cause effects that lead to affecting will, but they also need information and knowledge to think, plan, decide, execute, and check the validity of their own decisions. Thus they need an information-collection plan that is in operation before events occur and after events are over. The Internet has incredible amounts of useful information, but they cannot obtain all they need from it. Terrorists need to conduct personal reconnaissance and surveillance of potential targets. They also need to conduct HUMINT operations of sorts—asking questions and seeking guidance and advice from people who have information and knowledge they need. They will be forced to buy commercial satellite photography for their planning and execution. They will seek SIGINT-type info from cell phones and wireless computer communications. They will use cyberspace to attempt to obtain information and knowledge of their targets—and they will do so surreptitiously.

Terrorists need communications to talk, coordinate, and orchestrate events, they need computers to support planning and rehearsals, and they need encryption mechanisms to protect their nefarious deeds. Once they make their decisions and act, the terrorists need feedback on the "effectiveness" of their effects. The news media helps them immensely in the task, but often they will require information and knowledge from the Internet and from close interactions with people personally in the know to ascertain the true value of their actions and effects.

Terrorists need to synchronize their activities. The obvious example lies in the attacks on the World Trade Center and the Pentagon. How did the terrorists plan? How did they synchronize? How did they rehearse?

What were their backup plans? Why did they go at the south side of the Pentagon? These questions make it clear that orchestration and synchronization of events will become larger issues in the future. Because of globalization and the shrouds of relationships encompassing the world's tapestry of systems, terrorists will seek to orchestrate and synchronize in many locations, at the right time, and to create the right effects that lead to acceptance of their will.

If we anticipate their needs and efforts to collect data, information, and knowledge, it will help us anticipate and deny their attacks. Unfortunately, when we try to think through possible terrorist courses of action, for some reason we have trouble predicting their movements. Yet it is critical to do so if we ever hope to anticipate, thwart, deny, confuse, or deceive terrorist attacks.

When we attempt to anticipate terrorist attacks, there is one fundamental thing we must do as a starting point in our thought processes: think about what is important *to us*. What is our center of gravity, or the essential element or elements around which all else revolves? What would shut down our businesses? What would stop the flow of troops, tanks, or airplanes, or hamper logistics? What would stop us from communicating or collaborating or making decisions? In knowledge organizations certain intangibles—such as a formula or an algorithm—could be the most important thing. In service organizations knowledge workers themselves could be the most important component. In some organizations it may be the knowledge itself, or even the machines that enable knowledge workers to produce that knowledge. Then we have to make a logic shift, considering what the adversary would think is important from our perspective and identifying the underlying rationale for this terrorist thinking process. If we can logically reason out what the terrorist thinks is important to us and why, we can begin the process of anticipating and predicting the ways that terrorists—using the strategies, tactics, techniques, and tool kits they have in their possession—will set about to affect that which we hold dear or critically important. Once we understand the potential hows, we can work backwards to consider the actions that would lead to those effects; once we consider the effects that lead to outcomes, we can consider what the terrorist would need to know and understand to create those actions. To create the greatest second- and third-order effects the terrorists will have to seek, find, process, analyze, and synthesize information and knowledge.

We have to consider where the terrorists will find such information and knowledge. Some, obviously, will come from open sources. Other bits and pieces they might need could very well come through cyberspace. In addition, the terrorists could use both SIGINT (such as intercepting wireless communications among cell phones and computers) and HUMINT (such as conducting reconnaissance, "buying" people to obtain information and knowledge, or using surrogates to collect and report) processes. Once we have considered what terrorists need, and why and how they will obtain those pieces, we must then consider the indicators associated with such activities. When we understand and can spot the indicators, we can counter with our own traditional and nontraditional information-collection activities to confirm or deny the hypotheses that underpin our thinking. We then will have in place an effective terrorist indications-and-warning system.

Analysis and Synthesis

As the human race progresses into the future, computers will become smarter yet better able to replicate many activities heretofore accomplished only by human beings. In the short term, though, computers will not be capable of engaging in certain types of thinking—humans are still better at thinking that involves synthesis, nuance, and abstraction. Engaging in knowledge war will require excellent thinkers who can analyze and synthesize rapidly. Synthesis involves finding and exploiting obvious and seemingly disparate relationships, pulling bits and pieces of data and information together, and then combining them into a sensible whole. People can learn to synthesize, and they need to do so to enter and win in a knowledge war. While automation can help, it cannot do what humans can, given sufficient education and working in an environment conducive to thinking. Consider just a few of the many fundamental thoughts about analysis and synthesis needed to compete in knowledge war.

Human beings will need to improve their thinking capabilities to cope with the increasing complexities of the world. Improvements can occur through training, mentoring, and individual learning. Primarily, people need to learn to synthesize bits and pieces of information into greater, more powerful, and more meaningful wholes. They must learn to search for and find both obvious and disparate relationships. Intelligent agents and fast processors can help in such endeavors. Human beings will also need cerebral enhancements to enter and win a knowledge war. That is, a man-machine symbiosis will have to occur with the two operating as one, each capturing the strengths and minimizing the weaknesses of the other. Such symbiosis can occur through the growing competence of intelligent agents and other learning software. It can also occur through computer chips embedded in the bodies of people—implants designed to help people think well or to enhance their intellectual capabilities in language, math, computer science, and so forth.

Automation can help people think by collecting, collating, and visualizing the data, information, and knowledge needed to measure effectiveness (MOE) and measure performance (MOP). MOE tells people how effective their actions have been in creating the outcomes (effects) they seek; MOP tells people how well the actor or action that creates the effect has performed. Since many of the elements of importance in a knowledge war are nuance laden and intangible (invisible to the naked eye), computer software can be programmed to search for, relate, and interpret meaning. The ability to monitor and measure effectiveness while conducting information operations will be critical. Traditional intelligence collection, for the most part, will not work. But nontraditional ways of collecting data, information, and knowledge will assist in determining the fluctuations in these MOE; that is, cybercollection operations will quickly find data, information, and knowledge, form relationships, form tentative conclusions, and present possibilities (potentialities unbounded by constraints) to the human handler. Sensors need to "sniff" body chemicals to ascertain the psychological and physical states of people. Enhanced hearing and memory will greatly aid operatives engaging in HUMINT within a culture to determine the effectiveness of an IO campaign. Agents must be able to discern obtuse connections between a leader's decision making, the activities to execute the decision, and the effects (outcomes). The agent must then be able to relate them to measures of effectiveness. Agents also must be able to discern an opponent's MOE and MOP, particularly the methods for obtaining feedback as to the quality and timeliness of its decisions. Automation can be programmed to search for indicators of the strength of an individual's, an organization's, or a country's will, and how it fluctuates through the act–react–counteract cycles of involvement.

Creating conditions that are conducive to thinking will ascend in importance. Along with improving intellects through learning and hardware-software implants, people will place increasing emphasis on using technology to enhance the thinking prowess of knowledge workers and leaders. Conditions must be right to maximize human thinking and learning. Some people, for example, think best in quiet. Others believe listening to Mozart helps them think. In the future people will enter into altered states of consciousness to enhance their thought processes. Computer software will diagnose conditions most conducive to achieving maximum effectiveness of thought. Computer software will also quickly discern the best combinations of intellects and software to arrive at maximum effectiveness in decision making. The technology to bring enhanced thinking forward will range from simple sensory deprivation tanks to sophisticated hardware-software implants to sound and smell stimulants. All organizations must have thinking rooms where knowledge workers can adjourn to think. They will enter a world of virtual reality in which the way they think, create, and learn as individuals is constructed for them using smart software and powerful computers.

Information superiority is firmly connected to making decisions that are superior to an adversary's and combines information technology and intellectual power to create conditions with which to make those better decisions. With decision advantages comes superiority in tempo, initiative, and momentum at the time and place of the commander's choosing—not at the enemy's choosing. Information superiority brings additional advantages—intellectual (intellectual superiority), knowledge (knowledge management), technological, positional (physical and in cyberspace), and action (that lead to effects or outcomes that influence will). The best warriors of the new century will be able to seek, find, and sustain all of these advantages in every encounter with a foe, since assuredly the foe will be attempting to do likewise.

Information superiority cannot be all the time and everywhere. When faced with an adaptive, intelligent, and learning adversary, antagonists will find that the condition of information superiority evaporates quickly and isolates in pockets of physical space or cyberspace. Thus, any viable competitor of the future must absolutely ensure that his knowledge workers and leaders constantly improve and advance intellectually, and that technological enhancements occur constantly. These improvements offer the best hope of successfully maintaining skills sufficient to compete with and win against an adaptive, learning, and changing foe. With complacency or a tilt of emphasis to the kinetic area of conflict they will be left behind in the race for competitive advantage and the condition of information superiority, even if achieved.

Two new domains of conflict—the cyberworld and the cerebral world—will be added to the traditional domains of air, land, sea, and space. The world is so dependent upon cyberspace for the interaction of its macrosystems that we must prepare to compete in this domain even though it is invisible, its soldiers are cyberbots moving at the speed of light, and the outcomes of struggles in it are often nonkinetic.

With the advent of a knowledge war comes the ascendancy of the knowledge worker who turns information into knowledge for a decision maker's use in making effective decisions. Thus knowledge is highly dependent upon the intellects, experiences, intuition, and curiosity of knowledge workers' minds. With the increasing importance of precise, nonkinetic, intangible, and interwoven effects comes the need for advanced thinking and planning. Smart adversaries of the future will take their opponents into both cyberspace and the cerebral domain to achieve advantages, particularly in thinking and planning as they relate to decision making and with deception to create paralysis and shock in an adversary's decision-making apparatus.

Some of these things will occur soon, other things will not occur for decades. Clearly, future conflict and the changes in technology and thinking are not totally in evidence today. But it does not take much foresight to anticipate this type of conflict, which is enabled by technological and intellectual developments that are advancing exponentially. Superb intellects and constantly improving technologies combine to enable one side or the other to seek, find, and sustain for short periods the condition of information superiority. Since both sides are competing to create this condition, the side that doesn't have it will go to great lengths to recapture it. Thus, a sine wave of peaks and valleys will exist between adversaries who are adaptive, adjusting, and aggressive in seeking knowledge and using it for superior decision making. Advantages will accrue to the side that best melds mind and machines to provide a superhuman intellect in which it is impossible to differentiate between men and machines.

With a tilt toward intangibles and the morale domain of war, humans become more important than the machines that produce kinetic energy. This will not be an easy shift since the current power structures, money flows, organizational design, and leadership development remain firmly affixed to the conduct of a conventional, force-on-force, kinetically oriented operation. But the asymmetric foe is forcing us to pry ourselves from that schema; we must develop and eventually accept a new paradigm in which human beings, aided by information technology machines within flexibly designed and acting organizations, produce the knowledge that leads to victory. This trend will not occur without debate, though. Part of the discussion will involve how people think, how they process information, how they turn it into knowledge, and how they interact with computers. Part of the discussion will involve the leadership skills people need for accomplishing tasks in a virtual, cyber environment. Part of the discussion will involve using the minds and the right people to make the best decisions fastest in the environment. Part of the discussion will involve how important machines will become and if they will eventually be smarter than human beings. Further, much discussion and debate lie ahead when it comes to organizational design in an era in which knowledge is king.

Know the enemy and know yourself; in a hundred
battles you will never be in peril.

Sun Tzu, *The Art of War*

Information Operations   5

The Operational Environment

Information operations of the future will rely on both a philosophy and a
suite of tools. The philosophy of IO acknowledges the emergence and importance of knowledge war and its direct connection to knowledge-based
strategies and knowledge-based operations. The philosophy acknowledges
the preeminence of information and knowledge as they contribute to effective decision making, as well as the importance of thinking, planning,
and perceptions from two points of view: ours and our adversary's. The IO
suite of tools will be used to attack the thinking, planning, decision-
making processes, the mechanical turning of data into information and information into knowledge, and the machinery supporting the decision cycles of our adversaries. In addition, the IO suite of tools will protect our decision processes and information tools against an adversary's attacks. IO
tools will also serve as gatherers of data, information, and knowledge in
cyberspace, that is, information-collection operations in cyberspace.

Any useful discussion of IO must have as its central theme the exami-
nation of friendly and adversarial decision making. To begin, please consider the following working definition of information operations: "Activities in physical settings and in cyberspace purposefully designed to in-

fluence or directly affect the decisions of an adversary and the processes that support it, and to protect and enhance one's own decision making and the processes that support it."

It should be obvious to all serious students of future conflict that IO have become a major consideration when contemplating the art and science of war. Why is this so? First, IO have as their raison d'être the influencing of decision making of individuals, people in general, organizations, and political and military leaders. Affecting decisions is one of the ways that knowledge war experts will seek to influence will. Second, IO have pervasive influences and effects—and the potential to affect large numbers of people, civilian and military, and the way they live, think, perceive, and act. Third, IO cause enormously important and powerful second- and third-order effects that cut across social, political, military, economic, financial, informational, and ecological spheres of influence, particularly when they cascade into each of the systems composing a particular society. Smart IO warriors purposefully create such effects to manipulate or devastate the decision-making processes, actual decisions, and supporting actions of an adversary. The adversary who fails either to plan for or anticipate second- and third-order effects will be at a disadvantage compared to the adversary who acknowledges their existence, anticipates their use, and contrives ways to use their energy. Fourth, the soldiers of cyberspace are cyberbots moving at the speed of light. Cyberbots, the sophisticated software programs that operate in cyberspace, learn with experience, perform complex tasks, and roam cyberspace performing missions ensconced in their "brains" as devised by their programmers and directed by their human handlers (who, by the way, could be physically located thousands of miles away but instantly connected). Cyberspace terrain lies in the nooks and crannies of fiber-optic cables, communications channels, databases, and hard disk drives; its inhabitants move, think, and act at the speed of light. The interaction of cyberbots on digital terrain to create actions and effects (outcomes) has incredible potential for affecting decisions, and hence for affecting will, in military conflict and civilian competition for resources. It is difficult to anticipate the activities of cyberbots or to even discern their energy and activities. Yet this often invisible activity in cyberspace has the potential to influence the outcome of conflicts without the firing of even one artillery round, the moving of even one aircraft carrier, or the flying of a single aircraft.

Soon the importance of IO will surpass that of the ships, planes, and tanks that have served the U.S. military forces so ably over the past two hundred years. Though the tools of a kinetic, force-on-force conflict will continue to have their place, eventually IO will become a first consideration when planning military operations. Adversaries will use the many faces of IO to attack the United States; their capabilities and potential outcomes represent great potential for winning quickly and decisively. The important consideration for such victory on either side of a conflict, though, lies in maintaining superior intellects with which to think and plan, and developing continuously superior technology to exert direct influence as technical tools that assist human beings in operating successfully in cyberspace.

The Changing Face of Conflict

The spectrum of conflict has broadened significantly. Though the United States must still ensure that its armed forces can quickly defeat kinetic challenges from any nation or force on earth—agrarian or Industrial Age forces—all militaries are going to use whatever high technology they can harness to help attain any advantage in their particular situations. With the proliferation of cellular phones and computers, even agrarian and Industrial Age military forces will become hybrid forces wielding strange combinations of anachronistic weapons with modern technology. With the firm integration of the Internet in future conflict, the realms of kinetic and cyber are merging and the country must prepare to deal with both. Indeed, conducting operations in cyberspace is a great way for a weak force to create offsets against a stronger, more technologically advanced force.

War and conflict, indeed competition in general, will occur at speeds defying logic and understanding. Distance and time will become increasingly irrelevant, even though force generators still have to conquer the limitations of physical distance and the time one has to devote to it. With the tyranny of distance dissipating in importance, decisions will be even more important than they are now. There will be a distinct need for shorter, more concise decision cycles, with the added requirement of speed being relative to what an opponent's decision cycle is. Along with the physical spectrum of conflict mentioned above, the country's military will find itself immersed in the new, invisible, deadly, and abstract conflict of cyberwar because its actions will occur in the invisible, light-speed venues of cyberspace.

IO owe their growing influence to the ascendancy and pervasive presence of the trappings of the information revolution, as well as to the growing dependence of all aspects of the U.S. society, indeed of the entire advanced world, on information technology and the strengths and weaknesses that cascade from such dependence. IO will continue to thrive because of their invisible nature and the difficulty in ascertaining the identity, affiliation, or location of global perpetrators of attacks. IO, becoming a "weapon of choice," provide a way to offset the capabilities of a more powerful foe or to minimize the capability of weaker foes by using indirect approaches to attain goals and objectives related to decision making and information support systems. In addition, IO are a tremendous additive tool for waging knowledge war; that is, when all or even some of the elements of IO work together to create a synergy, outcomes will be far greater than if one element acted alone.

With its vast potential as a weapon coupled with the certainty that the country's adversaries will employ IO, the U.S. national security apparatus has no choice but to accept IO as a viable method for engaging in conflict. The country must prepare now to enter the arena of IO competition, engage capable opponents, and possess the will to persevere and win in this type of future conflict. Interestingly, IO are a strength with which to engage others in conflict while at the same time they are a vulnerability that all participants in conflict can use to attack and seek advantage. This unity of opposites (strength and vulnerability) is one of the reasons that IO remain such a mystery to many people.

Adversaries will use IO as a way to compete with the United States by attempting to affect its decision-making support processes and the actual decision making of individuals, of aggregates of people, and of leaders responsible for making decisions. IO are a critically important offensive tool. As such they represent a way for twenty-first-century opponents to wage knowledge war. IO are easy to conceal; moreover, the element of surprise becomes harder to anticipate and yields greater outcomes, particularly in the morale domain of war. IO are abstract and complicated, to the point that adversaries can easily assume that a country preoccupied with kinetic, force-on-force warfare will not be prepared to enter into a full-scale knowledge war. IO are also an effective defensive tool. With focused thinking and planning, and prudent, insightful development of technology and brainpower, IO can protect vital information systems and decision-making processes from adversary information operations attacks.

The Current Status

With the rise of IO as a major consideration in the art and science of war, it is quite obvious that the U.S. Department of Defense and its military professionals must begin thinking differently about war in the twenty-first century. Yet, from personal observations and interactions with people engaged in information operations, it appears that in a macro sense the DOD has yet to accept fully the viability of IO, to prepare adequately for conducting both offensive and defensive IO activities, or to commence the training of its officer corps to operate effectively in such an environment.

If the department was serious about preparing the military establishment for this kind of conflict, one could see a trail of "real" money leading to significant changes in doctrine, organizations, training and education, materiel development, leadership, people, and facilities—or, at the very least, see signs of serious experimentation with the concept of IO as a major factor in future conflict and the organizations responsible for national defense. These changes would be happening now. For example, it will take twenty years to "grow" an officer, from an intellectual or leadership perspective, to become a cyberwarrior in the existing system of training and education. Efforts sufficient to cause needed advancements in intellectual development and leadership only occur in isolated and sparse shreds of activity in the defense establishment's education system. Evidence of spending and "money trails" to prepare for knowledge war and IO is sparse or nonexistent. The real money in DOD still lies in the kinetic weaponry and traditional systems that support such a philosophy of conflict.

Part of the problem lies in the institution's macro view of IO. In a general sense, national security experts remain "confused" about this abstract but critically important subject, and, though understandable, this confusion has a cascading, debilitating effect on the important fields of thought concerning doctrine, organizations, training and education, materiel development, leadership, personnel, and facilities. These fields of thought are vital to the progress of change and are necessary for coping with IO, for learning what they are about, for learning how to employ them ourselves, and for learning how to protect us from our opponents' use of IO.

Currently cyberstrategists work the pieces on the margins of IO, some of them very well. But national security planners could unleash a tremendous force by planning and executing IO more skillfully in a holistic sense. As a first step, people need to understand fully how warfare is changing and recognize the power residing in information operations. People must combine the pieces of IO to find synergistic and combinatorial powers, and recognize that IO constitute a principal tool of asymmetric warfare that future opponents will use. They are equally viable as a weapon and as a means of defense for the United States.

Another reason for the basic misdirection of focus afflicting the defense establishment and precluding it from taking a great leap forward lies in the sheer complexity of IO. Each part of IO is extraordinarily complex in its own right. Then, for IO to work people must know and understand each part and learn how the parts of IO relate to other parts. They need to perform the mental function of combining things to achieve a planning and execution synergy—synergy is the only way to use IO to the fullest extent possible. Through physical and cyber combinations a tremendous force comes forth that enables either side in an adversarial relationship to win decisively with minimal costs.

Resistance to the changing face of conflict is not unusual, and most defense institutions have been conservative and slow to change. While several of the DOD's senior leaders desire change, the bureaucracy itself is slow to change. Such is the nature of bureaucracy, particularly one as massive and important as the Department of Defense. Yet, the dark cloud foreshadowing knowledge war and information conflict looms over the horizon. Looking closely one can feel its menacing nature, hear the rumblings, see the lightning bolts, and sense its amazing power and energy. One can sense the static electricity in the air connoting change and energy. If armed with such knowledge and understanding, people responsible for the country's future security would willingly focus the intellectual energy and fiscal resources required to use the power of IO to conduct offensive and defensive actions, and prepare the United States to be an unassailable bastion of information operations.

Thus the military establishment must spend the resources (time, intellectual, and fiscal) now to allow its brightest people to study the phenomenon of knowledge war and its principal enactment tool, IO. We must learn to use IO to impose our will on any adversary who will attempt to do the same thing against the United States. Recognition of IO as an important arbitrator of the outcomes of future conflict and the cascading implications that flow from such recognition are particularly important, owing to the speed of change that characterizes the age and the country's increasing dependence on information and knowledge in virtually every aspect of society.

Future Conflict

Interestingly, man has not exactly chosen knowledge war and IO as a preferred way to compete. Instead, other forces, particularly those involving survival and triumph of will, are driving adversaries of the United States to accept the precepts of knowledge war and IO. Our adversaries also learn from our example of engaging in conflict. In the Gulf War the U.S. war machine used IO to attack Iraqi command and control (C2) and Iraqi decision making, which considerably reduced casualties. The United States had the smarts and technical wherewithal to accomplish this transformation in roles and missions. Since that time U.S. adversaries have increasingly become aware of their conventional warfare weaknesses and the asymmetric advantage that the United States can wield over these forces. Out of necessity, therefore, future adversaries either have migrated or soon will migrate their thinking into the realm of knowledge war and IO.

Everyone, including the "have-nots" of the world, has access to information and knowledge no matter where they are located. People of all lifestyles can talk on cellular phones or access data on the Internet. Commercial firms lay fiber-optic cables above ground, under the ground, and under the ocean at staggering rates. Fiber has such great bandwidth potential that in many respects, twentieth-century constraints of bandwidth will diminish in importance as we move into the twenty-first century. From a military perspective the Internet, computers, and modern telephony greatly empower individuals and small groups, who can now orchestrate, coordinate, and enact activities around the globe with the stroke of a key. For hackers, some organized and state-sponsored militaries, terrorists, drug gangs, and sundry criminal elements, time and distance have reduced in importance. Basically, adversaries can

> Learn from the Internet,
>
> Coordinate across time and distance with encrypted communications, and
>
> Have access to the outputs of the high-tech world, such as high-resolution imagery heretofore attainable only by residents and organizations of rich, high-tech Western countries.

Owing to the nearly universal access to the Internet, communities of interest will spring forth as significant influences on all global security issues. Terrorists and other treacherous elements will connect, communicate, and share ideas in countless physical and virtual ways, thanks to the Internet. Alliances and conflicts will occur that transcend traditional notions of nationalism and state boundaries—and conflicts will occur between and among interest groups. People will have an increasingly difficult time differentiating among friend, competitor, opponent, adversary, and enemy. When violence and assaults occur against communities of interest, people won't immediately know whether the assault is coming from a criminal group, from a traditional armed force, or from deranged hackers who want to "play." All of these changes portend dramatic changes in our national security outlook, our strategies, and our use of resources to enact our strategies.

The Nature of IO

A first point of order in becoming more specific in our thinking about IO must answer the question, What are some of the pertinent functional characteristics of IO? In a general sense IO are much more than a mere process. They are too powerful and complex to be relegated to that. To better understand what IO are, we must think about what they do relative to knowledge war.

A significant part of IO exerts influence on how an opponent perceives, thinks, plans, makes decisions, initiates action, and measures the effectiveness of his actions. In this respect the goal is to affect an opponent's incoming or existing data, information, and knowledge, which fuel his thinking, decision making, and technological configurations to support those decisions. At the same time, astute cyberwar strategists, always wary and paranoid, must ensure that their sources of information, power, and decision making are protected. As plans transition into action, cyberwar operators must seek to manipulate the aggregate psyches of people they face and the individual psyches of the opposing decision maker. Moreover, they must attempt to affect an opposing leader's observe, orient, decide, and act—the "OODA loop"—cycle of decision making.[1] With an increased need to anticipate, the OODA loop that served us so well in the past could very well become the "AWOODAM2 loop" by adding anticipate, wargame, and measure and modify. This simple change in focus suggests the absolute imperative to anticipate and wargame and, after acting, to measure effectiveness and modify future actions accordingly.

In another sense people must be able to seek, find, and retain relative advantages in decisions, intellects, knowledge, technology, position, and actions while either shackling or diminishing an opponent's quest for similar advantage. Effective IO create entropy (in this context a gradual slowing down of decision making), which of course has value only if it enhances the friendly cycle. The object in such interaction is to accelerate and improve quality and timeliness of friendly decisions because one possesses better information and knowledge than one's foe.

IO consciously blend many capabilities to achieve synergy. Relational thinking drives the force and pervasiveness of IO by searching for and finding both obvious and disparate relationships. IO combine tangibles (cruise missiles) and intangibles (ideas) in time and space to create effects leading to relational advantage over opponents at the right place and right time to create desired effects (outcomes).

IO involve computing hardware, software, communications, and people that provide the data, information, and knowledge to help people think, plan, and decide. In this respect IO involve information systems—both ours and our opponent's. In a more technical sense, technology experts provide the pipes and engineering to blend various information technologies to provide data, information, and knowledge for people to make decisions. Information systems involve thousands of computers, channels of communication, software, and the architectures that organize them. Though marvels of modern engineering and technology, these machines, the communications pipes that support them, and the thinking that organizes their behavior together present lucrative openings for adversaries waging a knowledge war. This vulnerability holds true for both sides in future conflict. Information systems also involve the multitude of software engineers and contractors who maintain computing and communications equipment. These knowledge workers are at once a strength and a vulnerability: they are purveyors of information and knowledge who keep machines of information operating and improving, yet they represent targets that asymmetric opponents can attack or influence. If the opponent affects the "vessel" of knowledge and supporting information apparatus, the people who find or create knowledge and the decision makers they support suffer accordingly.

IO are an interactive mental chess game: they involve anticipating an opponent's assaults or attempts to manipulate our information centers of gravity while at the same time attempting to protect our COGs. Thus, IO are dynamic—we attack an opponent's information COGs knowing they will attempt to protect them. The enemy attacks the friendly information centers of gravity knowing that friendly forces will protect themselves and conduct similar attacks on enemy centers of gravity. The dynamic becomes more intense when both sides recognize that their opponents seek capabilities to affect information COGs, fully aware that the other side will have anticipated such assaults and are attempting to protect their sources of informational power. This dynamic constitutes an invisible struggle whose outcome has immense implications in accomplishing military and national objectives.

IO are founded on perception management. Twenty-first-century leaders will have to know and understand their own perceptions and those of their opponents. Perception is a critical aspect of IO, though it has a bifurcated meaning. While friendly perception remains important in the equation, probably the most important aspect of future conflict lies in understanding the opponent's perceptions—their perceptions of friendly capabilities and vulnerabilities, their perceptions of themselves, their perceptions of their environment, their perceptions of friendly perception of their perceptions, the nature of conflict, life and death, success, and so forth. This is indeed a complicated interaction, but it lies at the heart of conducting successful information operations against a skilled, intelligent, and adapting organism. To engage successfully in this dynamic means we must avoid, at all costs, mirror imaging (imposing our perceptions, thought processes, and value sets on the thinking of opponents).

IO influence and can periodically dominate the frequency spectrum (at a time, location, and effect of our choosing). The frequency spectrum is becoming increasingly crowded, and its scarcity will make it even more valuable. The United States must have the capability to protect the use of all pertinent frequencies and the capability to affect an opponent's use of frequency spectrums. All people associated with national security must take into account the spread of global communications and the dependence of these communications on the frequency spectrum. As such, people must understand what effects these systems will have on the future of warfare, particularly the system of IO. People must negotiate through their own perspectives and those of their adversary, then determine what the global spread of communications implies for manipulating, controlling, dominating, and interfering with the commercial frequency spectrum to permit a gain of temporary advantages in decision making.[2]

Before IO can affect, manipulate, or dominate an opponent's information systems military planners must understand several things: the adversary's decision cycle and how and what the opponent seeks by way of information, why he seeks that information, and how he intends to use it. People must also know how an opponent collects, moves, and processes information, and how he obtains feedback on the efficacy of his decisions. Planners and operators must know how the opponent engineers his communications, collection, and automation architectures to gather and use information to the fullest extent possible for making effective decisions. Our military and civilian analysts and decision makers must make the technical details of communications pathways, processing capabilities via automation, modems, encryption, routers, switches, firewalls, intrusion-detection systems, virtual private networks, antivirus software, and alternative routing support effective decision making.

IO are founded on knowledge management, which translates into an asymmetric advantage for the side receiving, processing, sharing, visualizing, managing, and using knowledge the fastest and most effective way in relation to the adversary. In the future all competitors using data, information, and knowledge will have to manage them quickly and efficiently using established human organizational processes, software (like collaborative tools), system managers, server administrators, systems engineers, and the like (see chapter seven). Military planners have to know and understand how the opponent manages his data, information, and knowledge and, as a corollary, figure out how to intrude upon and manipulate the flow of information and knowledge over his communications, collection, and automation architectures. These same planners need to know and understand how the adversary perceives the friendly system of knowledge management. IO will influence and periodically dominate the opponent's knowledge-management systems specifically to achieve advantages in decision making.

Information operations primarily involve an invisible war. As such, IO will be primarily a war of digits and wits—it will be a struggle for initiative, tempo, momentum, advantages, and mental agility. As a war of digits, digital information becomes a subset of knowledge maneuver. What, then, is knowledge maneuver?

When considering the maneuver of knowledge, in the future commanders will weight their battle space priorities with knowledge.

Knowledge is moveable—commanders can intensify their efforts in providing knowledge and the resources to produce it in a manner that is consistent with their needs, regardless of constraints inherent to time and space. Knowledge is fungible—it can cross levels of war, and it can exert multiple influences on the decision making of leaders in functional areas such as battlefield operating systems (air defense, fire support, and aviation). Knowledge is invisible and moves at the speed of light. Thus, a force can direct or manipulate knowledge much more quickly than it can relocate artillery fire or generate air sorties.

Knowledge maneuver enables commanders at all levels of war to acquire more and better information and knowledge. With knowledge maneuver people come together in collaboration in cyberspace to pool their intellects to solve complex problems. Knowledge maneuver also has more of a mechanical connotation—it is the manipulation of pipes and connections to provide the best bandwidth, data flow, and spread of information and knowledge. Knowledge maneuver can directly affect decision making. In a more specific view, commanders will weight future battle space efforts with the means of obtaining information and knowledge—collection, bandwidth, and processing priorities will be used to support the main effort.

Successful IO occur when people set conditions to achieve a relative, transitory advantage over their adversaries—at a time and place of their choosing—owing to timeliness, relevance, specificity, and accuracy of information and knowledge, and the skill that leaders use to make decisions. Leaders must be intellectually capable of making good decisions, or the very best information and knowledge will have no impact. This proposition has immense implications for the way the Department of Defense trains and educates its officers throughout their careers.

IO and Battle Command

Commanders and other leaders use information and knowledge not as power but instead as the means to power that one can only create through the art of battle command—but someone must first make a decision to act or not to act before that information and knowledge can become power. IO directly target and attempt to affect the decisions and decision-making apparatus of the adversary. IO can also be defensive—to protect commanders' and other leaders' decision making, to protect the knowledge workers who turn information into knowledge, and to protect the information technology apparatus that supports them.

Intelligence
Collection

Defense intelligence has numerous important functions in this endeavor of a knowledge war. For example, it must focus on and collect data, information, and knowledge for making better and faster decisions; this collection effort must include what the intelligence community does well now—collect in the kinetic world—and what it must do in the future—collect in cyberspace. Material developers must develop cyberbots to perform specific functions and missions in cyberspace.

What are some of these functions and missions? Cyberbots must surveil selected locations in cyberspace and reconnoiter pathways that data traverses as it moves from hub to hub and person to person. Cyberbots must act as sentinels to provide the warning that an adversary's offensive cyberbots are approaching and seeking entrée to friendly networks. These defensive cyberbots must parry and thwart the adversary's cyberbots and provide warning. Cyberbots must act as human intelligence agents in cyberspace and surreptitiously collect data, information, and knowledge from adversary databases residing on hard disk drives. The greatest coup of all, though, for these HUMINT-like cyberbots would be to steal the adversary's decision logic and the most trusted form of seeking and finding feedback on the veracity of decision making, without the adversary being aware of the theft.

All collected data, information, and knowledge must be transmitted back to human handlers who provide analysts and other knowledge workers with the capability to resource commanders' efforts using advanced information and knowledge sufficient to make decisions faster and better than other human beings. Some of it could be processed in cyberspace by virtual processors, mostly with data. But humans must still turn that information into knowledge, and thus the information and knowledge that cyberbots collect must be sent to a location where a human can receive and use it. Decisions in cyberspace will have to be made by cyberbots, because they move and change too swiftly to allow human intervention and guidance.

Positioning

Cyberbots are effective when they have positional advantage over opposing cyberbots, such as at a key switch or in the vicinity of a key database where they may hide either near to or in an adversary's key decision-making database, ready to awaken and disrupt or destroy the adversary's activities in cyberspace.

Intelligence Analysis and Synthesis

IO require fine-grained, reliable knowledge that comes through the intellectual work of analysis, particularly of the opponent's thinking, perceptions of reality, decision making, feedback mechanisms, and information support apparatus. Current conventional wisdom is that intelligence analysts will perform this function. Unfortunately, they are not adequately prepared to perform this type of analysis and synthesis, and such a shortfall isn't their fault. Cyberstrategists and other practitioners of IO must instead find country experts who have studied the opponent's thinking and decision making for years, and use them for learning how the adversary thinks and what influences his thought.

Analysts need to learn to think better, develop broader knowledge and understanding (particularly of foreign cultures), and tap into expertise available in the commercial, academic, or Reserve Component communities of expertise in a collaborative way. Large reservoirs of intellectual and cultural expertise exist throughout communities of interest and communities of practice in this country. The challenge is to find those communities, tap into their collective intellect, and shape the information and knowledge they have to meet commanders' decision-making requirements. The defense establishment must work with these communities, connecting virtually and physically with them to gain the level of analytic detail and synthesis they need to plan and execute IO. To perform IO functions with any degree of success we must either develop within or find this level of mental sophistication outside the Department of Defense.

Knowledge Weapons

A nascent implication that comes forth when thinking about a knowledge war and information operations concerns developing and using knowledge weapons against adversaries, and protecting people, organizations, and the country from an adversary's knowledge weapons. Kinetic knowledge weapons include actual weapons, terrorists' physical acts of violence, and bombs (truck, car, and human). Nonkinetic weapons include technology (such as laser and microwave weapons and other purveyors of energy), radio frequency and cyberspace electronic countermeasures (jamming), and software tools that hackers use to influence the data, information, and knowledge resident in computers or the flow of data, information, and knowledge. Knowledge weapons can include some aspects of terrorism (activities designed to influence knowledge workers or system administrators) and programmed cyberbots armed with viruses or other ways of affecting data, information, and knowledge (such as denial of service).

Knowledge weapons attack and manipulate data, information, and knowledge that an opponent may possess. Knowledge weapons disrupt the flow of data, information, and knowledge. Knowledge weapons attack databases that an opponent relies on for data, information, and knowledge inputs. Knowledge weapons interrupt the flow of a competitor's information to degrade the timeliness of arrival of the information. As in the case of all information that has value, timeliness generally is one of the variables that enable it to have and retain value. Thus it is safe to conclude that knowledge weapons will interrupt, manipulate, or affect the flow of information, thereby rendering its relevancy questionable. Continuing along this line of thought, knowledge weapons affect the meaning or coherence of information, also causing its relevance to be affected. Knowledge weapons will "fuzz up" the information an opponent seeks and finds. Thus these weapons will be a direct contributor to denying the principle of specificity that all competitors seek. Moreover, knowledge weapons will interrupt the flow of digits and affect them so as to distort accuracy. By distorting the accuracy of an adversary's precision-guided munitions, for example, they will have difficulty accomplishing what they are intended to do.

Knowledge weapons could be as simple as a piece of tin foil or as sophisticated as software programs possessing chemical releases that "eat" silicon (the right parts of chips at the right time and right place in an adversary's information-gathering and processing system). Since information is so important that it is safe to surmise that our opponents will attempt to develop and use knowledge weapons against us, we have no choice but to protect ourselves and learn to use knowledge weapons against them.

Transforming the
Face of Battle—Doctrine

The transformation of the face of battle isn't something that we can put off—the absolute need for this transformation is pressing upon us now.

Asymmetric adversaries, including terrorists, will use IO against the United States, and the United States must use IO against its adversaries. Much remains to be done in preparation for these approaching struggles. No preparation, though, is more important than changing our doctrine. Doctrine represents the collective thought of many bright people and helps people understand complex concepts like IO. IO doctrine applies useful mental processes to turn theory into practical application. It provides direction and guidance to trainers and educators, materiel developers, people who develop future organizations, and even the type of people the services recruit.

IO are a terrific tool for waging a knowledge war. In fact, knowledge war and its IO tools will revolutionize the way people engage in conflict. Doctrine is evolving, enormously complex, interactive; it possesses a constant dynamic between friend and foe struggling for advantages, and should highlight the importance of friendly and enemy information centers of gravity, to name just a few. Doctrine defines the phenomenon of existing in competitive environments that involve the invisible world of IO and should explain the many aspects of IO and how they relate among themselves and to other things. Military intellectuals need to think about the military applications of cyberspace and explain their thoughts and goals in doctrine. As doctrine writers put their fingers to the keyboard they must acknowledge the presence of a determined, ruthless, technologically adept, adaptive, and smart adversary. Doctrine writers must help prepare people for engaging in this wits-driven, often invisible competitive environment where cyberbots are the soldiers and cyberspace is the terrain. The military needs an intelligence aspect to its preparation for cyberspace in which competition will involve human beings, cyberbots, and the digital terrain.

In the cyberdomain a different type of key terrain exists: the collective intellects of the adversary's knowledge workers and decision maker(s). The intellect of the adversary will be a critical element in the thinking and planning that occur when knowledge workers turn information into knowledge, and when a decision maker thinks about the knowledge, turns it into understanding, and makes decisions. The psyches of individuals and groups of individuals condition their thoughts and perceptions. IO seek to influence thinking, planning, and decision making. Thus, people need to know and understand not only the individual thinking of the adversary but the collective thinking they follow as organizations and even as societies. When armed with this knowledge and understanding, planners have the potential to execute deception and psychological operations that slow or adversely affect the quality of decisions.

The social processes that IO planners will be increasingly interested in involve the intangible aspects of conflict (morale, moods of the populace, how opponents think and perceive) and the tangible aspects of conflict (such as location and condition of communication pipes, nodes, switches, automation, collection, and architectures). People need to spend "deep think" time to determine what is needed for creating this new type of doctrine; they must then write it down and ensure that it influences those responsible for developing the curricula of national defense schools and acquiring the materiel to support operations in cyberspace.

Activity in a knowledge war will involve time and movement in the boundless domain of cyberspace. What kind of activity? Cyberbots will attack the decision-making apparatus of the adversary. Cyberbots will defend against attack. Cyberbots will deceive opponent cyberbots and their human handlers. Cyberbots will act as sentinels warning of the probes and nefarious activities of opposing cyberbots. Cyberbots will surveil key targets and reconnoiter fiber and radio frequency pathways. Cyberbots will collect information from packets moving in cyberspace and from the code intrinsic to the memory of computers, cell phones, and personal digital assistants. Cyberbots will collect inside an adversary's databases, searching for decision logic that is buried in the code of software or that populates fields of databases.

Any substantive thought about IO doctrine must encompass aggregation theory, for it lies at the heart of IO. Aggregation theory, at its most basic level, involves the interaction of things (individuals, organizations, networks) and relationships (physical equipment, activities, thinking, and the like) in which combined outcomes are much more powerful than when not combined. Outcomes are the result of combinations coming together and emitting energy. With aggregation, second- and third-order effects occur—some intended, others not. If people understand aggregation theory and learn to create or anticipate second- and third-order effects, instead of coping with each unexpected effect separately as it comes along, the door of IO will open. Some useful examples help illustrate what is meant by aggregation theory.

First, psychological operations (PSYOPS) are used to affect the collective intellects of people enough to cause perturbations in morale. Such perturbations have the first-order effect of influencing the fighting quality of troops. A second-order effect is the dissonance caused in the minds of commanders whose decision making is being attacked or manipulated. The struggle in these competitive endeavors lies in the invisible realm of manipulating the collective-aggregate intellects of fighting people, of their leaders and politicians, and of aggregates of people—some friendly, some adversarial, some neutral. A third-order effect is the influencing of aggregate psyches of an adversary's soldiers to the point of creating dissonance in the psyche of the commander, thereby affecting his decision cycle.

As a practical example, the Gulf War experience can provide some insight into multiple layers of effect. On numerous occasions Iraqi soldiers capitulated, owing to the wear and tear on their collective psyches caused by the mental and emotional fears of death, the totality of a very effective PSYOPS campaign, and the follow-on physical pounding caused by B-52s and other aircraft. This emotional drainage and perception manipulation occurred over time and disturbed their commanders' ability to make effective decisions and thus their will to fight. It directly affected commanders' thinking and soldiers' willingness to fight. Ultimately, Iraqi commanders experienced the malaise of defeatism because, from poor psychological readiness and emotional perturbations in the ranks, they did not believe their soldiers would stand fast. Consequently, many leaders and their troops decided to surrender instead of fight, which compelled the Iraqis to capitulate to the will of the coalition. Thanks to the cumulative power of second- and third-order effects that come with successful IO, many Iraqi troops surrendered and enabled the coalition to avoid large numbers of casualties.

Second, think for moment how advertising works. If an advertisement affects just a few people it is meaningless. But if it affects a million people it will have a significant impact. Aggregation and the spread of ideas among people, encouraged by repetition and reinforcement by famous people and buzz, can influence individuals, groups of individuals, collections or organizations, communities of interests, and aggregates of nations. Once an idea or influence spreads to aggregates, the results are stupendous.

Imagine the potential aggregated effects arising from the use of terrorist-placed weapons of mass effect in key financial centers stretched across the United States and synchronized by cell phone and the Internet. The immediate effect would be the physical damage and destruction that can overwhelm the infrastructure's capability to respond. The second-order effect is felt when people—ordinary citizens and political and military leaders—realize the magnitude of loss and despair and respond with fear, suffering, and outrage. The third-order effect is then the erosion of trust, confidence, belief in sanctity of our way of life and the belief that law enforcement officials can provide safety. One event of this nature is a tragedy, but a multitude of events sequenced to create the greatest second- and third-order effects are a disaster of such immense proportions that the will to punish or continue to compete is diminished.

First-, second- and third-order effects become more important with aggregation. For example, if a killer application (app)—an idea and technology such as the stirrup or the cell phone that has an immense and far-reaching effect—takes hold it does not influence just one or a few people. For the innovation to be a true killer app it can affect millions of people and collectively change their way of life.[3] The second- and third-order effects coming from the impact and implications of killer apps often are unintended—such as the discovery of the simple horse saddle stirrup, which led to easier reconnaissance, cavalry maneuver, and so forth. The person who invented this killer app would have been surprised at all the second- and third-order effects that resulted. The same holds true for other killer apps: television, telephone, wireless communications, and computers. But the theory isn't the interesting part. Rather, killer apps take hold owing to the cumulative effects they have on aggregates, which in turn serve as the catalysts for second- and third-order effects that always seem to come.

Information operations make up a strange but fascinating world, at once exhilarating, dangerous, and difficult to comprehend because of the high level of abstraction involved and the fact that digital terrain involves the largely invisible maneuver of cyberbots inside fiber-optic cables and databases. It is safe to conclude that competing successfully in the domain of IO will demand ever-advancing technology, flexible and innovative organizations, and, most important, great intellectual prowess that comes neither cheaply nor without laborious and continuous expansion of mental capabilities.

The military establishment must acknowledge that, owing to the information revolution and the ascent of the digital age, the face of battle is changing. Cyberspace as a dimension of conflict and competition has vaulted to the forefront of importance of the future national security landscape, and now ranks equally with the air, ground, sea, and space dimensions of conflict and competition. The Department of Defense must act now to cope with this major change in the face of battle to shape the future national security environment. If inaction and inertia rule, the distinct possibility exists that twenty-first-century adversaries will leave us behind in the race to the future, a possibility that increases exponentially with the passage of time and stasis on the part of the DOD bureaucracy.

How can we best prepare for engaging in competition in the strange, exhilarating world that is the cyberdomain? People must adjust their thinking to acknowledge that a process involving people, machines, and time turns data into information, information into knowledge, and knowledge into understanding. A logic trail exists involving knowledge, understanding, decisions, actions, effects, and will. This logic trail forms the foundation of any effective architecture or knowledge-management system. It also provides a way to attack such a system.

Future cyberwarriors will readily understand that information isn't power, as the old adage claims. Instead, information is the *means* to power and it will take the art of battle command or decision making to turn knowledge into actual, *usable* power to create the activities and actions that cause effects. Understanding that information is the means to power lies at the conceptual heart of IO. Commanders and other leaders must put forth much intellectual energy and combine the intuitive art of battle command with IO to create conditions conducive to creating information superiority.

Before proceeding further, though, one must take note that the best and most timely information and knowledge are insignificant and have little value if they are not used worthily. Thus, developing the intellect and requisite thinking skills to think about, plan, and execute IO with the goal of influencing decisions will loom as increasingly important in the future of the country's security apparatus.

To be effective in planning and executing IO practitioners must receive extensive knowledge support. Individuals deploying to foreign countries often don't have the requisite knowledge and understanding of those countries to even begin to contemplate the intricacies of information operations. They need help from experts connected to them in a virtual, collaborative support environment.

How do we develop the talent, systems, methods of collaboration, and leadership suggested above? The Defense Department needs to "grow" officer and enlisted area experts, invest in their intellectual development, country knowledge, and language expertise, and then place them where they can readily help deploying or already-deployed forces. Along with country-specific expertise, these experts must know how to think using synthesis and analysis, how to communicate, how to manipulate data, and how to present information in the way their users need it for facilitating fast, effective decision making.

Since no one can predict the future with complete accuracy, the Department of Defense must posture itself now for the distinct possibility of the ascendancy of information operations as the principal tool of knowledge war, hence the key to manipulating an adversary's or our will. Posturing, though, is a multifaceted endeavor. Posturing calls for making the intellectual adjustments to think through, accept, and prepare to use information operations to win in any type of conflict. Adaptive thinking incorporates flexibility sufficient to accept and use rapid change for achieving and sustaining advantage. Military personnel must seek and use relationships, both obvious and disparate. They must be able to blend combinations of kinetic and cyberrelationships to achieve synergy because of the additional benefits that accrue to those wise enough to create it. Mentally adept people engaging in future conflict must be intellectually agile enough to recognize that they can sow the seeds of destruction even as they create—nothing will be permanent, everything will be in a state of chaos and flux, and "truth" will fluctuate constantly.

Posturing involves developing a coherence (unified, sensible, successful, cooperative, and complementary efforts) among materiel development, academic research, intellectual development, and software and hardware development to find existing technologies or create new ones that enhance decision making, while adversely affecting the adversary's. In addition, part of the posturing the DOD needs to prepare for engaging in offensive and defensive information operations means that people must stay abreast of research developments that come from academia, industry, and governments.

IO are the consummate asymmetric tool for both sides, since asymmetry can be an advantage for either the strong or the weak. If done correctly IO can provide great advantages to a stronger force, owing to the sophistication of machines and the decision processes they facilitate. IO also provide a way to bring the strong to their knees. If an asymmetric foe discovers and attacks or manipulates the stronger's information centers of gravity or hubs of power, he stands a great chance of being the modern David bringing down Goliath.

Unfortunately, IO may continue to receive short shrift owing to their complexity, to the tendency to keep the parts organizationally separated, and to the corollary difficulties in understanding what they are and what they do. Without the proper changes in the relevant elements, IO will remain diffused, confused, and often ignored by our leaders, mostly owing to the heavy burden of daily tasks these leaders already experience. As such, they will not ante up sufficient resources for enabling IO to be a defining aspect of the "American Way of War" of the twenty-first century. We must find a way to encourage decision makers to make wise choices to prepare themselves now, so they can indeed shape the future of the evolving environments of conflict.

All warfare is based on deception. Hence, when able to
attack we must seem unable; when using our forces,
we must seem inactive; when we are near, we must
make the enemy believe we are away; and when far
away, we must make him believe we are near.

Sun Tzu, *The Art of War*

The Art of Deception 6

All people involved in military activities need information and knowledge
to make effective decisions. Since deception is the manipulation of information and knowledge that people need to make decisions, it follows that
effective deception lies at the heart of successful information operations.
Deception can enhance the broad advantages of tempo, initiative, and momentum, as well as the more specific functional advantages of decisions,
intellect, knowledge, technology, position, and action. Deception, which
can be used to affect the information and knowledge that an adversary
uses for making decisions, is therefore fundamental to targeting the mind
of the adversary decision maker, his decision cycle, and his information
support systems. The following few thoughts about the art of deception
are in no way comprehensive, quite the contrary. Only some of the more
salient points of the cerebral aspects of deception are considered, and then
only briefly. But some of the issues surrounding the art of deception must
be thought through, all the while attempting to expunge some of the intellectual fog surrounding it so that the complexities of this great tool can
be better understood. Consideration must be given not to *what* to think
about deception, but rather *how* to think about deception. We must open
the door to the advanced "thinking about thinking" that we need to plan
and execute deception successfully.

Wise people will always attempt to use information and knowledge to gain a decision advantage over their opponent. It is just as important for people to manipulate the information and knowledge that their adversaries use for making decisions. Deception helps people think through complex situations and provides a way to dominate an opponent in a cerebral sense. In deception, an intensely cerebral form of competition, the issue is the mind—the thinking, planning, and perceptions of the opponent.

The cerebral struggles of deception are, for the most part, invisible. They involve a contest of wits rather than of might. Though physical things certainly still have a place in deception, the wise deceiver hypothesizes how physical effects will influence the mind of the adversary, how to manipulate these physical things, and, of course, how to pick the right time to deceive. The deceiver must then commit intelligence assets to confirm or deny the success of the deceptive measures.

Deception as a strategy is nothing new—wise men have been using it for centuries. At the root of all deception lies the attempt to affect how opponents think, plan, decide, act, and gather feedback on their actions. To think people need information and knowledge. To decide people need understanding that comes from knowledge. Information and knowledge help explain the world and relationships comprising it. People need information and knowledge to make decisions and then test the validity of those decisions. People think about, seek, and use information and knowledge coming at them from a variety of sources. Moreover, the validity of information and knowledge (that is, what people believe or don't believe) lies in their cultural backgrounds, social mores, education, religions, traditions, and so forth. Thus people using deception absolutely must have a good idea of how and why an opponent thinks, how he makes decisions, and what he uses as sources of feedback to measure the effectiveness of his decisions. If they don't use these three pieces, deception will always be sophomoric in design and no better in execution.

The Germans in World War II, for example, relied on human intelligence sources in Great Britain to keep them appraised of the intent of Allied invasion plans in 1944. Long after they had received information sufficient to discredit those sources, the Germans continued believing them. The Allies, on the other hand, believed their ULTRA transcriptions and protected with great vigor their undetected access to that source. At the time even a mere semblance of decent wargaming, in which some degree of respect for the foe comes forth, would have allowed the Germans to exploit the Allies' dependence on ULTRA by using deceptive inputs themselves. Fortunately for the Allies, the notion never seemed to cross the minds of German planners.

Information inputs people use for making decisions and confirming or denying the validity of decision making are always susceptible to manipulation. Opportunities to deceive abound because of the way humans seek and process information and knowledge. People making important decisions often have an insatiable appetite for information and knowledge, because conventional wisdom says that the greater volume of information and knowledge one has, the less risk one will endure. Risk fuels the continuous quest for more and better information and knowledge. Within "quantity," though, the seeds of deception can stay hidden, grow, and flourish.

Deciding always involves some degree of risk. Cautious people like to minimize risk, while those a bit more daring are more comfortable with increased risk of failure or unexpected chance events. Theoretically, risk decreases proportionately the more one is able to accrue timely and valid information and knowledge, because unknowns causing friction or surprise dissipate accordingly. As the availability of timely relevant information and knowledge decreases, risk increases. To make decisions with no information or with limited or questionable information is often tantamount to turning an acceptable risk into an unacceptable gamble.

Because each human is wonderfully unique, every individual thinks in a unique way. All people see the world differently, and their perceptions of reality vary widely. Information and knowledge influence their thinking, and thinking is critical to how people engage in the process of deciding on a course of action or a specific solution to a problem. Even more specific, the art of deception involves knowing to the greatest possible extent how an adversary processes, filters, and uses information and knowledge for either understanding the battle space or making decisions.

Information and knowledge seriously affect the views and perspectives of theoretical adversaries. Thus the modern deceiver has as a prerequisite for planning and executing deception (whether in the physical world, the mental world, or the cyberworld) the "thinking about the thinking" that an opponent uses for viewing and understanding the world and making decisions. Once people make this intellectual leap—knowing and understanding how an opponent views the world and the information making up that world—they can seriously start approaching other aspects of deception, some of which are indeed mechanistic.

Often people find themselves particularly susceptible to deception, owing to a proclivity to engage in reductionist thinking, that is, thinking that seeks simplistic solutions to or understanding of complex phenomena. Unfortunately, reductionist thinking results in a consistent failure to look for and find relationships, as well as inadequate thinking about aggregation theory and its "stepchildren," second- and third-order effects.

In particular, the operational environment for future conflict will have two sides. Thanks to modern information technology and the flood of available information and knowledge resident in the world's databases, decision making will be potentially easier, faster, and better than in the past. However, because of the connectivity of the world and the need to consult with allies and take their thoughts and preferences into account, constraints will be more restrictive, particularly in the use of violence and its offspring, collateral damage. Once constraints increase, risk increases and this increasing risk fuels the need for more information and knowledge. Ironically, the same technology or access to information that also could allow a commander to make very fast and good decisions could make him or her feel paralyzed, thanks to the need to obtain "truth" before acting.

All that is required to engage in complex deception activities are minds sufficiently agile for thinking through the thought process of synthesis; the ability to focus on relationships to conceive deceptive mechanisms; the ability to perceive the world through our opponent's senses; and, when required, the ability to think similarly to our opponents. Of course, people entering the realm of deception need to avoid mirror imaging their own value sets onto those of the opponent; if they don't they will have a distorted view of how the opponent thinks and what he will or will not believe. In this respect our country's security people must learn how to think like the terrorists who rammed airplanes into the World Trade Center and the Pentagon. If we cannot learn to think this way we need to hire and use asymmetric opposing forces to exercise and wargame with and against.

Besides being able to recognize how an adversary views the world, people engaging in deceptive activities must know their own particular view of the world. To engage successfully in the art of deception people must view the world as interrelated from both macro and micro perspectives. Traditionally, as citizens imbued with vestiges of Western culture, Americans have viewed the world as unrelated pieces or spheres of influence. Schopenhauer, the German philosopher, said, "But the sight of the uncultured individual is clouded, as the Hindus say, by the veil of Maya. He sees not the thing-in-itself but the phenomenon in time and space, the principium individuationis, and in the other forms of the principle of sufficient reason. And in this form of his limited knowledge, he sees not the inner nature of things, which is one, but its phenomenon as separated, disunited, innumerable, very different and indeed opposed."[1]

The world is and always has been interrelated in a wide variety of ways. Occasionally the human mind has failed to comprehend, seek, or create these relationships, which lies at the heart of the average person's inability to engage in quality deceptive activities. Advancements in information technology have enabled people to have the potential to understand the multitude of complex relationships making up the tapestry of the world. People can now use an increasingly plentiful and sophisticated variety of software to seek, find, and view relationships. People can now visualize information and knowledge to help them think in a more sophisticated way and enter into automated, adaptive wargaming for examining the "what if" and "could it be" scenarios and associated variables. Now, thanks to the immense variety of information and knowledge available, people can mentally view the world as giant spiderwebs, consisting of hundreds of interrelated cells and linkages, or as an endless string of pearls, all tightly connected but each containing a part of the others.

One positive element of the human condition is that each human perceives the world through a unique prism, but more precisely as a kaleidoscope. As he or she views the world, the kaleidoscope's patterns constantly rotate and change. Within their personal kaleidoscopes people search for patterns and linkages among patterns, but experience only varying degrees of success, depending on how that individual mind works. Minds that work easily with patterns, relationships, and a holistic sense seem to design deception tactics well and are less susceptible to being deceived than those that search for black and white answers, nonrelational reductionism, or absolute truths.

To be good deceivers and protect themselves from their opponent's deceptive measures, people must seek, find, and interpret information, all the while acknowledging that the opponent is or could be attempting to do the same to them. Smart people involved with future conflict must realize their near-total dependence upon information technology to function and, more important, realize how their perceptions, experiences, and values influence their ability to think, plan, and make decisions. While looking at themselves in this way, they must also look through the eyes of their opponents to see how they view such dependencies and vulnerabilities. This dependence leads both sets of people involved in a deceptive contest to be vulnerable to making grotesque mistakes in reasoning.

Perceptions, cultural backgrounds, and thinking patterns approximate the sum total of a human psyche. Psyches are tremendously important in deception, and involve the sense of "apperception" advocated so eloquently by renowned early-twentieth-century psychologist William James. James believed that education, culture, upbringing, religion, history, myths, and customs condition a person's mental outlook and identity, and generally influence how that person thinks, views the world, and solves problems—it is the essence of one's perceptions that guides one's thoughts about truth and worldview.[2] Thus James's notion of apperception could be used as a guide to train people to successfully perform deceptive operations.

The process of deception becomes easier to understand when realizing that what a person sees or perceives on the surface could be different than what it actually is. Typically, though, people involved with deception assess situations using black-and-white lenses and viewing them from easily discernible angles. To an expert deceiver the world instead becomes relationships characterized by hue. In the untrained deceiver's world solid lines and right angles are difficult to discern and ambiguity often clouds meaning. When people can view a world with hue and a healthy respect for ambiguity, avoiding clear-cut "truth" or black-and-white situations, they can start deceiving. It also helps to consider aspects of deception as a Rubik's cube: to enter the cerebrally intense world of deception people must observe, reflect, and think about thinking in terms of a slowly moving and rotating Rubik's cube that shows differing perspectives of the same situation (an adversary's perceptions).

Consciously and systematically people must think about the thinking of others to plan and execute successful deceptive activities. Others' perspectives and perceptions guide their view of the world—what is right and wrong, even what truth is. Such mental work doesn't mean acquiescence or agreement with those other perspectives, it simply suggests a definite need to respect others' points of view and outlooks. A planner must consider a theoretical adversary's perceptions, but it must go even deeper. The deception planner needs to consider how the adversary perceives his world and his opponent, as well as how the adversary perceives the perceptions that the opponent has about him. Furthermore, the skilled deceiver must look at self, at the adversary, at the adversary's view of self, and at the adversary's view of a friend's view of the adversary's view of the situation and the world. This mental exercise becomes more difficult as one realizes that the smart foe is doing the same thing. In addition, a planner contemplating deception must consider the adversary's goals and objectives and his constraints, since his fears and hopes might be consistent with the apperception mentioned above. Additionally, the planner of deception must consider what his adversary believes to be valid sources of information and knowledge, and how he seeks, finds, processes, and uses that information and knowledge. The deception planner must also consider the adversary's decision cycle, and he must consider how the adversary seeks feedback on the decisions he makes.

Armed with this information the deception planner can develop a deception campaign plan designed to manipulate the adversary's images, thought processes, and decision cycles. The deception planner must commit resources to the effort and use his own collection assets to track the adversary's efforts to seek feedback on his decisions, then use that feedback mechanism and ensuing flow of information to manipulate the thinking and follow-on decisions of the adversary.

Interpretive insight into the value of information from our perspective and from the perspectives of others who are not necessarily thinking, feeling, or intuiting like us is necessary to plan and execute deception. Undoubtedly the deception planner will need the assistance of knowledge experts—psychologists, linguists, country experts, and intelligence analysts—to get to the intellectual depth suggested here. This assistance is an absolute requirement for minimizing the impact of mirror imaging.

Another prerequisite for engaging in successful deception is information wargaming. People must be able to deftly slip back and forth between their minds and the minds of opponents to see what truth means to them in relationship to the opponent's point of view. Such mental gymnastics are complementary to the wargaming methodology presented earlier. They provide insight into an opponent's perspectives, goals, and objectives. Moreover, such mental work opens the door, so to speak, and allows an understanding about what information and knowledge are valuable, and why, from the perspective of each side in a contest, that situation has not yet come into being.

Determining the *value* of information and knowledge from the adversary's point of view is important and fundamental to successful deception.

But it is difficult to discern and manipulate because it is dynamic and relational to how the opponent views the world and how the friendly deception planner views the opponent's view. To add to the complexity, the deception planner must know very well what are important information and knowledge to the friendly decision maker, then be able to contemplate how the adversary judges and seeks confirmation and feedback of his own perceptions about what the friendly decision maker believes about the value of information and knowledge. The implication is, of course, that to engage in a viable deception operation deception planners must provide some of what the adversary considers to be valuable while still protecting the information and knowledge that are truly valuable, other than what is purposefully provided to affect the adversary's decisions. Deception planning and execution evolve into a quickly changing mental and physical landscape and place great demands on the intellects of both sides.

With the insights gained through information wargaming, people can estimate what their opponents know and understand, the realm of possibilities for deceiving opponents, and how to fool opponents with noise, ambiguities, false or partially true information, and disrupted hues. Wargaming helps people advance significantly toward winning the struggle for valuable information and knowledge so desperately needed to make good, fast decisions.

Here we must stop to consider the notion of truth. People entering the world of deception have to possess an outlook about information and knowledge without being burdened by "knowing truth." Let us face facts about truth: what makes information true or untrue lies in the eye of the beholder. Generally a given truth is not a universally accepted fact. Perspectives and apperception cloud the issue. Truth suffers further obfuscation by the "noise" coming from the mass of information deluging human beings daily and the rate of change that the information revolution brings to our lives. Human minds, though, have a distinct way of seducing themselves into thinking they know truth or believing that one truth exists, or, even more important, that one's inner self recognizes truth.

People take information and correlate it with information residing in their minds' "databases," resulting in messages of correlation, noncorrelation, and truth or "nontruth." People often seize upon "the correct answer" because they know "the truth." Minds typically relate informational inputs to their apperception and all that has been put into their minds over time. This is quite all right, as long as one recognizes that apperception presents one point of view and not necessarily the point of view that enables successful deception.

A final point about the issue of truth and deception: arrogance can be the ruin of anyone entering the realm of deception. Arrogance is an unstated belief of superiority or advantage of some type over an adversary. For example, one football team could have an unspoken belief that its skills in a passing attack far outshine its opponent's ability to defend. This could either be true or false. If false, and the passing attack fails against a better passing defense, the arrogant side will suffer an immense blow to its collective psyche when the team members discover they were wrong about the perceived advantage that led to their arrogance in the first place. Arrogance in a security setting could be, for example, as simple as believing that one's technology cannot be matched by the opponent, a belief that can be shattered when the opponent counters with an even better technology, or is able to use low technology or no technology at all to achieve an unexpected advantage. Or arrogance could come from the belief that an advantage has come from superior thinking, planning, or execution of tactics. The wise planner knows that the other side in a contest can come up with superior thinking, planning, and execution, too.

The implication here is simple and easy to grasp: we cannot let arrogance cause us to believe that we have an advantage which we don't have, or that we always know the truth. Such arrogance will cause our minds to be affected by what I call the "*Ozymandias* syndrome." In his epic poem *Ozymandias*, Percy Bysshe Shelley explains the original arrogance and past grandeur of an immense statue now lying in ruin on the desert floor in some far-off place. In the great game of deception, people cannot allow hubris to contaminate their thinking to the point that their minds become ruins lying in an intellectual wasteland. The wise deceiver avoids this arrogance at all costs and maintains respect for the adversary.

As an example, if we were to simultaneously input identical pieces of information into the minds of a native of Japan and a native of the United States, the resultant conclusions undoubtedly would be dramatically different between the two. The Japanese person has an upbringing, an accepted and practiced way of thinking and planning, education, morals, values, a gene pool, and an environmental context very different from the American's. His or her background—brought to bear on the type of input meant here—could very well cause the Japanese person to recognize as an "untruth" what the American would believe to be a truth. Each of us believes we can recognize truth. But in reality there is no absolute truth because truth depends on us—unique human beings and supreme individuals tempered by surroundings and backgrounds, culture, genes and chromosomes, soul and subconscious.

Let's take another approach. How many times have you thought you knew the truth absolutely and proceeded to make decisions from that absolute, only to find out that you were wrong and that inexplicably "the truth had changed"? Perhaps one shouldn't be so sanguine and positive about this kind of thinking that leads to the conclusion of either knowing or understanding the truth to begin with. Perhaps instead one should approach the concept of truth as being multifaceted; that is, there are many different perspectives of truth, each capable of fluidly changing many times and being influenced by many different and complex variables. Changing variables and factors eat away at the fabric of our perceptions of truth until we can gain the insight that "truth has changed."

This is not an assault on the fundamental truths that people in the United States believe to be true: inalienable rights, equality before the law, and so forth. But people from different cultures and backgrounds have views of truth that are different than ours. If the deception planner fails to recognize those differences his or her efforts to deceive will fail. As an example of my own tendency to mirror image "the truth," a few years ago I received an object lesson during a multicultural seminar on creative leadership. In that seminar I made a very strong point about the universality of integrity. I made this assertion because I thought the concept of integrity was inviolate and that it presented a black-and-white mental construct that was universal across all cultures. How naive! I found out quickly that I was wrong. Quite simply, people from other countries and other cultures did not think integrity was that important, nor did they think that it meant a person was truthful, nor did they think that a person with integrity necessarily had the moral courage to stand up for what he or she believed. In fact, they believed just the opposite, from my point of view. It was a powerful and painful experience, but it brought forth for internal consideration some valuable conclusions that seem apropos to the art of deception. I concluded that indeed truth changes, and we have to realize what I call the "flux of truth" to be able to enter into deceptive engagements with a chance of attaining and sustaining advantage. In addition, the lesson reinforced my belief that apperception is indeed important. Without question each person is unique and has a particular prism from which he or she views the world, and through which he or she filters information. The

structure of this prism strongly influences the shape of patterns and relationships inherent to situations. The conclusions that might be drawn all
depend on apperceptions, perceptions, and the variables involved in the
situation. Along with recognizing and dealing with apperception in a positive way, deceivers have to manipulate the variables that influence the
prism and thus the emergent patterns and thoughts about relationships of
the person who is the target of our deception.

With mental flexibility about the idea of truth people can begin to understand the many faces and uses of information and knowledge. They
can gain an appreciation for the perspectives of their opponents in the
competitive, dynamic quest for valuable information and knowledge that
will always be intertwined with conflict. It is possible and desirable to leap
into the opponent's mind for a review of his perception of truth, his worldview, and the inputs of information he seeks for verifying something as
the truth.

It is particularly important to understand the information and knowledge inputs an opponent seeks, uses, and trusts. Why? If people understand what information an opponent seeks, uses, and trusts, they can manipulate or create informational inputs specifically designed to confuse,
trick, or alter the opponent's perceptions. This is a twenty-first-century
view of deception—information and knowledge are fundamental parts
of it. Additionally, people must understand how an opponent thinks,
how he uses automation to process and sort information, and how he
employs communications to receive and send information. Moreover,
people must understand an opponent's informational feedback system
and mechanisms to manipulate the adversary's feedback to achieve advantage.

It seems only logical that deceivers must give much thought to how
their adversary seeks feedback for determining the effectiveness of decisions. Typically people have a source of information or knowledge that
they are willing to rely on in their quest for feedback. If the deceiver gains
insights into the feedback process, that person can create inputs that affect
the opponent's thinking and decision making.

It is also important for deceivers to understand their opponent's entire
decision cycle. If the deceivers do not know the cycle, they should use a
generic process for making decisions. Such a generic process would describe the desired outcomes and what information and knowledge would
be necessary to make decisions. A generic decision cycle theoretically
sketches an adversary's information-collecting, processing, analyzing, and

synthesizing processes. It also describes how collaboration works, how the opponent transmits decisions to turn them into actions, and how the opponent's ways of seeking feedback exert influence on ensuing outcomes. While variations may exist, all players in a conflict have decision cycles that resemble this generic cycle.

A further consideration is that deception planning will occur in cyberspace. In cyberspace intelligent agents or cyberbots will have many roles and missions. They will act as sentinels to warn of intrusion into networks and databases. They will engage attacking cyberbots and deny them entrance into valuable databases and networks. They will act as intelligence collectors. They will enable the intelligence analyst to "move" through cyberspace and see, hear, and perhaps someday touch the desired information and knowledge in virtual reality (reality in cyberspace). They will perform surveillance and reconnaissance duties along communications paths, and in switches, routers, and databases. These cyberbots will also be a major performer of deception. Deception in cyberspace will revolve around the duality of the struggle for valuable information and knowledge—both sides will need valuable information and knowledge to make fast, effective decisions. In some cases the struggle will involve surreptitious entry into databases to steal valuable information and knowledge. In other cases the struggle will involve the destruction of hardware and software to create an advantage involving either the use of or the veracity of information. How one defends, attacks, seeks, finds, and uses information in cyberspace will dictate how to deceive and attempt to hide the deception. The human mind will plan deception, but cyberbots will execute deceptive activities at machine speed. To help counter deception activities in an invisible realm, cyberbots will have programming that is sufficiently sophisticated to warn a distant software analyst that an opponent's deceptive activities are occurring, or that the friendly side's efforts to deceive are in danger of compromise.

Information, knowledge, relationships, and flexible thinking make the art of deception possible and plausible. As such, the process of deception involves thinking about and then denying or manipulating information and knowledge, causing relationships to be more obscure and thinking to become inflexible. Deception involves denying, manipulating, or inundating opponents with information we want them to receive and believe, rather than allowing them access to unaltered, valuable information. Armed with good information and a clear understanding of perceptions of truth, it could become possible to create conditions of truth in the minds of opponents in the forms, shapes, and shadows desirable for achieving advantage in decision making. This process could induce our opponents to use the type of information we want, when we want it, and with the purpose we have in mind to achieve some type of relative advantage for us. The principles remain the same. Nevertheless, people must be sufficiently motivated to develop plans that have sophisticated deception as their cause célèbre. Deception planning has to occur at the front end of a planning cycle, not as an afterthought.

Only a minute portion of the art of deception has been discussed here. Yet certain points must keep our attention: People cannot engage in deception with sophisticated foes without acknowledging the differing backgrounds of others, how they view the world, how they view information, how they think, and how they plan. They must have more than a rudimentary knowledge of the decision cycle and supporting information-collection, processing, and communications systems of an opponent to have any chance of influencing that opponent's thoughts and plans. Interestingly, to deceive we have to know how the opponent views our view of what is important, why it is important, and our conjecture about how the adversary could affect that which we deem important. The same holds true for our foray into the world of the opponent, as we see through his eyes into the recesses of our minds. Moreover, and probably the most important subelement of deception, is determining what the opponent believes or will have a difficult time believing, relative to information and how he receives it. Additionally, experts in deception have to learn to synthesize information, look for or create relationships, and engage in higher-level thinking and planning. These are learned skills, and, as a preparatory step, military leaders must enable the people populating their organizations to know and understand how to discern these relationships and how to use higher-level thinking skills.

Last, people most susceptible to deception are those who engage in reductionist thinking. They see the world too simply, as a world devoid of relationships. They, after all, know "truth" and thus conclude that others see the world as they do. How could it be any different? says the reductionist thinker. They are susceptible to deception by other moresophisticated thinkers, and will never have a clue about how to think through to conjure and then implement a successful deception program.

What is uniquely human is the application of
knowledge—recorded knowledge—to the fashioning of
tools. The knowledge base represents the genetic code
for the evolving technology. And as technology has
evolved, the means for recording this knowledge base
has also evolved from the oral traditions of antiquity to
the written design logs of nineteenth-century craftsmen
to the computer-assisted design databases of the 1990s.

Ray Kurzweil, *The Spiritual Age of Machines*

7   Knowledge Management

What is the essential tie between the subjects of knowledge war and
knowledge management? Knowledge war involves struggles between adversaries to use information and knowledge better than their opponents in
making fast, effective decisions. One of the ways to make decisions faster
and better than an adversary is to affect, in a negative way, the adversary's
decisions and information apparatus while protecting one's own decisions
and support apparatus. Information operations serve as a principal tool for
doing so because by design IO affect an adversary's decisions and supporting apparatus while protecting against similar incursions.

Another way to achieve advantage is to manage information and
knowledge better, since an edge in this type of struggle almost always goes
to the side that can acquire data and information, manage it, and turn it
into knowledge quickly and thoroughly. The masterful knowledge manager shares the acquired or developed knowledge with other interested
parties through a dynamic database, easily accessible applications, collaborative software, and communications with assured bandwidth for operating within and among organizations and disseminating information,
knowledge, and decisions. The goal, of course, is to blend the best intellects and information technology to turn information into knowledge for
making better decisions faster than an adversary.

As a start, a working definition of knowledge management is "the purposeful and systematic quest for quality, efficiency, and effectiveness of knowledge to support decision making brought about by retrieving, processing, organizing, analyzing, synthesizing, and sharing data, information, and knowledge among knowledge workers, decision makers, and organizations." What, then, is a knowledge worker? "A person working in the information society who adds value to information and existing knowledge products by contributing his or her thoughts, intuition, knowledge, understanding, experiences, and skills."

Most people are well aware of the dramatic changes that the United States and other societies around the globe are undergoing in information technology. It is difficult to miss, since we, as humans, are recipients of its amazing energy every day. The United States is just beginning to experience the swirl of rapid change that characterizes the information revolution. With the ascendancy of information technology (IT) in all aspects of life, a significant shift has occurred that involves knowledge as the most important commodity of value (now and far into the future) for any group engaged in conflict, particularly when competing in knowledge war. Even the perception of just exactly what is considered value is shifting, from the physical to the intangible—to knowledge, its generation, and its application in making better decisions faster than adversaries.

Arguably, knowledge is the crucial variable for making effective and timely decisions. How each side uses knowledge will determine the outcomes of knowledge war engagements. Thus, knowledge management is one of the advantages people seek because it opens the door to other advantages in making decisions, maintaining position, and taking action. Knowledge, though, does not appear without putting forth tremendous effort to find, process, sort, think about, develop, produce, share, and use it to make effective decisions.

Because of its importance, the functional area of knowledge management will become a battleground in its own right—effective knowledge management yields effective and timely information and knowledge and gives decision makers the potential to make better decisions faster than their adversaries. The methods and procedures organizations and individuals within organizations use to make decisions provide an asymmetric advantage to one side or the other. The wise adversary will expend energy to find the essential elements of an opponent's knowledge-management system, including its knowledge workers, then manipulate those methods, procedures, and machinery as an indirect approach.

The words "knowledge" and "management" sound simple, and they are simple to understand when thinking about them separately. But the intellectual underpinnings of knowledge management are not. How does one manage knowledge? How does one extract knowledge from the minds of knowledge workers, or encourage them to share what occurs and resides in their minds? How does one use automation and communications to manipulate the data and information that provide conditions for knowledge to come forth? How does an organization place value on something as abstract and intangible as knowledge, and how can it be managed to promote efficiency, intellectual growth, profit, security, and synergy among people and organizations or, indeed, among countries.

Knowledge management relies on human beings, their intellectual capabilities, and their expertise. Information technology enables the process though which humans provide or add value to products, but it clearly has a secondary role in knowledge management. Additionally, knowledge management involves synergy, that is, pooling people's minds (through IT) to find innovations or solutions to problems that dwarfs what either a single person or single machine can do. Moreover, knowledge management involves three types of knowledge: tacit knowledge (what people know and understand in their minds—intuition, beliefs, and gut feelings), explicit knowledge (what is written down—in databases and as standing operating procedures, policies, and the like), and embedded knowledge (what people do automatically in their work processes—it is fundamental to the business processes of organizations). The challenge to any organization is, of course, to be able to tap into the tacit knowledge residing in the minds of people and to be able to share it with others in a coherent and meaningful way (especially since what a person learns, knows, or understands constantly changes).

What are some of the characteristics of an ideal knowledge-management system?

> Dynamic database: An ideal system has a dynamic database in which data, information, and knowledge are constantly updated so that knowledge workers have access to the most relevant and timely material possible. A dynamic database has flexible, modifiable fields. Any knowledge worker in a collaborative environment should be able to access, manipulate, and input data from any location, even the most remote. Only a certain number of people, though, should have "write to database" authority.

(See more on these read-write rules below in "Organizational business rules.")

Knowledge rheostat: Knowledge workers can focus on a sliding scale of problems, from very specific to very broad. A dynamic database in such a system should have data, information, and knowledge in varying degrees of specificity, accuracy, relevancy, and timeliness to promote effective decision making at multiple levels.

Integration as a central theme: An effective knowledge management system has the capability to integrate data, information, and knowledge from many different sources or many different people. Integration, though, has many faces, such as between classified and unclassified information sources, between and among coalition partners, between state and local governments and national agencies, and so forth. Effective integration must be considered by system architects, decision makers, and knowledge workers at the start of the design process.

Collaborative environment: An ideal knowledge-management system is functional yet easy to use, has collaboration as its cornerstone, and incorporates cutting-edge multimedia collaborative software.

Knowledge attribution: Knowledge workers should be able to track their requests for data, information, and knowledge and know the status of their requests for assistance at any time in the cycle of data, information, and knowledge acquisition. This tracking information should be displayed in a format that is visually, audibly, or textually compatible with the knowledge worker's intellectual inclinations.

User friendly: The ideal knowledge-management system operates efficiently and effectively to the point that its processes are invisible to the knowledge worker. Such system transparency allows knowledge workers to spend their time adding value to information products instead of struggling with software or hardware glitches.

Visually oriented: Knowledge workers must be able to engage in many types of visualization. Visualization helps people understand complexity and, most important, the relationships among seemingly disparate elements of data, information, and knowledge.

Ample storage, fast retrieval: An ideal knowledge-management system has ample storage, easy and fast retrieval, and effective search engines that minimize the time knowledge workers spend searching for data, information, and knowledge.

Pattern recognition: An ideal knowledge-management system has the technology to search for and find relationships. Complex data mining and relational software should assist knowledge workers in thinking about, searching for, finding, combining, and exploiting both obvious and seemingly disparate relationships.

Information and knowledge veracity: An ideal knowledge-management system is well protected; that is, the information and knowledge within it are accurate and secure. Security is an absolute prerequisite for helping knowledge workers collaborate and decision makers to decide. If questions arise about the veracity of information and knowledge, forthcoming decisions take longer and often are more difficult to come by. Intruders cannot be allowed to enter the cyberspace of the knowledge worker or decision maker. Knowledge workers and decision makers must have an automated, fast, and effective way to verify the veracity of information and knowledge coming from other sources or repositories. Guaranteeing security is not an easy task, as threats to information systems and knowledge workers change quickly. Moreover, information security workers can overreact to perceived threats, thereby shutting down collaboration and exchange of information and knowledge and limiting the intellectual growth of knowledge workers and leaders. Any viable organization of the future must balance the contradictions posed by the quest to make fast, effective decisions with the need to ensure veracity of information and knowledge.

Knowledge workers of the future will know where to search for and find knowledge and expertise within and outside their organizations, and how to share that expertise. Sharing, however, is not an easy task, even when knowledge workers are sitting next to one another. Today most organizations don't have a knowledge repository or a database of expertise, nor do they have the collaborative software that knowledge workers need for easily finding and retrieving a particular piece of knowledge or expertise or for easily connecting with knowledge experts residing elsewhere. Besides being a difficult process to manage, issues of privacy seem to surface in any and all conversations regarding such a knowledge aid. Regardless of difficulty, though, in future security operations people will need to collaborate with experts within particular bodies of knowledge throughout the world. The world, specifically military missions in the realm of knowledge war, will be too complex for any one person to know enough to single-handedly plan or execute a successful operation.

The value that knowledge workers apply to data and information increases over time because people continuously learn—seeking, obtaining, reflecting upon, analyzing, synthesizing, evaluating, and internalizing information and knowledge. Learning, a preeminent factor within an effective knowledge-management system, involves interacting with and sharing ideas with other knowledge workers or converting explicit knowledge to tacit knowledge. Thus, learning, whether automated, distance, or face-to-face, provides a catalyst for improving value by enhancing the knowing, understanding, and thinking skills of knowledge workers.

Clearly, the power to exert influence lies in the tacit knowledge that knowledge workers have in their heads. But these people must have a management system to bring them information and knowledge so that they can share, manipulate, and visualize it, it must protect them, and it must allow them to eventually disseminate knowledge products and decisions. Effective knowledge management does not occur by happenstance. In fact, if people don't manage information and knowledge aggressively, effectively, and with great foresight, it can be dysfunctional and actually degrade the quality and timeliness of decision making.

From a forward-thinking leader's point of view, knowledge management first of all requires the proper perspective—a perspective of the value that lies in people and the knowledge they possess and produce. The wise decision maker recognizes the importance of knowledge workers and nurtures their well-being, intellectual growth, and dedication to the organization's success. In addition, future leaders must recognize the importance of information technology that "feeds" the brain of the knowledge worker to solve problems and contributes to good, quick decisions. Future leaders will expand their intellectual horizons to blend knowledge workers, information technology, and organizational processes that help transform data into information and information into knowledge. They will come to regard this blending with the same reverence they regard the historical, tangible assets of war—the ships, planes, and tanks that have served the country so well in the past. Future military leaders' efforts must be purposeful, aggressive, and continually renewing. They must recognize the distinct threat emanating from emerging, competent adversaries who want to seize upon the United States's dependence on information technology and attack its knowledge-management processes and the knowledge workers it supports.

It is a well-known fact that many organizations, civilian and government alike, have trouble adapting to change; these organizations cling to traditional ways of accomplishing their missions. With this clinging come staid processes for examining core competencies, and, if learning occurs at all, it is most often concentrated on past practices instead of creative views of the future. Typically, organizations that don't change easily fail to recognize the tremendous capabilities residing in the minds of knowledge workers of all ranks. With this failure they risk their ability to turn information into knowledge and their ability to use knowledge for making decisions faster than adversaries.

In less-than-forward-thinking organizations, data, information, and knowledge either are not shared at all or are shared begrudgingly, slowly, and awkwardly. These organizations often possess such rigid hierarchies that too many great, creative ideas languish and go through a filtering and smothering process before they reach decision makers. These organizations neither recognize nor foster innovation, creativity, or constant change—the hallmarks of the Internet age.

Knowledge management can be an asymmetric advantage. However, it can also serve as a target for adversaries to attack: perturbing an adversary's knowledge-management system can lead to decisions that are slowed and less effective. In fact, the effectiveness of an organization's knowledge-management system will be instrumental, indeed central, to how knowledge develops and how leaders make decisions in any type of conflict.

Even though the relationship between knowledge management and a knowledge war is somewhat abstract, without question knowledge management could very well be the primary determinant of success in future conflict. The elements composing the whole of knowledge management are these:

Seek   Share
Process   Anticipate
Manage   Protect
Apply expertise

A brief discussion of each element will demonstrate the point of importance of knowledge management.

Seek Information and Knowledge

Any successful knowledge organization of the future must have an organized way of seeking, finding, and gathering data, information, and existing knowledge so that knowledge workers can add value to what already exists or so that they can create value in what doesn't. This is a gathering process that "drinks" from the trough of plentiful data, information, and knowledge. Seeking is the paramount process of the Information Age because it is a first step in turning data into information and information into knowledge. But it also involves gathering what is relevant, specific, timely, and accurate from unbelievable amounts of data, information, and knowledge residing in databases throughout the world. Seeking must be easy to perform, heavily automated, and driven by intelligent software that can continuously and adaptively sort, categorize, bin, and perform statistical analyses on data, information, and knowledge.

From a philosophical perspective a viable data-, information-, and knowledge-seeking system has to serve knowledge workers actively, thoroughly, and quickly in their quest to provide leaders with the knowledge they need to attain and sustain knowledge and decision advantage. An effective knowledge-management system facilitates any competitive advantage that knowledge workers and the knowledge organization possess. Seeking will retain its preeminence, but it will constantly change in identifying ever-better ways to find, display, and share information and knowledge ever-more intelligent ways to help people make decisions better and faster than today.

Communications

The ways humans exchange and process data and information are crucial in knowledge-management processes, which means that communications paths will be a first-order element of any effective knowledge-management system. Communications must be fast, secure, and capable of doing what knowledge workers and decision makers require and expect. As people demand use of increasingly sophisticated software, the demands on communications systems rise simultaneously. Security forces, for example, will need the communications bandwidth to provide sophisticated graphics, imagery, visualization, and streaming video to users ranging from the first responder (such as a firefighter, police officer, or system administrator) to a senior decision maker. As two points of fact, often the best initiatives in knowledge management become stymied or stopped because of a lack of money to buy expensive communications equipment or because communications pathways offer lucrative targets for asymmetric foes using the tools of IO. Any communications systems of the future will have to take these vulnerabilities into account and add the necessary protections. That being said, reliable communication mechanisms are and will be an absolute necessity for any successful knowledge-management system.

Search Engines

Knowledge workers must have at their fingertips the means to seek, sort, find, and display the information and knowledge they consider relevant. The human being cannot now and will not in the future be able to single-handedly perform this function well. Unfortunately, there is too much data, information, and knowledge to sort through, particularly what is available to people who can use a browser to access the Internet. Computer search engines can search for and find information quickly and efficiently. Search engines have undergone great improvements over the past few years, and there will be even greater capabilities arising in the future. The world will always want and need faster, more thorough, and much smarter search engines. In fact, intelligent agents or cyberbots will eventually run search engines and learn from interacting with humans and from experiences in the digital world. Eventually intelligent agents will be able to identify relationships, explain nuances, and find "epiphany" relationships among disparate elements of data, information, and knowledge. In addition, future search engines will be able to present information befitting the way people think.

Storage

As more people grow to depend on data, information, and knowledge, and as visualization becomes increasingly important, the issue of storage will loom as a critical aspect of knowledge management. Storage must be interoperable, easy to understand, accessible anytime of the day, and possess high levels of security. Storage will continue to reside on increasingly more capable hard disk drives, transitioning into holographic storage devices and increasingly smaller peripherals. Eventually storage will occur in virtual storage areas, away from the user's location but easily accessed through the Internet and Internet2. Because of potential problems posed by hacking and cracking, as well as those caused by typical mechanical and human errors, all storage systems will be built with adequate backup and redundancies functions.

Retrieval

Retrieval goes hand in hand with storage. To be efficient and effective knowledge workers around the world must be able to quickly and comprehensively reach for and grab the data, information, and knowledge they require. Speed and accuracy will be of the essence in all knowledge-based activities. Smart agents can help in this effort, particularly when entrée requires elaborate passwords or if data sets and procedures are incompatible. Retrieval must be available from anywhere, including from space and cyberspace. Increasingly the absolute and reliable authentication or proof of identity authorizing human access, as well as sophisticated biometrics, will rise in importance.

Process

Data Marts and Data Mining

Data warehouses and data marts serve as a way to obtain, transform, manipulate, and organize data. Data warehouses and data marts shelter and facilitate data mining. Data mining identifies patterns and relationships present in available metadata (data about data) through complex algorithms. Data mining is particularly important because its execution helps people find relationships and patterns in the activities of people and provides value to existing information and knowledge. Data mining also helps knowledge workers find uniqueness that only machines can discern in seemingly meaningless metadata. Knowledge workers will increasingly depend on patterns and relationships to provide more value to information, to existing knowledge, and to knowledge they either find or create.

Data marts differ somewhat from data warehouses in that typically data warehouses are quite large, expensive, and difficult to operate. Data marts are small and contain data from a single department of an organization. They often make up the larger, more complex data warehouses. Warehouses allow for the analysis of data from disparate databases, and allow for sophisticated queries ("what if" questions, for example). Most organizations have only one enterprise data warehouse, but often have many smaller data marts. All data held in a data warehouse is analyzed and fed into the warehouse environment for use by analysts and managers. Data marts are usually smaller, more agile, and easier to access and manipulate. In addition, data marts often have a narrow focus and necessary applications built into the data structure.

Knowledge workers—who must be trained in the use of data mining starting early in their careers—will come to depend on the patterns and relationships that data mining produces in their continuing quest to remain ahead of adversaries. Data marts and data warehouses will reside in servers and storage areas scattered around the world and in cyberspace.

Integration

Some of the most significant challenges of the future lie in integrating data, that is, merging or bringing together differing elements, facts, data sets, and categories, and presenting them in a unified picture or whole. Integration serves as a facilitator for the mental act of creating synergy.

Information, knowledge, and understanding in various forms must be integrated to provide maximum value to knowledge workers. Integration involves sorting and embedding activities, functions, categories, and even data sets within existing activities, functions, categories, and data sets. The goal with integration is to enhance intellectual or physical performance. Examples include integrating national knowledge products into local and state government systems, integrating software, hardware, and telephony with computers, integrating imagery with geospatial data, integrating SIGINT and Measurement and Signature Intelligence (MASINT) with imagery, integrating coalition partner activities with U.S. activities, integrating classified with unclassified information, and integrating sound and images. Many integrating functions will occur over the Internet and in databases scattered throughout cyberspace and in the physical world.

Networks

Over the next several years data networking will drive how the economy functions, how people work, how they play, how they live, and how they wage war. Networking has many different connotations, some of which have application to our discussion here. First, networking can be viewed from a human perspective; that is, people network with others to find solutions to problems, make contacts, or share information, among other reasons. Second, networks provide an effective way for people to exchange data, information, and knowledge throughout local area networks (LANs), wide area networks (WAN), or by accessing the Internet. Third, networks describe what companies do within the organization proper. Of critical importance is that networks provide knowledge workers the capability of collaborating to share knowledge or to work on problems and projects. Interestingly, networks provide a conduit for cyberattack from cyberwarriors attempting to affect decisions by affecting operations or data, information, and knowledge residing within networks.

Manage

Database Management

Database management will continue to be increasingly important to help people find, manipulate, and share timely, specific, relevant, and accurate data, information, and knowledge. Databases will be dynamic, which means continuous data, information, and knowledge will feed into pre-established or forming fields. Snapshots taken in time will become passé. With automation of feeds into databases for continuous updating comes the responsibility to ensure data, information, and knowledge are accurate and valid. Databases will have common interfaces, enabling many databases to work together seamlessly.

Intelligent agents will assume an increasingly important role in maintaining the operation of dynamic databases. Intelligent agents, for example, will help tailor database-management systems with guidance provided by users. Increasingly intelligent agents will help people by searching for relevant data, information, and knowledge, and populating databases automatically. As intelligent agents become smarter they will take over much of the work of designing fields and inputting various types of data, information, and knowledge into those fields. They will ensure fast and effective integration, and alert knowledge workers when problems occur or as critical updates appear. Through dynamic database management a man-machine complementary relationship enhances a knowledge-management system to quickly and efficiently become more adept at combining pieces, things, or ideas into one powerful collage.

Knowledge Mapping

Knowledge mapping, which enables people to find the knowledge they need for adding value to products, improving one's knowledge and understanding, and for making rapid decisions, will increasingly exert influence on personal, business, and governmental activities. The complication is, of course, that any knowledge mapping is only as good as its inputs, logic trails, access rules, and procedures. Further, it takes effort by humans and machines to keep knowledge mapping up to date and to support knowledge workers effectively. Knowledge mapping helps organizations share and use information and knowledge efficiently and effectively.

Knowledge mapping will employ intelligent agents, particularly when knowledge workers attempt to identify and use relationships among data, information, and knowledge. Knowledge mapping will use extensive visualization so that workers will know at a glance where to find critical information and knowledge they need. Knowledge mapping also includes identifying the critical knowledge and skills of knowledge workers within a particular organization or even if located elsewhere. With the Internet it will be increasingly important for knowledge maps to depict not only the knowledge and expertise residing in physical organizations but also that which resides in virtual locations scattered about the world.

Knowledge Preservation and Transfer

Organizations of all types are experiencing increasing difficulties in retaining knowledge, specifically the tacit knowledge of their knowledge workers. This problem exists in the commercial marketplace and in organizations involved with security. Knowledge workers' minds hold value—in their thoughts, experiences, intuition, training, education, expertise, and visions. Unfortunately these skills are largely invisible, difficult to quantify, and almost impossible to convert into a database. Databases, after all, have fields and fields have rules of entry. Tacit knowledge often doesn't follow strict rules and procedures for entry into databases. Regardless, how an organization captures the tacit knowledge of its employees is crucial for that organization's viability. Otherwise new workers constantly have to "reinvent the wheel" when they arrive. Moreover, the richness of the synthesis that occurs in the knowledge worker's mind often becomes exclusionary to the problem at hand, because such thinking doesn't fit into preestablished fields of data.

What can be done to bridge this gap? For one thing, people need to put their ideas, thoughts, and insights into a repository that ties to a vision of the future. Software must be able to organize the material into categories and bins of knowledge so others can use it. Knowledge transfer should relate closely to knowledge mapping. People in an organization should know how to seek and find the knowledge they need, which means that fields of databases need to be flexible and capable of change. Changes in fields should go to users automatically and transparently— only intelligent agents should have to be concerned with such changes. Additionally, people need a good knowledge base to start with. Knowledge workers should develop their own knowledge bases and automated continuity files—a daunting task, particularly for new people or those who are extremely busy. Software can help overcome this challenge; that is, it seems obvious that a good, usable, and powerful wizard could go far in helping knowledge workers set up their own knowledge bases for the transfer of knowledge.

Apply Expertise

Training and Education

Training and education are the lifeblood of any knowledge-management system. Knowledge workers have to continuously learn to increase their knowledge and add more value to existing knowledge. They must ensure that the knowledge they possess remains broad and relevant. Because of constant change and technological development, people won't be able to go to an institution of higher learning for all their intellectual developmental needs. Instead, learning will occur in three ways: in the workplace (as peer-to-peer interactions or over the Internet), in a learner and mentor relationship, and as the responsibility of the individual. The Internet and distance learning will be primary sources of learning, experimenting, and receiving feedback about thinking processes.

Investments in Expertise

An important part of knowledge management is knowing what core competencies an organization has and why it has them. But that organization must also know what its vision of the future is and what core competencies its people need. With this type of thinking people can manage the differences between skills already on hand and skills needed for the future. In this sense organizations can and should pay for the advanced education work of young people in areas that relate to the needed core competencies—as a quid pro quo the person receiving an advanced degree should either work for the sponsoring organization (physically or virtually), or become the intellectual "sleeper agent" of the future in which an awakening can occur should the sponsoring organization need assistance. At that time the knowledge worker can respond and collaborate with others in the teaming effort to provide expert knowledge and intellectual support.

Knowledge Expertise Database

Knowledge workers need current, constantly updated, and easy-to-use databases that enable them to find the right people with whom to exchange information and knowledge. Key fields include life experience, education, work experience (including individual projects), thinking predilections, personality indicator scores, and interests. Such a database will facilitate easy consultation among workers either in a physical sense or from distant locations. In addition, such a database will facilitate the effective use of individuals and groups of knowledge workers in organizations and businesses.

Man-Machine Symbiosis

Man and machine will increasingly learn from each other until eventually machines will assume a more aggressive role and actually make verbal and textual recommendations for learning. These interactive tutors–learning assistants will learn from people with each interaction, and they will come to know the operator's strengths and weaknesses. The day will come when mobile software is critically important to individual learning, as the software will "know" what the individual needs in the way of information and knowledge and how and where to find it.

Man and machine will evolve to complement each other's weaknesses and strengths. As most people realize by now, computers perform some functions much better than people do, such as searching through databases, performing mathematical functions, and sorting through immense amounts of collated information, all at high speeds. People, on the other hand, at least for now have many skills computers do not have, such as judgment, seeing nuances and subtleties, and synthesizing abilities. Thus, man and machine are an inseparable team whose bonds will grow stronger with the passage of time. When people can actually talk with machines and machines can talk with people—a true interface between the two—the last obstacle will fade from our memories. We must consider now how machines will increasingly take on human characteristics such as talking, expressing, questioning, and perhaps even feeling emotions.

Share

Collaboration

A working definition of collaboration is "groups of individuals or organizations with varying degrees of knowledge, expertise, and thinking capabilities coming together physically or virtually to provide or share information and knowledge for solving problems or confronting issues to ultimately support decision making. Keeping this definition in mind let us examine why people collaborate:

With collaboration or the sharing of ideas come additional ideas.

Collaboration allows many minds to focus their intellectual power on solving problems.

Collaboration allows many people to understand a leader's intent and long-term vision.

Collaboration allows for savings in time and money, since people can be separated from each other yet continue to share ideas.

People can work issues "around the clock" by collaborating among and between multinational networks and organizations having a presence in countries around the world.

Collaboration allows people to form virtual and physical matrix groups specifically designed to solve complex problems.

Collaboration allows others to understand one person's information and knowledge requirements.

Collaboration allows people to learn from each other.

Collaboration fosters intellectual synergy. With the capability to bring many diverse minds to bear on problems, it becomes possible to achieve "intellectual superiority" that cannot be matched.

Collaboration allows for the instant sharing of an idea within thousands of minds.

People pool their knowledge (both tacit and explicit) to solve problems or add value to new or existing situations. With collaboration knowledge workers create intellectual synergy. The virtues of collaboration haven't changed in hundreds of years—when people blend what they know with what others know to add value to information, help solve vexing problems, or contribute to decision making the outcome is greater than the sum of the parts. Yet something significant about collaboration has changed—the tools facilitating it. With the Internet and modern browser technologies, Hypertext Markup Language (HTML) and Extensible Markup Language (XML) tagging of data, information, and knowledge, and visualization, high-speed computing, and collaborative software, collaboration has reached new levels of capability. One example is the multimedia aspect of collaboration—individuals or groups can now see and discuss what others are doing by viewing graphic images presented on their computer screens. People can even hear what others are doing. Moreover, people, from a few people up to hundreds of people, can participate in interactive events. Another example is the collaboration that takes place over the Internet, enabling people in global networks and communities of interest to work on projects and share ideas simultaneously or in sequence of time zones around the world. Projects can pass from one time zone or continent to another, enabling the project's completion by the time the originator sleeps and reports for work the next day.

In the future collaboration will include other senses, such as touch and smell. The welcome addition of these senses will round out the full multi-sensory, multimedia environment that is progressing now so beautifully in its ineluctable march toward assisting human understanding of complexity. The future also holds intelligent agents playing a much larger role in finding, sorting, categorizing, and displaying information. People with similar intellects and thinking predilections will gravitate toward each other for possible inclusion in teaming efforts, thanks to advances in intelligent agents and computers sufficiently powerful to discern a participant's emotional state of being as well as profiling intellectual and emotional intelligence.

People will enter virtual reality environments as they wish and roam the "terrain" of the digital world by themselves, with avatars, or in a collaborative environment with others. Thus, groups of people will experience the same thing, be capable of discussing the nuances of what they saw, and possibly learn to recant previously held beliefs. As a group they will solidify existing notions or come up with new ideas, in virtual reality. Moreover, because of the chance to roam digital spheres of being, people will collaborate vicariously by living in physical reality while taking advantage of the digital experience of the person experiencing virtual reality.

Instant Messaging

People want information and knowledge for a wide variety of purposes. Perhaps the most interesting reason lies in the continuing human need to reduce risk in decision making. People want relevant information at the location they want, at the time they want, and in the format they want. Soon, people will be managing their sources and types of information over the Internet, particularly when security and privacy issues become important (in a historical sense).

In the future people will receive information from any physical location around the globe. In this respect, people will respond to a sender immediately and enter into collaborative environments with others using instant messaging that is voice activated and voice driven. In addition, people will receive and send instant messages using graphics, imagery, music, and text. Soon olfactory and physical messages will affect the receiver's state of equilibrium, too. These messages will come across the person's PDA or cellular phone, fully capable of being annotated and returned. In the long term these messages will feed directly into the receiver's mind, as fast as that person desires. Moreover, instant messages will be fed into microcomputers embedded in people's bodies. Mobile instant messaging will come about primarily through voice directives.

Intelligent agents will package messages in the format and time span that the sender desires.

Visualization

People are increasingly visual. Several reasons exist for this trend. First, there is too much information for people to process, sort, and use. Second, people make faster decisions using visual information and knowledge. Third, IT is one of the reasons people are becoming "lazy" in an intellectual sense—they often grasp the meaning of the picture without thinking about the layers of complexity that, when taken together, comprise its meaning. This isn't a bad thing unless people think that the image portrayed is the fullness of its complexity.

People in the near future will visualize information and knowledge to complement the ways their minds work. Intelligent agents will interact with humans, becoming digital "pets" for people. These digital pets will interact with the humans and become increasingly smart about what a particular person needs and how that information can be optimally displayed. The increasingly smart intelligent agent will reside in the mind of the machine, so to speak, or when the "master" is mobile it will reside on a CD or in a similar medium. Visualization of the future will range from two dimensions (2-D) through six dimensions (6-D), depending on a person's predilections. The six dimensions include the traditional dimensions of height, width, depth, and time, but also the nontraditional dimensions of mind and cyberspace.

Visualization will be individualized to the point that the user will possess an internal rheostat for changing visual images on PDAs, televisions, and computer screens. Visual images will come to the person through wireless networks via enhanced retinal devices and heads-up displays that blend with a person's skin, hair, and neural enhancements such as neural implants. Changes to images will come via the miniature computer a person wears all the time that is embedded in the person's skin, or by the computer and signaling from a host machine.

The future holds numerous technical wonders, such as giving users the ability to paste pertinent data sets onto digital maps or images, or draping imagery over terrain features. Visual images will be enhanced with sound, olfactory, sight (color), and tactile properties no matter where the person might be located, even if that happens to be a virtual environment. Visual monitoring of a person's heartbeat, blood pressure, and pulse, presented on enhanced retinal displays or imported as images directly into a person's brain, will enable people to know what is happening with their bodies at all times.

Knowledge Engineering

Knowledge engineering is the organizational use of knowledge-based systems, processes, decision making, and business rules. Knowledge engineering uses an object-oriented methodology coupled with advanced algorithms to perform knowledge relationships that can be structured automatically by a computer. An advanced extension of this methodology will allow computers to intercept raw data and derive relevant knowledge from it for sharing with others. The computer determines the semantic relationships from the data and, with the aid of cognitive processing, allows the computer to distill the raw information in a way similar to the way a human brain does.[1] Advancements in software and processing power must use strengths of the computer and associated software to ease the life of human beings. The more transparent the functioning of knowledge engineering, the more people will find it useful. It is easy to foresee the day when a person has knowledge engineering built into the personal computer that is a part of the person's persona. The knowledge engineering software will organize, sort, interpret, display, and store information and knowledge automatically as the person goes about his or her daily life. The computer will "blow in the ear" of the human being when it has information and knowledge already sorted and organized and ready for use.

Business to Business and Peer to Peer

Much of the current exchange of information and knowledge lies in the realm of business-to-business (B2B) and peer-to-peer (P2P) collaboration. B2B and P2P are merely examples of a trend that has been obvious for quite some time and that will continue. The future will also bring a variety of other types of technology-manifested collaboration: commander-to-commander, knowledge worker–to–knowledge worker, physician-to-physician, hospital-to-physician, patient-to-physician, and researcher-to-physician. All forms will occur as direct interactions through peer-to-peer technology and digital philosophy. Much of this activity can and will take place through the use of increasingly dynamic websites (or, better yet, reside with people no matter where they are located). Such a movement makes people less dependent on servers and a central organization or location inherent in the server-client philosophy that has served humankind so well. When Internet2, with its great, wide-band potential, becomes more than a figment of imagination, leaders and knowledge workers, people and physicians, and businesses and organizations will communicate without having to call up websites or wade through webpages. Collaboration will see dramatic improvements and enhancements with the advent of tactile interactions, even when far apart physically. Additionally, the future will bring the ability to connect with experts in a collaborative environment to seek and use opinions, a situation in which true power lies in the synthesis of several different expert opinions. Intelligent agents will assist people in finding the right experts and the right times to bring people together in a collaborative, virtual environment.

Anticipate

Technology Trend Mining

Knowledge workers must stay abreast of changes in technology, particularly information technology because of the capabilities that improvements in IT can bring to human cognition. Unfortunately, most people don't have a systematic way to discern changes in technology or to reflect on what the changes mean for work, play, war, and so forth. Thus, knowledge workers and knowledge organizations of the future will engage systematically in searching for, sharing, and using technological changes.

Intelligent agents will provide part of the answer, and increasingly powerful search engines will provide yet another way. Additionally, knowledge workers could have explicit missions to mine for technological change in specific areas of concern, such as advancements in a particular technology. Eventually people will conduct this mining on a routine basis because technology is changing so rapidly. The more systematic and methodical the search and categorization, the better the situation becomes for the knowledge worker (given that time is not a pressing and persistent constraint).

Intelligence Gathering

Competition will grow in scope and intensity owing to the ubiquity of information technology, the plentiful nature of information and knowledge, human beings' innate desire to learn, rapidity of change, increased stakes commensurate with higher risks, and diminishing resources. A world that is totally interconnected will undoubtedly create a witch's brew rivaling that of the opening scene in Macbeth.

Thus, a blinding flash of the obvious leads to the conclusion that it is increasingly important for people to know their opponent as well as or better than they know themselves. To do so has become easier with the advent of websites, webpages, and the increasingly intrusive capabilities of information technology. People and organizations need to know what their competitors and opponents are thinking, planning, and doing. Intelligence collection, both traditional and in cyberspace, and the information and knowledge products aiding in decision making will become increasingly important.

The old cycle of intelligence operations involved collecting, processing, analyzing-synthesizing, presenting, deciding, and communicating. Perhaps the new intelligence cycle won't be a real "cycle" all the time. Perhaps it will be a simultaneous coming together of the many aspects of seeking, gathering, blending, analyzing-synthesizing, collaborating, and sharing.

Smart sensors will be added to the arena of intelligence, as well as intelligent agents, strong processors, and specific software programs written to accomplish intelligence-gathering, analysis, and synthesis roles. These collection, processing, and communications activities will occur in physical reality as well as in cyberspace.

Surrogate Travel into Human Intent

"Surrogate travel into human intent" is another way of describing virtual reality and cyberwargaming. Surrogate travel into human intent, though, concentrates on the cognitive part of wargaming and uses smart software to "travel" to the opponent's mind via his databases, hard disk drives, business logic, decision-making tools, past activities, budgetary information, and the like, to gather data, information, and knowledge that the friendly human wargamer can use.[2] Leaders and knowledge workers will increasingly need to "wargame" courses of action to gauge opponent actions, counteractions, and reactions, to identify possibilities, and variables and the sensitivity of both, and to focus scarce collection assets to confirm or deny hypotheses coming from wargaming activities. They will enter graphic and virtual environments to discern differences between real and virtual environments. Knowledge workers will put forth hypotheses in this virtual reality environment and confirm or deny the hypotheses based on indicators or intelligence information they collect. Human opponents in this wargaming environment will be aided by computers and software at the beginning; eventually opponents will be software avatars (intelligent agents) that take on the appearance of and act like human opponents (or machine opponents, or a combination of the two), assimilating their thought processes and actions. In particular, wargaming against avatars programmed to think, plan, and act like human beings will offer a way to mitigate the effects of mirror imaging, and will offer glimpses into ways to mitigate risk and how certain combinations of activities and machines could work under particular environmental situations. In the surrogate travel environment people will see, hear, smell, and even touch. It will be a different and strange world that enables people to travel digitally through fiber-optic cables, over wireless radio waves, enter databases, peruse various types of media, and examine the internal components of switches, telephony, and computers that control them.

Protect

Fundamental to any successful knowledge-management system will be the capacity to know and understand what knowledge is valuable and if that knowledge is reliable. Total protection all the time is unattainable—in the Internet age with Internet time, "truth" changes quickly (including the "truth" of expertise). Nevertheless, certain things will be important to the success of a particular operation, but only for a limited time. Thus, knowledge workers must know the value of the information they have, know that it needs protecting, and know that people will try to "pirate" their important knowledge and how to protect it. These knowledge workers must always ask, What is important? Why is it important? How could someone affect that which I deem important? What would the adversary think important, and why? How would the adversary attempt to affect my important things? With what tools? What would the adversary have to know to affect my important things? How can I anticipate and deny his activities? What will signal that something is coming? What sources of information and knowledge can help me know the adversary is planning to come, and how can I thwart his activities? How can I avoid mirror imaging?

Information Security

Since competition will increase in intensity and resources will be harder to come by, it follows that people will be vying for the knowledge having the most potential for exploitation, and turn it into money. Thus, business logic, internal decision processes, research and development projects, experimentation, and ongoing studies will be important. Just as important, however, is realizing that intellectual property will be valuable, though its value will be limited owing to its half-life in Internet time. Intellectual property will be valuable unto itself and in combinations with other technologies. The next killer apps will come from people combining technologies and sharing the results in a new, higher-order intellectual property (IP) domain. Cyberthieves will steal IP from individuals and from organizations in traditional ways—physically stealing blueprints, ideas, reports, and diagrams and the like—and in nontraditional ways—digitally stealing valuable information and knowledge. Nontraditional theft will be a much more dangerous and insidious activity than we have previously known, owing to the difficulty in discerning presence, activity, and origination of cyberbots trained to be thieves. Technology has allowed the theft of information in small pieces which can later be aggregated into a whole. Since it can be obtained in small, unobtrusive ways, it may never be recognized as theft. Thus, organizations must be wary, and they must catalog and tag IP, identify rules to govern its sharing, and protect it.

Intelligent agents in the form of cybersentinels will form the main line of defense against the assault of information pirates that is certain to continue. However, traditional means of protection—intrusion detection systems, firewalls, and so forth—can be put to work. In addition, knowledge workers must protect IP from alteration and malicious destruction. One example, bio-hacking, is starting to raise its ugly head—hackers have become bored with electrons and are switching to biotech and gene work to practice their sordid trade. Our people must develop an expertise in creating cybersentinels as well as in designing sensors to help protect from physical intrusion. They must design miniature sensors for identifying and denying intrusion in the physical and cyberworlds. They must design virtual reality environments for cybersentinels and to aid in understanding the digital terrain, so to speak, wargaming against cyberbots that are guided by lower- and higher-order machine intelligence.

Massive preparations in the area of knowledge management is critical for winning engagements in knowledge war in the future. Knowledge management is particularly important because the country's military relies on information and knowledge for its effective functioning and as the means to achieve information superiority and decision dominance. Adversaries of the United States will surely see such reliance as dependence and will therefore attack it with sound and fury. In fact, knowledge management will constitute one of the most important "battlefields" of future engagements in a knowledge war. The U.S. military must prepare for such conflict now.

Making decisions better and faster than an adversary means that the U.S. military must have smarter knowledge workers, better thoughts, better research and development, and better technology than any adversary, and then continuously demonstrate the capability to learn, adapt, and change rapidly. The country's military and civilian security forces also must learn to use generated knowledge purposefully to make better decisions faster. The U.S. military must realize that in the future almost nothing will be "stable," least of all the knowledge in a knowledge worker's mind or information technologies. Thus, knowledge workers must continue to learn and accept the change that learning brings. In addition, organizations must continuously update the information systems that support their knowledge workers and decision makers, and track the IT their adversaries use.

In future conflicts, how each side manages its data, information, and knowledge, and how it adds value to existing information and knowledge could be the key variables in determining outcomes of engagements. People and organizations need to define what knowledge management means in their particular contexts, they must break it into constituent parts and understand those parts and how they function in a whole, and they must constantly upgrade and improve their knowledge-management processes.

Knowledge management is much more than simply buying the latest in information technologies. The important goal of a successful knowledge-management system is the expansion or creation of value that comes from the minds of organizational knowledge workers. These workers have to learn constantly to improve their ability to add value to information and turn it into knowledge. They also have to understand the criticality of knowledge in the process of making decisions. Interestingly, adversaries of the future will see knowledge workers as a great vulnerability and they will target and attack them.

Knowledge management can be an asymmetric advantage for either the weaker or the stronger force. If this country's armed forces want to possess an asymmetric advantage in knowledge management, a great deal of mental and physical work has to occur. As a first step, though, leaders must recognize the importance of the process and system of knowledge management, the reason it is important, and the fact that knowledge workers who provide value to information and knowledge have to be protected and nurtured (not unlike the nurturing and caring for the weapons of war that occurred in the Industrial Age). Leaders in all organizations will have to learn to anticipate their adversary's intense efforts to perturb their knowledge-management systems as an indirect approach to influencing decision making. Indeed, this is a changed world!

8 |Triumph in Knowledge War

By now the reader has probably concluded that the United States has no option other than to prepare for knowledge war. Knowledge war is inevitable, considering how the conventional fighting prowess of the U.S. military has forced adversaries to contrive new ways to compete and have any hope of winning. Knowledge war gives such adversaries a way to compete, perhaps even win, with an excellent chance of surviving.

Knowledge war will cut across the tapestry of world systems to which the United States is inextricably linked—the intertwined systems of social, political, economic, financial, ecological, informational, and military networks. Though the military system has been a prime focus here, what follows can be applied to all systems comprising the world tapestry, from the highest levels to the lowest. These potential solutions should be a part of the intellectual preparation of leaders in all systems that make up America's fabric. Specifically, four commingled pathways into the future for improving the nation's capabilities to cope with knowledge war assaults from adversaries and to create advantage for the country are discussed: knowledge advantage centers (KACs), a joint asymmetric opposing force (OPFOR), a joint information operations proving ground (JIOPG), and developing and using an Internet replicator. Of course, the challenges to preparing for waging a knowledge war against terrorists and other future enemies are easy to describe but difficult to accomplish. That is, the security forces of the United States must prepare for *both* kinetic warfare and a knowledge war.

Pathways to the Future

The following four main ideas are essential to what should be done to prepare for knowledge war:

> Knowledge Advantage Centers: A knowledge advantage center (KAC) system of collaborative networks would connect people and machines in all fifty states and four territories and possessions for the express purpose of improving the homeland security of the United States. If we believe that knowledge warfare is either upon us now or will be soon, we must agree that making better decisions faster than our adversary is crucially important in attaining an advantage. One way to make better decisions faster than an adversary is to think better by using the combined intellects of hundreds or thousands of smart people linked in collaborative environments. The goal, of course, is to provide to the people making decisions the intellectual advantage (intellectual superiority) that enables them to think, plan, execute, and adjust better than their adversaries.

> Opposing Force: The country's military forces must develop and use a joint asymmetric opposing force (OPFOR) that operates against U.S. joint, service, or civilian security forces, one that speaks a foreign language, thinks like opponents from a foreign country, and acts in ways appropriate for people who don't think like Americans. Such an OPFOR must have a vision, goals, objectives, and the ruthlessness of a real opposing force instead of one that simply mimics the thinking and actions of U.S. security personnel.

> Joint Information Operations Proving Ground: The country needs a joint information operations proving ground (JIOPG) where collaborating joint and component commanders and their staffs can enter the world of information operations being planned and conducted in cyberspace. In addition, the research and development of warfare information software, lethal weapons, and nonlethal weapons would occur here, as well as the testing and evaluation of equipment in an IO environment. An IO proving

ground would provide a place for joint doctrine and experimentation to flourish.

Internet Replicator: The country needs an Internet replicator to train its military forces. Cyberspace is difficult to imagine, let alone train in. Moreover, numerous laws, rules, and social conventions preclude U.S. military forces from entering and training on the Internet since people and organizations around the world use it every day. As a consequence, it is difficult to train people to think about and operate in cyberspace and to experiment with new concepts and doctrine there. The country's military and state and local government security forces need a place to develop individual and organizational processes to meet the foe of the future. The country also needs "a place" for security people and scientists to meet and develop the machines of cyberspace—the multifaceted capabilities of cyberbots. Development of cyberbots to the extent needed to engage in and win a knowledge war will not happen without training, education, research and development, and experimentation within an extensive and elaborate Internet replicator.

Knowledge Advantage Centers

Knowledge advantage centers are the modern version of centers of gravity—they are *information* centers of gravity. Knowledge advantage centers are "the confluence of collection, communication, automation, thinking, planning, and decision making where data, information, and knowledge fuse, where knowledge workers collaborate to turn information into knowledge, and where leaders make decisions." In knowledge advantage centers people and machines collaborate to integrate data, information, and knowledge. They also provide expertise to support security operations, particularly since security operations of the future will be knowledge-based operations.

Sharing seems like a normal, everyday occurrence, but in reality people and organizations, particularly bureaucracies, tend to hoard information and knowledge. Regardless, we must be open with the data, information, and knowledge that people need to make lifesaving decisions. For example, local law enforcement people need access to national-level intelligence if it answers their needs for timeliness, accuracy, relevance, and specificity. On the other hand, national agency people need firsthand information from first responders since this information, though not yet knowledge, provides pith in a national overlay of situational awareness that can only come by digesting on-the-scene videos, sounds, voices, smells, observations, and location information.

Americans are aggressive—it is a great national strength. Thus, we must put this aggressiveness to work in the war against terrorism. We cannot just be consequence management experts, we must become anticipation and denial experts, too. As such, we need a philosophy that militates against waiting for the terrorists to bring the fight to us. By then it will be too late, particularly if terrorists use weapons of mass destruction. Two of our country's greatest strengths in the fight against terrorism lie in our collective intellectual power and technology development to achieve advantage. These two assets—our brains and our technology—if used effectively will allow people to make better decisions faster. Decision and intellectual advantages lead to other types of advantage: technological, knowledge-management, positional, and action. These are the essence of knowledge warfare, and clearly we are at the start of a long knowledge war.

We as a country need to do what many corporations are already doing, that is, collaborate across bureaucratic boundaries and turfs in an extensive network with the explicit purpose of sharing data, information, and knowledge to enable decision makers (corporate, local, state, first responder, and national) to make better decisions faster than the terrorists. In all cases our goal in this collaborative network must be to "be there waiting" when the terrorist arrives to do his deeds. We need to tap the creativity of knowledge experts and pool the collective intellect of thousands of people to help leaders at all levels and in all types of organizations make informed, anticipatory, and effective decisions.

Implied in this concept is a convergence of brainpower, information, knowledge, automation, and communications together operating as a whole. Learning, intellectual inquisitiveness, and technological improvements provide the energy for this operating whole. We must be able to help people improve their abilities to collect, report, analyze, synthesize, collate, collaborate, and visualize. Moreover, we need to push our scientists and technologists to develop software to help us do the things every situation demands, especially things that we aren't universally adept at, most notably identifying relationships among disparate but related elements.

This collaborative network has many other networks that have many other networks, and so forth. We must design a system that will take us through a difficult, zero-sum struggle that may last a hundred years. We need to design the system now and adjust as we learn and change.

The knowledge advantage center (KAC) system of collaborative systems is expansive: it is a nationwide, collaborative civilian and military computing and communications environment, both virtual and physical, encompassing the entire continental United States and its territories. The KAC system enhances knowledge workers' and analysts' support of decision makers and staffs (national, state, local governments and private companies); it provides knowledge to decision makers for making rapid, effective decisions in both contingency and homeland security operations (military and nonmilitary). A KAC system promotes a continuous expansion of knowledge workers' knowledge and mental capabilities. This collaborative system of systems spreads across the country, indeed to many parts of the world; it responds to the information and knowledge requirements of America (at all levels of business and government); and it capitalizes on our greatest strengths—our intellects and technology—to deal with insidious threats. This collaborative system of systems connects businesses, local and state governments, law enforcement agencies, national government agencies, coalition partners, intelligence agencies, and military forces in a huge, collaborative network. Its purpose is to exploit collaboration to allow decision makers to make better and faster decisions than adversaries. Why do we need this network when it will take so much effort and so many resources to create?

First, terrorists are clever, adaptive, and learning. They operate in small and elusive groups. They know their susceptibilities to being caught. They are careful with what they do, and they are deceptive. They will never come out and say "here we are." No, we will have to find and eradicate them by using good anticipatory thinking, recognizing centers of gravity, using traditional and nontraditional information collection, and, most important, sharing information and knowledge to form relationships. We need to develop "*aha!*s" on a continuum rather than as epiphanies.

Second, much is happening in the country. We are being watched and surveilled by terrorists. We cannot stop this activity because to do so would require that we give up some of the liberties that make our country great. So, how do we proceed? We can observe, share information and knowledge, and form relationships with the help of computers and visualization. We can consult with knowledge experts. We can anticipate via hypotheses, and confirm or deny with the help of a systematic way of collecting, processing, analyzing-synthesizing, and presenting information, and communicating conclusions. We are not able to do these things now, but we should learn to. We have to keep in mind the "gold nugget" of critical information buried in "two tons of sand." This analogy holds true for our endeavor: we are searching for nuggets that lead to relationships that enable us to anticipate and deny. Could the local cop in Fremont, Nebraska, have the information that leads to halting an anthrax assault on Washington, D.C.? We must design a system in which people can share what they observe, collaborate with experts, push data and information into a relational database, search for and form relationships (data mining), visualize a solution, and make informed, fast decisions.

Third, our smart people have to learn continuously to sustain an intellectual advantage over the terrorists. All people have the opportunity to learn over the Internet and from other repositories of information and knowledge. We have to respect the terrorists' abilities to learn. It would make sense to enable our knowledge workers to continue to learn every day in a systematic way. A collaborative network would help in such an effort.

As we proceed with our thinking about knowledge advantage centers we need to keep in mind differences in information and in knowledge requirements. People working at a local level often have different information needs than people working at a national level. Private corporations have different information requirements than the military. Specificity, timeliness, accuracy, and relevance—four helpful terms for understanding the challenge—mean different things to organizations and at different levels of action. Any system we develop should keep the information needs of a wide variety of people in mind since all levels and all people will become involved in the war against terrorism.

Knowledge advantage centers should exist at the national, regional, state, local, community, and business levels of organizations. These centers should stretch and connect across the country and vary in size. What they won't vary in, however, is function and form. Every KAC will be a fractal or representation of the whole. Each will have integration, visualization, collaboration, and modeling and simulation at the center of its conceptual heart.

Knowledge advantage centers are confluences of communications, automation, information collection, thinking and planning, and decision making. They are part physical and part virtual. They provide leaders with an environment to make fast, effective decisions. Their entire purpose is to provide knowledge to decision makers to make better decisions faster than the adversary. The knowledge advantage center construct has to be operated by permanently assigned knowledge workers, by people from the interagency community (IAC), by regional, state, local, law enforcement people, by technologists, by functional knowledge experts (such as chemists), by intelligence analysts, and by operational people in functional areas (disaster–consequence management people). These people must have continuity for the purpose of continuously building knowledge; we must invest in their continuous training and education.

Exact numbers of people in each knowledge advantage center will vary based on mission, complexity, and the expanse of that center's information requirements. Some of the manning will take place in the physical world and some will take place in the virtual. Moreover, knowledge project managers will pull together virtual project teams to work on specific issues and projects. Typically when the issue is resolved or the project is finished the virtual team will disband only to have another community of interest be born. As an aside, this has definite implications for leadership: How does one lead people when they are physically located thousands of miles away and in collaborative sessions are connected only by electrons in virtual reality?

Each knowledge advantage center will focus on its decision maker's information and knowledge requirements. As such, decision makers will know and understand what constitutes their ever-changing information and knowledge requirements, and they will articulate those requirements in clear and concise language to the KAC's knowledge workers. A secondary role and mission would be for knowledge workers to focus on the information and knowledge issues of other knowledge advantage centers. They must always be on the lookout for data, information, and knowledge that other KACs need. These needs will be displayed visually and textually. Moreover, knowledge workers at each KAC will have the ability to display the needs of other KACs, to show the relationships among key pieces of information in each of the KACs, and to show how they relate with other key pieces of information in other KACs.

Eventually a dynamic, self-adjusting, constantly pulsing (because of data, information, and knowledge inputs) database will service the information needs of knowledge workers. Knowledge advantage centers will not be just a repository for huge volumes of data, information, or knowledge. Instead, they will be "couplers" or "conduits" for seeking, finding, gathering, analyzing, synthesizing, visualizing, and transferring knowledge. Central to the KAC idea lie the blurring or merging of four functions: collaboration, integration, visualization, and modeling-simulation. Problems are so complex that we need knowledge experts and the minds of hundreds of people working together to create solutions and provide advice—collaboration. We have to merge, fuse, relate, and converge data, information, and knowledge from a vast number of sources. Data and information from these sources have to be inserted into a dynamic, relational database so knowledge workers, with the help of intelligent agents, can sort, sift, find, relate, and apply experience, intellect, and intuition to turn information into knowledge—this is integration. We must have visualization to help people think. Because people think in different ways visualization software has to adapt to how individual people think, not a normal population. Moreover, we must be able to search for, find, and discern patterns and relationships that we didn't know were present—visualization can help. We have to use models to find possible friendly and enemy second- and third-order effects so as to avoid surprise for us and to create surprise and shock effects for terrorists. In addition, we need to be able to represent the reality we cannot create, that is, cyberspace, physical attacks, concentric rings of protection, and webs of information in 3-D for the purposes of training and educating knowledge workers and decision makers—modeling and simulation.

At the heart of the KAC concept lie the concepts of infrastructure, knowledge management, philosophy of technology, and technological enhancements. Most important, we have to define and think through the constituent parts of knowledge management. These parts must be updated constantly and protected. They must work together in a synergistic way. We must be rigorous in our approach to developing the concepts and technology of knowledge management, as a multitude of definitions and approaches exist. But the design of these complex functions must support knowledge workers and decision makers and facilitate integration, visualization, modeling and simulation, and collaboration. If we hold design and knowledge process engineers to these vectors, we will build capable knowledge-management systems.

Our country needs to develop regional knowledge advantage center constructs. Each region will have a core of knowledge workers having focused expertise that is particular to the region but generally cutting across the political, military, economic, social, information, and infrastructure realms. Regional KACs will connect to a national knowledge advantage center (NKAC), to federal agency KACs, and to state and local KACs.

Functional areas and knowledge workers in each KAC will have communities of interest and practice in their collaborative environments. Information and knowledge will constantly flow among and between elements in the collaborative environment of each community of interest.

Some people may disagree with the need for regional KACs. Nevertheless, the underlying rationale is this: 1) Each region has distinct centers of gravity and a focused view of information and knowledge; 2) If we seek to link all fifty states to a national center it will be very difficult to manage; it might be more effective for layers of knowledge centers to interoperate; 3) We need fast decisions, and a regional decision maker will often have a larger perspective than a state or local decision maker. It makes good sense for the president to delegate decision making to decision makers at the regional level of a KAC organization; 4) Regional KACs will focus on regional problems quickly and effectively owing to breadth and depth of limitations and perspectives. While keeping the needs of the suprastructure in mind, their missions will dictate a focus on regional decision making all the while pushing and pulling information and knowledge to local KACs and to the national knowledge advantage centers.

Regional KACs will have pertinent representation. As you think about the wedges of the knowledge advantage center pie, imagine knowledge workers, liaison people, technical people, and machines doing much the same as a national knowledge advantage center construct. Then, enlarge each wedge and see to whom it is connected. The connections will be many collaborative communities of interest feeding information to and sharing with each other to search for, gather, and use the best data, information, and knowledge with which to support decision makers.

We must have the means to integrate open source data and information with classified information. We have to fuse all classifications of information and, if classified, make informed decisions about sanitizing or fuzzing up the elements causing sensitivity so that we can provide it to people who need it the most to thwart the actions and effects of terrorists. Governmental agencies at the national, state, and local levels will integrate their information requirements and must share much better than they have in the past to achieve the rapid recognition of patterns and relationships desired.

Knowledge advantage centers purposefully search for and determine, in cooperation with many other connected KACs, second- and third-order effects. These effects spring forth from hidden relationships composing the world; when perturbed they cause the effects of surprise and shock to launch forth. Knowledge advantage centers will work through modeling and simulation communities (government and private) to anticipate second- and third-order effects (from the view of terrorists, as they plan to strike our societal tapestry).

The "form" piece is always the hardest to predict—many issues come up because of money, bureaucratic turf, people, influence, and the like. In this case, form follows function. If the country develops a national knowledge advantage center as it should, that center will serve two roles that cannot be pried apart: homeland security and contingency deployments. The "pointy" end of the spear now has two heads—one that points to an overseas deployment location and one that points to the U.S. homeland. Even as troops deploy, attacks will occur against military posts, camps, and installations, against lines of communication, against sea and aerial ports of embarkation, and against intermediate staging bases. In addition, our future adversaries will attack centers of gravity at all levels in the United States—physically, perceptually, with WME, and with cyberbots attacking through cyberspace. Terrorists will surely be attacking even when a deployment is not underway, which will have a heavy impact on the capability to deploy and sustain military operations.

For homeland security, knowledge advantage centers will exist at the national, operational, and tactical levels of war, and at the national, state, local, and commercial business levels of functions and organizations. They tie people in collaborative networks of relationships involving individuals, machines, organizations, and communities of interest and practice. These KACs will operate together in a huge collaborative network designed specifically to share data, information, and knowledge in an effort to improve homeland security. Each regional combatant commander of a joint unified and specified command (U&S command) will have a knowledge advantage center that connects to knowledge advantage centers at the national and component levels, with state and local governments, and with communities of expertise, knowledge, and interest.[1]

The homeland security (HLS) portion of the NKAC will have permanent representation from the interagency community (IAC), intelligence agencies, corporations, medical community, technologists, and liaison officers from military organizations and coalition partners. As events unfold, more representation will be added as required. These knowledge workers will connect in a collaborative environment with knowledge workers at other KACs, with knowledge experts, and with communities of interest scattered around the world but connected through and in cyberspace. A chief of staff will orchestrate the activities of knowledge workers. Knowledge workers will not make decisions—they will devote themselves to collaborating, thinking, sharing, and improving their knowledge. Decisions will be made by decision makers who are fed expert information and knowledge that come from the merging of man and machine in multiple knowledge advantage centers.[2] Each KAC will have a group of people who are cross-functional; that is, these people have knowledge skills that apply to everyone—chemical, biological, radiological, nuclear, IO, and crisis responders.

The notion of knowledge advantage centers needs a bridge into an organized, coherent way of seeking, finding, gathering, analyzing-synthesizing, and presenting-packaging tailored information and knowledge to leaders and decision makers. An organization designed to do these things could be an integral part of the NKAC and KAC concepts—some currently exist in the commercial world. We need to consider the functions of seeking, finding, gathering, analyzing and synthesizing, and presenting-packaging information and knowledge at all levels of the construct. The intellectual and technical strength of each cell and link contributes to the strength of the overarching web of information and knowledge that, in a conceptual way, enshrouds the United States and connects it to other parts of the world. We need to think this aspect through carefully—considering roles and missions, conditions, functions, form, technology, and training and education and tailoring information and knowledge support to meet the diverse needs of local, state, regional, national, and coalition organizations.

Security will be a critical aspect of any knowledge advantage center. Each center will have physical security specialists and guards, an elaborate sensor network, multiple biometric means of entry, individual machine monitoring (to deal with insider problems), intrusion detection software, firewalls, antivirus software, information assurance specialists, and counterintelligence specialists, to name a few. Veracity of data, information, and knowledge will be crucial in an era demanding speed in decision making. Protection against cyberterrorists, whether they are trying to damage our information systems or trying to collect our valuable information and knowledge, will have to occur. We will need a way to replicate attacks against our KACs over the Internet. Knowledge advantage centers will be susceptible to physical attack, too, so training (physical and simulated) specifically designed to thwart difficult physical attack must occur. Of course, any KAC is an information center of gravity. As such, operating under the concentric ring of protection and web of information theories will always be of paramount importance to their effective functioning. Terrorists will also focus on key knowledge workers and their families. Thus, organizational training has to develop a new paradigm to thwart attacks of terrorists aimed at knowledge workers' or leaders' families. Both the families of knowledge workers and knowledge workers themselves will have to be participants in continuous homeland security training as a condition of their employment.

Asymmetric Opposing Force

U.S. security forces need a joint asymmetric opposing force to help friendly security forces prepare for combating a multitude of terrorist attacks against key people, physical landmarks, and symbols, and, of course, to conduct operations in cyberspace. An asymmetric opposing force exists to look, act, think, plan, make decisions, and seek feedback like an adversary who hates America would, an adversary who will use any asymmetric tool to win, who will sacrifice his life to win, and who thinks and plans differently from the people who live in Western democracies.

Currently, military services conduct excellent training using programs that have served the country well. But will these programs serve us in the future as well as they have in the past? No, for several reasons: conflict is changing from an emphasis on kinetic warfare to knowledge war, but the country's military training emphasis remains first and foremost affixed on force-on-force and kinetic operations; in most circumstances the opposing force in training activities consists of people from the United States who think and act like the forces they are opposing (when they should think and act like the foreign country that the U.S. military forces are training against); current military force training gives only lip service to the need to work against forces that speak a foreign language; when conducting wargames U.S. military forces operate with force-on-force, attrition, and movement models and simulations that generally replicate where the country's military leadership would like to fight instead of training in places that asymmetric foes of the future will undoubtedly choose. In these models and simulations very little progress has occurred to bring in the intangibles of warfare such as fear, surprise, deception, PSYOPS, fatigue, and the nonkinetic, often intangible aspects of IO to merge with the kinetic aspects of warfare; U.S. military forces have no place to go where they can become immersed in an information operations environment and be forced to deal with the future soldiers of cyberspace—cyberbots—and the terrain of cyberspace—fiber-optic cables and radio frequency airwaves.

Because our adversaries of the future plan to use knowledge war against us, how can our military forces say they are preparing for future conflict if they do not train to engage true asymmetric foes waging a knowledge war? Unfortunately, most of the current training focuses on what people have done well in the past: kinetic engagements in force-on-force scenarios and modeling and simulation wargames that only involve attrition, movement, and weapons effectiveness and efficiency. While some training in force-on-force operations needs to occur, asymmetric warfare, ranging from the strategies and tactics of conventional forces that are forced to gain offsets against the U.S. military, to the decentralized terrorist waging a knowledge war, need to be a part of the training proposition. Without the addition of a realistic asymmetric warfare, complete with role players having the language, technology, and thinking and planning traits of asymmetric adversaries, current military training will continue to inadequately prepare the U.S. armed forces for future conflict.

A brief discussion of the principal forces driving the need for a joint asymmetric OPFOR is in order. One of the most significant lies in the enhanced capabilities of asymmetric foes to attack or manipulate the multitude of relationships that characterize a globalized world. The United States, for example, cannot go it alone in the world anymore—what even one tiny country does affects many other countries immediately. That immediate effect spreads rapidly outward to other systems and second- and third-order effects surge forth. What happens in the world economic system affects the world political system, while what happens in the global political system has a direct impact on how military operations occur, and so on. Constraints such as rules of engagement come from domestic and international political environments and dictate how much force can be used, how many casualties can occur, and how much collateral damage a country can endure. The United States has increasingly demonstrated a penchant to finish conflicts quickly and to get on with life—the immediacy phenomenon influences all that happens during a conflict. The media, armed with highly capable computers and telephony, comb the earth broadcasting to and from political capitals on the events that make news, influence populaces, and cause politicians to make decisions. The media has a great deal of influence on will, the obvious center of gravity for knowledge war. The point is that the environmental context of globalization and the tapestry of systems comprising the world's body politic present lucrative targets for asymmetric adversaries.

Future conflict will involve both asymmetric and traditional force adversaries competing in traditional and nontraditional battle spaces. The U.S. military is well prepared to compete against traditional forces but less prepared to operate against nontraditional ones. The U.S. military needs to train against both types of threats. Moreover, the country needs to train to meet the asymmetric threat in the continental United States, in places that are en route to objective areas, in objective areas, and in cyberspace. Right now the United States doesn't have an asymmetric OPFOR that can present a credible challenge.

Opponents waging a knowledge war will use the Internet and computers to their advantage. From the Internet they will learn, adapt, and change. The Internet also serves as a medium for worldwide coordination and the means for sending encrypted messages from great distances at extremely rapid speeds. It also enables people to collect information and intelligence ranging all the way from simple Reuters news reports to purchasing half-meter resolution imagery on the commercial marketplace.

Opponents of the future will have access to weapons of mass destruction and they will choose to use those weapons—in the U.S. proper, where they can achieve the greatest second- and third-order effects—when it serves their purposes. One of the terrorists' principal targets will be the air and sea points of embarkation using weapons that will be small and difficult to find (to facilitate a terrorist's individual emplacement and detonation). Timing will be of critical importance while the Internet will enable adversaries to employ their weapons at the right time and right place to create the right effects in the minds of the general populace, in the military people, and in knowledge workers such as people operating the computers that operate the seaports. Interestingly, opponents of the future will also use nonlethal weapons, such as incapacitants that make people sick. In those kinds of scenarios it will take a long time to learn that their incapacitants either are or are not biological weapons. They will also use nonlethal weapons, such as chemical lubricants or foam that immobilize or lasers that blind, striking the objective areas of conventional forces and special operations forces. They will use these weapons to counter what they perceive as U.S. asymmetric advantages. They will, for example, use high intensity light and lasers to blind helicopter pilots using night vision goggles. Asymmetric foes will likely employ off-the-shelf technology to jam the global positioning system on which we have grown dependent.

Adversaries of the future will have access to changing technologies; it is folly to believe otherwise. In fact, some future adversaries will have the access and money to enable them to acquire technologies approaching or even surpassing the quality of technology the U.S. military possesses. Nevertheless, these technology-rich foes will find themselves at a disadvantage, at least for the time being, because of the stupendous potential of the United States in the functional areas of knowledge management, networking, visualization, and the building of dynamic databases. It makes sense not only to practice against adversaries who have access to the very best technology, but to practice against opponents who know how to use such technology to achieve their goals and objectives.

U.S. military forces face future adversaries that have very different outlooks, goals, and objectives; it will prove difficult for military thinkers to anticipate our foes' activities and intentions with even a modicum of accuracy because they will be neither stylized nor lock-step in their approaches to problem solving. Thus, U.S. military forces will find themselves preparing for future conflict and using intelligence collection in battle space and in cyberspace to confirm or deny hypotheses about the enactment of adversary possibilities. Keep in mind the difference between possibilities and capabilities: possibilities have a much greater applicability to thinking, planning, acting, and anticipating in asymmetric warfare than capabilities. The forces of asymmetric adversaries will adapt and flex according to what they see and hear on the Internet and television and through their own collection means. These forces also will learn and change in response to U.S. force activities.

Asymmetric adversaries will usually speak a language different than English. Thus, U.S. military forces must cope with and train against foreign-language foes in all their exercises. Language presents unique and difficult training challenges since it is expensive to mimic the language of an opposing force. Moreover, language causes perturbations in force structure, command and control, and the accuracy, quality, and timeliness of intelligence. In short, if an OPFOR uses a foreign language the way a real foe would use it, training exercises could become messy and other training objectives could go by the wayside. Intelligence becomes less timely and less precise causing exponential increases in risk. Yet if commanders are serious about undergoing realistic and challenging training, their opposing forces have to use the native tongues of foreign countries. When military forces train against opponents using foreign languages, commanders should take linguists along on deployments, should train with interpreters, and should be able to cope with less-than-perfect knowledge of what an opposing side is doing, saying, and intending, owing to the difficulties in interpreting the meanings that lie hidden in the nuances of a foreign language.

U.S. military forces must work against a realistic, determined, ruthless opponent having goals and objectives difficult to surmise and definitely antithetical to those of the United States. Such an opposing force must possess leadership with sufficient intellect to think, plan, and orchestrate events, activities, and effects far beyond the simplicity that dominates today's training scenarios; moreover, such opponents must have patience and be comfortable watching linked events occur over time as they move either slowly or quickly toward goals and objectives.

Such opponents will attempt, at all costs, to strike perceived vulnerabilities in U.S. military forces and against the U.S. and coalition partner societies. Such forces will neither fight fairly nor always operate in the conventional realm. They will operate as a conventional force when it suits their purposes, but they will use cyberspace and knowledge war as well, because of its low risk and high payoff. They will learn to seek offsets against U.S. military forces by using indirect approaches to create actions that lead to outcomes that influence decision making. Thus, to the greatest extent practicable the U.S. military must insert these types of asymmetric activities into its training, it must develop a changing and capable asymmetric OPFOR for conventional force-on-force and cyberspace conflicts, and it must use OPFOR organizations in simulations and live field exercises. The civilian security realm needs to train against similar threats.

Objectives

An ideal training asymmetric OPFOR will help U.S. military armed forces prepare to operate in future environments against smart, unpredictable, and unscrupulous foes. This asymmetric OPFOR represents a global threat and connects with other asymmetric forces through the Internet and modern telephony. An ideal asymmetric foe learns, adjusts, adapts new technology, and becomes more formidable over time. These improvements are not happenstance—they take effort and resources. The asymmetric OPFOR uses foreign languages in all its operations, and it is schooled in the tools of the modern knowledge warrior, including weapons of mass effect, nonlethal weapons, information operations, cyberterror, and traditional terrorism. Asymmetric OPFOR leaders must be thinking people who can learn and adjust rapidly. They must work relationships and seams for advantages. They also must use the global media as "allies" to manipulate public opinion and attack fissures in coalitions. Believers in good, accurate, and timely intelligence, these capable asymmetric leaders establish their own human intelligence networks and purchase commercial imagery when needed. They actively think about their opponent's minds and engage in sophisticated act–react–counteract wargaming with the specific intent of winning at any cost. Our capable asymmetric adversaries possess their own version of campaign plans including traditional ways and means, strategies, tactics, tools, phases, centers of gravity analyses, vision, intent, goals, objectives, and actions. They aim their efforts at the decision making of U.S. military leaders and politicians to influence the will of the populace of the United States.

The U.S. military must invest serious money in buying the type of information and weapons technology that asymmetric adversaries have access to, are proficient with, and use to further their aims. It does little good to arm an OPFOR with the same technology that the U.S. armed forces use because it is often obsolete. Arguably, in real situations sometimes the technology an asymmetric OPFOR uses will be inferior to that used by U.S. military forces; at other times said technology will be superior to the technology that U.S. military forces use. An asymmetric OPFOR should have the resources to request imagery from the National Imagery and Mapping Agency (NIMA) and other commercial providers of satellite imagery having the quality, quantity, and timeliness that are available to anybody with a computer, an Internet service provider, and money.

Essential Elements

A training asymmetric opposing force must be difficult to beat, it must be smart, and it must use the thinking, customs, and cultural mores of the country in question. How will an asymmetric OPFOR look, think, and act? They must be tough in their thinking, in how resolutely they accomplish their objectives, and in how they execute activities to create effects against their adversaries. A useful OPFOR must be smart, do smart things, and be creative in its approach to competing with U.S. armed forces. Their goals, objectives, tasks, and subtasks must come from particular thinking and planning of the country being emulated. The OPFOR must portray absolute determination and steadfastness while accomplishing its missions. It must portray an adversary whose will to win surpasses all costs, including life. These opponents must be unpredictable and continuously operate unconventionally in ways contrary to group thinking and the easy-to-anticipate, mainstream thinking that characterize mental activity in most Western democratic countries.

In all scenarios, whether simulations or live exercises, extensive information operations must occur from the start and continue through exercise completion. IO cannot be one-sided—U.S. military forces cannot prepare for conflict in cyberspace by employing cyberforces but not having anyone or anything to employ them against. All exercises must have an asymmetric "cyber OPFOR" that is highly skilled and uses the latest information and knowledge available in the open press, in commercial markets, and, most important, on the Internet. Asymmetric IO cyberoperations must use a foreign language, complete with said language populating all adversary databases.

The asymmetric OPFOR leadership must participate in all campaign planning, but such planning only occurs to the extent that people from the representative country or group of terrorists would or could plan. As such, an asymmetric OPFOR must have a plan that realistically represents a worldly, relationship-intensive system of systems view. Any asymmetric OPFOR campaign plan should call for intense engagement in interactive wargaming to discern the adversary's act–react–counteract cycles and its attempts to achieve feedback on decisions via its information-collection systems. A modern, capable asymmetric OPFOR leader must attempt to gain insight into the adversary (friendly) leadership's perceptions of himself and of the adversary (friendly) perceptions of the asymmetric OPFOR's perceptions of the adversary. An asymmetric OPFOR leader will use the most modern technology possible, but constrained by reality (such as money available, technical proficiency, and availability on the black or open market).

Challenges

The creation of an OPFOR faces many challenges. One is the tribal wisdom that presents a formidable obstacle to initiating something as "out of the box" as an asymmetric OPFOR. Tribal wisdom, passed from generation to generation and from promotion board to promotion board, lives in the past and clings to traditions and past ways of doing things. Its conservative tendencies are resistant to real change. Many military people will resist the notion of an asymmetric OPFOR because its enactment could lead to failure, certain aspects of expensive exercises could get disrupted, and the all-important collective mind-set of the officer corps remains firmly set on preparing to compete against conventional, kinetic opposing forces.

Many people will worry that an asymmetric OPFOR will cause perturbation in training objectives. In the past, commanders have often been overly concerned about interjections—such as radio communications jamming—hindering the accomplishment of training objectives. These same people could very well be inclined to worry about asymmetric OPFOR operations. These leaders, however, have to introspectively ask themselves about the meaning of realism, about what type of conflict they are preparing for, and if they are training for peace or training for war.[3] We have to find common ground and consensus about realism and how best to prepare to operate in and win in such an environment. This view may be only one of many, and others may have equally valid arguments. Since the future is so difficult to foresee, the vast array of choices will be fraught with error. But it would be better for the country's military forces to err on the side of overstating an OPFOR capability in training because then it would be easier to adjust to less-difficult problem sets. It is much more damaging to prepare for a weak adversary only to be faced with a capable and ruthless opponent.

It will be difficult to replicate training exercises in which communications and intelligence occur just as they would during an actual conflict. At best, nothing is ever perfect in the two disciplines—they are too complex, they rely on too many interconnections, and both are governed by the thinking and activities of human beings prone to error. Nonetheless, when intelligence and communications are so important in producing the precise and timely information and knowledge needed for making effective decisions, it follows that our leaders must prepare for the perturbation and occasional demise of such processes. This is what realism is all about when preparing to engage in modern conflict.

Without a standing, capable, and representational joint asymmetric OPFOR, commanders should rest uneasy about whether or not they preparing to square off against a realistic threat or simply preparing for a threat they hope to face. There is little room for discussion in this matter. If people want to train for conducting future conflict and win against capable, asymmetric foes, they must train against a truly representative foe. U.S. forces walking around in different clothes cannot satisfy this requirement. Instead, a capable asymmetric OPFOR must consist of many different sizes of units, must train to act as a realistic foe would act, must be interspersed with native linguists, and must use the weapons and technologies that a realistic opponent would use. Such an OPFOR is adaptive, learns from its mistakes, and anticipates how U.S. forces will react. Such a force engages in activities to take away or offset the projected advantages of U.S. military and civilian security forces.

Finding role players who speak the right languages and have the right skills to replicate our future adversaries will be expensive. In addition, role-playing, rehearsing actions, finding and using the country expertise, and obtaining the appropriate weaponry and technology will be expensive. But what constitutes "too" expensive when it comes to preparing our country's youth to operate in both a kinetic world and in cyberspace against a formidable foe?

Benefits

The benefits of a joint asymmetric OPFOR are incalculable. U.S. officers and organizations will operate against foes approximating the capabilities that reasonable people anticipate for future conflict. Thinking and planning will improve because of the challenges put forth for dealing with a formidable and technologically capable adversary. Generations of officers will develop and mature, imbued with the notions of an interrelated earth, an unpredictable and ruthless foe, and the connection between decisions and will, actions and decisions, and effects and actions. Along with determining how asymmetric foes will seek advantages, U.S. planners, operators, and commanders must learn to wield asymmetric advantages a strong force possesses while attempting to offset the asymmetric advantages weak foes have over strong foes at the same time. In addition, by being forced to deal with constraints such as imperfect intelligence and the vagaries of an unfamiliar foreign language, U.S. officers will have a much better approximation of how the changing technological environment influences time, distance, environmental variables, and, most important, risks and decision making. As a final point, cyberspace is an abstract subject that the typical officer of today spends little time contemplating. By undergoing repeated brushes with an asymmetric OPFOR who engages in IO as a "tool du jour" with which to wage knowledge war, U.S. officers will start giving cyberspace its due consideration. From such recognition our brightest and most forward-thinking officers will surpass previous limitations in abstract thought and start to contemplate the conduct of operations in cyberspace and how a smart, adaptive, technically competent foe of the future will attempt to win in such an environment.

Joint Information
Operations Proving Ground

Preparation necessary to succeed in future conflicts in cyberspace can only occur through realistic training and education for military personnel, through effective research and development with IO hardware and software, through test and evaluation of information systems and weapons (lethal and nonlethal), through experimentation with new concepts and technology, and through a thorough evaluation of joint information operations doctrine within the context of people interacting with machines in living and virtual environments.

Information and knowledge constitute the most important aspects of any future field of strife because using them wisely enables people to make good, quick decisions. Information and knowledge also enable people to expand their minds continuously, thereby making a specific attempt to maintain an intellectual edge or advantage over future opponents. Information and knowledge are also critical in competitive engagements because knowledge is the means to power, implying a distinct and potent relationship between information and knowledge and the art of battle command, effective decisions, and the purposeful and intelligent creation of actions that lead to meaningful effects.

Information operations focus on affecting thinking, planning, perceptions, and decisions. IO attempt to manipulate information support systems that enable decisions, while protecting from similar incursions and attacks. There is very little question that the United States will use information and knowledge to engage in conflict; other countries will do likewise not out of a position of strength but rather from a position of weakness.

To reemphasize some of the main ideas from chapter five (Information Operations), it follows that the country's military must train to engage successfully in this type of conflict. Such training will help its officers learn to plan, wargame, and conduct operations in cyberspace. Actions in cyberspace lead to effects (outcomes) that will often be invisible because they are digital and occur in cyberspace. Actions in cyberspace occur rapidly because cyberbots move at the speed of light. Moreover, cyberspace does not house terrain military people are accustomed to—in cyberspace the terrain is fiber-optic cable, hard disk drives, and databases. This is the world of information operations.

This new arena of conflict is distinct from the physical world (other than how actions occurring in cyberspace affect the physical world and vice versa). It is a world so dramatically different from what the country has experienced in past conflicts that it requires very different thinking, planning, decision making, feedback, intelligence operations, and execution of actions leading to effects. This new arena of conflict cannot be replicated in the physical world.

The defense forces of the United States must start preparing now to defend against asymmetric attacks certain to come from foes using information operations, particularly in computer network attacks, computer network exploitation, and cyberdeception. Such adversaries will attack U.S. vulnerabilities in the continental United States (CONUS) en route to and in objective areas. They will work very hard, for example, to affect decision making at all levels of command. They will work to disrupt command, control, computers, communications, intelligence, surveillance, and reconnaissance (C4ISR) systems, which provide much of the data and information leaders use to make decisions. Future adversaries will use the Internet to orchestrate and coordinate activities of individuals and small groups around the world at the speed of light. Information warfare weapons add to an already impressive quiver of arrows that asymmetric foes possess.

To prepare adequately for cyberwar, U.S. military and security forces need a Joint IO Proving Ground. It is important to note that the U.S. military and security apparatus will suffer a severe degradation in overall capabilities to wage operations in future conflict without a JIOPG. What, specifically, are some of the negatives one can expect if civilian and military leaders fail to establish a JIOPG? Without a JIOPG it will be difficult to conduct research and development or realistic experiments. It will also prove difficult to provide feedback and guidance to materiel developers, to doctrine writers, to trainers, and to organizational developers to enable them to improve to the point that they are able to enter into and win future engagements in cyberspace. The processes in place for conducting research and development, acquisition, and training worked very well for preparing for conflict against conventional, Industrial Age forces. These processes, however, will not work to prepare U.S. armed forces for engaging in conflict in cyberspace. Additionally, laws and organizational structures have not kept up with changes created by the information revolution and its effects on the ways to conduct conflicts. Military intervention, for example, requires increasingly close cooperation with other agencies, governments, corporations, nongovernmental agencies, and private volunteer organizations. To cope with threats in the continental United States, such cooperation and coordination will be even more important here than abroad. A JIOPG would allow experimentation, which will identify shortfalls in law or in procedures among relevant agencies.

Since the DOD does not have a proving ground that deals with digital conflict, it lacks both a place and the processes needed for proving or disproving (testing and evaluating) the effects of digital conflict on people, organizations, or materiel—computers, fiber, databases, and information warfare (IW) weapons. Continuous research and development are the lifeblood of innovation. R&D at a joint information operations proving ground should involve a symbiotic relationship between civilian industry R&D and government laboratories. R&D would be fueled by experimentation at the proving ground and would be a place to prove or disprove concepts and interactions.

A serious contradiction exists between what DOD officials say they can do and what they actually can do in regard to engaging in information operations. Quite frankly, the way the defense institution thinks, develops its materiel, organizes, and develops doctrine is not adequate for conflict in cyberspace. Please consider four thoughts: First, in a mechanical sense the digital transformation process (data to information to knowledge to understanding) and fusion process (of digits) remain locked in a state of disrepair owing to legacy systems, interoperability problems, and an anemic basic research situation. Until the pathway lights up and "jointness" dictates the absolute requirement for interoperability, this situation will continue. Second, in a thinking sense many officers and DOD civilians continue to think in a linear, reductionist, predictive way, even while Internet realities and forthcoming digital engagements in cyberspace argue for nonlinear, synthesis-inducing, and possibility-considering ways. Without improving the thinking capabilities of our people, it is quite likely that the U.S. military establishment will not discern the relationships that make up the environment of the future. Third, the military needs a way to experiment with IO developments in software, such as using cyberbots for collecting information, for deceiving, for denying, and so forth. The military needs a place to test, evaluate, and validate the systems and weapons that support IO in physical reality and in cyberspace. Fourth, the military needs processes and procedures to experiment with the materiel, doctrine, and organizations that any force will need to engage successfully in information operations.

Military officers need realistic cyberspace simulations and modeling that approximate the environments that future conflict will resemble. Such activity is the cornerstone for preparing officers' minds to operate in "a place" they neither see nor feel. Indeed, cyberspace is even difficult to imagine; virtual reality and simulation will greatly assist in understanding the environment, the relationships comprising this new domain of conflict, the nature of conflict, and potential outcomes, and in developing the mental skills to compete. Moreover, without sophisticated simulations and modeling that truly replicate the reality of cyberspace and operations in this environment for training of the officer corps, officers won't easily move out of the here and now of *capabilities* (potentialities bound by experiential constraints) to a future of *possibilities* (potentialities unbound by constraints).

Without experimentation using information operations–related equipment, organizations, materiel, and people, it will prove difficult, if not impossible, to provide guidance and feedback to those needing it. Specifically, materiel developers, doctrine writers, trainers, and organizational developers need focus, guidance, and the proper priorities for enabling U.S. forces to use IO with a sophistication sufficient to engage cyberopponents successfully. Since the DOD does not have a viable proving ground dealing with digital conflict, it is unable to prove or disprove the effects that digital conflict has on people, on organizations, or on materiel (computers, fiber, databases, and IW weapons).

Keeping these challenges and issues in mind, consider the following logic: The country is deficient in figuring out how to protect itself against a determined IO-oriented adversary that will undoubtedly attack the country's military and civilian decision making and information sinews, from the lowest to the highest levels of command. Efforts to defend against an organized cyberattack are not fully coordinated, and it is difficult to replicate an environment in which human beings interact with cyberbots to operate in a place they cannot see. Military defensive IO efforts don't correlate with civilian efforts, and homeland defense issues involving IO are not nationally orchestrated, coordinated, or practiced. When, for example, was the last time the country replicated a nationwide cyberattack complemented by simulated terrorist physical attacks, to exercise with crisis management, decision making elements of state, local, national, and civilian governments, and military leaders?

The country has no way to approximate its own capabilities to attack in a conflict environment the decision-making apparatus, information processes, and minds of a full range of capable opponents. It is difficult to replicate the Internet environment that will host future conflict. Yes, the country has brilliant people working on offensive information operations. But they often work in isolation, and they have limited ways to practice what they have learned or developed in isolation. Too many laws, rules, and regulations exist to allow for the type of training the country's military and security forces need to hone their intellectual skills and "weapons" of cyberspace.

Benefits of a JIOPG

A primary intangible benefit is the promotion of change within the military culture. A joint information operations proving ground would provide a tremendous learning mechanism and it would dramatically change the way people believe and think, thereby affecting institutional thinking in the aggregate. Moreover, a JIOPG would exert far-reaching influence on doctrine, organizations, training, maintenance, leadership, facilities, and people. Such a program would go far in preparing the structures and processes of organizations and their human and machine inhabitants to start thinking seriously about what it will take to execute national security operations in the unpredictable and invisible conflict in the twenty-first century. Additionally, a JIOPG would provide the means to conduct reasonably cheap experiments, and enable the intellects of scientists and military personnel alike to develop advanced, sophisticated concepts about information operations and how to leverage their inherent power. As a final point, a JIOPG would provide great insights into the creation of informational synergy and its principal ingredients: information collection, automation, communications, thinking and planning, and decision making. With such insight the U.S. military establishment might develop the capability to create advantages, some tangible some not, at times and places for purposes of their choosing.

Internet Replicator

Currently U.S. forces do not have an acceptable way to replicate cyberspace or to train seriously to operate in a military sense "there." If training occurs at all, it usually does not include the correct spoken language, a worthy opponent, or rudimentary defensive tools such as intrusion-detection systems, firewalls, antivirus software, honey pots, and the like. Moreover, in many cases commanders do not employ a realistic "red team" to penetrate their information systems or to manipulate the flow of information and data going into and out of their headquarters. Commanders across the joint spectrum often do not have a good view of how their units stand in relation to the totality of the asymmetric threats they face. In the current state of affairs commanders cannot use the existing Internet to train because of legal issues, issues involving the importance of the Internet, and the consequences if people made mistakes with offensive and defensive operations in cyberspace.

Leaders responsible for homeland defense know that enormous, complex, and ominous dangers loom in the distance. For starters, the country's infrastructure is at risk. The crux of the issue lies in four general areas of consideration: first, the very word "infrastructure" is ill-defined and often misunderstood; second, though much good work has been done in the area, the constituent elements of some aspects of our critical infrastructure have not been identified or properly defended. Moreover, even when we can define in great depth what constitutes "critical" at the state, local, and national government levels, we need to spend much more time thinking through when, where, and how a worthy adversary will attack and why they might attack. We must consider how the parts of our country's infrastructure (military and civilian) work as a whole so we can anticipate how the adversary might view the creation of second- and third-order effects through infrastructure attacks. We have neither sufficiently identified our infrastructure's vulnerabilities nor have we delved into sufficient intellectual depth to accurately predict how an able adversary could exploit those vulnerabilities. In addition, much of the nation's infrastructure lies in the control of people loath to share relevant information, thereby rendering the notion of a holistic, sharing, knowledge-oriented networked defense system moot. Finally, the country has few ways to prepare for cyberattack; that is, the civilian populace and the communications, electrical, air traffic control, banking, finance, and news infrastructures are vulnerable to cyberattack in isolation, but such attacks would not count for more than a sound bite on television. However, if an entire state or region blacked out owing to an extensive cyberattack emanating from within the country or without, it would have extraordinarily negative first-, second-, and third-order effects. Such attacks would be particularly devastating if timed to coincide with orchestrated kinetic bomb attacks, weapons of mass effects

attacks, or nonlethal warfare attacks. Such vulnerability and unprepared-
ness affect all governments (local, state, and national) and much of the
commercial marketplace. The country is not ready for the inevitable conflict of the future.

If military leaders believe that the nature of conflict is changing to one
that is a combination of kinetic mechanisms operating in physical space
and new mechanisms operating in cyberspace, then they must have the
tools to train their subordinates to enter and win such conflicts—but they
do not have that capability today. If senior military leaders believe that operations in cyberspace are going to occur owing to the ascent of knowledge war and IO, they must also accept the fact that their forces are only
partially prepared for conflict in the twenty-first century. The country's
twenty-first-century leaders need to look at this issue seriously and give
to leaders across the spectrum of government, commercial organizations,
and the military the means to train, experiment, develop software and
hardware tools, and make decisions in an environment that approximates
the reality of tomorrow.

The only conclusion is that the country needs an Internet replicator to
simulate cyberspace and the real Internet. An Internet replicator, complete
with a cyber information operations opposing force, should be a part of
every training exercise across the spectrum of joint and individual service
training. Without an Internet replicator the country's military leadership
cannot prepare for the type of warfare and conflict certain to be in the off-
ing.

What should an Internet replicator do? First and foremost it must operate like the real Internet, that is, be transparent to people operating it
whether it is the real Internet or a simulated Internet. Thus it must have
"real" Internet service providers, enclaves, and clusters of servers, real Internet protocols, time delays inherent to the real Internet, and normal security requirements and procedures one usually finds on the real Internet.
Moreover, an Internet replicator must be flexible in size; that is, it must be
able to simulate a few servers and enclaves or it must represent huge networks comprised of thousands of servers and enclaves. A realistic Internet
replicator must simulate the peer to peer connections and instant messaging that people use today. It must also replicate the internal networks, intranets, wide area networks, and local area networks that all organized
forces use. Moreover, an effective Internet replicator must replicate databases just like those an adversary would use, that would be populated
with data, information, and knowledge in a pertinent foreign language.

Some of the data, information, and knowledge should be relevant, but most would be irrelevant, just like on the real Internet.

An Internet replicator must have a cyberwar OPFOR that can replicate opponents that face our military, government, and civilian organizations. This OPFOR should be composed of the best and the brightest hackers in the country. They should construct cyberoperations in both offensive and defensive modes. This OPFOR should conduct intelligence-collection, denial, surveillance, counterintelligence, and communications operations. It should employ cyberbots as entities or in swarms for attacking, defending, alerting, denying, deceiving, collecting intelligence, or communicating. All OPFOR operations should evolve around the capabilities of the country or organizations in question. Undergoing a thought process that replicates how a real opponent thinks should precede all OPFOR actions. Language should be a part of any Internet replicator OPFOR since future opponents will not be sending their e-mails or populating their databases using English.

An Internet replicator should be an integral part of a joint asymmetric OPFOR and a joint information operations proving ground. These three ideas work best if they integrate synergistically. The day will come when joint commanders, state and local government officials responsible for defending the infrastructure, and commercial enterprises come to a knowledge advantage center and immerse themselves in a simulated cyberenvironment on an Internet replicator to prepare for conflict. Facing them would be the joint asymmetric OPFOR that thinks, speaks, and acts like the adversary in question and that operates its "soldiers" in cyberspace; the joint homeland force would have to deal with simulation that combines the kinetic world with cyberspace and its digital environment.

Scientists and technologists would use the Internet replicator to develop the cyberbots, intelligent agents, and avatars proposed throughout this book. The Internet replicator and simulation that would accompany it would provide a way to experiment, develop, test, and evaluate mechanisms for warning, denying, deceiving, collecting, and using cyberbots handled by human beings. Scientists and technologists would have a terrific R&D program with the Internet replicator envisioned. As a result, the country would experience an upswing in technological progress that would influence the entire country, perhaps the entire world. Moreover, research and development and experimentation with the military would prepare cyberwarriors to enter and win conflict in cyberspace at home, around the world, in space, or in cyberspace itself.

The notion of an Internet replicator will face many challenges. An Internet replicator will be costly, and the hardware, software, and qualified scientists and technologists working with military leaders will add to the cost as will accessing the necessary bandwidth to connect the enclaves and knowledge centers. The price to support modeling and simulation sufficient to simulate this environment will also be expensive. Replicating networks of potential foreign adversaries will be complex and time consuming. Keeping up with technological progress and ensuring the Internet replicator represents the latest technological changes (software and hardware) will take time and money. Accomplishing tasks and populating databases in the language of a foreign adversary will take time and money. Simulating the digital environment will take time and money. Many bureaucratic turf fights will erupt over the control and development of an Internet replicator.

So what is a realistic way to move ahead and develop this needed training and research and development tool? As a first step, the Internet replicator should develop in one of the federal government's national labs. Much developmental work will need to occur to replicate the Internet and the connectivity that an Internet replicator will need for its operation. An Internet replicator will need to start small and grow, probably starting with a few existing networks, servers, and connectivity within DOD until sufficient funding becomes available to enable the full-scale vision to come into being. Eventually the Internet replicator will grow to include several sites with many enclaves of servers connected to many protective gateways in many states.

A second step would be to expand the ideas of an Internet replicator with the development of a joint asymmetric OPFOR and a joint information operations proving ground. These three mechanisms are inextricably related and will provide the best preparation for the country as it faces the challenges and ravages of knowledge war. The DOD, other national government departments, state and local governments, commercial businesses, and congressional support are all important. The Internet replicator will be operating in many states which means that many congressional leaders will have a vested interest in its development. Many leaders in Congress are concerned about the state of defense for the homeland and the cyber-readiness of the country's military forces. An Internet replicator, complete with a capable asymmetric OPFOR, will go far in ameliorating those concerns. Moreover, since homeland defense will occur in all the states the congressional leadership will want their constituents to train on the Internet replicator against the asymmetric OPFOR that could attack the U.S. infrastructure. The joint staff should take leadership of the project and use the replicator, the joint asymmetric OPFOR, and the IO proving ground to ensure that joint forces are trained for knowledge war, that people have a place to experiment and conduct research and development, and that joint IO doctrine has a place to mature and become more meaningful. The Joint Forces Command or the newly formed North America Command should and could easily become the command to whom the joint staff delegates responsibility for operating the Internet replicator. Either of these commands will focus the attention of electrical engineers, computer scientists, and psychologists on developing the software and hardware to engage in conflict in cyberspace, on training joint commanders and their staffs on operations in cyberspace, and on experimenting with organizational design specifically in preparation for conflict in cyberspace.

The country needs a vast networked system of knowledge advantage centers that are at once large and small, that are linked, and that operate twenty-four hours per day seven days a week. These centers should be collaborative, they must build on previous work, and they must create new knowledge that leaders of all walks of life and status need for making effective decisions. The country's defenses need a joint asymmetric opposing force that thinks, plans, acts, and speaks like the foes we will face in the twenty-first century. In particular, we must replicate terrorists, conventional forces using asymmetric strategies, tactics, and tools, and loose confederations of criminals, drug gangs, terrorists, and conventional forces. These foes may speak a foreign language and may think differently than we do. These adversaries will have motives and perceptions far different from our own. They will operate in the physical world and in cyberspace. These forces will have asymmetric strategies, tactics, and tools that will challenge U.S. military forces to the point of forcing our officers to think, plan, and act differently.

Our country needs a joint information operations proving ground and the Joint Staff should have responsibility for its inception and survival. Joint Forces Command or North America Command should be responsible for its operation, training, experimentation, research and development, and testing and evaluation. Such a place will give commanders the opportunity to train their forces in a cyberspace environment against the types of foes they will face in a future knowledge war and will give scientists and technologists a way to conduct applied and basic research experimentation, and a way to develop spin-off technologies that will help the country in the long term. Such a place will provide ample opportunity to conduct the necessary testing and evaluation of knowledge weapons, kinetic weapons in support of IO, and nonkinetic weapons. Such a place will provide the means to evaluate evolving IO doctrine and to put teeth into the insightful vision articulated in "Joint Vision 2020."

The country needs an Internet replicator to train leaders and their staffs to operate successfully in cyberspace. This replicator needs to approximate the real Internet and the networking a real opponent would employ and the language of the people we are training to operate against. Moreover, the Internet replicator must provide a capable OPFOR against U.S. cyberwarriors in the invisible world of cyberspace. This replicator program needs to build slowly, allowing for the development of simulation software that will make training more realistic. Slow growth will also allow for the integration of the knowledge advantage center system of collaborative systems, a joint asymmetric OPFOR, and the Internet replicator into the joint IO proving ground. Taken in isolation each idea is powerful, but when one combines all four, the country will have a capability for preparing the intellects and technical capabilities of its teams of knowledge workers and leaders for years to come.

[A]ll these complex systems have acquired the ability to bring order and chaos into a special kind of balance. This balance point—often called the edge of chaos—is where the components of a system never quite lock into place, and yet never quite dissolve into turbulence, either. The edge of chaos is where life has enough stability to sustain itself and enough creativity to deserve the name of life. The edge of chaos is where new ideas and innovative genotypes are forever nibbling away at the edges of the status quo, and where even the most entrenched old guard will eventually be overthrown.

M. Mitchell Waldrop, *Complexity*

Cyberwarriors 9

Far into the future human beings will be the most important aspect of our national defense. Human beings will engage in future conflict and human beings will direct the cyberbots that will inevitably appear. It is important that leaders from all walks of life carefully consider how to nurture and develop the intellects that its cyberwarriors will need to engage our asymmetric adversaries, who will be waging a knowledge war through terrorism, IO, weapons of mass effect, and the other nefarious tools of the modern knowledge warrior. Unfortunately, the U.S. Department of Defense does not place equal emphasis on developing intellects as it places on developing the physical machines with which to wage attrition-based warfare. Yes, we spend a lot of money on our schools, but the learning that goes on in those schools is focused on the training necessary to think about, plan, and execute attrition-based operations. In the types of conflict described in this book, our security forces will be engaging in both an attrition-based and a knowledge war. Knowledge war is different than traditional, attrition-based warfare because it requires far greater intellectual skills, a broader understanding of the social, political, economic, financial, informational, and military spheres of human intercourse, and the existence of a far greater symbiotic relationship between man and machine. We need to change the way we recruit, train, educate, and retain people that we want to engage in asymmetric conflict, specifically in a knowledge war. In addition, a cyberwar is just over the horizon. It will take a long time and a lot of money to prepare the training and education that we need for developing the human beings to effectively conduct information operations (both offensive and defensive) in cyberspace against formidable foes (human, software, and computer-guided) who intend to destroy our cherished way of life.[1]

The DOD must purposefully develop information warriors to operate successfully in a knowledge war environment. The country's military leadership has to find the most capable people in America to be its warriors in cyberspace. The most capable people are those with the best computer skills, best thinking, and best planning capabilities; they are the best leaders; and they are the best future conflict strategists to represent the country in this new arena of conflict. These people have to be the most capable talent that our country has because they will face people with gifted intellects and strong networks and resources who couple their capabilities with computers in their attempt to impose their will on the United States.

As currently structured the U.S. military has no chance of finding, attracting, or developing the ideal kind of person we need for waging a knowledge war: the twenty-first-century cyberwarrior. Our current ethos of egalitarianism, attention to rigid processes, well-defined standards, a hierarchical leadership structure, a belief that achieving higher rank correlates with more advanced thinking and knowledge, and an Industrial Age emphasis on kinetic warfare that influences all organizational processes won't meet the requirements we have for developing, nurturing, and retaining the ideal cyberwarrior who will defend our country's cyberspace.

An interesting book by Michael Lewis, titled *Next*, describes today's strange "marriage" of man and machine, and the young people who engage in a different kind of learning. They have become knowledge experts, they share ideas with people they don't know in peer-to-peer collaboration, they freely create new knowledge, they easily develop expertise without formal schooling, credentialing, or licensing, and they have been able to develop a wonderful symbiosis with both the computer and the Internet. If one believes that the information revolution has caused the United States to depend on computers, telephony, and the Internet, and that they all involve operations in cyberspace, it follows that some important aspects of future conflict will occur in cyberspace. As mentioned earlier, the soldiers of cyberspace will be cyberbots. They roam a terrain of fiber-optic cables and electromagnetic waves in hard disk drives, along switches and routers, and within databases. One cannot see, feel, or sense this space though it exists nevertheless, and it is critically important.

In an ideal sense these young people have the skills and learning abilities to serve as our cyberwarriors (with just a bit of taming, of course). But how in the world could they ever adjust to the military and how will the military ever adjust to them? The answer to this question won't be easy to find, but our military needs to think about meeting this challenge head-on. Serious leaders have to start asking some serious questions: How will the country defend itself against cyberattacks by very skilled foes? How will the country use cyberspace to conduct offensive operations? Who will program the cyberbots and who will control them? Who will match wits with machines moving at the speed of light, while being controlled by the best and brightest hackers and cyberwarriors in the world? How does one go about training and educating people to enter conflict in this new "terrain" against the new "enemy" yet feel confident about winning? Will the country's military school systems meet the stimulation, intellectual, and emotional needs of a rapidly learning and maturing cyberwarrior, as they currently stand?

Finding the answers to these questions will not be easy and will require some serious introspection and eventually a serious commitment of resources to do what is right. The answers will not be found in a hodge-podge of continuing dependence on the status quo and its choices in the intellectual development of cyberwarriors, in how to attract the right kind of person to operate in cyberspace, or in retaining such talent once it experiences the military's way of life. No, the answers lie in recognizing the changing landscape of future conflict, the changing nature of future adversaries, and purposefully selecting then grooming, stimulating, and intellectually preparing the country's future cyberwarriors.

Perhaps the country's military needs to acknowledge that it cannot or will not be able to develop ideal cyberwarriors. Perhaps it will be forced to hire its cyberwarriors—a different kind of mercenary not unlike what great societies in the past have done when their militaries could no longer meet the challenges of the age. This tantalizing subject will be left for another time and another book. Instead let us concentrate on defining the characteristics of future cyberwarriors knowing that such identification will certainly cause people to think about doctrinal, organizational, training and education, materiel, leadership, and personnel systems to support this kind of person. Clearly the right kind of people that can thrive in cyberoperations do exist and they can develop the skills those situations demand. The cyberwarrior is not the kind of person currently functioning in the country's military system—it may be decades before the military can truly transform itself to the extent that it can find and nurture such individuals. But the cyberwarrior will be an extremely formidable American fighting person in the great tradition of American creativity and commitment to ideals.

The Internet is developing into a biological organism—pulsing, growing, and constantly reproducing and changing in shape, influence, and implications. Its edges are growing in size exponentially, constantly gaining in intelligence and knowledge. It will never stop changing, and it is growing an amazing intelligence all its own. It is a collective intellect—a strange attractor of combinations of human and machine intelligence, collective intellects, and even the intertwining of, or symbiosis between, man and machine. Peer-to-peer computing has empowered the edges of the Internet to develop this shared intellect. Empowering the edges of the Internet and providing the power of networking has begun the process of rendering anachronistic all emphasis on the center, with its tedium and bureaucratic lethargy. Along with this high degree of humanlike intelligence resident in an inanimate object, people (primarily young people) are developing truly incredible intellects, depths of knowledge, and understanding of relationships as they surf the waves of this marvelous learning, sharing, and problem-solving tool.[2]

It takes no imagination to understand that such a stunning intellectual capability, such as is currently developing on the Internet, is value neutral; that is, all people everywhere, believing in any doctrine and having any goals and objectives, have the capability to bring their native intelligence into play and learn and develop ideas and programs. Unfortunately, the results of their work could become killer apps that change the way military planners think about, plan, and execute operations in future conflict. The Internet and its "intelligence" recognize no correct way of thinking, no ideology, no people as superior, and make no judgments about right or wrong. Indeed, to the collective intelligence of the Internet all activity is neutral and exploitable. The Internet is a means for implementing an electronic enabler of meritocracy—if Thomas Jefferson could only see how the Internet brings his notion of meritocracy into being!

All branches of service must purposefully start now to develop "information warriors," as it can take twenty years to "grow" and groom an officer and at least fifteen years to "grow" and groom noncommissioned officers to become competent information warriors within the military's current system and hierarchical establishment. If the country's military chooses to develop its cyberwarriors along traditional lines within the current framework of training and education, the twenty-year timeframe will hold true.

What is needed? The military needs a core of cyberwarriors to help plan and think about the future and to put offensive and defensive IO into effect. Cyberwarriors will not be just information specialists, nor will they be "geeks." Instead, they will think, plan, lead, and they will be master technologists. As such, they will be able to combine technical expertise with twenty-first-century leadership skills and a mastery of the age-old craft of warfare. What, then, are some of the more important characteristics that typify the cyberwarrior of the twenty-first century?

U.S. military cyberwarriors are proud of who they are: they are different; they think, plan, and perform differently than what ordinary people do to defend the United States; and they therefore require training, education, and nurturing that is different than that given to mainstream military. Cyberwarriors first and foremost are twenty-first-century leaders. They have the ability to operate in person-to-person interactions, in peer-to-peer chat rooms in cyberspace, or to lead virtual projects with people and machines they have never met or seen. They believe in the ability of people to learn throughout life and are life-long learners themselves. It is through continuous learning and application of that learning that cyberwarriors ensure their intellectual superiority over any foes. They will ensure that future foes won't outsmart the country's cyberstrategists and develop "leap ahead" advances in technology.

The cyberleader of the future believes in mentoring subordinates and fellow knowledge workers by helping them learn. Leaders of the future develop and execute vision. People need a plan to think about, debate, and follow. The cyberwarrior of the future learns not only to develop and impart vision but to enact that vision's essential elements. As a leader of the future the cyberwarrior motivates people to do things well and correctly without actually being physically close to supervise.

The cyberwarrior of the future sets up environments for the pursuit of heuristics, or discovery learning. These environments are free of fear and retribution for people coming forth with creative or off-the-wall ways of viewing the environment, accomplishing tasks, or thinking about life. The cyberwarrior of the twenty-first century is in great physical, intellectual, emotional, and spiritual condition. As such, he or she understands how the Aristotelian whole (quadrants of life forces focusing on spiritual, intellectual, physical, and emotional spheres of growth and balance) changes with time, operates, fluctuates (as to which quadrant dominates at a particular time), and functions as a unified entity.

The twenty-first-century cyberwarrior, with the assistance of technological enhancements, endures fatigue, pain, fear, and anguish with no degradation of either intellectual or physical performance far beyond normal human capabilities or what ordinary people can endure.

Cyberwarriors possess and demonstrate great intellectual skills: they analyze quickly, relate obvious and disparate entities and events, and, most important, synthesize data, information, and knowledge into an amazing relational intellectual environment that is somewhat physical and somewhat virtual. It is through synthesis that cyberwarriors search for and create relationships that lead to combinations that in turn lead to the creation of synergy. Synergy is essential for gaining dominant knowledge advantage and information superiority. From this knowledge and understanding, developed through superior intellectual and machine skills, the cyberwarrior seeks or creates conditions for finding hidden relationships that heretofore have escaped the attention of the conscious human mind.

The cyberwarrior views the world holistically or as totally interrelated, with the intertwining of systems and relationships comprising human interaction among themselves and with the environment. Once people understand the interrelated nature of the world, a logical antecedent allows them to understand or at least intuit why things happen. If people understand why things happen they can also understand why activities and actions affect some things while other things appear unaffected. Once the interrelated nature of the world becomes a way of life and a holistic view of the world dominates one's thought processes, the potential to manipulate variables sufficient for affecting perceptions and shaping the future surges forth as conscious thought. Such an intellectual transformation enables cyberwarriors to position themselves to deal with the rapidity of change inherent to the information revolution.

The twenty-first-century cyberwarrior is a master of technology, keeping abreast of changes in technology and understanding the implications of new technology on the task of national defense. It is a constant search for new technology and ideas about technology on the Internet. Cyberwarriors frequently inhabit peer-to-peer chat rooms involving many aspects of technology, including hacker thoughts and software. The cyberwarrior recognizes technology as a tool that helps people learn better, perform psychomotor tasks better, and think exceptionally well. The cyberwarrior never forgets that no matter how glitzy technology is, it is merely a tool for helping human beings accomplish physical and mental tasks better.

Cyberwarriors possess a worldview different than most people's—it is broad rather than narrow, relational instead of isolated, and changing instead of stagnant. They have a philosophy of life, they articulate it, and they live what the philosophy mandates. They are inquisitive and lifelong learners who help subordinates approach learning in a similar way. The cyberwarrior believes in the doctrine of holism, in which wholes constitute a greater force than the sum of constituent parts. As such, cyberwarriors believe in human potential for growing spiritually, physically, emotionally, and intellectually. The cyberwarrior possesses a strong belief in values that make our institutions and country great: dedication, integrity, loyalty, selfless service, strong work ethic, and so forth.

America's cyberwarrior creates organizations but sows the seeds of their destruction as they come into being. Thus, along with the creative process of building organizations and relationships, cyberwarriors are destroyers. The process of destruction and creation facilitates adaptation to change. Otherwise the future and vicissitudes of change will do the destroying and the serious malaise of passivity will set in to stay.

Cyberwarriors loathe the status quo. They believe in perturbation, change, and chaos and purposefully create them if the condition of stasis becomes evident. As such, cyberwarriors have sufficient intellect to understand the unfolding and enfolding of variables well enough to manipulate them, thus effectively shaping the future. Additionally, cyberwarriors believe in the interaction of opposites and view the Hegelian dialectic as a medium of change and creativity—thesis, antithesis, and synthesis strike a clarion call for creativity and provide a resonate tone in their minds.[3]

America's cyberwarriors of the twenty-first century understand how strength can be weakness. As such, they comprehend the nature of interacting relationships. Sometimes these relationships surface as interactions among opposites, but to cyberwarriors they are perfectly logical and present no intellectual contradiction.

America's cyberwarriors save lives through asymmetrical or asynchronous advantage, through technological advantage, or through the greatest of all assets, the human mind. Their goal in any competitive endeavor is to operate and win convincingly with minimal costs in human lives. As such, cyberwarriors of the twenty-first century are masters of deception and the thinking that deception requires. As Sun Tzu said, to deceive we must know ourselves as well as know the enemy. Thus cyberwarriors do not fall victim to the intellectual ravages of "mirror imaging." Instead they know and understand much about foreign cultures and differences among people.

Through technology and collaboration these inquisitive and always thinking cyberwarriors obtain the detailed knowledge expert support needed to conduct skillful, undetected (until too late to respond) deception operations, while being constantly wary of the deceptive efforts of an equally skilled foe.

One of the things the country's military can do to develop cyberwarriors with the attributes and characteristics discussed above is by changing its training and education systems. This change must aid in the development of future cyberwarriors as they progress through the military's rank system. Along this line of thought, the military needs to train and educate the best and brightest computer people and military strategists and allow them to expand their horizons, learn, create, innovate, and "romp" in a cyberspace-like environment and have multiple ways to facilitate intellectual growth and improve requisite psycho-motor skills to compete against formidable foes in cyberspace. Our country needs these people to protect our way of life. The adversary of the future and the environment in which he will engage the United States is coming into view—and it will incorporate the strategy of knowledge war and a variety of asymmetric tools, one of which is information operations. Learning will demand coherent, exciting, and stimulating learning experiences. It will involve some training, and it will require lots of education carefully choreographed to help satisfy the intellectual capabilities of the future cyberwarrior.

The military must devote sufficient resources to develop the intellects of young people and advance these intellects into increasingly higher levels of thinking as they progress through their careers. Simulation and modeling, plus constant doses of twenty-first-century mentoring by leaders and learning resource mentors, will be an important combination to consider for aiding this process of intellectual growth. Military learning institutions and organizations have to encourage a new kind of intellectual development: inquisitive people empowered to supplement their institutional and organizational learning with what they learn on the Internet by exchanging ideas in chat rooms, exploring concepts, theories, new ideas, and approaches to problems, all on the edge of the living, pulsing, biologic Internet, will produce the kind of constantly learning cyberwarrior-philosopher this country needs.

The military has little choice other than to develop cyberwarriors. Adversaries will eventually adopt this way of thinking to further their own agendas and goals. Some of them have been bold enough to explain their intentions in sufficient depth for us to understand (the Chinese, for example). Others are like a penumbra—they are there, growing, changing, learning, becoming one with both Internet and machine, but we can't see them or learn anything about them until they decide to come forth or make a mistake. These types of people will constitute our adversaries of the future—they are adversarial cyberwarriors and the country's military must prepare to battle them and win.

If the military fails to take a bold step and purposefully develop our own cyberwarriors, the country writ large will be at a serious disadvantage relative to the adapting, learning organisms that will surge forth and retreat, be aggressive and be passive, be a friend one day and an adversary the next, and sometimes remain an implacable foe, all the while, however, reconnoitering, sensing, probing, planning, and deceiving in cyberspace on the Internet and over electromagnetic waves. The country's military needs a core of cyberwarriors to meet this adversary on this new kind of "terrain" in different conditions, and then defeat it soundly and repeatedly. Thus the military establishment must put as much time, thought, effort, and resources into developing the minds of cyberwarriors as it places on developing machines of war. People must think through the implications, make the necessary changes in the military establishment, and purposefully develop the new cyberwarrior of tomorrow.

What should the solving of Nature's secrets be? . . .
If you wish to advance into the infinite, explore the
finite in all directions. If you desire refreshing
contemplation of the Whole, you must discern the
Whole in the smallest of things. In the infinite the same
events repeat themselves in eternal flux, and the
thousandfold vault of the heavens powerfully conjoins
with itself, and then the joy of life streams out of all
things, out of the smallest and out of the greatest of stars.

Goethe Selected Verse (Edited by David Luke)

10 | The Way Ahead

Knowledge war is upon us, and we must act now to prepare for this type of conflict because it will dominate our future security environment. Knowledge war is the struggle for control of the data, information, and knowledge with which to make better decisions faster than the adversary. When people make better decisions faster they can create actions that lead to effects that lead to the most important center of gravity of the twenty-first century: will. Knowledge war has knowledge-based strategies and knowledge-based operations as its lineage. Knowledge war has information operations as its principal tool, which attacks and manipulates decision making, thinking, planning, and perceptions. IO also attack and attempt to manipulate information systems that provide the data, information, and knowledge that people need for engaging in knowledge war.

Clearly the United States has entered a new era of defense. Just as clearly the country has to change the way it views threats to the state of security that we all desire. The events of 11 September 2001 provided a glimpse into the future. It wasn't a pretty glimpse, as the adversary's willingness to sacrifice his own life and the lives of thousands of innocent people to further his cause suddenly brought the struggle to the forefront of every American's consciousness. The ferocity of the terrorist attacks should cause all of us to shudder because of their implications for today and far into the future.

How, specifically, does the country need to change and adapt to meet this new threat to our long-standing sense of security? First, people cannot extricate the Department of Defense from homeland security, nor can they isolate it from this challenge. People cannot isolate parts of this whole anymore—homeland security is a challenge that must be dealt with holistically; that is, local, state, and national governments and commercial organizations are part of the same security tapestry. Second, the struggle will undoubtedly be a "Hundred Years' War." Thus we have to cast off our narrow perspectives and short-range viewpoints and take on broader intellectual vistas and long-range viewpoints as we design our twenty-first-century security system. Third, we have to prepare for combating asymmetric adversaries who think and perceive in ways very different from the way we do. These adversaries have strategies, tactics, and tools that can cause great damage to the country's physical icons, infrastructure, and treasures, as well as great harm to the physical well-being and psyches of American citizens. Fourth, we have to learn to think differently and become more anticipatory to stop terrorist attacks before they happen. Fifth, we have to enter collaborative environments that connect across the breadth and depth of the country, with the explicit goal of seeking, finding, or creating valuable knowledge of sufficient quality to enable decision makers to make better decisions faster than any adversary. Sixth, the country's security forces must train against a joint asymmetric opposing force that thinks, plans, and executes like a true terrorist or other asymmetric foe would, not like an American would. Last, we have to prepare now to engage in cyberwar. We need to develop cyberwarriors, a joint information operations proving ground, and an Internet replicator for our scientists, operators, planners, and commanders to work with and train on.

On a philosophical plane the world in which we are privileged to reside has several contexts. There is the physical world—that which we can see and touch and that we know well because of our physical senses—and there is the wonderful world of our minds—that which we can imagine, dream, create, and think about (or be constrained in our thinking about). With a twitch of our bodies or a toggle of a neuron switch in our minds we can overlay the tapestry of our minds onto the physical context of our existence. Then and only then can we start to understand the magnificence of the world in which we reside, that is, the merging of our minds with matter (our bodies, our physical context, and so forth). In particular, our minds have to conjure patterns of relationships and tapestries of linkages, networks, and the connectedness of the world. We also must consider a third context: the invisible context that affects us all. We cannot see the atoms or the cells that make up the universe. We cannot see or touch quarks or shadowy photons. We cannot see the digits that transform our world so dramatically. We cannot quantify our consciousness or our unconsciousness. Yet who can deny their existence? In our quantify-driven world we tend to discount what we cannot see or touch. Thus we fail to understand and appreciate a piece of the brilliance our maker put forth into the wonderful world in which we live. This failure causes us to lock the door of opportunity, and we fail to open it even when the knocks are loud.

Rest assured, knowledge war will eventually dominate conflict. This is not to say that kinetic warfare will go by the wayside. But the struggle for information and knowledge and their effects on making decisions faster than an adversary can will become so important that this struggle will dominate all other kinds of conflict. The logic underpinning knowledge war is the conversion of data into information by machines, and turning information into knowledge by the application of intellectual powers of knowledge workers. Leaders take knowledge, form relationships with other knowledge, and gain understanding. With that understanding leaders can make decisions—they decide to act or not to act. If they decide to act someone provides the action that creates effects or outcomes. Effects in turn influence will—of individuals, of groups of individuals, of leaders, and of countries.

Adversaries of the United States will choose to engage in knowledge war for several reasons. First, these adversaries know they cannot compete in a force-on-force way—they cannot win, and they know it. Thus, they seek and will find other ways to compete. We must anticipate these moves and be waiting for them when they come. In this respect they are asymmetric foes that seek offset against a much larger and wealthier opponent: the United States. Second, the United States has magnificent technology upon which it is extraordinarily dependent. This dependence presents vulnerabilities that future adversaries will attack using a far-reaching variety of strategies, tactics, techniques, and tools. Third, the information revolution has greatly empowered weaker adversaries to synchronize and coordinate across time and space to accomplish their objectives. Along this line of thought, information and technology are neutral—they know neither good nor bad. People and organizations with money can buy cutting-edge technology to help them think, plan, and act; they can purchase information and in some cases even valuable knowledge. As an example, anyone can buy information and knowledge products such as high-resolution satellite imagery to use for purposes beneficial to their cause; this knowledge and the effects it can create could very well be inimical to the goals and objectives of the United States. This trend will continue and become more difficult to counter.

Competition will become even more intense and with this growth in intensity comes the need to acquire valuable information and knowledge for making fast and effective decisions along with a situation in which valuable knowledge is increasingly difficult to find, obtain, and safeguard. Information and knowledge and the intangible effects they create will help destroy, blind, manipulate, deceive, or retard an adversary's means to acquire, analyze, synthesize, and move information and knowledge. Future conflict will not be like the conflict human beings have known and grown accustomed to. Instead, future conflict will involve conducting sophisticated mental warfare and psychological operations against both military and civilian populaces.

Information superiority is an important condition in this type of struggle. It provides conditions conducive to achieving advantages over an adversary. These advantages include decision, intellectual, knowledge (knowledge management), technical, positional, and action. Information superiority will continue to be difficult to measure, but, regardless of difficulty, scientists and technologists should work with communications, operations, and intelligence personnel to measure it. Information superiority will fluctuate, sometimes slowly and sometimes rapidly, because if one side has it the other side will seek it through change, adaptation, and counteraction.

Valuable knowledge provides commanders with the means for making faster, better decisions than adversaries. Thus, it is important to create a combination of technology and people to achieve the greatest efficiency and effectiveness all the while producing knowledge—this is knowledge management. Knowledge management will be a critical aspect of future conflict because how one side or the other manages knowledge will go far in determining the outcome of knowledge war engagement. Knowledge management can be an asymmetric advantage, but it also presents a lucrative vulnerability through the principal perpetrator of action in knowledge war: information operations.

Information operations affect thinking, planning, perceptions, decision making, and decision support systems. IO are tools that can be used by weak and strong forces alike. IO can involve largely invisible assaults, in indirect and often subtle ways, against stronger, more technologically capable opponents. IO can also represent a way to win by maintaining and using aggressive, offensive maneuvers (computer network attack, deception, PSYOPS) while staying protected through computer network defense, OPSEC, and deception. (In this respect IO can allow one side or the other to stay strong, minimize costs, and enjoy a definite asymmetric advantage.)

As human beings we run the risk of becoming so mesmerized by a location, region, or the here and now that we lose sight of the greater whole, the greater purpose, or a broad view of the enemy. In a philosophical as well as in a physical sense the best way to describe the world is to say it is composed of closely intertwined relationships—globalization has a meaning that is significantly relational and increasingly sensitive to perturbation. Our enemy has sensed what "global" means and plans to use its workings—its webs of information and enshrouding social systems—against us. What happens in one location affects other locations (in sometimes obvious ways, at other times in unobservable ways) owing to the instantaneousness and ubiquity of communications and information.

The world is made of relationships. This is no revelation, but for some reason we don't think as much about relationships as we should. We must do so to enable our minds to compete with our adversary's. Relationships can be both obvious and disparate: some we can easily notice and understand, but we often experience difficulty with the hidden or seemingly disparate ones that make up everything in the world.

To counter adversaries using asymmetric strategies, tactics, and tools such as information operations, the United States has the implied task of developing cyberwarriors. This breed of warrior will do battle on the new frontiers of conflict and the battle space of the future: cyberspace. Cyberwarriors combine the best characteristics of the warrior ethos from years gone by with the capabilities that only people living in the Information Age and experiencing firsthand the information revolution can possess. Military cultures must purposefully nurture and grow cyberwarriors because it will not occur by happenstance. It will take a concerted effort to develop these people while they are young and as they progress through the military's rank and military training and education systems. Though there will not be a need for a huge number of them, cyberwarriors will prove critically important for winning and attaining advantages in decision making, for only they will have the brainpower and technical skills to perform the sophisticated cerebral functions that IO requires, such as synthesis, holism, analysis, deception, futuristic planning, and conceptual creation.

The twenty-first-century competitors will project their minds into the minds of their opponents to understand what opponents want and what they have planned. Relational thinking will be mandatory, so cyberwarriors of the twenty-first century must be expert in analyzing information. Even more important, they must be expert in synthesizing information that is gained through analysis into a higher level of meaning. In addition, twenty-first-century cyberwarriors will always seek intellectual, emotional, and physical synergies because of the immense potential and power they offer. Twenty-first-century cyberwarriors will also think and plan holistically searching for, finding, and creating relationships among large and small systems and obvious and discrete events and things. The cyberwarrior will create and destroy, creating and designing new organizations and processes with ease and planting the seeds of destruction in that which he or she creates. These creations will get changed either by outside forces (giving the creator no chance to shape the future) or through internal changes (which gives ample opportunity to shape it).

Two activities in knowledge war stand out in importance: those aimed against the opponent in an offensive way and those the opponent will attempt to do in return to create the effects he believes will influence friendly decisions and our country's will. The country's military forces must respect the intellectual and technical capabilities of adversaries and acknowledge that they will continue to learn, adapt, change, and become more capable. We must agree that making better decisions faster than the adversary is crucially important in attaining advantage over him. One way to make better decisions faster than an adversary is to think better by using the combined intellects of thousands of smart people linked in collaborative environments

We must learn to share and collaborate on a scale heretofore unknown in our daily lives. We must develop a system of knowledge advantage centers that stretch across the country connecting people everywhere—at all levels of state, local, and national governments and commercial businesses—with the purpose of sharing data, information, and knowledge with which to anticipate and deny terrorists the capability to attack our homeland. Though this idea is simple in concept it will be difficult to accomplish owing to our human tendencies to hoard data, information, and knowledge and our basic mistrust of the motivations of others. But develop it we must, to collectively thwart future terrorist attacks.

The country's security forces need to train against a capable joint asymmetric opposing force to shake off the vestiges of the technological arrogance that have come with being a huge, rich, and technologically gifted country such as ours. Arrogance can deceive and even blind. The truly asymmetric OPFOR will help keep our armed forces firmly rooted in reality and it must be a thinking, adapting, learning, and changing entity. OPFOR members must use foreign languages with ease and think like foreign adversaries; during their training they must be armed with the latest technology and possess goals and objectives different from the friendly force. The asymmetric OPFOR should not be constrained by the idea of fighting fairly. The OPFOR must have capable hackers who face U.S. cyberwarriors in cyberspace on an Internet replicator. Asymmetric opposing forces will have a global view and not be confined to thinking about narrow, regional perspectives.

Currently, America's military and government agencies (local, state, and national) do not have a viable way to train against asymmetric foes. The training of our military forces and preparations for the country's homeland defense are constrained. The country needs an organized place where preparation for knowledge war can occur: a joint information operations proving ground. Several activities will occur here. First and foremost, commanders and their staffs, state and local leaders, the people who support them, and even important commercial organizations will conduct exercises and train to operate in cyberspace. The overarching proposition of the JIOPG is the creation of a learning environment in which after-action reviews and critiques using the most modern simulation and learning facilitation techniques are of primary importance. People from all types of organizations concerned about security and from all levels within those organizations need a place to train to meet the looming digital onslaught that our future foes will undoubtedly bring. Scientists and IO operators and planners also need a place to experiment with technology, to anticipate the actions of asymmetric foes, and to allow staffs to work with the stability or fragility of operating doctrine in a realistic but expectable environment.

Cyberspace is the "playground" where information operations will occur. It is the boundless extent where computers, computer networks, communications, and people exist and blend into a whole. Though it is real in one sense, in cyberspace people, software, and hardware interact in an invisible, digital, and virtual world where no sanctuary exists anywhere at any time. The country needs a scalable Internet replicator, also known as a Fishbowl, complete with an opposing force of competent cyberwarriors to train military, government, and civilian leaders, and knowledge workers for intense information warfare struggles in cyberspace. People role-playing as cyberstrategists and cyberwarriors in the Internet replicator training environment must use relevant foreign languages, thought processes, and decision-making devices. The government, the military, and civilian institutions together must learn to recognize, cope with, and succeed in defending against ruthless, highly skilled, and determined cyberattackers. The Internet replicator will help to meet this challenge because such a replicator portrays an environment that looks and acts exactly like the real Internet. For example, it will have some time delays built in, it will have enclaves with Intranets, local area networks, and wide area networks, it will have a skilled cyberopponent or opponents who act as cyberattackers, and it will have authentic opponents representing adversarial countries. Moreover, all activities on an Internet replicator must occur in a secure environment. Without practice on an Internet replicator, people will never have the opportunity to prepare adequately for defending themselves against the nefarious activities of cyberbots and their handlers, let alone against common hackers and script bunnies. In addition, our country's scientists and technologists need an Internet replicator for research and development in a secure environment with which to experiment to develop constantly new technologies for warning of and denying cyberattacks by any foe. Local and national governmental agencies responsible for the well-being of the citizenry of the country will not be able to train people nor possess adequate technology to defeat cyberattacks without an Internet replicator.

What are some of the main thoughts that can serve as a conceptual framework for developing a strategy for planning and executing knowledge war operations? First, people need to be able to describe the purposes of the information and knowledge they are seeking. Generally, these uses involve adding value to information, using knowledge to understand, and using understanding to make decisions. Strategy helps us think through how to enhance decision making, degrade the opponent's information and knowledge, or enable some kind of an advantage to spring forth. More specific elements of a knowledge war strategy involve gaining knowledge and understanding about the information centers of gravity or knowledge advantage centers of both sides. Both will have critical places where collection, automation, communications, thinking and planning, and decision making merge. Any knowledge war strategist must identify information centers of gravity, anticipate how each side will defend them, and design ways to influence the centers in a positive or negative way.

The knowledge war strategist knows and understands the details of the opponent's decision-making apparatus—how the adversary thinks, plans, makes decisions, and perceives the world. In addition, the strategist knows how the adversary develops information requirements, and technically how data, information, and knowledge are collected, processed, analyzed, synthesized, and packaged for decisions to arise. When the decision becomes action that leads to effects, the strategist understands how the adversary seeks feedback on the decisions and actions that he has let loose.

Antagonists in an information-operations context know and understand the conceptual and actual architectures (operational, technical, information) flows and connectivities. In addition, strategists in a knowledge war think through the computing hardware and software necessities that support decision processes. The knowledge war strategists also think through their own hardware and software needs in relation to the adversary's. The strategist visualizes communication paths and attendant bandwidths and anticipates what the effects will be when one or more paths are affected. The strategist also identifies the databases, decision-making software, and visualization tools his adversary uses. Finally, the knowledge war strategist understands the adversary's technical means of seeking, obtaining, and processing data and information for the purposes of receiving decision-making feedback. It is easy to see that the technical requirements for supporting the knowledge war strategist are immense and will continue to expand into further complexity, especially for cyberdeception.

Predicting and controlling effects, both tangible and intangible, are crucial to the knowledge war strategist. Effects in this context are outcomes the strategist seeks to bring into being through actions in kinetic and nonkinetic ways. Conditions are necessary to achieve maximum effects through action. Conditions in the context of knowledge war primarily deal with perceptions, minds, thought processes, and psyches of adversaries. The knowledge war strategist shapes the adversary's thinking and perceptions through both kinetic and nonkinetic means.

Information strategists must define their information requirements relative to specificity, timeliness, accuracy, and relevancy. They must be able to describe the information they believe opponents will seek, what is valuable to the opponent, and why an opponent needs it. Information strategists also take the time to think through constraints and restraints, designing goals and objectives relative to knowledge war outcomes. Ultimately, a viable knowledge war strategy identifies resources required to reach informational goals and objectives. Strategists must "sow" the seeds of destruction in the system that they create to stay abreast of changes in the continuing revolution of information technology that they use for gathering and using information and knowledge.

A knowledge war strategy will face resistors—for every action planned strategists must anticipate resistance from a foe and the environment and develop plans to overcome those resistors. If strategists are able to learn the nature of resistors they will be able to devise ways to eliminate or at least reduce the source(s) of power the opponent possesses to deny our will. Reducing the strength of resistors empowers.

In a broad sense, the context, constraints, resistors, and will (of an individual, of groups of individuals, of politicians and other leaders, and of our allies) all help define the boundaries of a conflict's intensity, enabling the strategist to plan effects that fit the situation. A conflict's context is the environment in which people operate—the physical terrain, cyberspace, weather, culture, nature of information, degree of technological sophistication, and so forth. Constraints, imposed either externally or by one or both antagonists, have a direct and powerful impact on intensity by limiting the activities opponents choose to use to create the effects that influence decisions, hence will. (The self-imposed ban on chemical weapons in World War II by both sides provides an example of self-imposed constraints.) The strength of resistors and an analysis and synthesis of second- and third-order effects help to define intensity of conflict. Will, the critical intangible in any conflict, is the determination, perseverance, and cost (moral and physical) that opponents in a conflict will endure to win.

We have no choice but to set about designing philosophical, cerebral, and technical solutions to the ominous threat that will haunt our sense of safety and well-being for the long term. We have to design a national security system that has an outlook of one hundred years. If we don't proceed with this kind of thinking and vision, the country will be in grave danger from our asymmetric foes who have sworn to destroy our way of life. It will not be enough to be good at consequence management of horrific mass casualty and destruction events that terrorists have unleashed upon us. Instead, we must anticipate the terrorists' actions through good, creative thinking that is bonded to superior technology and enables us to "be there waiting" when the terrorist comes knocking at our door. We can make this happen, and we have to get on with it right now. The onslaughts are coming and we must prepare ourselves to anticipate and deny those attacks regardless of their nature, location, and ferocity. The attacks will be physical and intangible, and we must prepare for both. Unfortunately, asymmetric warfare and the knowledge war will not go away. Thus, we need to master them now and maintain superiority in everything we do in the way of national security.

Notes

Preface

1. This scenario was developed at the behest of Lt. Gen. George Fisher, U.S. Army (Ret.), at Oak Ridge, Tennessee, in October of 2000. Lt. Col. Tim Vane, U.S. Army (Ret.), helped in the creation of the scenario.
2. Timothy L. Thomas, *Like Adding Wings to the Tiger: Chinese Information War Theory and Practice* (Fort Leavenworth: Foreign Military Studies Office, 2000), 6.

Chapter 1. The Setting

1. *Webster's II New Riverside University Dictionary* (Boston: Houghton Mifflin, 1984), 669.
2. *Webster's New Riverside Dictionary,* 1175.

Chapter 4. Information Superiority

1. Joint Chiefs of Staff, "Joint Vision 2020" (Washington, D.C.: U.S. Government Printing Office, June 2000), 3.
2. "Joint Vision 2020," 8.
3. *Webster's II New Riverside University Dictionary,* 396.
4. "The Promise of Internet2," *The Futurist* (July–August 2001): 12.
5. "The Promise of Internet2," 12–13.

Chapter 5. Information Operations

1. Col. John Boyd, U.S. Air Force (Ret.), lecture to U.S. Army School of Advanced Military Studies, spring 1985.
2. Personal communication with Col. David Michael Tanksley, U.S. Army (Ret.), in late December 1998.
3. Larry Downes and Chunka Mui, *Unleashing the Killer App* (Boston: Harvard Business School Press, 1998), 11.

Chapter 6. The Art of Deception

1. Arthur Schopenhauer, "The World as Will and Idea," *The European Philosophers from Descartes to Nietzsche*, ed. Monroe C. Beardsley (New York: Modern Library, 1960), 682.
2. William James, *Psychology*, ed. G. Allport (New York: Harper & Brothers, 1961), 120.

Chapter 7. Knowledge Management

1. The concept of knowledge engineering came to the author by way of Col. David Michael Tanksley, U.S. Army (Ret.), in December of 1999 at Knoxville, Tennessee.
2. Conversation between the author and Capt. Michael Maston, U.S. Navy (Ret.), at Oak Ridge, Tennessee, in March 2001.

Chapter 8. Triumph in Knowledge War

1. The notions of knowledge advantage centers and information centers of gravity first came to me as I studied with my contemporaries at the Army's School of Advanced Military Studies (SAMS). The notion of Knowledge Advantage Centers supporting CINCs and spreading across the levels of war came from discussions with retired U.S. Army generals Gary Luck, Anthony Zinni, and Charles Wilhelm, among others.
2. The notion of knowledge workers sharing information and knowledge in a collaborative environment surfaced during numerous conversations and interactions with retired U.S. Army generals Gary Luck, Anthony Zinni, and Charles Wilhelm, among others, in the spring and summer of 2001 while working for Joint Forces Command. The idea that knowledge workers do not make decisions but instead supply knowledge for decisions surfaced in these same conversations over the spring and summer of 2001.
3. The readiness question, "Are we training for peace or are we training for war?" came from personal communication with Col. Duwayne Lundgren, U.S. Army (Ret.), at Fort Riley, Kansas, in the summer of 1977.

Chapter **9.** Cyberwarriors

1. I have been exploring the thoughts expressed in this chapter for well over eighteen years. They first began careening around in my mind when I was a student at SAMS in 1984–85, and in particular when I served as a G2 of the 82d Airborne Division from 1987 to 1989 and as a brigade commander and J2 in the Republic of Korea from 1994 to 1998. I have spent numerous hours pondering the future and the intellectual capabilities that the country's military will need to compete in the future environment. Portions of this chapter were previously published in the spring and fall 2001 issues of the Joint Staff periodical *Cyber Sword*. I hope to encourage the Department of Defense to consider making long-term resource and intellectual commitments to developing cyberwarriors for future conflicts.

2. Michael Lewis, *Next* (New York: W. W. Norton, 2001), 27–133.

3. Albert Hofstadter, "On the Dialectical Phenomenology of Creativity," in *Essays in Creativity,* Stanley Rosner and Lawrence E. Abt, eds. (Croton-on-Hudson, N.Y.: North River Press, 1974), 116.

Index

About the Author

Brig. Gen. Wayne M. "Mike" Hall retired from the U.S. Army in 1999 after thirty years of active military service. His last assignment on active duty was as study director for the Army's Intelligence XXI study.

Hall holds a bachelor's from the University of Nebraska, a master's of science from Kansas State University, an MMAS (master's of military art and science) from the U.S. Army Command and General Staff College, and a doctorate from George Washington University. While in the military, Hall attended the Command and General Staff College, School of Advanced Military Studies, and the National War College.

Hall is currently the senior executive vice president for homeland security and future conflict at MZM, Inc. in Washington, D.C. He and his wife, Sandy, live in Suffolk, Virginia.

The Naval Institute Press is the book-publishing arm of the U.S. Naval Institute, a private, nonprofit, membership society for sea service professionals and others who share an interest in naval and maritime affairs. Established in 1873 at the U.S. Naval Academy in Annapolis, Maryland, where its offices remain today, the Naval Institute has members worldwide.

Members of the Naval Institute support the education programs of the society and receive the influential monthly magazine *Proceedings* and discounts on fine nautical prints and on ship and aircraft photos. They also have access to the transcripts of the Institute's Oral History Program and get discounted admission to any of the Institute-sponsored seminars offered around the country. Discounts are also available to the colorful bimonthly magazine *Naval History.*

The Naval Institute's book-publishing program, begun in 1898 with basic guides to naval practices, has broadened its scope to include books of more general interest. Now the Naval Institute Press publishes about one hundred titles each year, ranging from how-to books on boating and navigation to battle histories, biographies, ship and aircraft guides, and novels. Institute members receive significant discounts on the Press's more than eight hundred books in print.

Full-time students are eligible for special half-price membership rates. Life memberships are also available.

For a free catalog describing Naval Institute Press books currently available, and for further information about joining the U.S. Naval Institute, please write to:

Membership Department
U.S. Naval Institute
291 Wood Road
Annapolis, MD 21402-5034
Telephone: (800) 233-8764
Fax: (410) 269-7940
Web address: www.navalinstitute.org

THE COLD WAR AND THE COLOR LINE

THE
COLD WAR
AND THE
COLOR LINE

American Race Relations in the Global Arena

THOMAS BORSTELMANN

HARVARD UNIVERSITY PRESS
Cambridge, Massachusetts, and London, England 2001

Library of Congress Cataloging-in-Publication Data

Borstelmann, Thomas.

The Cold War and the color line : American race relations in the global arena /
Thomas Borstelmann.

p. cm.

Includes bibliographical references (p.) and index.

ISBN 0-674-00597-X

1. United States—Race relations—Political aspects.   2. United States—Foreign relations—
1945–1989.   3. United States—Foreign relations—1945–1989—Social aspects.   4. Cold War—
Social aspects—United States.   5. Southern States—Race relations—Political aspects.
6. African Americans—Civil rights—History—20th century.   7. Civil rights movements—
United States—History—20th century.   8. South Africa—Race relations—Political aspects.
9. Blacks—Civil rights—South Africa—History—20th century.   10. Civil rights movements—
South Africa—History—20th century.   I. Title.

E185.61 .B728 2001

305.8'00973—dc21   2001024861

For Lynn, John, and Daniel

CONTENTS

PREFACE

IN 1972, WHEN I WAS fourteen years old, the ultraconservative Jesse Helms was first elected to the United States Senate from my home state of North Carolina. That a figure so closely associated with racial segregation could become our senator in the 1970s astonished me, even at that young age. By this time, however, Helms had mostly dropped the rhetoric of white supremacy and spoke primarily in the idiom of anti-Communism. With my friends and especially my older brother Michael, I made much counter-cultural fun of the senator's concern about "subversives," both red and black. Little did I know that the seeds of a lifelong fascination with race relations and the Cold War were being sown.

Three years later, as juniors at Phillips Exeter Academy in New Hampshire, my classmates and I were required to choose between two courses for fulfilling the second semester of the year-long U.S. history sequence. One course focused on the domestic side of American history since 1865, while the other paid greater attention to U.S. foreign relations in the same period. There was, it was thought, simply too much material to cover both stories adequately between January and June. A similar division shadows the entire field of American history, as I later found out, from the training of graduate students to the journals that reflect the research of professional historians. In 1975, I opted for the international track but was never happy—then or since—with having to make that choice. This book intends to contribute to a growing movement to reunite the internal and external sides of the American past.

The story told here ranges widely across U.S. and international landscapes in putting the history of white supremacy on the same page with the history of the Cold War. Particular attention is paid to Africa and to African Americans, both of which figured prominently in the era of the Southern civil rights movement and Third World decolonization. Recent scholarship has revealed the ways in which black Americans linked their struggle against discrimination at home to the quest for national independence in Asia and Africa. This book focuses on how the white men in Washington who wielded enormous global power after 1945 wrestled with the simultaneous demands of people of color at home and abroad for full equality, and with the parallel responses of Southern segregationists and white authorities in Africa. It was during the Cold War that one of the central tensions of American history was largely resolved, at least in the public sphere: the conflict between an older vision of an America founded on hierarchies such as race and gender, and a newer ideal of a society free of legalized discrimination and enforced inequality.

A WRITER ACQUIRES many debts on the road to publication. I am grateful to several people who took time from their own important work to read parts of early drafts of the manuscript and provide generous and critical comments, as well as much needed encouragement. They will recognize their suggestions in the stronger parts of the book. For its faults, they bear no responsibility; they truly tried to tell me. They include David Cecelski, Bill Chafe, Mary Dudziak, Cary Fraser, George Herring, Lisa Hoffman, Michael Hogan, Michael Hunt, Walt LaFeber, Mel Leffler, Gary Okihiro, David Painter, Dick Polenberg, Andy Rotter, Keith Taylor, Peter Wood, and Jonathan Zimmerman.

The Cornell History Department has provided a wonderfully supportive as well as challenging environment for carrying out this project; I am particularly indebted to two wise and helpful department chairs, Itsie Hull and Sherm Cochran, department manager Judy Burkhard, and the entire Americanist phalanx. Financial assistance from several sources at Cornell eased the book's completion: the Return Jonathan Meigs Fund of the History Department, the Robert and Helen Appel Fellowship for Humanists and Social Scientists, and university sabbatical and study leaves. Audiences at various forums provided helpful feedback on earlier versions of several chapters: the Ohio State University Graduate History Workshop, the Southern New England Foreign Policy Seminar at the University of Connecticut, the Porter L. Fortune, Jr., History Symposium at the University of Mississippi, and the Cornell Comparative History Colloquium. The editors of two journals kindly allowed the reuse of materials first published with them: "Jim Crow's Coming Out: Race Relations and American Foreign Policy in the Truman Years," *Presidential Studies Quarterly*, 29 (September 1999): 549–569; and "'Hedging Our Bets and Buying Time': John Kennedy and Racial Revolutions in the American South and Southern Africa," *Diplomatic History*, 24 (Summer 2000): 435–463.

Two literary professionals eased the publication process with consummate grace and skill: my agent, Lisa Adams of the Garamond Agency, who placed the manuscript; and my editor, Joyce Seltzer of Harvard University Press, who improved it immeasurably. Manuscript editor Elizabeth Hurwit provided superb polishing in the final stages. Long-term sustainers of the author include John "JB" Borstelmann, Tim and Barb Beaton, Dub and Libby Gulley, John Platt, Suzanne Mettler, and Wayne Grove, all of whom helped preserve my perspective on the larger journey. My deepest gratitude goes to those to whom this book is dedicated, my wife and our sons. Patient with me, they are my best teachers.

T. B.
Manlius, New York

The past is the present. It's the future, too.

—Eugene O'Neill, *Long Day's Journey into Night*

PROLOGUE

"WHEN I ASKED for coffee, the good woman said she could not serve me." Malick Sow, the ambassador to the United States of the newly independent African nation of Chad, was explaining to a reporter in 1961 about his reception in a Maryland restaurant. "She said, 'That's the way it is here.' I cannot say how I felt. I was astonished. I was so angry. President Kennedy himself has made deep apologies, but these humiliations are bad." When asked for her side of the story, Mrs. Leroy Merritt of the Bonnie Brae Diner replied with what she considered evident common sense. "He looked like just an ordinary run of the mill nigger to me. I couldn't tell he was an ambassador."[1]

An "ordinary run of the mill nigger" or an ambassador? Malick Sow and fellow African diplomats created a dilemma for white Americans long accustomed to discriminating against people of color. Americans determined to preserve their leadership of the non-Communist "free world" could no longer ignore the world's nonwhite majority, whose representatives now flowed into the United States as their peoples gained independence from European colonial rule. "How can we persuade these Africans and these Asians, whose skins range from dark to black," State Department official Pedro Sanjuan asked Maryland legislators, "that we believe in human dignity when we deny our own citizens the right to this basic dignity on the basis of skin color?"[2] American foreign relations could not be insulated from the nation's race relations in an era of maximum U.S. involvement abroad.

In the twentieth century the global movement toward racial equality and self-determination gathered speed and finally broke upon the bulwarks of white supremacy with irresistible force. This happened around the world and within the United States at the same time, and the purpose of this book is to trace the ways in which the U.S. government responded to demands for an end to racial discrimination both at home and abroad.[3] Colonialism— mainly European but also Japanese and American—retreated from Asia, Africa, the Caribbean, and the Middle East in the decades following World War II; legal racial segregation in the United States was exorcised in the same years. The two world wars gravely weakened the European empires, and from the ashes of each global conflagration arose a new revolutionary state to inspire and encourage colonial independence: the Soviet Union in 1917 and the People's Republic of China in 1949. Decolonization became, in a sense, the twentieth-century equivalent of emancipation for slavery, doing for nations what had been done a century earlier for individuals. The unfolding of national self-determination across Asia and Africa, in turn, nourished the struggle for equality in America.

The twin efforts of the anticolonial and civil rights movements, and the resistance they encountered from Europeans, white settlers in Africa, and white Southerners, presented the U.S. government with a dilemma. How could it defend equality and the rule of law while not alienating its allies across the Atlantic and local authorities south of the Mason-Dixon line? This conundrum was immensely complicated by the onset of the Cold War. The American practice of the Cold War was grounded in the central belief that the liberal, democratic, capitalist order of the United States represented a more open and humane society than that of Communist states. With the Soviets and the Chinese competing for the friendship of the new nonwhite nations, American policymakers had to thread their way through a minefield of potential disasters in dealing with race and its international implications after 1945.

The essential strategy of American Cold Warriors was to try to manage and control the efforts of racial reformers at home and abroad, thereby minimizing provocation to the forces of white supremacy and colonialism while encouraging gradual change. They hoped effectively to contain racial polarization and build the largest possible multiracial, anti-Communist coalition under American leadership. This effort proved generally more successful at home than abroad. The relatively small percentage of nonwhite Americans could be reasonably accommodated within the flexible structure of American democracy. But the more revolutionary situations in much of the Third World proved harder to control, as even deeper racial divisions of wealth and power alienated nonwhite majorities from their colonial and white settler overlords.[4] The potential for racial conflicts that might derail the American pursuit of a First World–Third World alliance was most evident in southern Africa, where white settlers dug their heels in for a last-stand defense of white privilege. White violence against civil rights organizers in the American South also threatened to limit Third World sympathies with the West in the Cold War.

The story of domestic race relations and American foreign relations over the past half-century has had a particular geography. The narrative encompasses most of the globe, as does the reach of U.S. power, but the states of the old South loom especially large. It was there, south of the Mason-Dixon line and east of the hundredth meridian, that human relations along the color line conflicted most sharply with the nation's pursuit of a "free world" abroad. Much changed in Dixie during the decades after 1945, and the region was always more complicated and diverse in its social patterns than its harshest critics understood. But the states of the former Confederacy did emerge from World War II as a region of abiding racial hierarchy, coercion, and violence, whose white authorities remained firmly entrenched. This set it at odds with most of a broader world that had long endured the inequalities of predominant white power but was now moving rapidly up the road to at least formal racial equality.

The racial discrimination required by Southern state legal codes represented one pole in a growing debate between two visions of sovereignty and human rights. Traditionalists adhered to a doctrine of local authority in managing social relations and maintaining order. This doctrine took the form of defending states' rights to manage their own internal affairs. It also bolstered national sovereignty against inquiries or incursions from international organizations. The second vision embodied the opposite view: that in the wake of the Holocaust, the international community could no longer fully trust any local sovereignty to protect its own citizens. Innocent people should no longer be allowed to suffer and die owing to the sheer misfortune of which government they happened to live under. Instead, standards for human rights needed to be made globally applicable. The United Nations embodied elements of each of these visions, with clauses in its charter declaring both the common rights of all people against discrimination and the sovereignty of nations over life within their borders. In the early postwar years, the U.S. government cited the latter clause in rejecting requests for an international investigation of American racial discrimination—a position shared by European colonial powers and South Africa. A few years later, white Southerners used precisely the same arguments in making their own case against federal investigations of Southern racial practices.[5]

The South's centrality in struggles over American race relations engendered a debate in the early postwar decades about the region's precise relationship to the rest of the country. At issue was the very definition of Americans' core values and the essential identity of the United States. Non-Southern liberals tended to view Dixie as a foreign culture and almost a different country. Less troubled by social hierarchy, less committed to racial equality, and more impressed by the South's religiosity and general conservatism, moderates and especially conservatives saw the region instead as somewhat distinctive but ultimately in line with the rest of the nation. From the opposite end of the political spectrum, radicals agreed; they rejected the idea that non-Southerners should feel superior to the South. Attuned to the reality of de facto segregation and discrimination in the North and West, they tended to agree with Malcolm X's assertion that "Mississippi is anywhere south of the Canadian border."[6]

Critics of Southern race relations, including black Southerners, had abundant ammunition in a culture so marked by rivalry with the USSR. The Cold War developed after World War II as a new bipolar U.S.-Russian competition that encompassed the globe, and strident American anti-Communism presented critics with the image of a totalitarian Soviet Union that bore some resemblance to white-controlled Dixie. A short list of common features of Russia and the American South included the language of justice and the reality of inequality before the law; involuntary labor (peonage and prison labor in the South); kangaroo courts and summary executions; arbitrary imprisonment; the denial of human rights through the use of inhumane prison conditions and even torture; and the state's use of spies and terror to intimidate potential dissidents, including occasional "disappearances." Above all, the role of the government—whether in Moscow and Soviet provincial capitals or in Jackson and Philadelphia, Mississippi—was that of a crucial supporter and perpetrator of such practices.

The American South and the Soviet Union had different histories, of course, and different forms of inequality and injustice. But the parallels between political totalitarianism and racial totalitarianism could occasionally be striking. Each seemed to serve an ideological "higher truth" in its legal system rather than pursuing questions of actual innocence or guilt. Harvey McGehee, the chief justice of the Mississippi Supreme Court, unintentionally illustrated this point soon after that state executed Willie McGee, an African American, in 1951 for the supposed rape of a white woman in Laurel. McGehee's court had upheld the conviction, and he responded revealingly to a journalist's suggestion that evidence of the woman coercing McGee into a sexual relationship might have been suppressed in the case: "If you believe, or are implying, that any white woman in the South, who was not completely down and out, degenerate, degraded and corrupted, could have anything to do with a Negro man, you not only do not know what you are talking about, but you are insulting us, the whole South. You do not know the South, and do not realize that we could not entertain such a proposition; *that we could not even consider it in court.*" Truth, as in the Soviet Union, was often established long before anyone walked into the courtroom. Like many others, Willie McGee paid for that higher truth with his life.[7]

The other primary site for official U.S. anxieties about color and subversion during the Cold War, besides the American South, was Africa. The swift retreat of European colonialism from the Middle East and most of Asia after 1945 left Africa as the center of colonial contention. In contrast to the confusing "browns" and "yellows" Americans saw in the rest of the Third World, the "black and white" racial demographics south of the Sahara fit with their assumptions about their own society, though in reversed proportions. The continent even offered its own Deep South, its own centers of white supremacist resistance that would outlast less determined white colonial authorities farther north: the apartheid regime in South Africa, the white settler government in Rhodesia/Zimbabwe, and the dogged Portuguese rule in Angola and Mozambique. While the rest of sub-Saharan Africa emerged from white control between 1957 and 1964, the region below the Zambezi River remained frozen under white totalitarianism until late in the Cold War.[8]

The African and African American freedom movements encouraged and reinforced each other in ways remarked on at the time and since. The appearance in Washington and New York of dignified nonwhite leaders from the Third World, combined with their respectful treatment by white American officials, made a profound impression on segregated black Americans.[9] Conversely, the model of successful, nonviolent organizing in the American South spread powerfully outward from American shores, its influence washing ashore in freedom struggles in places as disparate as China, Poland, northern Ireland, Palestine/Israel, South Korea, and South Africa.[10] In a world increasingly linked by televised imagery and intercontinental jet travel, American policymakers could not preserve a clear distinction—much as they sometimes wished to—between the social upheaval they faced along the color line at home and in certain parts of the world abroad. The movements for racial equality and self-government that arose among the world's non-white majority during the Cold War were destined to succeed or fail, for the most part, together.[11]

DISCERNING THE ROLE of race relations and racial attitudes in American foreign policy after World War II is not a simple diagnostic exercise. The construction of U.S. relations with the outside world does not work like a mathematical equation, in which a single factor can be removed and the resulting difference in outcomes then calculated precisely. Indeed, racial thinking itself is merely one of several ways to divide the world in hierarchical and pejorative categories; others like ethnicity, class, religion, and gender frequently overlap with race in the minds of responsible officials, making it difficult to isolate completely any one variable.[12] Yet there is abundant evidence that racial considerations were regularly present in the minds of U.S. officials throughout the Cold War, as should hardly be surprising in light of the turmoil surrounding race relations in the United States and the world in those decades. In a nation increasingly absorbed by racial problems while also engaged abroad with a mostly nonwhite world, it seems only reasonable to ask how these two themes were related. The focus here on race should not, of course, suggest a downgrading of the salience of other well-known factors in American foreign policy, like economics or geostrategy, but rather encourage the inclusion of an aspect not previously well understood.

The high stakes of the nuclear-tipped contest between the Soviet Union and the United States promoted institutional and intellectual rigidity while discouraging creative thinking about the period's international history. In retrospect, the conflicts between the great powers of the Northern Hemisphere after 1945 distracted attention from the period's perhaps more significant long-term development: the emergence of the world's nonwhite majority from white colonial rule into national independence. With the democratic elections of 1994 in South Africa, the long era of legalized white rule over people of color—much, much longer than the period of competition between Communism and capitalism—came finally to an end. Continuing differences in the racial distribution of power and wealth confirm the ongoing relevance of this theme to contemporary international history.

Thomas McCormick has observed, "People do not think one way about their national society and a different way about the world society. Instead, they tend to project and internationalize conceptual frameworks first articulated at home."[13] Historians are nearly unanimous in their placing of the civil rights movement "at the center of their history of the postwar United States," according to a review of recent major U.S. history textbooks.[14] Yet scholars of that movement have paid scant attention to its international context and significance.[15] Conversely, even eminent observers of U.S. foreign policy sometimes display a notable lack of concern for domestic realities.[16] Harold Isaacs, by contrast, called attention nearly four decades ago to "the peculiar consonance that has existed between the rise and fall of Western white power and the rise and fall of white racism in America."[17] The issue of race is one of several that can illuminate the ways in which American history in the second half of the twentieth century was also international history.

Some clarity about the use of the word "race" is required. The absence of quotation marks around this term throughout the book reflects an effort at literary felicity, not an acceptance of the still common belief that race refers to a biologically significant characteristic of human beings useful for classifying them in distinct populations. In a society as suffused with anxiety about racial identity and race relations as the contemporary United States, it is worth emphasizing how much race is a social construction. Americans throughout the Cold War and after, like their predecessors, have conflated several distinct physical features—skin color, hair type, various facial features—to create a system of absolute racial identities: white, black, yellow, red, and sometimes brown. Race has become so accepted a part of "common sense" in this culture that it bears repeating here that race is a historicized construction, which other societies in other eras have built differently.[18] The tendency of North Americans and Europeans to wield racial thinking with considerable force and impact in modern history should not be understood as a permanent condition. There was, after all, a time when "civilized" peoples of the Mediterranean region viewed northern Europeans as "deficient in skill and intelligence," in Plato's words, and Britons as "utterly stupid and incapable of learning," according to Cicero.[19]

In the course of the twentieth century, American racial classifications tended to become more rigid, although Americans remained uncertain sometimes about how to apply them. Ironically, the scientific community moved in the opposite direction by refuting previous assumptions of significant biological differences.[20] The status of Jews suggested continuing malleability about race, from the 1940s—when they were viewed primarily as a separate race, having an interfaith marriage rate of 3 percent and suffering intense discrimination—to the end of the century, when they had been thoroughly assimilated into the category "white" with an interfaith marriage rate of over 50 percent.[21] The legal status of immigrants from India remained unclear until the Supreme Court ruled in 1923 that they were Caucasians but still not "white persons," and thus not eligible for naturalization.[22] So few blacks lived in Hawaii in the early twentieth century that the census classified them as "Puerto Ricans," while Puerto Ricans were in turn defined as "Caucasians" and thus—unlike Indians—as whites. In other words, along Waikiki Beach, blacks were white.[23] Russians troubled American racial categorizers throughout the Cold War, appearing to some as Europeans and to others as Asians or half-Asians.[24] The anxiety produced by such uncertain identities perhaps helps explain the sign over one east Texas town: "The Blackest Land—The Whitest People."[25] And it is crucial for understanding the American institution of the "one-drop rule," whereby having any African heritage at all came to define a person as black.[26]

The absurdities of the one-drop rule were legion. Walter White, long-time president of the National Association for the Advancement of Colored People (NAACP), had fair skin, fair hair, and blue eyes. Like his parents, he appeared white, but like them, he chose to identify with the tiny percentage of his genetic ancestry that was not European. When he traveled abroad with his second wife, a white brunette, and they were introduced as an interracial couple, puzzled foreigners often asked him how he happened to marry a "colored" woman.[27] African American diplomat Ralph Bunche headed the 1960 United Nations mission in the Congo, where Americans assumed his color would be an asset. But he was so light-skinned that the distinction was lost on the Congolese, who tended to consider him just another European. Black in America, he was white in central Africa.[28] Even Malcolm X, the exemplar of black pride and, for most of his career, black separatism, embodied such ironies. His relatively light skin tone resulted in the nickname Red in his early years. During a visit to Kenya in 1964, Malcolm accompanied President Jomo Kenyatta to the Nairobi racetrack, where he was noticed from afar by U.S. ambassador William Attwood as "a white man" in the official entourage. Only as Attwood approached the presidential box could he see who the guest was—and thus what color he "really" was.[29]

Race, then, means "race." But the artificiality of the term and the often invidious meanings imputed to it do nothing to reduce the importance of those meanings to Americans in the postwar era. Most Americans, including policymakers in Washington, viewed the world in explicitly race-conscious terms, and U.S. officials made decisions that affected the lives of hundreds of millions of people partly on the basis of racial assumptions. This is the story of how those racial lenses helped shape U.S. relations with the outside world in the era of American dominance in the international sphere.

1

RACE AND FOREIGN RELATIONS BEFORE 1945

ALONG A RIVERBANK in Panama in 1577, Spanish soldiers hunting for their piratical English rivals finally caught up with Sir Francis Drake's lieutenant John Oxenham and his fifty comrades. The English had allied themselves with the considerable local force of Cimarones, enslaved Africans who had escaped from their Spanish masters. Together they raided the gold and silver shipments bound for Spain from the mines of Peru. These Protestant Englishmen were intensely hostile to the Catholic Spanish, who returned the sentiment by hauling them off to be executed, but they evidently enjoyed the company of the Cimarones. What the Spanish interrupted that day along the river was an African-English encampment and barbecue, replete with large quantities of roasting pork and abundant good humor and fellowship. The Spanish described the black and white allies as "amusing themselves together"—in contemporary terms, they were partying.[1]

The foreign relations of the United States, the country that emerged two hundred years later from the colonies established by Drake's countrymen, have always involved relations between peoples of different skin colors. Natives of three continents—Europe, Africa, and North America—gathered in the land that became the United States, later to be joined by immigrants from Asia. White appropriation of black labor and red land formed two of the fundamental contours of the new nation's development and its primary sources of wealth. Like the enslavement of African workers for two and a half centuries, wars of Indian removal helped define American society and American foreign relations from the early seventeenth century to the late nineteenth. The savagery inherent in the institution of human bondage and in the frequently exterminationist campaigns on the frontier assured that their poisonous legacy would continue to afflict American life throughout the twentieth century. Slavery and westward expansion wove together issues of race relations and foreign relations from the very beginning of American history.[2]

The subjugation and destruction of people of color rests at the center of the early American experience, in great tension with the egalitarian, democratic tradition of the American Revolution. The dominant power of white Americans in most of their interactions with people of color has meant that, for most of American history, white political and military superiority has been an imposing and aggressive force in the shaping of these relations. Francis Drake himself, after all, was involved in the African slave trade. But scenes like those of the multiracial barbecue in Panama suggest that from the earliest days of its colonial past, the United States has harbored competing inclinations among its white majority: toward racial hierarchy and domination, on the one hand, and toward equality and color blindness on the other. The oscillation between those two poles has helped define the manner in which racial perceptions and attitudes have always been a factor in the American experience.

The Color-Coded Expansionism of the 1800s

During the first half of the nineteenth century, American expansionism acquired an increasingly explicit sense of racial destiny. White Americans in these decades shared the growing European scientific fascination with the meanings of human difference and racial classification. Popular American conceptions of the nation's future in the decades immediately after the Revolution focused on the ideological power of republican liberty. Although assumptions about "Anglo-Saxon" superiority and the inevitable disappearance of "inferior" peoples were present in 1800, they did not become dominant in the language of expansion until the middle of the century. Only gradually did a specifically racial view of human destiny overcome a more egalitarian political perspective in post–Civil War America. As North and South reconciled, black and white became increasingly segregated. As renewed racial hierarchy set in at home, it weakened the perception of the United States as a model of liberty for the rest of the world to follow.

A racial perspective pervaded the debates about annexing Texas and going to war with Mexico in the 1840s. Most white Americans saw Mexicans as a people degenerated by a racially mixed heritage of Spaniards and Native

Americans, who could neither manage nor improve the vast territories they held. They stood in the way of U.S. settlers eager to occupy new lands to the west. The annexation of Texas and the rapid conquest of the land that became the American Southwest came to be seen by most white Americans as the "manifest destiny" of a racially superior people.[3]

Profound disagreements over whether to carry slavery into new territories and states in the West helped bring on the Civil War in 1861, as domestic race relations and national expansion remained intertwined. After the brief egalitarian hope of Reconstruction, white Americans put their regional divisions largely behind them in rebuilding racial hierarchies in the form of Jim Crow segregation and renewed campaigns against Western Indians. This other reconstruction—of unified white authority over nonwhite peoples—placed the United States squarely within a broader global pattern of the expansion of white power. The final decades of the nineteenth century and the first decade of the twentieth marked the zenith of European and North American imperial might, at the end of which only a fifth of the world's land remained outside the control of Europeans or the descendants of European settlers. Racial theorist Lothrop Stoddard in 1920 referred back to the turn of the century as "the high-water mark of the white tide which had been flooding for four hundred years." The reestablishment of white dominion in the American South and its final assertion in the American West came in precisely the same years that Europeans at last brought the interior of Africa under their subjugation. American immigration policies in these years reflected the broader trend of white settler states, including Canada, New Zealand, Australia, and South Africa, to define citizenship in racial terms, with non-Europeans ineligible for naturalization. Racial homogeneity seemed a defining characteristic of great nations like England and France.[4]

The growth and consolidation of white power at home and abroad did not seem accidental or unfortunate to most Americans. The legitimacy that white Americans accorded to notions of white supremacy was reflected in the rise of Social Darwinist thought, which proclaimed Europeans and their descendants as fittest to survive among the races, and Anglo-Saxonist thinking, which posited the preeminence of an even narrower British race. The common tendency was to rank racial groups by their supposed relative merit, with northwestern Europeans and their descendants at the top and others trailing off through darkening skin shades to Africans and their descendants at the bottom. Such a view of racial categorization around the world precisely reflected domestic attitudes.[5] The trend toward dividing up the human population went so far as to threaten the concept of a single white race, as Anglo-Saxonist proponents conceived of several white races: Anglo-Saxons, Teutons, Latins, Celts, and so on, down to Slavs, who were seen as partly Asian. This racialized vision of humanity emerged in the rising nativist sentiment by the turn of the century against immigrants from southern and eastern Europe. "They are beaten men from beaten races," announced the president of the Immigration Restriction League in 1894.[6]

With many Europeans not even qualifying as fully white, native Americans and African Americans fell much further down the scale. Indians west of the Mississippi River faced an onslaught of white settlers backed by the military power of a reunified federal government. The alternative to the U.S. Indian Bureau's reservation system, with its emphasis on the eradication of Indian culture, was warfare.[7] Regardless of the extent of antislavery motivation among Union troops in the Civil War, soldiers in blue were shipped west immediately thereafter to vanquish native Americans. "The more we can kill this year, the less will have to be killed [in] the next war," declared General William T. Sherman, conqueror of much of the South and the first postwar commander in the West in 1868, "for the more I see of these Indians the more convinced I am that all have to be killed or be maintained as a species of pauper."[8] Theodore Roosevelt summarized the common assumption that underlay the often exterminationist thrust of anti-Indian warfare: "The most vicious cowboy has more moral principle than the average Indian."[9] The subjugation of native Americans and their treatment as dependent wards of the state laid the groundwork for parallel U.S. actions in the Philippines a few years later. Each represented a form of colonialism: conquering and controlling culturally different peoples, and ruling them as subjects rather than incorporating them as citizens.[10] Domestic and foreign policies regarding people of color developed as two sides of the same coin.

The African American experience of segregation after Reconstruction reflected the importance of black labor in the Southern agricultural economy, in contrast to the resentment of Indian presence on valuable Western lands. But the goal was control of both nonwhite workers and nonwhite lands, just as it was in European overseas colonies in the same years.[11] While most white residents of the Confederate states accepted their defeat and reincorporation into the Union, some plantation owners refused and left. Of these, many settled in Brazil, where slavery still existed and expatriates could reconstruct racial hierarchy and the plantation economic system in their fullest form.[12] White supremacists thus indicated an international consciousness, just as they would a hundred years later in looking to Rhodesia and South Africa for support. Those white Southerners who did not leave after the Civil War succeeded in establishing a rigid system of racial segregation confirmed by the Supreme Court in the *Plessy v. Ferguson* case of 1896. In conjunction with the expropriation of Indian lands, this decision laid the groundwork for the treatment of nonwhite inhabitants of the overseas territories that the United States seized two years later in the war against Spain. Only Hawaii and Alaska among those territories with dense populations of people of color would become states, as the U.S. government sought to preserve the whiteness of its citizenry.[13]

Two prominent figures of the late nineteenth century represented to the American public the meanings and importance of white authority. Henry Morton Stanley and Theodore Roosevelt each understood race relations in international terms, tying the story of black and native Americans to that of Africans. Through the lives of these two men, Americans could link the exploration and conquest of two of the last great mysterious wilderness frontiers: the interior American West and central Africa. The fame of Stanley, the lesser known of the two, derived from his successful effort to track down missing British missionary David Livingstone in 1871 near Lake Tanganyika, and his epic three-year transcontinental exploration of the headwaters of the Congo River, which brought him, exhausted, to the Atlantic Ocean in 1877. He was subsequently involved in the creation of the Congo Free State, a colony infamous for its brutality toward the indigenous inhabitants. A Welshman by origin, as a young man Stanley came to Louisiana, where he observed plantation slavery at close range. He fought in the Confederate army and was taken prisoner at Shiloh. After the war he became a star reporter for the *New York Herald,* covering first the Indian wars in Colorado in 1866–67 and then the British punitive campaign into Abyssinia later in 1867. Stanley's subsequent reputation for ferocity among Africans—his nickname in the Congo was Breaker of Rocks, and he left his bloody mark across the continent—built upon earlier experiences with American slavery and frontier wars. There he had first learned the place of people of color in an era of white dominion, and he applied that lesson in explicit fashion, even quoting to Africans from speeches that General Sherman had made to Native Americans. Stanley represented the truly international sweep of white authority in the decades after the Civil War.[14]

Like Stanley, Roosevelt built on experiences with Jim Crow and frontier warfare in the American West in understanding Africans as an amalgam of black and red Americans. Roosevelt's life span of 1858–1919 coincided with the era of global white supremacy, and his fascination with race and racial differences reflected the Victorian world in which he lived. Though a New Yorker by origin, he learned sympathy for white Southerners and even for slavery from his Southern-born mother, as well as hatred for Indians from his years in the Dakota Territory in the early 1880s. He opposed lynching but shared common white assumptions about the subversive possibilities of African Americans and the importance of restricting non-European immigration. By the time he landed in Africa in 1909 at the head of a Smithsonian-sponsored safari, the former president embodied white American views of "natives"—whether red or black—as both dangerous and less interesting than exotic animals. Roosevelt, like Stanley, helped teach Americans about the international character of white rule over people of color.[15] In a popularized version, the Tarzan novels of Edgar Rice Burroughs that began to appear in 1912 did the same. These stories of white heroism in the jungle paralleled tales of white courage on the American frontier. Tarzan's identity as a British lord did not diminish the sense of a Davy Crockett adventure transported to central Africa in these quintessentially American stories.[16]

Issues of race pervaded the American acquisition of an overseas empire after the defeat of Spain in 1898. Most imperialists and anti-imperialists shared a common view of the world as divided between civilization and barbarism, with those possessing lighter skin in the former camp and those with darker skin in the latter.[17] Where they differed was over how, precisely, the United States should engage with the lands it had seized in the course of the fighting—the Philippines, Cuba, Puerto Rico, and the Virgin Islands, as well as Hawaii. In winning the argument regarding the Philippines, annexationists succeeded in convincing enough white Americans that the domestic racial status quo would not be upset. Anxieties about increasing the nation's non-European population had contributed to the initial refusal to annex Hawaii to the United States after the white planters' rebellion in 1893. The final decision five years later to make Hawaii an American territory reflected a growing determination to control Asian immigration to the islands to put them on a path toward someday forming a predominantly white state.

One major issue of the congressional debates of 1897–98 demonstrated the unstable and malleable definition of race, even as U.S. policymakers used this idea in powerful ways: were Hawaiian residents of Portuguese origin or descent white? According to the 1896 census, Americans made up 3 percent of the islands' population, non-Portuguese Europeans 7 percent, and Portuguese 15 percent; the rest were Japanese, Chinese, or native Hawaiian. Annexationists said yes, of course the Portuguese were European and thus white, boosting the whiteness of the territory. Anti-imperialists said no, they were not really white, lowering the whiteness quotient and the attractiveness of annexation to most white Americans. Americans spoke forcefully to one another about race, even if they were not always referring to the same category of distinction.[18]

U.S. policy toward its new colonies abroad after 1899 derived directly from its prior policies toward Native Americans. The administration of racially and culturally different peoples as dependent wards of the state without political representation marked an extension across water of long-standing governmental practices in the American West. The unanimity of white American opinion about the rightness of subjugating Indians made this precedent a powerful argument for annexationists in Congress in 1899. If those opposing the acquisition of the Philippines were right, Senator Henry Cabot Lodge declared, "then our whole past record of expansion is a crime." Roosevelt wrote to a friend that if whites were "morally bound to abandon the Philippines, we were also morally bound to abandon Arizona to the Apaches." Apaches had few white defenders at the turn of the century, and the American frontier with indigenous peoples pushed westward again, into the far reaches of the Pacific Ocean.[19]

Filipino resistance to American annexation confronted U.S. policymakers with a repetition of the problems they had just finished facing in Indian lands at home, and their response was the same. Both officers and soldiers of the American army had considerable personal experience in Indian fighting, and their racially charged language and behavior in the Philippines reflected the belief that they were engaged in a similar struggle.[20] President William McKinley spoke of Filipinos needing to be disciplined "with firmness if need be, but without severity so far as may be possible."[21] The ordeal of insurgent and counterinsurgent warfare soon eliminated most restraints, however, and produced considerable severity. From the ready use of torture and the killing of prisoners, some American soldiers descended into sheer racial slaughter. They did so at the command of their superiors, such as General Jacob Smith, who ordered his men to kill all hostile Filipinos over the age of nine: "The interior of [the island of] Samar must be made a howling wilderness."[22] A. A. Barnes, an American infantryman, expressed a racially focused rage in a letter home to his brother: "I am probably growing hard-hearted, for I am in my glory when I can sight my gun on some dark skin and pull the trigger."[23]

Some white Americans were troubled by the course of the war in the Philippines, but many openly accepted it as one of the unfortunate burdens of civilization. Charles Francis Adams, Jr., the former president of the Union Pacific Railroad and president of the Massachusetts Historical Society, explained this point of view in a public address in Lexington in December 1898. Reviewing the violent history of white interactions with people of color on the North American continent over the previous three centuries, Adams called it "a process of extermination." But, he concluded, that process was also "the salvation of the race. It has saved the Anglo-Saxon stock from being a nation of half-breeds."[24] U.S. actions in the Philippines fit into a broader global pattern of imperial severity as European powers pressed further into Africa and Asia and encountered resistance. The massive slaughter of Congolese by the Belgians, the attempted extermination of the Hereros in Namibia by the Germans, and the destruction of the Boxers in China by an international force all unfolded in these same years. Like other American elites of the time, Roosevelt understood American behavior in this global context: "The rude, fierce settler who drives the savage from the land lays all civilized mankind under a debt to him. American and Indian, Boer and Zulu, Cossack and Tartar, New Zealander and Maori—in each case the victor, horrible though many of his deeds are, has laid deep the foundations for the future greatness of a mighty people."[25]

Unlike Germans and Belgians who engaged in large-scale racial violence beyond their homogenous homelands out on the fringes of their empires, Americans constructed their campaigns against the Filipinos out of a set of experiences at home in a multiracial, hierarchical society. Indian wars provided only part of that domestic framework; the ongoing campaign of subjugation of African Americans provided most of the rest. "We do not have to go to Luzon for American barbarities," one Brooklyn minister noted in 1899, pointing to the chartering of an excursion train to carry aficionados from Atlanta to the well-advertised lynching of Sam Hose in the town of Newnan, Georgia, from which some returned with souvenirs of his charred and mutilated body.[26] The years of U.S. warfare in the Philippines, 1899 to 1902, encompassed a wave of mob violence against African Americans at home. In addition to lynchings in the rural South, the terror included race riots in New Orleans, New York City, Akron, Ohio, and Wilmington, North Carolina. Black soldiers stationed at bases in the South drew the particular ire of white soldiers and civilians who, just thirty-three years away from slavery, were enraged by the authority implied in the uniforms and weapons. The racial coding of deadly conflict on the periphery bore similarity to that in the center of the American empire.[27]

African American soldiers fighting in the Philippines found themselves caught between powerful, conflicting currents of identity. Despite their cultural and linguistic differences with Filipinos, many of them felt some admiration for the islanders' nationalist resistance to American power. Some also acknowledged a sense of racial commonality as people of color, an inclination that grew as they listened to their white colleagues refer to Filipinos as "niggers" and as the insurgents appealed to them to stop fighting in a "white man's war." A small number even deserted and went over to the other side, finding the elements of racial conflict in the war ultimately more significant than their common ground with white Americans' economic and military expansion. The vast majority of black soldiers stood strong in their American identity, however, even as they evinced little enthusiasm for the idea of the "white man's burden" and the more gruesome forms of racially tinged counterinsurgent warfare. Hoping their courage and patriotism on the field of battle would translate into better treatment of African Americans at home, they swallowed their anger at the spread of segregated facilities in the American-controlled areas of the Philippines and the other territories taken from Spain.[28]

In this process, black soldiers took encouragement from small signs of progress among their paler countrymen. Captain W. B. Roberts wrote home from Santiago, Cuba, in the fall of 1898 to report an incident where the American proprietor of a hotel restaurant had tried to eject him on racial grounds. "The dining room was full of officers and others, and you could have heard a pin fall" when Roberts loudly declared that he and other black soldiers had spilled their blood on San Juan Hill to protect American businessmen like him and would accept no such discrimination. Before the restaurateur could respond, a white general got up from his table, walked over, and shook Roberts's hand, saying, "Come, Captain, take my seat; and you, Mr. Hotel Proprietor, get it quick; and I don't want to hear any more of this damn foolishness with these officers of mine!"[29] Such outcomes were unusual in a period when the outward expansion of the American empire usually served to encourage more intense discrimination against Americans of color. But the logic of color-blind equality—that is, the logic of democracy—persistently shadowed the reality of Jim Crow, articulated by darker Americans determined to make real the promises of the Declaration of Independence. The growing American engagement with a mostly nonwhite world pointed toward an era when many more white Americans would be forced to confront and resolve the contradiction between liberty and racial hierarchy.

White Dominion and Its Challengers in the Early 1900s

The U.S. war against the Philippines overlapped precisely with the British war of 1899–1902 against the South African Boers, the farmers of Dutch descent seeking to preserve their independence from the expanding British empire. Americans' response to events in southern Africa revealed their ambivalence about race and its meanings. Most Americans sympathized with the Boers, seeing parallels to their own revolutionary past as white settlers on a new continent far from Europe. Some even volunteered to fight alongside the Boers, especially Irish Americans resentful of continued British control of their own homeland. The U.S. government, however, strongly supported the British. The administrations of both William McKinley and Theodore Roosevelt viewed the British empire as a civilizing and modernizing force, bringing law, liberty, and commerce in its wake, in contrast to the white South Africans, whom they saw as traditional, hidebound rural folk—figures from the past rather than the future. Both presidents spoke of the conflict as a racial one between more advanced Anglo-Saxons and more retrograde descendants of the Dutch. A European or even a northwestern European identity was not sufficient to put one in the front ranks of whites. The Dutch and Irish origins of men with the names "Roosevelt" and "McKinley" added a particular irony to the presidents' emphasis on the supposed biological superiority of the British; Roosevelt did at least argue that the Dutch "down at the bottom have the great basic virtues of the Teutonic races." Sure that race was a critical method of distinguishing among peoples, Americans nevertheless demonstrated ongoing uncertainty about precisely who fell into which category.[30]

Arguments about the relative genetic merits of Europeans paled, of course, beside confident assertions of non-European inferiority. Almost completely ignored in American discussions of the Boer War were the vast majority of South Africans, who were black. Social Darwinism offered a vision of weaker peoples of all shades inevitably disappearing before the relentless march of stronger peoples. A supposedly unavoidable course of history offered a clear case of ends justifying means, when it came to matters of tactics in pacification campaigns on the fringes of empire. Professor Washburn Hopkins of Yale asked in the pages of *Forum* in 1900, "Is there not here a counterpart in the moral world to the inflexible severity of physical laws, whereby the maintenance of the race is upheld at the sacrifice of individuals?" However excruciating the forms of that sacrifice might be, it accorded with "the higher law of racial superiority" that Hopkins and so many other white Americans saw at work in the Philippines and in southern Africa—and in the American South—at the dawn of the twentieth century.[31]

In the same years that the United States and the European powers were extending the reach of white authority around the globe, signs of antiracist and anticolonial organizing began to twinkle through the imperial gloom, pointing the way to the coming decline rather than triumph of worldwide white supremacy. The first dramatic breakthrough came on a battlefield at Adowa in Ethiopia in March 1896. There, in the same year that the U.S. Supreme Court put its stamp of approval on the South's system of segregation in the *Plessy v. Ferguson* decision, soldiers of King Menelik II's Ethiopian army destroyed an invading Italian army. This first modern military defeat of Europeans by non-Europeans preserved Ethiopia as the one great exception in Africa in the early twentieth century: an independent nation. Out of one part of Africa came hope for people of color everywhere, and fear for white supremacists.[32]

The second sign of trouble for white dominion appeared in northeast Asia. The Japanese destruction of the Russian fleet at the Battle of the Japan Sea in 1905 ended the Russo-Japanese War and represented the entry of the first nonwhite nation into the elite ranks of industrialized great powers. Some Europeans had reassured themselves after Adowa that Italy had had only a second-rate army, but the Russian empire was an entirely different matter. Observers around the world interpreted Japan's success in racial terms, as a victory of Asia over Europe. On a boat crossing the Suez Canal soon after this news arrived, Chinese revolutionary Sun Yat-sen found himself besieged by enthusiastic Egyptians eager to celebrate their anti-European solidarity, even after they realized he was not Japanese.[33] Politically attentive African Americans took vicarious pleasure in the defeat of the overconfident white Russian troops in Manchuria.[34] Most Europeans, however, heard the news from the Sea of Japan with foreboding. A French reporter in Japan described the arrival of captured Russian soldiers: "Whites, vanquished and captives, defiling before those free and triumphant yellows—this was not Russia beaten by Japan, not the defeat of one nation by another; it was something new, enormous, prodigious . . . it was the awakening hope of the Oriental peoples, it was the first blow given to the other race, to that accursed race of the West."[35]

Even the British, formally allied with Japan since 1902, were uneasy about an Asian nation defeating a fellow European power. Roosevelt helped mediate peace terms between Japan and Russia in part to preserve the balance of power in northeast Asia and thereby prevent what he called "the creation of either a yellow peril or a Slav peril."[36]

The military defeats of Europeans in Africa and Asia fit into a broader pattern of rising resistance to the spread of white power in the first years of the twentieth century. Those ruled by colonialism or otherwise discriminated against increasingly sought national, ethnic, or racial self-determination. In the intellectual realm, the publication in 1911 of *The Mind of Primitive Man*, by American anthropologist Franz Boas, signaled the coming retreat of scientific racism, with its rejection of biological theories of white racial superiority. In the political realm, anticolonial agitation in India and elsewhere accelerated, and the first of a string of Pan-African Congresses was convened in 1900. Eastern European Jews, stigmatized as racially distinctive, responded to continuing anti-Jewish violence by organizing the Zionist movement for liberation. African Americans and their white supporters formed the National Association for the Advancement of Colored People in 1909 in reaction to intensifying racial discrimination in the United States. And Africans surveying the grim realities of the new dispensation of joint British-Afrikaner control in South Africa created the African National Congress in 1912. The outlines of a global struggle for racial equality were visible by the time World War I exploded across the European landscape.[37]

The carnage on the fields of Europe between 1914 and 1918 struck a mortal blow against the certainty of Europeans that theirs was the highest civilization to which all others should aspire. World War I appeared to many as a civil war among whites, especially in light of centuries of assertions about Europe as a unified culture superior to the non-European world it had colonized. Not only were Germans, French, and British slaughtering one another in unprecedented numbers, but they were even using colonial troops from Asia and Africa against each other. The "white race" seemed to be engaged in a kind of suicide, many observers noted, some with sadness and some with hope. Nation clearly ranked above race, revealing huge cracks in the edifice of white authority. Soldiers of color paid close attention, whether they came from India, Senegal, or Mississippi. People who had lived all their lives under the power of white supremacy could not avoid being impressed by the deaths of millions of white people. Most nonwhite troops who survived the horrors of modern warfare returned home less willing to be intimidated by white civilians. Young African American veterans, the *Chicago Defender* observed, "are not content to move along the line of least resistance as did their sires."[38]

Waves of new immigrants over the previous generation had combined with the processes of industrialization to create a more diverse and more contentious American society. Ethnic tensions in the United States swelled as Europe descended into war. People with ties to Ireland, Germany, Eastern Europe, and Russian Jewry observed the conflict with different emotions than did those with links to England, France, Italy, and gentile Russia. The administration of Woodrow Wilson kept the United States out of the war for the first three years, in part because of such conflicting public sentiments. Wilson's sympathy for the British side of the conflict was no mystery, however, and the start of U.S. belligerency in 1917 led to an outbreak of intense repression of domestic dissent. Wilson's ambassador to England, Walter Hines Page, responded with rage to the hopes of many Irish Americans that a defeat of England would lead to Irish independence: "We Americans have got to hang our Irish agitators and shoot our hyphenates and bring up our children with reverence for English history and in the awe of English literature."[39] Wilson's Southern roots and strong segregationist commitments also did not incline him to sympathy for the major revolutionary movements of the World War I decade, in Mexico, China, and Russia. He tended to view Mexicans as somewhat childlike and in need of American tutoring to learn how to govern themselves responsibly. His Yankee secretary of state, Robert Lansing, wrote to a friend about the loathsomeness of having to "swallow one's pride and to keep ladling out soothing syrup to those Greasers [Mexicans] while they smiled sarcastically and kept on with their insults."[40] Lansing blamed much of the Bolshevism that took power in Russia on the supposed Asian influences in that vast, bicontinental nation, equating Lenin's rule with "the Asiatic despotism of the early Tsars."[41]

American entry into World War I led to an extraordinary outbreak of violence against African Americans. The delicate balance of domestic race relations was upended by the war hysteria of thousands of white Americans, who were troubled by black migration to Northern cities for defense industry work. From the summer of 1917 through 1919, white mobs across the nation beat, shot, burned, and mutilated black Americans with impunity. The incidents ranged from rural lynchings across the South to full-scale riots in cities around the country, including the ferocious slaughters in East St. Louis and Chicago.[42] The contrast of these actions with Wilson's proclaimed cause of expanding democracy abroad was stark. Howard University professor Kelly

Miller asked in an open letter to the president: "Why democratize the nations of the earth if it leads them to delight in the burning of human beings
after the manner of Springfield, Waco, Memphis, and East St. Louis, while
the nation looks helplessly on?" Eight thousand African Americans marched
down New York's Fifth Avenue in protest.[43] America's closest allies in the war,
England and France, experienced smaller-scale versions of the same pattern.
Whites assaulted blacks in Liverpool and Cardiff, while nonwhite workers
imported from the colonies to help in France's war industries were carefully
segregated and frequently abused. White male anxieties about social contacts
of men of color with white women figured prominently in the patterns of
discrimination in all three countries.[44] The French government, however,
readily used African colonial troops against Germany, including a large number in the occupation of the Rhineland in 1923. Germans, whose imperial
army had engaged in the near extermination of an entire African people in
Namibia just nineteen years earlier, apparently did not appreciate this small
bit of historical justice. They bitterly resented black authority and called for
the removal of African soldiers not just for Germany's sake but also "for civilization and the white race."[45] By dividing the white nations and drawing in so
many non-European combatants, the war threatened the survival of global
white rule.

The extensive participation of African Americans in their nation's cause in
World War I created novel problems for U.S. commanders. The army con-
scripted a slightly higher percentage of blacks than whites, and then sent
many of them off to France, a country that U.S. commanders feared would
breach American racial boundaries. As a result, U.S. Army authorities campaigned strenuously to convince French soldiers and commanders not to
treat darker-skinned U.S. soldiers as equals—especially not to eat, shake
hands, or socialize with them and not to praise them too highly in front of
lighter-skinned U.S. soldiers. American authorities exuded particular anxiety
about the likely liaisons between black GIs and French women, given the relationships that commonly develop between male soldiers and some female
civilians during wars. Such links, it was feared, could lead to the breakdown
of army morale abroad and segregation at home after the war. Wilson himself
worried in 1919 that decent French treatment of African American soldiers
"has gone to their heads." The American campaign to extend Jim Crow
across the Atlantic met with only limited success, as most French were delighted to have Americans of any color fighting to defend them against Germany. And the problem turned out to be French men more than women:

French commanders generally treated African American soldiers as equal to other soldiers, as they did their own colonial troops. This was a novel experience of equal treatment for black Americans, precisely what their white commanders and colleagues had feared.[46]

Black veterans returned home from Europe with guarded hopes that their patriotic service and sacrifices would help initiate a new era of diminishing discrimination. The white hostility they encountered back home suggested the opposite. Walter Jones recalled many years later, "They told us to 'go catch the Kaiser and everything'll be all right.' We went over there and fought and the first thing I heard when I got back to Waco, Texas, was a white man telling me to move out of the train station. He said, 'Nigger, you ain't in France no more, you're in America.' He didn't even give me time to take off the uniform."[47] During the war years the urban black working class had grown considerably, as African Americans left the rural South for better opportunities in booming war industries. This brought them into volatile situations as a wave of labor conflict and political radicalism swept through the country from 1918 to 1920. Conservative whites viewed blacks as especially inclined to subversion or to manipulation by revolutionaries because of their secondary status. Army intelligence officers kept a close eye after 1917 on "Negro unrest," and Southern leaders warned African Americans against seeking equality. "This is a white man's country," warned South Carolina Congressman and future Secretary of State James F. Byrnes, "and will always remain a white man's country."[48]

Wilson's determination to create a League of Nations at the postwar settlement at Versailles revealed the close ties between domestic American race relations and U.S. foreign policy. Japan responded in the spirit of Wilson's Fourteen Points by proposing a clause in the league's charter that would establish the equality of all people regardless of race. Despite having the approval of a majority of delegates at the peace conference, the proposal died when the president, supported by Australia and Britain, declared that there were "too serious objections on the part of some of us" to include it in the league covenant. Racial equality was still a radical concept for most Americans in 1919, especially Southern segregationists and Westerners opposed to Asian immigration, who lobbied Wilson against the proposal.[49] Philip Kerr, the secretary to Prime Minister David Lloyd George, reminded U.S. diplomats of American vulnerability on matters of race when he directed them to ignore Irish American lobbying efforts regarding Irish independence; the United States had no more right to ask that Ireland speak at Versailles, Kerr declared, than a "delegation of Negroes from America" could demand from England the right to appear there and argue for relief because "they had been disenfranchised in the South."[50]

Japanese leaders resented the rejection of their proposed clause, just as they did the ongoing discrimination against Japanese immigrants in the United States. Former prime minister Shigenobu Okuma warned that "the problems of population and race will in the future form the hardest and most important issues between nations." Echoing W. E. B. Du Bois's famous dictum about the color line being the central problem of the twentieth century, Okuma concluded: "The satisfactory solution of these baffling problems is the responsibility of twentieth[-]century diplomacy."[51] People of color everywhere shared Japan's disappointment in the U.S. government's commitment to racial discrimination and the support the United States received from Britain and the white-ruled Dominions. Organizers of the Pan-African Congress in Paris in 1919 ran into opposition from the State Department, which denied passports to many Americans trying to participate in what was seen to be a radical, international, antiracist movement.[52]

For many opponents of the League of Nations, however, Wilson's restraint did not adequately protect white supremacy from international pressure. Senator James Reed of Missouri ridiculed the idea of international sovereignty: "Think of submitting questions involving the very life of the United States to a tribunal on which a nigger from Liberia, a nigger from Honduras, a nigger from India . . . each have votes equal to that of the great United States."[53] America's emergence from World War I as a global power had brought the country's internal conflicts into the full daylight of the international arena.

In the two decades between the end of one world war and the beginning of another whites attempted to reassert traditional racial hierarchies while there began a movement toward racial equality that would flower after 1945. The domestic violence of World War I slid over into the 1920s, with lynchings and race riots continuing long after the guns had fallen silent in Europe and workers had retreated from the barricades of 1919. As African American calls for greater civil liberties escalated, segregationist forces increasingly resorted to violence. Entire black communities reeled from the onslaught of armed whites, often instigated by the newly revived Ku Klux Klan: more than a hundred and fifty black residents of Tulsa were murdered by marauding mobs on 1 June 1921, and the all-black town of Rosewood, Florida, burned to the ground. The deaths of some fifty whites in the Tulsa riot illustrated the determination of African Americans to defend themselves—"I ain't looking for trouble, but if it comes my way I ain't dodging," one war veteran said—but they also suggested the risks whites were willing to run to assert their racial dominance.[54] Three thousand whites, including many undergraduates at the University of Georgia, savored the torture and slow burning of an African American in Athens in February 1921, going so far as to hunt for souvenirs among the charred remains—one of more than fifty lynchings in Georgia during the previous four years.[55] Yet the antilynching legislation introduced in the U.S. Congress in 1937 never overcame the opposition of powerful conservatives and Southern lawmakers, who viewed it as an opening wedge of federal intervention in the "Southern way of life."[56] Federal law-enforcement officials such as those in the Federal Bureau of Investigation (FBI) showed little interest in the South's peculiar forms of justice.[57]

New restrictions on immigration after World War I reflected broad white anxieties about controlling people of color and culturally distinct European newcomers. Sentiment against blacks in the South, Asians and Mexicans in the West, and immigrants from southern and eastern Europe in the Northeast flowed together in a new nativist spirit in the 1920s. The Immigration Act of 1924 created a system of "national origins" to reduce radically the number of new Americans and to ensure that none would be Asians and few would be Europeans from south of the Alps or east of Berlin. U.S. policy fit into a global pattern in the 1920s of white settler states erecting walls against nonwhite immigrants, as was done in South Africa, Australia, Canada, Southern Rhodesia, and New Zealand. The victory of fascism in Italy and the growth of the Nazi Party in Germany further illustrated the enthusiasm and vitality of white racism beyond American shores in the wake of the Great War.[58]

Nowhere in America's relationship with the outside world did racial attitudes loom so large as with Africa. American films in the interwar years revealed a fascination with African wildlife and topography and a disdain for the continent's peoples. Movies of the 1930s like *Trader Horn* and *King Solomon's Mines* bolstered white views of supposed African savagery. The first of a wildly popular series of films, *Tarzan the Apeman,* appeared in 1932, confirming casual American assumptions of white superiority and cultivation. In the minds of almost all white Americans, there could not have been a wider gap between the virtues and refinement of European civilization and the primitive and savage lands of Africa. Such a vision of ancestral continents was closely tied to the assumptions of white cultural and technological progress in the United States; Americans of African descent, it seemed, could hardly rank intellectually or morally with Americans of European descent. The most popular film of the interwar era, *Gone with the Wind* (1939), marked the end of this period with the same stereotypes of black venality and foolishness as *The Birth of a Nation* (1915) had done a quarter of a century earlier.[59]

While reprising certain familiar American attitudes, however, the 1920s and 1930s gave rise to an ultimately more important change in thinking about race. These were the years in which American and British anthropologists and biologists began to repudiate the scientific basis of racial hierarchies. Anthropologist Franz Boas helped lead this rejection of race as a biological category meaningful for understanding differences within the human population. Not coincidentally, Boas worked with U.S. Congressman Emanuel Celler of New York in unsuccessfully opposing the national origins system of immigration restriction. The decline of the view of race as a scientific designation and the progressive recognition of race as primarily a social grouping found perhaps its foremost statement in the publication in 1942 of Ashley Montagu's *Man's Most Dangerous Myth: The Fallacy of Race,* which described the biological limitations of racial categorization. The political sphere offered its own parallel to this scientific shift, as Eleanor Roosevelt and other mainstream liberals began openly to oppose racial discrimination and segregation. Her husband was not yet willing to take that position in the Oval Office, but his successor, Harry Truman, would do so a few years later. As the Depression dealt a broad blow to European mastery and self-confidence, the European Left turned to the Soviet Union as a model, and Western radicals showed increasing interest in non-European cultures. The intellectual groundwork for a broad international movement for racial egalitarianism was being laid just before the racial claims and savagery of the Nazis brought "Aryanism" to the forefront of international relations and politics.[60]

Race and War in the 1940s

As the ruler of the powerful German state from 1933 to 1945, Adolf Hitler put race at the center of his worldview and forced the rest of the world to confront the consequences. For him German foreign policy had a racial mission: to expand the nation's sovereignty to the east, destroying or enslaving the inferior Polish and Russian peoples in the way. He wrote in *Mein Kampf* (1923) that in this pursuit he would not repeat the mistake of the Spanish conquerors of Latin America, whose intermarriage with Indians and Africans had led to what he considered racial degradation. Rather, he would follow the North American model of annihilating indigenous peoples to ensure "Aryan" or northern European racial purity and domination. After taking power in 1933, he responded to international criticism of Nazi racial legislation by having his ministers remind diplomats and the foreign press that Germany was hardly unique in this regard, pointing to the segregation and immigration laws of other predominantly white nations. He explained Nazi views of Slavic inferiority by using an analogy he thought Westerners would understand: "Russia is our Africa and the Russians are our Negroes."[61]

Italian actions leading up to World War II revealed a similar readiness to destroy people categorized as inferior. The senior statesman of fascism, Benito Mussolini, sent his forces into Ethiopia in October 1935 in a campaign of conquest, replete with the use of outlawed mustard gas, the bombing of clearly marked hospitals, and the indiscriminate slaughter of women and children. In revenge for the Italian defeat at Adowa four decades earlier and in quest of recreating the glory of the ancient Roman empire, Mussolini snuffed out the last truly independent African government. The European conquest of Africa was, briefly, complete, as the League of Nations refused to take meaningful action against Italy.[62] The brutal Italian campaign did have the effect, however, of electrifying a wide array of African Americans who had long treasured, from a distance, Ethiopia's peculiar prestige. "Almost overnight," historian John Hope Franklin has written, "even the most provincial among Negro Americans became international-minded."[63] W. E. B. Du Bois wrote in the *Pittsburgh Courier* that the status of Africans and African Americans was closely linked: "I do not believe that it is possible to settle the Negro problem in America until the color problems of the world are well on the way toward settlement." Many recalled the argument of Marcus Garvey at the end of World War I that black safety in America depended on African independence: "The Japanese in this country are not lynched because of fear of retaliation. Behind these men are standing armies and navies to protect them."[64]

When those Japanese armies and navies attacked U.S. forces in the Pacific in December 1941, American race relations and the racial dimension of U.S. foreign policy became prominent issues that would now be struggled over and forged anew in the crucible of war. Franklin Roosevelt had already laid out the basis of U.S. war aims in two forums earlier that year. In his January state of the union address he had enumerated the "four essential human freedoms" that the United States must promote, including the "freedom from fear" and the "freedom from want." In his joint statement of war aims with the British prime minister Winston Churchill in August, known as the Atlantic Charter, the president had emphasized the "right of all peoples to choose the form of government under which they live" and the need to "see sovereign rights and self-government restored to those who have been forcibly deprived of them."[65] Roosevelt and Churchill had Europeans in mind—French, Poles, Dutch—but most of the rest of the world seemed to qualify as well. Who more than colonized peoples had been "forcibly deprived" of their "sovereign rights and self-government"? Did black Southerners and Africans deserve "freedom from fear"? Filipino General Carlos Romulo recalled in 1945 how residents throughout the European colonies in Asia had been thrilled to read the brave words of the president and the prime minister, wondering only, "Is the Atlantic Charter also for the Pacific?"[66] Even in Latin America leftist political unrest was sometimes blamed on the subversive influence of the much-publicized four freedoms.[67] By framing their war propaganda as a struggle for democracy and against the Third Reich's racist tyranny, the Western Allies opened themselves to intensive critiques of their own colonial and segregationist practices.[68]

Roosevelt confronted European and American colonialism in World War II, as he had the domestic problem of racial discrimination during the 1930s. Both colonialism and racism represented distractions from his primary goals of ending the Depression at home and defeating the Axis powers abroad. The president worked closely with the British and therefore with their colonies in the war, just as he had constructed the New Deal in the United States through intimate cooperation with white Southern legislators who administered the rule of Jim Crow. Personally inclined against colonialism and racial discrimination, Roosevelt was foremost a pragmatist who accepted the necessity of working with reactionary allies for common ends. He was not about to weaken a common front in a dire global crisis, whether economic or military, to pursue what he saw as a more narrow issue of social justice. Such problems, from his perspective, could be dealt with only after the larger crisis was successfully resolved.[69]

Few places more fully embodied the racial contradictions and uncertainties of American culture than the islands from which the U.S. war in the Pacific was launched. The territory of Hawaii in 1941 was still in many ways a colonial plantation society, where a tiny white elite controlled vast estates of sugar cane, pineapples, and coffee on which a quarter of the mostly nonwhite population lived. But Hawaiian social customs, at least on the surface, did not include the pervasive segregation common elsewhere in the United States. When the war brought enormous numbers of American troops into the islands as a staging ground for operations against the Japanese, significant transracial friendships developed. Conflicts were inevitable. White Southerners urged islanders to practice more intensive discrimination, while African American soldiers defended themselves against white assaults in street confrontations with greater support from their commanders than they had received back on the mainland. Blacks' experiences in Hawaii during the war left them pleased with a degree of color-blind camaraderie but also bitter at the spread of some Southern-style discrimination to a place where it had not previously taken root. By contrast, many whites from Dixie were alienated by the absence of a racialized social order as they understood it. One wrote home to complain of African Americans that "the government is sending a lot of them out here and giving them all the privileges of a white man" and to warn that "we are going to have a war with the Negro[e]s when this one is over right in the states."[70]

The war in the Pacific following the surprise attack on Pearl Harbor seemed to draw all the racial bitterness of white Americans out into the open. World War II was not racial in its origins, but in the Pacific it became for most American soldiers a racially coded conflict. In contrast to most U.S. residents of German and Italian heritage, those of Japanese descent were stripped of their property and incarcerated because of what they looked like—not because of their actions or even beliefs. Anti-Japanese vitriol pervaded American life between 1941 and 1945, along with the assumption that "Japanese-ness" was an innate, racially determined quality that required guarding against by quarantining all its genetic carriers. This racial coding of the Pacific war troubled African Americans. The arbitrary nature of white racial thinking disgusted one Harlem resident: "All these radio announcers talking about yellow this, yellow that. Don't hear them calling the Nazis white this, pink that. What the hell color do they think the Chinese are anyway! And those Filipinos on Bataan! And the British Imperial Army, I suppose they think they're all blondes?" Fighting a war with nonwhite enemies and allies forced American authorities to try to educate the public about distinguishing "good Asians" (Filipinos, Chinese, Indians) from "bad Asians"

(Japanese). The problem of needing to supersede simple stereotyping would arise again in subsequent wars in Korea and Vietnam.[71]

The European theater seemed quite different to Americans. Lacking the exterminationist sentiment evident in the war against the Japanese, U.S. forces fought another white nation that was engaged in the organized process of actual genocide. While effectively seizing the moral high ground against the Nazi regime, the U.S. government downplayed the intensity of anti-Semitism in the United States, which reached a new peak during the war years. Jews were still categorized legally as members of "the Hebrew race" rather than "the Caucasian race." This system enabled the U.S. government to admit few Jewish refugees from Europe after the Nazis came to power in 1933. American officials hoped they would emigrate elsewhere; some even suggested various parts of Africa, an option that would have restarted the process of European colonization of that continent. This potential linking of blacks and Jews also appeared in U.S. opinion polls, which showed Jews ranked second only to African Americans as the most disliked group. Although few Americans approved of German racial practices, it was not clear that knowledge of the Holocaust would necessarily motivate U.S. soldiers in Europe to fight harder. The army magazine *Yank* decided in the fall of 1944, for example, not to use a detailed U.S. government report on the Holocaust for an article because of fears that a focus on Jews might stir up latent anti-Semitism in the army. Anti-Jewish sentiment pervaded Congress and the State Department, further hemming in Roosevelt's own limited interest in helping European Jews even after the initial reports of the Holocaust had been confirmed.[72]

Despite enduring prejudice, going to war with the most murderous racists in modern history certainly left its mark on American society. Most Americans were horrified by the pictures and stories of the concentration camps liberated by the Allies. Hitler forced the world to see where racist thinking ultimately could lead, thereby sharply reducing its legitimacy for the postwar era. This encouraged popular opinion in Western nations like the United States to begin to support the antiracist consensus that had been building in the scientific community since the 1920s. Fighting against the Nazis eased this transition for many white Americans. In part by downplaying America's own racial history, or even projecting it onto the enemy, they found it possible to protest racism—and became increasingly tolerant of the idea of integration in the process. The fierce prejudice on display in the Pacific theater suggested how complicated and incomplete this shift away from explicit racism would be. But the struggle between the contending forces of racism and antiracism continued after 1945 on a new terrain won by the sacrifices of those who conquered the Third Reich.[73]

The United States entered the war in 1941 with a thoroughly segregated military, representative of broader American society. Soldiers trained disproportionately in the South, where eight of the nine largest army training camps were built. Most black units were commanded by white Southern officers, who were widely believed to understand African Americans better than did other white Americans. White Southerners expressed deep anxiety about having large numbers of armed and uniformed black troops—and any black officers—in their region. To some whites the very idea of a black soldier with authority was literally fantastic. Edgar Huff was one of the first fifty African Americans accepted into the Marine Corps in 1942, for service in a segregated unit in North Carolina. On leave in late December to visit his ailing mother and still in uniform, Huff got off a bus in Atlanta and was immediately accosted by two Marine military police (MPs). The MPs tore up his furlough papers, arrested him, and threw him in jail for impersonating a marine. There he spent five days, including his first Christmas in the Marine Corps, before his commanding officer was able to bail him out. The MPs simply refused to believe there could be such a thing as a black Marine.[74]

The decision in January 1942 to segregate "white" and "black" blood plasma for transfusions illustrated the routine fashion in which military authorities preserved traditional racial practices. The secretary of war and the secretary of the navy, along with their respective surgeon generals and the head of the American Red Cross, approved this policy, despite agreeing that there was no scientific reason to do so. The purpose was solely to placate anxious segregationists, especially in Congress. African Americans noted how this policy gave credence to long-standing beliefs about the mystical qualities of blood as racially distinctive—as in such terms as "pure-blooded" or "Aryan blood"—and thus matched up perfectly with official Nazi doctrine. The critical role in blood plasma research played by an African American physician, Charles Drew, deepened the irony of U.S. policy.[75]

Even more controversial were the efforts of U.S. military authorities to expand segregation into areas where it had not previously existed in rigid, legal form: the North, the West, and overseas. The problem arose most sharply in determining racial policies for new army facilities being built in these areas. Military commanders needed uniform policies, which inevitably required imposing new practices on some parts of the country, as racial traditions varied across different regions. Black soldiers particularly resented what they saw as the infusion of Jim Crow rules into regions and places previously free of them. Puerto Rican recruits experienced an acute version of an expanded, bureaucratized system of racial classification, as an army officer walked through their ranks cursorily deciding who would be assigned to "white" units and who to "black" units.[76]

Efforts to extend the reach of Jim Crow overseas during the war provided one of the most visible intersections of race and U.S. foreign policy. Nowhere did this issue arise as dramatically as in England, which served as a staging ground for the vast American force that attacked the German-controlled beaches of Normandy on D-day, 6 June 1944. Of those U.S. soldiers 130,000 were African Americans, whereas England itself had only 8,000 black residents. Military authorities and white soldiers focused great attention on the relationships of male soldiers with civilian women in nearby communities. American commanders feared that Englishwomen might develop intimate relationships with darker-skinned GIs. Many white American males were determined to export their definition of racially appropriate socializing, seeing themselves as protectors of white womanhood in England just as at home, where most states in the South and West banned interracial marriages. When many British female civilians insisted on treating black and white Americans equally—allowing themselves to get emotionally and sexually involved—the results among American troops were often explosive.[77]

Both the United States and Britain tried to defuse such situations by working to isolate black soldiers as much as possible, but war had deprived British women of male companionship, and military perks and antiracist Allied propaganda limited these efforts. The British government found itself caught between its own civilians' desires and its sensitivity to two critical supporters: white Southerners in the U.S. Congress, who were enthusiastic defenders of England, and the large majority of U.S. troops on English soil, who were white. Its overriding goal was the defeat of Germany and the preservation of unity among the forces engaged in that task, regardless of discriminatory attitudes within the Churchill government or inclusive behavior among some civilians. General Dwight Eisenhower, the commanding officer of American troops stationed in England, laid down a policy of informally preserving racial segregation to minimize possible tensions. But he refused to order an explicit color bar, as he recognized the impossibility of preventing social contacts with civilians or trying to control the behavior of civilians of another sovereign country. Instead, he tried unsuccessfully to convince the Women's Army Corps to station more black American women in England.[78]

Eisenhower explained the situation in a letter to his superiors in Washington in September 1942. "Here we have a very thickly populated country that is devoid of racial consciousness," he argued, exaggerating British color blindness but revealing white American anxieties about other white people who did not adhere to strict segregation. "To most English people, including the village girls—even those of perfectly fine character—the negro soldier is just another man, rather fascinating because he is unique in their experience, a jolly good fellow and with money to spend." The problem, Eisenhower noted, arose with "our own white soldiers [who], seeing a girl walk down the street with a negro, frequently see themselves as protectors of the weaker sex and believe it necessary to intervene even to the extent of using force, to let her know what she's doing." Both those men and their supreme commander seemed to share the conviction that such an Englishwoman did not "know what she's doing."[79]

British civilians generally responded to African American soldiers much as they did to white American soldiers. When distinctions were made, in fact, they often ran in favor of blacks. Many English found black GIs more polite and less arrogant than white GIs, and less likely to complain about the relative lack of American-style comforts like refrigeration and abundant home heating. "I don't mind the Yanks," said one, "but I don't care much for the white fellows they've brought with them." Civilians leaned to the underdog's side, especially when confronted with unprovoked cruelties by white soldiers against black troops. White American insults and discrimination against nonwhite British colonial troops from the West Indies serving in England further offended civilians. The British government downplayed such incidents, seeking to avoid conflicts with U.S. forces. Indeed, one proposal by the Colonial Office—not actually enacted—would have made black Britons wear a badge to distinguish them from black Americans, presumably to enable white Americans to know immediately who was a "nigger" and who was not. Both British officials and civilians had long been annoyed by American criticisms of British rule in India, and they were swift to note how the American racial discrimination that had now washed up on their shores undercut American moral posturing. Many African Americans who had long criticized Britain for its colonial practices abroad found themselves divided between that view and a newfound appreciation for the wartime humanity of English civilians at home.[80]

When race-baiting and discrimination among U.S. forces occasionally exploded into large-scale violence, British civilians were as likely to side with blacks as whites. Racial violence among Americans abroad was rarely separable from race relations at home. The largest incident in England during the war unfolded just four days after the massive race riot in Detroit on 20 June 1943, which killed thirty-four people. A typical scenario spun out of control on 24 June, as white MPs from the nearby U.S. air force base came to a pub in Bamber Bridge, Lancaster, to arrest supposedly disorderly African American soldiers. White MPs were well known for their eagerness to punish black troops for socializing with local women. In this case, local white Englishmen sided with the soldiers, urging them to resist the MPs and helping unleash what evolved into several hours of uncontrolled violence between white and black troops. "The more I see of the English, the more disgusted I become with Americans," a black lieutenant named Joseph O. Curtis wrote to a friend in 1944. "After the war, with the eager and enthusiastic support of every negro who will have served in Europe, I shall start a movement to send white Americans back to England and bring the English to America."[81]

The unrest at Bamber Bridge, like the Detroit riot, was part of a broader pattern of intense racial conflict in American life during World War II stemming from major shifts and dislocations of people and places. Incidents of racial violence unfolded among civilians and soldiers in all areas of the country as well as at overseas bases. They reached a peak in 1943, as over two hundred racial battles flared in more than forty-five cities. The ferocity of the killings in Detroit forced Roosevelt to send in six thousand federal troops to reestablish order, an extraordinary diversion of military force in the midst of a war for national survival. Military installations in the South, where most African American soldiers were trained, served as the most common flash points. Military spending brought both economic growth and social dislocation to Dixie, as cities swelled with war workers, and army camps were carved from pine forests to hold hundreds of thousands of new troops. White civilians and white MPs and soldiers shared a determination to preserve the color line, even at the cost of beatings, shootings, disappearances, and riots on bases and in cities. Their dismay at the presence of black officers and armed black soldiers paralleled the fear and anger of black Northerners, unaccustomed to the deference expected under Jim Crow, who were sent south for training.[82]

The Roosevelt administration sought to dampen such conflicts. Presiding over a political party encompassing vast differences in regard to race relations, the president and his advisers focused on the overriding task at hand: winning the war. This produced an ambivalent approach to racial discrimination. When greater racial justice dovetailed with the needs of the war, it was pursued; when it did not, it was allowed to slide. Many officials privately and publicly rejected the idea of racial equality or integration. Attorney General Francis Biddle, for example, responded to the Detroit riot by calling for more rigidly enforced segregation. Roosevelt himself entrusted racial matters in the armed forces to two Southern aides, one of them the avowedly racist James Byrnes. Segregationist Southern Democrats dominated both houses of Congress, far out of proportion to their region's share of the nation's population. Nevertheless, the manpower needs of the military eventually breached the wall of segregation within the armed forces during the war. The sheer inefficiency of duplicating facilities resulted in an abundance of cases of greater, if unintended, equality of treatment and opportunity, including the first use of racially mixed infantry companies during the Battle of the Bulge. In the process, many whites experienced limited doses of desegregation that served to inoculate them against too easy a return to traditional assumptions about people of color after the war.[83]

While the United States tried to muffle the sound of racial conflict within American ranks, Germany and Japan strove to amplify it. German and Japanese propagandists eagerly brandished every report of racial discrimination in the United States as evidence of American hypocrisy and the hollowness of Allied rhetoric about democracy and freedom. Their lurid but often accurate stories of lynchings and race riots targeted the nonwhite peoples of the European colonies, especially those in Asia. The Nazis themselves were hardly good candidates for promoting the cause of racial equality, but their propagandists did not need to enter the realm of fiction to weaken American claims to moral superiority.[84]

As a nation of color in a region dominated by European colonialists, Japan had greater success in raising doubts in Asia about the true intentions of the Allies. Despite their own ethnocentrism and brutal treatment of conquered Asian peoples, the Japanese portrayed themselves accurately as leaders of the effort to sweep white power out of Asia. They buttressed their call to "join us in our fight for the liberation of Asia" in part by focusing attention on the anti-Asian character of U.S. immigration laws.[85] Nationalist leaders from India to Korea varied widely in their attitudes toward Japan, but they were all inspired by the possibilities created by early Japanese victories over the British and Americans, whose colonialist and segregationist practices were well known. Japan's success in highlighting American racial discrimination and violence—encapsulated in the persistent question, "Will white Americans give you equality?"—dismayed many liberal Americans. Popular writer Pearl Buck saw American racism as now too politically expensive a luxury, bemoaning "the persistent refusal of Americans to see the connection between the colored American and the colored peoples abroad, the continued, and it seems even willful, ignorance which will not investigate the connection."[86]

The Tokyo government used racial themes in making a particular effort to influence African American opinion about the war. Many black Americans had long appreciated Japan as the nation that disproved the idea that people of color were unable to compete equally with whites. Black communities around the country shared an element of vicarious pleasure in the victories of a nonwhite military over white troops in 1942. "I don't want them to quite win," went a typical response, "but to dish out to these white people all they can dish out to them." Elijah Muhammad, the founder of the Nation of Islam, and a small number of other black sympathizers with Japan were imprisoned for encouraging African Americans to abandon their national loyalty to the United States for a racial identity with an international colored community under Tokyo's leadership. But the vast majority of black Americans remained intensely loyal, sharing the national revulsion at the attack on Pearl Harbor. Few believed that any advantage could arise from an Axis victory. African Americans instead swallowed a full measure of harassment and discrimination in order to serve in the U.S. armed forces, with hopes that their sacrifices would lead to greater equality for all Americans after the war.[87]

The Pacific War at Home

The racial lenses through which the United States government viewed much of the world during World War II were most evident in the incarceration of some 110,000 American citizens and residents of Japanese descent in the West. Japanese Americans were not interned in Hawaii, where they were too important to the territory's economy. Despite the attack on Pearl Harbor, which supposedly justified the mass incarceration, neither military intelligence nor the FBI found any military reason for the policy. By contrast, few Americans of German or Italian heritage were imprisoned. European immigrants, in other words, could become Americans, but Americans born in the United States who had Japanese ancestors remained ultimately Japanese. It was hard to imagine a more racially exclusive definition of "American." Japanese Americans were not only rounded up like livestock but were also forced to live for months in buildings designed for animals, such as stockyards and fairground cattle stalls. Several of the internment camps constructed for them in the interior of the country were located on Indian reservations, as a way of avoiding provocation of local white residents who resented the presence of the internees. Confining Japanese Americans to reservations served to obscure a new concentration of people of color within an older one.[88]

Federal and local officials did not hide the racial logic of their policy. General John Dewitt, head of the Western Defense Command, explained that in this war "racial affinities are not severed by migration. The Japanese race is an enemy race." Dewitt believed his point to be obvious, as his hypothetical construction of the reverse situation revealed: "To conclude otherwise is to expect that children born of white parents on Japanese soil sever all racial affinity and become loyal Japanese subjects, ready to fight and, if necessary to die for Japan in a war against the nation of their parents."[89] Governor Culbert Olson of California did not shy from this kind of racialized thinking with any audience. "You know," he told a group of Japanese American editors in February 1942, "when I look out at a group of Americans of German or Italian descent, I can tell whether they're loyal or not. I can tell how they think and even perhaps what they are thinking. But it is impossible for me to do this with the inscrutable Orientals, particularly the Japanese."[90] The *Los Angeles Times* summarized white views more bluntly in an editorial that same month: "A viper is nonetheless a viper wherever the egg is hatched—so a Japanese-American, born of Japanese parents—grows up to be a Japanese, not an American."[91] The internment camps represented racial segregation with a vengeance. Dixie had no corner on the market of discrimination.

Even Japanese American war heroes could not immediately overcome this racial construction of American identity. In response to heavy Japanese propaganda in Asia about the United States fighting a racist war, the Office of War Information finally convinced Roosevelt in 1943 to allow military recruitment from the internment camps. American citizens of Japanese heritage, having lost their constitutional rights and freedom, were now expected to shoulder the citizen's duty to fight for his country. Reaction in the camps was deeply divided. Many Nisei (those born in the United States of Japanese immigrants) who did serve when drafted hoped that their sacrifices—which were considerable, especially in the Italian campaign—would help grant full citizenship to Japanese Americans after the war. They did, though it turned out to be a long process. Captain Daniel Inouye, the future U.S. senator from Hawaii, lost an arm in combat in Italy. Stopping in San Francisco on the way home, he tried to get a haircut. The barber looked at him, with his empty sleeve pinned to a U.S Army uniform covered with ribbons and medals, and said simply, "We don't serve Japs here."[92]

The attitude toward Americans of Japanese descent provided the domestic backdrop for the treatment of actual Japanese soldiers and civilians during the war. Historian John Dower has demonstrated how the surprise attack on Pearl Harbor triggered "a rage bordering on the genocidal among Americans." This response went beyond the wartime requirement of demeaning and killing the enemy to create a pattern of language and behavior permeated by racial thinking. The determination to slaughter as many enemy soldiers and civilians as possible did not characterize U.S. actions against Germans and Italians in the European theater of the war.[93] Even Hollywood movies reflected the acceptability of racially coded killing when it came to the Japanese: while cries of "hell" and "damn 'em" by American gunners shooting down Japanese planes were censored as obscene, "stinkin' Nips" and "fried Jap going down!" were not.[94] Racial considerations were not the primary factor in official policy and individual actions in the Pacific, but they did help determine the peculiarly ferocious manner in which the war there was fought. Japanese audacity in attacking U.S. territory had challenged the very structure of white supremacy that suffused American life in 1941. "The thesis of white supremacy," Undersecretary of State Sumner Welles pointed out in 1944, "could only exist so long as the white race actually proved to be supreme."[95] The Japanese challenge to white power threatened not just the political order of the western Pacific, but also the social order of the United States and the European colonies.

The fact that the United States had Asian allies as well as an Asian enemy in World War II complicated the picture considerably. Since the ubiquitous epithet "little yellow . . ." appeared to cover people on both sides of the Sea of Japan, including America's Chinese allies, who were keeping millions of Japanese soldiers occupied, U.S. officials and other prominent Americans encouraged their countrymen to narrow their typically broad anti-Asian sentiments. *Life* magazine published an article soon after the attack on Pearl Harbor that sought to help readers distinguish Chinese from Japanese, including pictures that made Chinese appear more like Europeans. One Chicago newspaper took this logic a step further, declaring China to be "our 'white hope' in the East." Despite strong opposition, Congress in 1943 amended U.S. immigration laws to allow a tiny but symbolic number of Chinese into the country each year, in an effort to undercut Axis propaganda. Indians received a similar allotment three years later. These breaches in the wall of immigration restriction pointed toward the racial egalitarianism that slowly overtook the U.S. legal code after 1945. The war in the Pacific, in sum, illuminated both the deadly consequences of American racial discrimination and the beginnings of a rethinking of U.S. relations with nonwhite nations.[96]

Africans, African Americans, and the Prospect of Equality

Like Asians who saw in Japan's initial victories an opportunity to reconstruct their national futures, many Africans experienced World War II as a loosening of their historic colonial ties to Europe. Major battles were fought on the northern littoral of the continent, and Africans served in Allied armies both there and abroad. British, Nigerian, and South African troops liberated Ethiopia from the Italian army in 1942, establishing a precedent for the process of decolonization that would sweep through all but the southernmost end of the continent over the next twenty years. From south of the Mediterranean, World War II appeared in large part as a vast European civil war. White people slaughtered white people by the millions, casting into further doubt the supposed superiority of European civilization to African "barbarism" that underpinned the system of colonial rule. Above all, Africans listened carefully as the major colonial powers on their continent—Britain, France, and Belgium—condemned Germany for its efforts to rule over other peoples. African nationalists were dismayed at the apparent surprise of Europeans at the success of fascism, which seemed not so different from colonialism. Aimé Césaire noted that before Europeans "were its victims, they were its accomplices": before it was inflicted on them, they had "absolved it, shut their eyes to it, legitimized it, because, until then, it had been applied only to non-European peoples."[97]

Africans by 1945 were unsure how to view the United States. Americans had no direct colonial connection to the continent, and they had helped defeat the racist Nazi regime by fighting under an officially antifascist creed. While still mostly segregated, U.S. forces did include large numbers of black soldiers. And American officials seemed eager to open up new avenues of trade with territories previously controlled by colonial offices in London, Paris, Brussels, and Lisbon. But behind their wealth, power, and fine rhetoric stood the shadow of Jim Crow. How the U.S. government and the white majority of Americans treated African Americans at home seemed the likeliest indicator of how they would deal with dark-skinned people abroad. Reports of American racial discrimination trickled into African colonies from a variety of sources: British and French radio broadcasts; German propaganda during the war and Communist Party propaganda afterward; African soldiers coming home from foreign battlefields; and African students returning from sojourns at American colleges. The close working relations of U.S. officials with colonial and white settler authorities in Africa also suggested something less than a firm commitment to equality. Few white Americans imagined African self-government as a viable option in the foreseeable future. Among whites racial attitudes combined with economic and strategic considerations to consolidate the primacy of European over African priorities.[98]

Black Americans, by contrast, strongly supported the independence of colonies in Africa and everywhere else. They understood during the war that the status of people of color in the United States could not be separated from that in other parts of the world. Independence in Africa would weaken the logic of continued white supremacy in the United States, as black people demonstrated their ability to govern themselves. Three major figures in African American political and intellectual circles—W. E. B. Du Bois, Walter White, and Rayford Logan—published books in 1944–45 that focused on the international character of racial inequality and the need for its demise.[99] Politically engaged Americans of color hoped in particular that the new international organization under construction in San Francisco in the spring of 1945—the United Nations (UN)—would promote both decolonization and full human equality. The problem for the UN was the inherent tension between respecting national sovereignty and defending racial nondiscrimination; the second had to be pursued without overriding the first, or else the United States and the European colonial powers would not be supportive. The UN charter thus compromised by including language supporting both principles. A divided U.S. delegation did agree to the principle of nondiscrimination—a significant symbolic step—but refused to take a strong stand against colonialism. Delegates like John Foster Dulles opposed the human rights clause in the charter out of fears that it could lead to an international investigation of "the Negro question in this country." What Dulles feared, others welcomed.[100]

Like African nationalists, black Americans emerged from the traumas of World War II with a new determination to liberate themselves from the burdens of inequality. Membership in the NAACP grew ninefold between 1940 and 1946. African American protests against discrimination escalated sharply before the United States entered the war, as the threat of a massive march on Washington organized by labor leader A. Philip Randolph led to the creation of a federal Fair Employment Practices Commission in the summer of 1941. While American troops were engaged on foreign battlefields, such activism was muted by a respect for the cause of defeating the Nazis and a determination to strengthen the war effort. But their experiences of fighting abroad for democracy rendered black soldiers less willing to return home and quietly accept their prewar status as second-class citizens. For many, the scale of wartime violence and destruction that they had survived made white supremacists at home seem less intimidating. For others, experiences of better treatment by whites abroad altered their expectations when they returned. Black Mississippian Wilson Evans remembered how quickly racial discrimination had disappeared under fire in the Battle of the Bulge: "For those what six, maybe eight, ten days there was no black or white soldiers. We was all soldiers. White was afraid of dying as blacks. And there was no color. During the [German] breakthrough I did see that Americans could become Americans for about eight or nine days."[101]

Whether Americans could remain Americans, in these terms, remained an open question as World War II wound down. On the one hand, many white Americans, particularly soldiers, emerged from the war with a more acute sense of the injustice of racial discrimination. Having faced down the weapons of Japan and Germany, some were quick to confront perpetrators of petty harassment at home. Some had seen firsthand the horrors of where Nazism led; most others had at least glimpsed the pictures of liberated concentration camps. Many soldiers had experienced intimacy and comradeship with darker-skinned colleagues, which challenged their society's traditional segregationism. Georgian Dean Rusk of the War Department, for example, invited his counterpart from the Office of Strategic Services, Ralph Bunche—an extremely rare high-level black official—to dinner at the previously segregated officers' mess. When Bunche tried to decline, Rusk insisted, "Come on, we'll change that rule right now"—and they did.[102]

On the other hand, white anxieties about returning black soldiers were high, especially in the South. Black Mississippian Henry Murphy came home to Hattiesburg with a Purple Heart earned on the battlefield. White police in the area, however, were randomly assaulting and searching uniformed African Americans; if they found pictures of white women, the assaults could turn deadly. Murphy's father therefore met him at nearby Camp Shelby with a pair of overalls for him to change into in the car, so the returning war hero could survive by posing as a simple field hand. Dabney Hammer's reception was no less menacing. He arrived home in Clarksdale, Mississippi, with a chest covered with ribbons and deeply proud of his performance in five battles in Europe. One of the first whites he met stopped to admire his medals: "Ow-w-w look at the spangles on your chest. Glad you back. Let me tell you one thing though, don't you forget." Forget what? Dabney asked. "That you're still a nigger."[103] "Niggers" or "Americans": the essential question remained.

No issue more clearly revealed white Americans' ambivalence about race and foreign relations in the 1940s than the contrasting treatment of African American soldiers and German prisoners of war, particularly in the South. Axis prisoners in transit often enjoyed access to hospitality in restaurants where black Americans—even those sometimes guarding the prisoners—were not welcome. This pattern reflected, in part, a belief in decent treatment of prisoners of war, especially with hopes that American prisoners in Germany would be treated with similar respect. But it also starkly illuminated the racial contours of many white Americans' understandings of their own identity. Dark-skinned U.S. troops, risking their lives to defend all Americans, could not sit at the same table in fellowship with white Americans, while white soldiers of the German Wehrmacht, which was engaged in killing as many Americans as it could, were invited to the table. African Americans during the war were often enraged by this confusion over who the enemy really was.[104] Nor was this pattern of discrimination limited to blacks or to the South. In Wyoming, where both German prisoners and American citizens of Japanese descent were incarcerated in separate camps, neighboring white residents frequently built relationships with the Germans that outlasted the war, while making no contact whatsoever with their fellow Americans imprisoned for the color of their skin and the shape of their eyes.[105] For many Americans of all regions, whiteness continued to transcend national borders and even enemy lines in conferring the ultimate form of status.

Nonetheless, as World War II came to a close, the long-standing tendency of white Americans and their government to see African Americans and other antiracists as potential subversives was beginning to ebb. The fears of the FBI and the military intelligence services that significant numbers of black Americans might support Japan on racial grounds proved unwarranted, as did their concerns about Communist Party inroads into black political allegiances. African Americans instead demonstrated their patriotism precisely as their white countrymen did—arguably more so, at least in terms of how they treated their fellow citizens of different hues. Governor Eugene Talmadge of Georgia did win reelection in 1944 in part by campaigning against "Moscow-Harlem zoot suiters trying to take over Georgia," and Senator Theodore Bilbo of Mississippi did ask Congress to pass a bill providing for the deportation of African Americans to Africa. But the tide was against them, as the Supreme Court's outlawing of the white primary in 1944 demonstrated. In his widely acclaimed study of U.S. race relations, *An American Dilemma* (1944), sociologist Gunnar Myrdal pointed toward an eventual eradication of discrimination. Even many Southern segregationists—like Afrikaners in South Africa—recognized that fighting a war against the Nazis was bad news for Jim Crow. "The Huns have wrecked the theory of the master race," complained Frank Dixon, former governor of Alabama. The U.S. government acknowledged this turn in quiet ways, such as the shift in policy by the Civil Service Commission's loyalty board in 1943 directing its investigators to cease asking applicants whether they attended racially integrated meetings. Patriotism no longer required segregationism. As Americans sailed forth into the uncharted waters of the Cold War, they carried with them a greater openness than ever before to reconstructing race relations along more egalitarian lines both at home and around the world.[106]

2

JIM CROW'S COMING OUT

"THE RACIAL PROBLEM has . . . become not merely a moral and social problem in American life," political commentator Harold Isaacs concluded in the summer of 1950, as American troops poured into South Korea. "It has become a political problem of the first magnitude in the making of American world policy."[1] The course of World War II had drawn the United States out of its relative isolation, and the early postwar years thrust the nation into a conspicuous position of world leadership and even dominance. Like a deep-sea diver surfacing too quickly, American society hurtled upward into the unprecedented attention of the open air. Without adequate time for decompression, Americans seemed almost to have a case of the cultural bends in the late 1940s, debilitated by certain vestiges of intolerance they carried with them. Systematic racial discrimination and violence were foremost among those vestiges and caused uncertainty among other peoples who admired much else about the new North American superpower.[2]

The ethnic and racial diversity of the American population made the United States a potentially positive model for the international community as it rebuilt from the ashes of World War II's racially charged destruction. Not only was the United States enormously wealthy, militarily supreme, and democratic in its creed, but it had also marched to victory in World War II with armed forces made up of the descendants of every continent on earth. If any nation looked like the world in all its human diversity, it was the United States. As international leaders pieced together a new multiracial organization to promote peace and "respect for human rights and for fundamental freedoms for all without distinction as to race," all eyes looked to the country that would host that United Nations.[3] "The possibilities for good or evil within the continental United States are precisely the possibilities for good or evil in the vexed postwar world," American writer Wallace Stegner declared in the early months of President Harry S Truman's administration. "The problem of making one nation from the many races and creeds and kinds, one culture from all the European, Indian, African, and Asiatic cultures that the promise of freedom has drawn to our shores, comes to a head in our time."[4]

Stegner and many others recognized that the reconstruction of race relations along more progressive lines would be a crucial task on both the national and international levels after World War II. Colonialism, in particular, would not remain the same. The most important colonial powers of Western Europe had been either defeated and occupied by Germany during the war, like France, Belgium and Holland, or at best militarily victorious but financially devastated, like Great Britain. By contrast, nationalist movements in Asia, the Middle East, and Africa—like racial egalitarianism in the United States—had been strengthened by the course of the war, and in the two decades after 1945 independent states emerged in almost every former colony. Harold Isaacs urged Americans quickly to grasp this sea change in international and interracial relations: "We are going to have to learn to see Europe not as the center of the world we know, but as a peninsula dangling on the western end of Eurasia."[5]

The struggles for national and racial equality in the world system and in the United States, which came to characterize the entire Cold War era, intensified in the Truman years of 1945–1952. The domestic civil rights movement and what might be called the international civil rights movement of anticolonialism moved on parallel tracks. African Americans sought to vote in the American South, while Africans and Asians strove for self-government. Black Americans struggled to overcome the use of states' rights to justify white supremacy, promoting instead the higher mandate of human rights and the U.S. Bill of Rights, while nonwhites abroad struggled to overcome the UN's reliance on domestic jurisdiction, wielded by racially discriminatory white authorities in places like South Africa. The tradition of white supremacy in the United States was embedded in a broader global pattern of white control of people of color, and both systems of racial inequality appeared to some to be directly related. They predicted that the two systems would survive or fall together.[6]

The Cold War developed after 1945 as a state of heightened tensions between the two great powers that emerged from World War II, the United States and the USSR. The ideological origins of the conflict dated from the

Communist revolution in Russia in 1917, which most Americans abhorred for its rejection of private property, religious worship, and multiple-party political liberty. Such differences in values took on greater political and military significance at the end of World War II, as the two victorious nations found themselves face-to-face in Central Europe, East Asia, and the Middle East. Their visions of the world order that should be reconstructed after the war were largely incompatible: command economies run by Communist parties versus global capitalism free from most state interference. U.S. policymakers were deeply concerned about Communist influences in the anticolonial nationalist movements in Asia and Africa. In Western Europe, home to America's most important allies, the Truman administration feared that growing Communist parties in Italy, France, and Belgium might use public dismay at grim postwar economic conditions to win popular elections and seize control of those governments.[7]

The major American Cold War initiatives of the late 1940s and early 1950s—the Truman Doctrine, the European Recovery Plan (Marshall Plan), the North Atlantic Treaty Organization (NATO), and National Security Council document 68 (NSC 68)—emerged against a background of mounting demands for racial equality and national autonomy. People of color at home and abroad were particularly sensitive to these policies' racial meanings. Winston Churchill's Iron Curtain speech of February 1946 represented a declaration of Cold War, but it also called for Anglo-American racial and cultural unity. The Truman Doctrine of March 1947 opposed potential "armed minorities" of the left but not those of the right, who actually ruled much of the world: European colonialists. The Marshall Plan (1948) and NATO (1949) bolstered anti-Communist governments west of the Elbe River, but they also indirectly funded those governments' efforts to preserve white rule against indigenous independence movements in Asia and Africa. NSC 68 (1950) laid out an offensive strategy for diminishing Soviet influence abroad, but it also revealed American anxieties about a broader "absence of order among nations" that was "becoming less and less tolerable," when the largest change in the international system was coming not from Communist revolutions but from the decolonization of nonwhite peoples.[8] This concern about disorder abroad paralleled the Truman administration's unhappiness with the volatility of race relations at home after World War II.

In designing his big house of anti-Communist democracy, Harry Truman faced the same fundamental challenge at home and abroad. He had to build it large enough to include people of all colors, while preserving his relationships with the British and French colonialists who ruled so much of the world beyond Europe, and with the Southern segregationists who through their seniority dominated much of the U.S. Congress. In an era of decolonization and rapid change toward greater racial equality, the president and his advisers needed to demonstrate that traditional white racism would not be a central element in the domestic and international anti-Communist coalitions they were constructing. In its quest for nonwhite loyalties, the Truman administration was confronted by a more racially inclusive vision deriving from the radical Left. The Soviet Union played on Third World experiences of European colonialism and Western racism, while Henry Wallace and the Progressive Party appealed to African Americans against the intransigence of white Southern authorities.[9] Both were external sources of pressure that encouraged the administration to take a more racially liberal stance within the limits imposed by its higher priority of containing the expansion of Soviet power. Racial issues and the management of racial change were central to the American experience of the early Cold War, and the nation's borders proved quite porous in this regard.

Harry Truman and People of Color

The relationship between race and U.S. foreign policy in the Truman years was grounded in the segregated culture in which the most influential members of the administration had lived before 1945. The president's family connections to the South are well known: his upbringing in the former slave-owning state of Missouri, at a time of fierce Jim Crow discrimination and violence; his slave-owning grandparents; and his mother's traumatic experiences as a child with Union Army raiders and subsequent lifelong hatred of blue uniforms and Abraham Lincoln. Historian David McCullough has pointed out how Truman's hometown of Independence was "really more southern than midwestern," especially in its racial practices. The attitudes about people of color that the future president imbibed from these sources were not surprising. His conversation and letters were littered with racial and ethnic epithets for Asians, Jews, southern Europeans, and African Americans, among others.[10]

People do not choose the circumstances into which they are born, but what they make of those in later life provides a measure of their character. In contrast to his siblings, Harry Truman chose to move away from the explicit racism of his childhood as his political career developed and his contacts in the world widened. He successfully courted black voters by treating them as a legitimate political interest group and following a fairly progressive path on civil rights issues. As a U.S. senator from Missouri from 1935 to 1944, he consistently supported antilynching legislation and the abolition of poll taxes. By the standards of later generations and of more liberated contemporaries, Truman remained a racist in his personal attitudes: he opposed what he called social equality of the races and continued privately to disparage nonwhite peoples on occasion, as he would to the end of his life. But for a man of his place and time, the Missourian made impressive strides on matters of race.[11]

Truman's ascension to the White House resulted to a significant degree from his record on racial issues. His selection as vice president in Franklin Roosevelt's last campaign of 1944 represented a compromise at the Democratic Party convention. The real struggle was between supporters of the incumbent vice president, Henry Wallace of Iowa, and of the former Supreme Court justice and close wartime associate of Roosevelt, James Byrnes of South Carolina. Wallace headed the reform forces of the New Deal and was strongly supported by African American voters and unions; Byrnes, a devoted segregationist and union opponent, was the candidate of the party's powerful white Southern wing. Truman proved acceptable to both sides, a border-state figure strong enough on labor and civil rights without being a complete racial egalitarian. The vice presidential contest thus previewed the struggle that would unfold four years later in the 1948 presidential campaign, with Truman's bona fides on civil rights again being tested by Wallace's stronger opposition to racial discrimination.[12]

In contrast to Wallace, Truman's primary foreign policy advisers all stood to his right on racial issues. There was no equivalent on the international side of the administration to political aide Clark Clifford's encouragement to promote civil rights more forcefully.[13] The elite white men who ran the State and Defense Departments and the intelligence agencies were comfortable with the world they had grown up and succeeded in, a world marked by European power, Third World weakness, and nearly ubiquitous racial segregation. To varying extents, assumptions of white racial superiority underlay their interpretations of the postwar situation facing Washington. Even George Kennan, the author of the containment doctrine underpinning the administration's entire foreign policy and an intellectual widely admired for sophisticated strategic thought, made no secret of his distrust of people of racial and ethnic descents different from his own. He saw African Americans and Jews in the

United States as potentially subversive "maladjusted" groups.[14] As late as 1938, Kennan had written privately that the U.S. government should be transformed into a "benevolent despotism" of elite white males, with women, immigrants, and blacks excluded from the franchise.[15]

Kennan, like his colleagues in the Truman administration, was a man of the north. A Soviet specialist with a Scandinavian spouse, he later observed that his "entire diplomatic experience took place in rather high northern latitudes."[16] His occasional contact with the world beyond Europe gave him the opportunity to indicate personal dislike and even loathing for peoples of Africa, Asia, the Middle East, and Latin America. He tended to lump them together as impulsive, fanatical, ignorant, lazy, unhappy, and prone to mental disorders and other biological deficiencies.[17] In the case of Latin America, Kennan specifically singled out generations of racial intermingling as a primary source of that region's supposed neuroses and delusions.[18] Third World neutralism angered him. Nonwhite leaders needed to be seized "by the scruff of the neck" and made to defend their newly independent nations from potential Soviet incursions, he told American officers at the National War College in 1952.[19]

Perhaps most revealing in Kennan's thought about race and international relations was the unself-conscious fashion in which he, like many of his colleagues in the U.S. government, attributed ideological and diplomatic failings to racial identity. He located a major—if not the major—root of Soviet despotism and tyranny in the Soviet Union's partly Asian identity. He considered the suspiciousness and inscrutability of Soviet diplomats and leaders "the results of century-long contact with Asiatic hordes."[20] The "Long Telegram" that he sent to the State Department from the U.S. embassy in Moscow in February 1946, which first put him on the upward path from obscure diplomat to major policymaker, attributed much of the Soviet government's behavior to its "attitude of Oriental secretiveness and conspiracy."[21] The Bolshevik Revolution of 1917 had stripped away "the westernized upper crust" of the old czarist elite, revealing Russians in their true form as "a 17th-century semi-Asiatic people."[22] It was Asia and "Asian-ness" that had done so much to corrupt the healthier, "European" elements of Russian life and character, according to Kennan, and that now made it imperative to contain the USSR within its own boundaries.

What is important about Kennan's perspective on race is not its singularity but its commonness within American policymaking circles. He was no fool and could be, in many ways, a strategist of great subtlety and even humility;

he was quite critical, for example, of what he saw as the material corruptions of modern American and European life.[23] But if the most reflective of Truman's diplomatic elite—the one specifically assigned the task of long-range thinking, as the head of the State Department's new Policy Planning Staff, and a non-Southerner to boot—could be so traditional in his assumptions of white superiority, his colleagues were unlikely to do much better. In fact, they did not. State Department adviser John Foster Dulles plumbed the depths of official racial insensitivity by making his famous 1951 comment that "the Oriental mind, particularly that of the Japanese, was always more devious than the Occidental mind" to Ambassador Wellington Koo of Taiwan.[24]

Truman demonstrated just how little the opinions of African Americans counted in American public life in 1945, and how much influence white Southerners had, by making James Byrnes his first appointed secretary of state. The South Carolinian's "cool formality of a sophisticated diplomat" masked the heart and mind of a typically reactionary racist of his time and place: "a chronic, absolute, unquestioning believer in the natural inferiority of the African stock."[25] Many Americans were appalled. W. E. B. Du Bois, the director of special research for the NAACP, traveled to the capital of Byrnes's home state in October 1946 to urge the Southern Negro Youth Congress to make the American South "the firing line" for the emancipation of people of color everywhere. Du Bois situated Byrnes in a long line of other prominent white supremacists from the Palmetto state, including slavery proponent John C. Calhoun and rabid segregationist "Pitchfork" Benjamin Tillman, all "men who fought against freedom." The secretary of state instead "must begin to establish in his own South Carolina something of that democracy which he has been so loudly preaching to Russia."[26] Byrnes's successor in the State Department, General George Marshall, was no less comfortable with traditional racial hierarchies. A white Virginian whose career in a segregated army inclined him to discount the abilities of African Americans, Marshall in his brief tour as secretary of state touched on the subject of race only to reject accusations of American hypocrisy about democracy.[27]

Truman's final and most influential secretary of state, Dean Acheson, came out of the elite, all-white world of the Groton School, Yale University, Harvard Law School, and corporate Washington. A sophisticated and worldly man of strong convictions, Acheson was a fierce Cold Warrior and a close and loyal adviser to the president. His oft-cited 1946 letter to the Fair Employment Practices Commission emphasizing the damage done to U.S. diplomacy by domestic American racial discrimination seemed to indicate a sensitivity to racial issues.[28] In fact, it showed his immensely practical mind. He retained deeply prejudiced attitudes typical of his generation and class. "If you truly had a democracy and did what the people wanted," he argued privately, "you'd go wrong every time."[29] Acheson was a passionate Europhile who disliked and disparaged Asians, Latin Americans, and other people of color. Despite his partial awareness of explicit racism's cost to American diplomacy, he presided over a segregated State Department.[30] The son of a British-born clergyman who emigrated to the United States, Acheson did not share the anticolonialism common among many Americans. He supported white minority rule in southern Africa, even sharing anti-Asian sentiments by letter into the 1970s with former Central African Federation leader Roy Welensky.[31] One of the most striking aspects of his thinking about race was how it grew less egalitarian through the 1950s and 1960s, when much of the rest of white America was moving in the opposite direction. In his 1969 autobiography, Acheson rued the failure of the United Nations to be housed in a serene European city like Geneva or Copenhagen, regretting that it wound up instead in New York, "a crowded city of conflicting races and nationalities." The most influential American policymaker "present at the creation" of the Cold War offered little enthusiasm for racial equality or national independence in the colonial world, as he pursued his primary objective of containing Soviet expansionism.[32]

If Truman's leading foreign policy advisers were ill prepared for a postwar world of growing racial equality, they appeared positively prescient when compared with the predominantly Southern leadership of the Congress, especially the Senate. The judicial branch of the federal government provided crucial support for desegregation after 1945, but much of the nearly all-white legislature set off into the new era of the United Nations ridden with racial attitudes from an earlier time. The late 1940s and early 1950s marked the apex of Southern influence in the Senate, as seniority rules and the one-party character of Southern politics had elevated Dixie Democrats into the chairs of a majority of that chamber's most powerful committees—the pattern in the House of Representatives as well.[33] Mississippi's Theodore Bilbo used his position in the Senate to promote violence against African Americans, calling on fellow whites of the Sunflower state in 1946 to "remember the best way to keep the niggers from voting. You do it the night before the election."[34] Bilbo represented an extreme position on Capitol Hill regarding methods, but his goal of preserving white supremacy was in no way unusual or disreputable. The majority of his most influential colleagues agreed. Their leader, Richard

Russell of Georgia, was the "most powerful man in the Senate," according to the *Christian Science Monitor* in 1951. Southern segregationists stymied the efforts of Northern liberals to pass an antilynching bill or any other piece of what Russell called civil wrongs legislation.[35] A few years later William S. White's book on the chamber summarized its character: "So marked and so constant is this high degree of Southern dominion . . . that the Senate might be described as the South's unending revenge upon the North for Gettysburg."[36] It is worth remembering how at the onset of the Cold War "democracy" and "freedom" for the majority of human beings were alien concepts to the bulk of the leadership of the United States Congress.

Changing American Race Relations in an International Context

Wars, by their very nature, create social dislocations and tensions, and World War II was no exception. The unprecedented scale of its destructiveness ensured that it would be followed by a contentious period of reconstructing damaged political orders. In many places, like China, Greece, and Korea, this process emerged as civil strife and even civil war. In others, like Eastern Europe, it appeared as an army of liberation transforming itself into an army of occupation. But in much of the world the conflict over postwar reconstruction took the form of anticolonial struggle and metropolitan resistance, with a veneer of racial distinction. Indonesian nationalists fought against Dutch troops; Vietnamese guerillas went to war with a French army; Indians prevailed relatively peacefully over retreating British authorities; and French colonial forces slaughtered tens of thousands of Malagasy insurgents and civilians in Madagascar.[37] Strikes and protests in the Gold Coast and Nigeria put British authorities on notice that Africa would be next, and by 1952 Kenyans were at war with white settlers and colonial soldiers. In each of these cases, the experiences of non-Europeans in World War II, especially those who fought in colonial armies, had helped embolden them to challenge white supremacy after 1945.[38]

So it was in the United States. The swelling tide of racial tension and violence that rolled through the American South in 1946 and 1947 was part of a global phenomenon of race relations being reconfigured in the aftermath of the defeat of history's most murderous racists, the Nazis. "Bringing the boys home" for most white Americans meant returning to the peacetime life of 1941. In the South most dramatically, but also above and beyond the Mason-

Dixon line, that meant a life grounded in segregation and white domination of people of color. African Americans, however, emerged from the war with very different hopes and expectations. Those on the home front were bolstered by better wages and job opportunities, and by egalitarian war rhetoric extending from Roosevelt's four freedoms and the Atlantic Charter of 1941 to the founding of the United Nations in 1945. Those returning in uniform from abroad, having risked their lives for their country and having survived the often unspeakable brutalities of the modern battlefield, were even less inclined to accept the harassments and indignities of traditional segregation. They had in mind the Double V campaign announced by the *Pittsburgh Courier,* for victory against fascism abroad and against racism at home, not a quiet return to a Jim Crow society of poverty, discrimination, disenfranchisement, intimidation, and violence. Close to a million of these dark-skinned veterans percolated back into American society in late 1945 and 1946, providing many of the shock troops for the racial confrontations that roiled American life in the early years of the Truman administration.[39]

For most white Southerners, the reestablishment of Jim Crow far surpassed other Cold War priorities like promoting democracy. Local officials in Jackson and Hattiesburg made this clear in their decisions to ban appearances of the Freedom Train in 1947, a "rolling showcase of democracy" sponsored by the U.S. government, to prevent unsegregated viewing of potentially subversive documents like the Declaration of Independence.[40] The efforts of blacks to vote and to be treated with respect enraged many of their white neighbors.[41] Following the lead of public authorities like Senator Bilbo, white Southerners unleashed a wave of intimidation, terror, and death upon African Americans that lasted for much of 1946 and 1947.[42]

The violence came from many sources and in a variety of forms. It ranged in scale from full-blown white riots with random shooting into black homes and businesses, as in Columbia, Tennessee, in February 1946, to the gruesome intimacy of the blowtorch and cleaver used to dismember veteran John Jones near Minden, Louisiana, six months later.[43] It varied in agency from unidentified civilians in the lynchings of two black couples near Monroe, Georgia, to police chief Lynwood Shull of Batesburg, South Carolina, gouging out the eyes of uniformed Army Sergeant Isaac Woodward (and then refusing him medical attention for a day)—and to the hundreds of state patrol officers and National Guardsmen arbitrarily destroying black-owned property in Columbia. It included the execution of eight African American prisoners at a state prison camp outside Brunswick, Georgia, by an inebriated chief warden and other guards in July 1947. And it incorporated untold numbers of "disappearances," near lynchings, and police beatings.[44] The best accountings suggest at least sixty known violent deaths of black Southerners at the hands of whites between the end of the summer of 1945 and the end of 1946, with abundant police collusion.[45] Most revealing, no perpetrators were punished, despite mounting international protests.[46] What Truman's assistant attorney general called "the tide of lawlessness" was evidently acceptable to much of the white South.[47] Few could miss the unhappy implications for America's reputation. Even the conservative California Republican senator William Knowland ("the senator from Taiwan") took time out from his Cold War priorities in East Asia to demand on the Senate floor—much to his colleague Richard Russell's annoyance—that "such things must not continue in the United States."[48]

This wave of "white death" in postwar America crested in the South in 1946, but it neither ended quickly nor was limited to the lands below the Mason-Dixon line. International attention returned repeatedly to events in Dixie during Truman's two terms in the White House, particularly the 1951 executions of Willie McGee in Mississippi and the Martinsville Seven in Virginia on charges of rape.[49] Racial segregation and discrimination had the force of law in the South but were almost as common in public life in the North and the West.[50] The veneer of greater liberalism outside the old Confederacy frequently proved to be quite thin, as the race riot at the military prison in Leavenworth, Kansas, demonstrated in the spring of 1947.[51] More dramatic was the crowd of three thousand whites, abetted by the police, who gathered in Cicero, Illinois, on 12 July 1951 to defend their community from invasion—by the family of an African American veteran, Harvey E. Clark, who had rented a unit in a previously all-white apartment building.[52] With the nation at war on the Korean peninsula, liberals and many others were dismayed. Francis Russell of the State Department called the ensuing Cicero riot, which was quelled only by the intervention of the National Guard, the equivalent at home of the early U.S. defeats on the Korean battlefield.[53] A letter to the *New York Times* reminded readers that the events in Illinois would bolster the conviction of many Asians that white Americans regularly abused black Americans and thus undermine the nation's security: "Every rioting participant in the Cicero incident has won for himself the Order of Stalin."[54] Many Northern whites in the early postwar years demanded more rather than less segregation, and specifically cited the Jim Crow South as a model of successful race relations.[55]

If the Mason-Dixon line failed to wall off injustice from the rest of the country, there was still no doubt of the brighter prospects on its other side that had drawn black Southerners northward in large numbers since World War I. What those migrants left behind was a society where in many states, as the President's Committee on Civil Rights reported in 1947, "the white population can threaten and do violence to the minority member with little or no fear of legal reprisal."[56] For millions of Americans, the police were not a source of safety, and courts did not provide a refuge from injustice. U.S. society was far from life under Soviet rule, but the two did have certain similarities. Even FBI Director J. Edgar Hoover, a fierce anti-Communist and a segregationist of well-known racial prejudice, recognized the resemblance. When his agents tried to investigate lynching cases in the South, they received almost no cooperation from local authorities or the white community. Instead, he admitted, "we are faced, usually, in these investigations, with what I would call an iron curtain." Hoover found the arrogance of many Southern whites "unbelievable."[57] Poverty and the absence of protection from arbitrary and cruel authority—that is, the absence of the rule of law—characterized the lives of millions of Southerners, just as it did the lives of so many Russians and Eastern Europeans. Likewise, dissent against that authority in either place was dealt with swiftly and brutally. African Americans in Dixie were not likely, then, to be deeply impressed by Truman's Cold War distinction between "freedom" and "slavery." The New York–based League for Freedom of the Darker Peoples of the World agreed with Hoover's analysis of Southern isolation and offered a novel solution in a 1947 telegram to the president: send General Douglas MacArthur and his forces, having successfully reconstructed Japan and taught its peoples "the fundamental principles of Americanism," to Georgia and South Carolina to do the same there.[58]

Despite totalitarian tendencies along the color line, the American South was not the Eastern bloc. With the 1860s now well behind it, Dixie was part of a much larger constitutional republic, just as the United States was part of a still bigger world moving fast away from officially sanctioned racial inequality. One of the most striking aspects of the late 1940s was the instability of traditional political and social structures, including racial hierarchies. The defeat of Nazism, the sharp decline in "scientific racism" since the 1930s, and the creation of the United Nations and its Human Rights Convention all encouraged the forces of egalitarianism around the globe. Within the United States, African Americans took the lead in organizing for a less discriminatory society. Their unwillingness to accept prewar white threats and practices helped lead to the white race riot in Columbia, Tennessee, in 1946. Black Columbians knew every detail of the lynching of a youth in their town thirteen years earlier and of the other eight lynchings in the state since, and news of a white lynch mob assembling downtown moved them to arm themselves and then open fire on white officers entering their neighborhood in the dark. "Most people here remember that [last lynching]," one black resident told a reporter, "and they didn't want another."[59] Even the destruction that followed could not diminish the greater confidence of Columbia's black community that grew from its demonstration of a will to defend itself.[60] Jackie Robinson's breaking of the color barrier in major league baseball in 1947 highlighted the vulnerability of Jim Crow across the country, encouraging black Americans as well as millions of white Americans who found themselves cheering for a person of enormous talent and grace under pressure.[61]

No institution in American life loomed larger in the changing race relations of the 1940s than the federal judiciary. In decisions that paralleled racism's declining international legitimacy, the U.S. Supreme Court outlawed the white primary in 1944, ruled against segregated public interstate transportation in 1946, and banned racially restrictive property covenants in 1948. Also in 1948, the California Supreme Court declared the state's antimiscegenation law unconstitutional. No judges were immune from the Cold War atmosphere that heightened international attention to American race relations, and the Truman administration made good use of these judicial breakthroughs in its diplomacy abroad. The U.S. embassy in Manila, for example, reported great public sensitivity among Filipinos to issues of color and considerable enthusiasm for the California decision.[62] Truman's Justice Department consistently intervened in civil rights cases in an effort to convince the country's highest court of the negative impact of officially sanctioned racial discrimination on American foreign relations. The Justice Department's amicus curiae briefs in the various school desegregation cases that would eventually be collated into the 1954 *Brown v. Board of Education* decision were especially revealing of the international significance of race in America in the early Cold War. The State Department's leading argument in dealing with foreign critics had been that segregation was a regional and declining phenomenon, which would soon be eliminated through the established American legal system. If instead the Court should rule against the plaintiffs in the school cases, finding that segregation was fully compatible with democracy and individual rights, the American position in a mostly nonwhite world would be devastated—and the United States revealed for the hypocrite its critics accused it of being. The ultimate decision in *Brown* followed the antidiscriminatory logic of the Court's previous civil rights decisions, and American diplomats were relieved that Jim Crow's legal sanction had disappeared.[63]

Like both the escalating racial conflicts and the rising tide of antidiscriminatory reform in the immediate aftermath of World War II, the Truman administration's civil rights policies were embedded in a larger international context. Truman's support for civil rights legislation and his 1948 executive orders to desegregate the armed forces and the federal civil service had their closest parallel in the challenges facing Britain's colonial administrators. Like Truman's Democratic Party, the vast British empire incorporated conflicting constituencies ranging from left-liberal egalitarians of all colors to staunch white supremacists. Like Truman's efforts to hold as much of his party together as possible, especially for the 1948 election, the Labor government of Prime Minister Clement Attlee sought to preserve as much unity as it could across three continents. Generally eschewing the French route of counterinsurgent warfare in Indochina, Madagascar, and eventually Algeria, Britain responded to the unyielding demands of nationalist organizers by granting independence to India, Pakistan, and Sri Lanka and setting the Gold Coast—and soon the rest of British West Africa—on the road to self-government.[64] Truman responded to similar pressures at home, stemming from the general decline in black deference that brought a violent white backlash and serious unrest across the South, and from A. Philip Randolph's campaign to desegregate the U.S. military. In both Washington and London, policymakers' relatively liberal inclinations regarding race relations were powerfully encouraged by deteriorating social stability brought on by nonwhite democratic organizing. At the same time, both governments sought to accommodate rather than alienate white supremacist forces that retained considerable influence: for Truman, the Dixiecrats, and for Attlee, archconservatives in England and white settlers in Africa, especially the apartheid government that took power in South Africa in 1948.[65]

Much like the next Democratic president sixteen years later, Harry Truman entered the White House in a period of rising political activism by African Americans and corresponding repression by white Southerners. Like John Kennedy, Truman had previously tried not to anger either side in civil rights disputes and did not wish to see racial issues rise to the top of the national agenda. And, like Kennedy, Truman was forced to choose more plainly between segregationists and egalitarians, stepping clearly into the latter camp.

Both men believed that they governed in times of global crisis caused by Soviet actions, and both were deeply concerned about the damage that racial discrimination at home was doing to their international priorities. Truman's earliest response to this problem—assembling a President's Committee on Civil Rights in late 1946—fended off criticism while establishing a baseline of information about racial discrimination in the country. The committee's final report in October 1947 placed great emphasis on the international reasons for ending Jim Crow, concluding that "our domestic civil rights shortcomings are a serious obstacle" to American leadership in the world.[66] That June Truman also became the first president ever to address the NAACP, telling the crowd of ten thousand and an international radio audience that "our case for democracy . . . should rest on practical evidence that we have been able to put our own house in order."[67]

The political crisis that Truman faced in his uphill battle for reelection in 1948 mirrored the simultaneous international crisis. At home he confronted competitors—Republican Thomas Dewey and Progressive Henry Wallace—with stronger civil rights stands as he fought for his own political survival, while abroad he faced the Soviet Union in what he saw as the nation's struggle for political survival in a world tilting leftward. Burdened with racist allies at home in his own party, he was weighed down with colonialist allies abroad; Dixiecrats were most obviously like Afrikaner nationalists in South Africa, but their political parallel to European colonialists on matters of race relations was also evident. That Truman would share a party affiliation with Senator Richard Russell of Georgia seemed no odder than his alliance through NATO with such white overlords of Africa as Portugal's Antonio Salazar. His domestic political advisers thus encouraged his own growing commitment to standing against racial discrimination. Clark Clifford's well-known memorandum of advice about the campaign emphasized that "the northern Negro voter today holds the balance of power in Presidential elections." Unlike the white South, which he wrongly believed "can be safely considered Democratic," Clifford argued that African Americans might well return to the party of Lincoln—much like Third World peoples possibly turning to the party of Lenin—if the president did not woo them more explicitly.[68] These were, of course, similar rather than identical situations: African American votes were clearly more immediately important to Truman than the friendship of impoverished nations in the long arc of new states stretching from Morocco to Indonesia. But the political dilemma for the president of racially charged conflict continued on both domestic and international levels.

The obviously political timing of Truman's initiatives on race in 1948, combined with his continuing effort to keep white Southerners on board his campaign, seem less significant in retrospect than the unprecedented fashion in which he placed the authority of the White House and the Democratic Party behind the goal of racial equality.[69] The Democrats, it must be remembered, had long been the party of the white South, of slavery and the lynching rope. It was Truman's rejection of racial discrimination that began to shatter that historic identification and started the seismic shift of a majority of white voters in Dixie to the Republican Party by the 1990s. His civil rights message to Congress on 2 February 1948 made explicit the international context of American race relations: "If we wish to inspire the peoples of the world whose freedom is in jeopardy, if we wish to restore hope to those who have already lost their civil liberties . . . we must correct the remaining imperfections in our practice of democracy."[70] Prominent among the pieces of legislation he called for was a federal antilynching law, as lynching—"the law of the jungle," as the NAACP so accurately and pointedly called it—probably did more to undercut American legitimacy as a leader of the "free world" than anything else.[71] Truman's executive orders for the integration of the armed forces and the civil service marked other milestones, as did the Democrats' strong civil rights plank and the president's campaign speech in Harlem to a crowd of 65,000 three days before the election. All helped him win a large majority of black votes and a narrow overall victory, to the delight of racial liberals both in the United States and abroad.[72]

One of the fruits of victory in 1948, ironically, helped limit Truman's pursuit of racial reform in his second administration. The Democrats regained control of Congress, assuring its Southern members of renewed political power through their committee chairmanships. The situation paralleled the problem of Truman's successful creation of a Western European alliance with the Marshall Plan and especially NATO, which bolstered the economic and military might of the colonial powers in Africa and Asia. The president's civil rights package ran into the buzz saw of segregationist opposition on Capitol Hill, where a liberal effort in early 1949 to weaken the filibustering option in the Senate was trumped by a successful conservative effort to strengthen it instead.[73] So Truman looked to the courts and the military for movement toward desegregation, where success was gradual but more forthcoming. But the twin crises of 1950—the onset of the Korean War and the meteoric rise of Senator Joseph McCarthy—brought a swift end to major efforts at social reform of any kind. Awash in the maelstrom of a military emergency in East Asia and a political emergency at home, the president sought above all to unite his party and the country behind his leadership. Potentially divisive issues would have to wait for calmer days.[74]

Diplomacy and Shifting Racial Identities

The most salient fact about both race and American diplomacy between 1945 and 1953 was how contested and in flux each of them was. African Americans battled for equality at home; people of color fought for independence from white rule across the colonial world; for the first time in world history, the great powers united in the UN Charter in opposing racial discrimination; and the very idea of race as a biologically meaningful way to distinguish between peoples was fast losing its scientific authority.[75] Over all of these shifts loomed the chilling rivalry between the United States and the Soviet Union, dividing the world into competing blocs, threatening it with nuclear destruction, and absorbing most of the attention of the Truman administration. One aspect of race in the United States that was not succeeding fast enough, however, was the struggle of African Americans to achieve full civil rights. Consequently, Americans remained in peril of forfeiting their claim to leadership of a mostly nonwhite world.[76]

Yet the crumbling of many long-standing hierarchies of color helped create some hope that racial equality could be achieved. New uncertainties about the racial identities of individuals arose as blacks and whites alike tested the color line. "Passing" as a member of a different "race" had long been an option for many persons, owing to the artificiality of race as a category of absolute distinction for people of diverse genetic inheritances.[77] After 1945, the phenomenon of passing moved in many directions. Traditional efforts by some light-skinned people of color to be treated as whites—that is, to be treated without humiliation—continued, as journalist Carl Rowan observed at a whites-only restaurant outside Atlanta where he was refused service. Rowan struck up a conversation outside with a man of dark hue, who explained that he was waiting for his lighter-skinned cousin inside: "Yeah, he's passing. And he ain't much lighter'n you 'n' me. When we get hungry in these country places he walks right in and orders some sandwiches. If they tell him they don't serve colored, he says, 'I don't blame you; I wouldn't serve no damn niggers, neither.' Them crackers get so flabbergasted they don't know what he is, and they serve him. They figure that anybody who don't like niggers is on their side."[78]

With visitors from the Third World arriving at the United Nations and touring the United States in growing numbers, some African Americans occasionally enjoyed passing as foreigners. Putting on a foreign accent and donning exotic attire like turbans, they could share temporarily in the status of nonwhite visitors as "honorary whites" in Washington's segregated restaurants and hotels.[79] Such adventures could be perilous, however, as white Americans commonly made the reverse identification of dark-skinned foreigners with black Americans, offering them instead the back of Jim Crow's hand. When four black students from the British West Indies sat down for a meal at a lunch counter in Washington, D.C., for example, they were refused service until they produced their British diplomatic passes. Then the embarrassed waitress apologized, explaining that she had not realized they were "not niggers." Being a "nigger" in America's capital in 1948 was evidently a function not simply of skin color but also of nationality.[80]

It was no surprise that some white Southerners were angered by this new wrinkle in policing the color line, but many did their best to accommodate it. After a sharp debate among the waiters at a restaurant in Montgomery in the summer of 1951, visiting Indian politician Rammanohar Lohia was finally allowed to have lunch. Afterward the waitress explained confidentially to Lohia's white escort, future U.S. senator Harris Wofford, "I knew he wasn't no colored man. We're getting good at spotting foreigners. Besides, no colored man would come in here."[81] Lohia's dark skin did not quite make him a "colored man." Even for segregationists, racial identities did not always fit neatly into boxes.

As racial hierarchies and distinctions came increasingly into question in the years after World War II, a few white Americans occasionally tried to pass as African Americans to see what life was like under Jim Crow. The best known case was John Howard Griffin, a white Northerner who had his skin chemically altered to appear black and then went south to investigate the other side of the color line.[82] Less dramatic was the example of Dick Sanders, a white Southerner and classmate of Carl Rowan at Oberlin College. Boarding a segregated bus together in McMinnville, Tennessee, in 1947, they sat down together in the black section at the back. Eventually alerted by a white woman passenger who kept turning around to stare at them, the driver pulled off the road and came back to tell Sanders that he could not sit in a "colored seat."

Without a pause, the young man responded in an embarrassed whisper, "My father was white." Embarrassed in turn, the driver returned to the wheel, explaining the situation to the woman en route. When they got off in Knoxville and the driver handed him his bag, Sanders smiled and said, "Say, I forgot to tell you back there that my mother was white, too." The driver responded first with an embarrassed look, then with a smile, and finally a loud laugh, which Sanders and Rowan joined in. Not all white Southerners failed to appreciate the absurdities of Jim Crow's racial classifications.[83]

Another kind of passing in the swiftly changing times of postwar America came with a crucial change in U.S. immigration law in 1952. For the most part, the McCarran-Walter Act was a conservative bill that reconfirmed the racially discriminatory system from 1924 for using national origins to limit entrance almost exclusively to European immigrants. Truman unsuccessfully vetoed the legislation because it preserved that system, which he called "a slur on the patriotism, the capacity, and the decency of a large part of our citizenry" and "a constant hardship in the conduct of our foreign relations."[84] But the State Department had actually urged the president to sign the measure because it also included a historic section ending the ban on the naturalization of Asians.[85] For the first time in American history, residents of the United States who had come from Asia could become American citizens. A crucial step had been taken along the road toward a nonracial definition of U.S. citizenship. Asians could now pass as—in fact, become—Americans.[86]

The indeterminacy of racial identities in the immediate aftermath of World War II also helped shape shifting American attitudes toward the state of Israel, created in 1948 in the midst of Truman's campaign for reelection. Traditional American anti-Semitism and the scientific orientation toward racial categorization had combined in the 1924 immigration legislation to define Jews as members of a distinctive "Hebrew race." Anti-Semitic sentiment among gentile Americans continued to build, reaching a peak during World War II, simultaneous with the Holocaust. Then came a sharp change. For several weeks after the German surrender in May 1945, U.S. newspapers were filled with stories of the Nazi gas chambers and horrifying pictures of the victims of anti-Semitic evil. These were Harry Truman's first two months as president, and he shared the shock, revulsion, and perhaps even guilt of most Americans about what had happened. When Zionist settlers just three years later won a stunning military victory in their war of independence, gentile American attitudes toward Jews continued to shift, leading to a steady decline in anti-Semitism and a turn away from defining Jews in racial terms. No longer impotent victims of Nazis, Israelis—and, by extension, Jews else-where—were increasingly seen by gentile Americans as tough, independent pioneers who had conquered hostile indigenous peoples and opened up pre-viously marginal lands to fruitful agricultural production, much like Ameri-cans' own ancestors. With Israel on its way to becoming a crucial American ally in the Middle East, and with anti-Semitic quotas disappearing in the United States, Jews could increasingly pass as fully privileged American citi-zens and Israelis as members of a modern, Western nation.[87]

The relative nature of racial designations also characterized American per-spectives on East Asia after the war ended in the Pacific. Long-standing white American prejudice against Asians and people of Asian descent had often been complicated by an opposing tendency to appreciate those who seemed to admire American ways. Unlike the old racist adage about Native Ameri-cans, there apparently was such a thing as a live "good Asian." This bifurcated view showed up vividly in World War II, when most white Americans viewed the Japanese enemy in explicitly racial terms—as inherently fanatical, savage, treacherous, and inhuman—but regarded their Chinese allies as courageous, democratic, and business-oriented. Between 1945 and 1949, however, the ra-cial traits attributed to the Japanese and Chinese were almost entirely re-versed. Occupied and friendly to Americans, Japanese became democratic, reasonable, and pro-business, while the newly Communist Chinese took on the role of fanatics and savages, especially after they entered the war in Korea in late 1950 and began killing American soldiers in large numbers.[88]

A similar shift occurred regarding Korea in the same period, as the postwar political chaos on the peninsula encouraged American observers to reverse their prior views of Koreans and Japanese. Brave former Korean nationalist allies became suddenly uncooperative, difficult, and even subversive, while brutal Japanese forces were now seen as cooperative and orderly—and were thus used by U.S. commanders for preserving order long after their surren-der.[89] The early Cold War revealed that white American racial thinking was quite malleable in the face of changing political and strategic realities. De-spite anti-Chinese sentiment that swelled after the People's Liberation Army crossed the Yalu River, there was no internment or even sustained harassment of Chinese Americans as there had been of Japanese Americans eight years earlier.[90]

The same flexibility, in light of changing political circumstances, about meanings assigned to racial categories could be seen in the federal govern-ment's relationship to African Americans. The federal employee loyalty boards set up by the Truman administration in 1947 assessed the loyalties of black and radical workers partly by the degree of their support for biracial friendships. It remained a suspicious activity to socialize across racial lines and carried the taint of radical political associations. FBI Director J. Edgar Hoover was particularly eager to link civil rights activity and mobilization to subversive organizations, and he flooded Truman's office with reports linking the Communist Party to civil rights organizing efforts. But the president's own growing support for an end to racial discrimination kept him at odds with Hoover on this issue. Truman and his advisers might have had no use for black radicals like Paul Robeson and W. E. B. Du Bois, but support from anti-Communist African American leaders like Walter White encouraged the president to avoid simplistic conflations of racial egalitarianism with political radicalism.[91]

White Southern political elites in the president's own party worked much of the same ground as Hoover in these years. Particularly from 1948 on, they responded to the new stance of the national Democratic Party in favor of civil rights by elevating the rhetoric of anti-Communism as a more widely acceptable cover for segregationism. If Communists and other radicals supported racial equality, they asked, what clearer evidence could there be of its subversiveness? They played on rising fears of domestic subversion in an effort to fend off Truman's liberal racial legislative goals. After the establishment of the Truman Doctrine and the federal employee loyalty programs in 1947, the president and other administration officials increasingly explained demands for social change within the international system—especially immediate decolonization—as influenced by a monolithic Communist conspiracy headquartered in Moscow. Dixiecrats and other white Southern leaders brought that argument home to their own region, attributing efforts at racial reform there to Communist agents.[92]

The smearing of liberal reform with Communist associations reached its apogee in the early days of McCarthyite political anxiety, as segregationists in North Carolina sought to defeat the most renowned Southern liberal of the era, U.S. senator Frank Porter Graham, in his reelection effort in the 1950 Democratic primary. A young Raleigh journalist named Jesse Helms played a critical role as campaign publicity director for segregationist Willis Smith, whose supporters catered to voters' anti-black and anti-red inclinations by labeling the incumbent as a dupe of Communists and a proponent of the "mingling of the races." The latter was the trump card, based vaguely on Graham's support for school desegregation but played out on fliers with pictures of black U.S. soldiers dancing with white Englishwomen in London during the war, under the captions "Remember these . . . could be your sisters or daughters" and "WAKE UP WHITE PEOPLE." On the very day North Korean troops poured across the thirty-eighth parallel into South Korea, white North Carolinians took the advice of Helms and his comrades and voted to defeat Graham.[93]

While white Southerners challenged Truman from the right by conflating anti-Communism and racial discrimination, Henry Wallace and his Progressive Party pressed the president from the left by declaring both to be bankrupt policies. The former vice president became the most prominent white supporter of full racial equality in the late 1940s, reflecting the growing openness of many white Americans after the experiences of the Depression and World War II to move beyond the simple boxes of racial identification.[94] Wallace threatened Truman, in part, by linking this more liberal domestic stand with a fierce critique of the president's anti-Soviet actions abroad since 1945. Harking back to the wartime Grand Alliance, Wallace sought to defuse the Cold War by recreating a reasonable relationship with Moscow in place of the containment doctrine advocated by George Kennan. His opposition to colonialism, and to the imminent NATO alliance that would partly support it, suggested a sensitivity to international racial hierarchies that matched his actions at home.[95] He and his running mate in the 1948 presidential campaign, Senator Glen Taylor of Idaho, each toured the South while refusing to speak to segregated audiences and endured threats to their physical safety as a result.[96] Daisy Bates, a black Arkansan and key figure in the Little Rock school desegregation crisis nine years later, recalled listening to Wallace speak: "I had waited all my life to hear a white man say what he said."[97] Truman's campaign was able to publicize Communist Party support for Wallace as an effective method of reducing his popularity within a largely anti-Communist electorate.[98] But Wallace had at least articulated the antiracist convictions shared by a growing number of white Americans, helping in the process to push Truman toward the more liberal racial stand he took in 1948.[99]

The Wallace campaign pointed to one final issue of uncertain racial and national identity that helps illuminate the unsettled relationship between race and diplomacy during Truman's presidency. African Americans of all political stripes had united during World War II in their opposition to colonialism and support for Asian and African independence. After the war they continued to view racial discrimination in the United States within that broader international context. But the onset of the Cold War between the United States and the Soviet Union forced them to choose between retaining an internationalist perspective that would leave them outside the bounds of mainstream debate in the United States by opposing America's European allies, or adopting a more nationalistic, anti-Communist stand that supported U.S. foreign policy while pursuing racial equality more narrowly at home. In a sense, Cold Warriors such as Truman and Kennan in 1946–47 implicitly raised the question of whether African Americans identified themselves primarily in racial or national terms. Did they represent a potential fifth column of the world's nonwhite majority within American walls, or were they loyal American nationalists fully committed to the country's dawning struggle against the forces of international Communism? It was a question, backed by a withering fire of political, social, and economic sanctions, that forced painful choices on black political leaders.[100]

Most African Americans in the political arena went with Truman and the Democratic Party. The president's unprecedented support of civil rights at home convinced them to tone down their criticism of American policies toward the colonial world to press on with the struggle for justice at home. Some were genuine anti-Communists, while most remained, on average, more doubtful than white Americans about the seriousness of a domestic threat of subversion; their experiences of power and prejudice in the United States left them unlikely converts to a campaign of repression. But they could read the writing on the wall. As the whole structure of the wartime Popular Front, uniting the Center and the Left, caved in under the weight of the Cold War, people like Walter White hustled out to the safe ground of the anti-Communist and increasingly antiracist president. Black leaders further to the left refused to leave that alliance structure to which they had committed their political lives and were destroyed along with it. Du Bois, Robeson, and William Patterson lost their passports for continuing to criticize American diplomacy and racial practices and had their careers crushed by federally sanctioned repression. A credible, coherent black Left sank below the surface of American political life, as the chilly waters of the Cold War closed over it.[101]

Truman, Colonialism, and Africa

Western European and North American colonialism had long been charged with racial tensions and meanings, which inevitably complicated the Truman administration's policies toward decolonization. The essential problem for the White House was how to create as large and strong an anti-Communist, "free world" coalition as possible. The alliance had to be not only international but also multiracial, just as did the liberal anti-Communist political coalition Truman was assembling at home. The difficulty came in the conflicting axes of global tension that emerged after World War II: the East-West axis of the Americans and Soviets, and the North-South axis of the European metropolitan powers and their vast colonial territories. To keep both paler North and darker South on the Western side of the horizontal axis represented a major challenge for American foreign policymakers. In a report to the president in September 1948, the Central Intelligence Agency (CIA) underlined the "serious dilemma" of trying to forge close ties with the anticolonial new nations of the Middle East and Asia, while preserving good relations with European powers eager to retain their last colonial territories in Africa. Over it all hung the issue of color, the agency observed: the "deep-seated racial hostility of native populations toward their colonial overlords," caused by centuries of imperial exploitation.[102]

The situation facing the Truman administration abroad after World War II had certain similarities to the problems confronting the U.S. government after the Civil War. In both cases Washington had to figure out how best to reconstruct a defeated enemy, whether the Confederacy or the Axis powers, to mesh with a desired postwar order. In 1945 reconstruction included rebuilding war-torn Western European allies, whose economies depended greatly on the labor of nonwhite workers in their colonies—much like the South's dependence on African American labor in 1865. In both cases, the U.S. government had to decide whether to force white overlords to grant greater liberty to those workers—whether emancipation from slavery or independence from colonialism—and if so, how to reconcile that liberation with the rebuilding of needed white allies. Well aware that Jim Crow rather than full citizenship had ultimately followed emancipation, racial egalitarians three generations later sought to avoid either neocolonialism in Asia and Africa or a form of neosegregationism in the United States. Robert E. Lee was not Winston Churchill, Frederick Douglass was not Mohandas Gandhi, and the cotton plantations of Alabama were not the rubber plantations of Southeast Asia or the Congo, but the parallel dilemmas of Reconstruction and Cold War decolonization suggested the enduring character of race as a central problem for American policymakers across national boundaries.

Truman and his advisers hoped for a gradual but steady process of national independence spreading across the Third World, with power passing peacefully from retreating European colonial officials into the hands of pro-Western, anti-Communist indigenous elites. The ending of colonial rule had created the United States, after all, and anticolonial sentiment and conviction had endured through America's own rise to global power and acquisition of overseas territories. The economic advantages of greater American access to others' colonial territories—that is, free trade—dovetailed with traditional political principle. The Truman administration believed its own grant of independence to the Philippines in 1946 provided a model for the rest of the West.[103] The president was especially pleased when Britain arranged to withdraw from India and Pakistan a year later, a momentous step away from the old hierarchical empire and toward a new multiracial commonwealth.[104] Many in Britain and in other metropolitan nations resented American pressures for decolonization, particularly in light of Jim Crow's continued existence. Randolph Churchill, for example, asked visiting American journalist Walter Lippmann over lunch, "Why do you always worry about our niggers? We don't worry about yours."[105] But the U.S. government recognized the importance of self-rule in the colonial world for the successful waging of the Cold War. The Truman administration even occasionally used serious economic threats to promote that process, as when it forced the Dutch to de-escalate their war against non-Communist nationalist guerrillas in Indonesia.[106]

Sluggish economic revival in Western Europe and American fears of Communist advances in 1946 and 1947 diminished the Truman administration's concern with decolonization. The consolidation of a robust, integrated, anti-Communist Western Europe soared to the top of Truman's agenda, tightening the American embrace of colonial regimes in London, Paris, Brussels, and Lisbon and deepening Third World skepticism of American motives. The Marshall Plan and NATO aimed to bolster the economies and military forces of the metropolitan governments but also served to strengthen them in their quest to retain control of valuable colonies abroad.[107] Critics pointed this out at the time: Senator George Malone of Nevada told his colleagues that "the North Atlantic Pact simply guarantees the integrity of the colonial systems throughout Asia and Africa."[108] As the Dutch discovered regarding Indonesia, the results of American aid were not always that straightforward. But it was clear that the president's famous anti-Communist distinction in his Truman Doctrine speech—between a way of life "based upon the will of the majority" and one "based upon the will of a minority forcibly imposed upon the majority"—skipped over the most common form of minority rule in the world: colonialism.[109]

Race remained a crucial factor in the association of the United States with Western Europe.[110] The Soviets were swift to tell Third World listeners that the implementation of the Truman Doctrine would inevitably bring American racial discrimination in the wake of American dollars.[111] Scholars and journalists like Harold Isaacs and Richard Deverall traveled extensively in Asia after World War II, returning to report, in Isaacs's words, that "American racial discrimination is one of the facts that identifies us with colonialism in the Asian and African mind. They recognize it for what it is and relate it to their own experience."[112] The administration's own Psychological Strategy Board agreed that in colonial and newly independent areas, "the memory or actuality of domination by the white man is a far greater psychological reality than the Soviet menace."[113] While many African Americans protested Winston Churchill's Iron Curtain speech in 1946, delivered before Truman at segregated Westminster College and emphasizing the "special relationship" between English and Americans, most powerful white Americans applauded. Senator Richard Russell even took Churchill's suggestion of eventual "common citizenship" a step further by offering to invite England, Scotland, Wales, and Ireland (and perhaps Canada and Australia) to become American states—a construction of the whitest possible U.S. imperium.[114] Few democrats missed the irony of Dean Acheson, the undersecretary of state, promoting U.S. financial assistance for "freedom and democracy and the independence of nations" abroad to an audience of elite white planters in Cleveland, Mississippi, in May 1947, while just down the road the poorer allies of those planters were busy crushing a nascent voter registration drive by darker-skinned residents. "White Supremacy Is in Peril," declared the *Jackson Daily News,* asking its readers, "[Do] you want a white man's government, or will you take the risk of being governed by Negroes?" Acheson's pleas to help the administration support "human dignity, human freedom, and democratic institutions" worldwide had a peculiar ring in the Mississippi Delta in 1947.[115]

The racial element in U.S. tolerance of European colonialism showed up most starkly in regard to Africa. The centrality of race in U.S. policy toward that continent was due partly to the European tendency to contrast "black" Africa with "white" Europe, which mirrored the bipolar racial thinking typical in the United States, and partly to Africa's status as the last major area of European overseas control. Rapid postwar decolonization and Communist-led insurgencies in Asia made the Truman administration grateful for an area with little visible anticolonial or anticapitalist organizing. What the senior administration official with responsibility for Africa, Assistant Secretary of State George McGhee, called Africa's "situation of relative stability" preserved both the continent's significant contribution to the economies of Western Europe and American access to certain critical minerals there, particularly the uranium ore of the Belgian Congo and South Africa.[116]

With less previous involvement in Africa than on any continent besides Antarctica, American policymakers after World War II relied heavily on the European and white settler authorities of the region in dealing with its problems. This policy resulted partly from traditional diplomatic practice everywhere: governments deal mostly with governments. But it also reflected the political, cultural, and racial ties that American elites felt with their Western European counterparts. American diplomats and visitors had almost no significant contacts with Africans in the late 1940s, admitting privately their nearly total ignorance of what they called "native issues." Sympathetic with the white officials whose segregated society they shared there, white Americans tended instead to view Africa through European eyes.[117] Occasional exceptions startled and encouraged African nationalists, while exacerbating the latent fears of Europeans that the United States ultimately planned to replace them as the dominant force on the continent. Journalist William Attwood, later a U.S. ambassador in Guinea and Kenya, recalled the surprise of Guineans when he shook their hands during a 1947 visit. French officials had apparently encouraged Africans to expect no white American to touch a person with dark skin, and the more egalitarian style of some Americans like Attwood complicated the picture of a segregated United States, especially in light of the Truman administration's growing support for desegregation at home.[118] But the enduring closeness of white Americans with colonial authorities, along with the disenfranchisement of African Americans at home in the South, left uncertain the answer to the increasingly common African question, paraphrased by the U.S. ambassador to Liberia: "Does the U.S. favor rule 'of' the majority 'by' the majority in Africa as it does in Europe, the U.S., or Communist areas?"[119]

The Truman administration essentially answered this question, "Eventually, but not yet." That dawning of democracy was closely correlated to the density of white settler populations. Where European settlers were few, as in British West Africa and in North Africa outside Algeria, self-government would come soon. Where they were many—in the temperate areas of eastern and particularly southern Africa—their resistance to majority rule would slow the process of decolonization. In most of Africa, in other words, the presence of white people corresponded directly to the absence of democratic practices. This correlation created a conundrum for white American policymakers, who were accustomed to associating the influence of whites on the continent with Western civilization and material progress.[120]

For the best possible resolution of this dilemma, the United States looked again to Britain to provide a model of progressive, orderly decolonization that would not play into the hands of radicals or Communists. U.S. diplomats in Salisbury cheered the organizing of Southern Rhodesia, Northern Rhodesia, and Nyasaland into the Central African Federation in 1953, despite grave African doubts, as giving "the white man . . . a golden opportunity to really make his concept of a Free World work successfully."[121] That effort failed miserably over the next several years, however, demonstrating the limits of what metropolitan governments could or would do to restrain white settlers.[122] Meanwhile, the extent of African anger at colonial rule and European settlers became clear in British Kenya, where the Mau Mau rebellion broke out in the final months of the Truman administration. The chief American official in Nairobi recognized that a deep "racial division of wealth" lay at the heart of the conflict in Kenya, as whites there understood. But for a U.S. government primarily committed to containing Soviet influence and deep at war in Korea the bottom line on the anticolonial struggle in Kenya was how it worked "to the benefit of International Communism in creating another focus of unrest in the Western sphere."[123]

The most extreme version of the Truman administration's dilemma of how to fit diplomacy and race together unfolded in the Union of South Africa. The electoral victory of the Afrikaner nationalists in 1948 established the system of apartheid, as white South Africans rejected the postwar global trend toward racial equality. It was as though the Dixiecrats had won in the United States: absolute racial segregation and discrimination became the law of the land, and the evangelically racist government of Daniel Malan looked to spread its gospel beyond its borders. As the wealthiest and most powerful state in the region, South Africa became the core of a southern African white redoubt that utterly rejected the principle of human equality. While Communist totalitarianism rolled across eastern Europe, racial totalitarianism seeped into every corner south of the Limpopo River. Indeed, the rule of apartheid seemed a perfect model for how to drive Africans into the arms of the Communist Party, which remained the only political group open to all races and committed to complete racial equality. The injustices of anti-

Communist apartheid threatened to legitimate Communism as the only real defender of democracy in the region. Here was a terrible dilemma for American Cold Warriors.[124]

Strong common interests linked Washington and Pretoria. Historic alliances in both world wars and the Korean War, mutual anti-Communism, and growing trade patterns tied them together.[125] Strategic minerals proved even more important, with South African manganese, chrome, and uranium becoming crucial elements for the postwar American armaments industry.[126] Underlining these tangible concerns, Secretary of State Dean Acheson reminded U.S. officials to avoid "taking hold of glowing principles [of racial equality and majority rule] and dropping these other important considerations."[127] White Americans were also strongly inclined by tradition to identify culturally with white South Africans, whose European ancestry and frontier past seemed so like their own: bearers of Western civilization and Christianity to a continent inhabited by less technologically sophisticated, non-Christian, darker-skinned peoples. The racial character of this common identity was occasionally made explicit, as when State Department intelligence officers emphasized the importance for U.S. policy of South Africa's "having the largest white population on the African continent."[128] George McGhee, later recalled his own Texan heritage as a source of the "sympathy" he felt for South Africans "for their extremely difficult racial problem." Which South Africans he was referring to was no mystery: "It was unfortunate that I was not able, during my visit to Africa in 1950, to meet any of the African leaders."[129] Enduring habits of racial identification limited the understanding of American policymakers—even those of considerable goodwill, like McGhee—about who the vast majority of South Africans actually were.

The contrasting results of the 1948 elections in South Africa and in the United States indicated that the window of opportunity for a rising apartheid and a declining Jim Crow to embrace each other comfortably would not remain open forever. Unlike the Afrikaner nationalists, the Dixiecrats were resoundingly defeated by both major party candidates, who each supported greater racial equality. In South Africa there was no Supreme Court interpreting the Constitution in increasingly color-blind fashion. Dixie was only part of the United States; South Africa had no equivalent to the American majority beyond the Mason-Dixon line.[130] The governments of the two countries could still follow mutual interests into the same bed in the late 1940s, but growing tensions between the two over race relations began to point to the parting of the ways that the decline of the Cold War in the 1980s finally brought.[131] In the meantime, the Truman administration did its best to fend off critics who emphasized the parallels between apartheid and Jim Crow.[132] Liberal officials pointed to accumulating evidence of racial reform in the United States. Traditionalists like Acheson struck a more apologetic note while making the same point. He explained to the South African ambassador that the U.S. position regarding any UN discussions or investigations of apartheid "must of course necessarily be conditioned to an undetermined extent by complicating factors of our own domestic and public opinion situation."[133] In sum, South Africa had begun its lengthy term as the most awkward member of the multiracial alliance that the United States sought to maintain against the Soviet Union.

Jim Crow on the International Stage

The elemental problem for America's first Cold Warriors in dealing with race was their inability to wall off white American racial attitudes and practices from the rest of the world and its nonwhite majority. The very idea of a containment of domestic racism cut directly against the grain of a major purpose of postwar U.S. foreign policy: to assert American leadership around the globe, thus exposing more rather than less of Americans and American culture to other nations. Jim Crow survived the inward turn of American society after World War I, but it fared much differently as an outrider on a sharply expanding American presence in the international arena after 1945. Competing with the Soviet Union in the postwar world meant, by definition, maximizing the amount that other peoples saw of American life. Hiding one's flaws was considered the telltale sign of a secretive, unfree country— precisely what American leaders in the late 1940s identified as the defining characteristic of a Communist nation.

The swift projection of American power onto every continent during World War II and Truman's assertion of U.S. leadership thereafter brought unprecedented international attention to American society. Other peoples had long been curious about this nation of immigrants, whose multicultural character connected it to almost every part of the world. The Cold War focus on the ideals of democracy and freedom assured that racial exceptions to the American practice of those principles would receive careful attention.[134] The long shadow of the Holocaust encouraged the postwar internationalism that also challenged unlimited national autonomy. This was the new era of the

United Nations and its Human Rights Convention for the whole world, a far cry from the American tradition of local government being solely responsible for the protection of citizens' civil rights. Acts of racial violence in obscure rural parts of Dixie changed almost overnight from events of mostly local interest to headlines splashed across newspapers around the world.[135]

The Soviet government and its allies, unsurprisingly, delighted in publicizing news of American racial discrimination and persecution. Racial events offered an irresistible opportunity to respond to American publicity about repression of individual liberties in the Soviet bloc. When Secretary of State Byrnes tried to protest the Soviet denial of voting rights in the Balkans in 1946, he was stumped by the Soviet retort that "the Negroes of Mr. Byrnes' own state of South Carolina" were "denied the same right"—"a checkmate of the first order," admitted a U.S. psychological warfare official. The U.S. Central Intelligence Group (precursor to the CIA) reported in 1947 that "the bulk of Moscow's criticism appears to be focused on U.S. discrimination against Negroes."[136] Moscow Radio framed U.S. military actions in Korea in terms of domestic American racism, calling bombing runs over North Korea and "the oppression of the colored people in the U.S. . . . links in the same chain."[137] Soviet propagandists zeroed in on evidence of double standards, such as the Voice of America's extensive coverage of the arrest of white South Carolinians accused of the 1947 lynching of Willie Earle, compared with its brief mention of their subsequent acquittal.[138] American diplomats in Moscow believed their hosts utterly insincere in these attentions to African Americans, but Soviet leaders consistently supported the overthrow of colonialism and the end of legislated racial discrimination. U.S. officials like Ambassador to India Chester Bowles admitted privately that the Soviets—despite their intense political repression at home and in Eastern Europe—in regard to racial equality "have done a better job than we."[139]

America's Western European allies found reports of racial persecution in the United States alternately scandalous and reassuring, offensive to their sense of decency but also helpful—especially to government officials—as they tried to fend off criticism of their own colonial practices. Some could not resist highlighting American double standards when the U.S. government issued anticolonial statements while allowing segregation to continue at home.[140] Certainly, though, the foreigners most interested in—even preoccupied by—what commentator Harold Isaacs called the infection of racism among white Americans were the nonwhite peoples of the new nations of Asia and the Middle East and the remaining colonies of Africa.[141] For them, the implications of how the world's most powerful country dealt with its own people of color were enormous. For them, the Klansman's knife, the police officer's baton, the senators' debates, and the president's actions had a particular immediacy. "Throughout the Pacific, Latin America, Africa, the Near, Middle, and Far East," the President's Committee on Civil Rights concluded, "the treatment which our Negroes receive is taken as a reflection of our attitudes toward all dark-skinned peoples."[142]

Truman and his foreign policy advisers were well aware of international attention to American race relations. Acheson in 1946 regretted with some annoyance being "reminded over and over by some foreign newspapers and spokesmen that our treatment of various minorities leaves much to be desired."[143] Two years later, Eleanor Roosevelt, a member of the U.S. delegation to the United Nations, called "our open discrimination against various groups . . . the one point which can be attacked and to which the representatives of the United States have no answer."[144] By 1950 Republican senator Henry Cabot Lodge was referring to race relations as "our Achilles' heel before the world," while the State Department noted that "no American problem receives more wide-spread attention, especially in dependent areas, than our treatment of racial minorities, particularly the Negro."[145] Not surprisingly, American policymakers sometimes tired of the issue and enjoyed turning the tables on critics such as South Asian Indians by pointing out their own color prejudices. Segregationists like Byrnes and Marshall were particularly inclined to respond in this fashion.[146] But the problem endured throughout Truman's presidency, as Bowles reminded an audience at Yale University in 1952: "A year, a month, or even a week in Asia is enough to convince any perceptive American that the colored peoples of Asia and Africa . . . seldom think about the United States without considering the limitations under which our 13 million Negroes are living."[147]

Perhaps nothing elicited world attention to American racial practices as effectively as direct appeals by African Americans to the international community.[148] Black Americans filed three major petitions with the United Nations during Truman's presidency, cataloguing the forms of discrimination under which they suffered and asking for assistance in pressuring American authorities to eliminate them. The appeals of the National Negro Congress in 1946, the NAACP in 1947, and the Civil Rights Congress in 1951 made clear that many Americans viewed civil rights in an international context.[149] These petitions were acutely embarrassing to the Truman administration; Attorney General Tom Clark publicly declared himself humiliated by such evidence that American citizens could not receive decent treatment in their own country.[150] There was little new in the petitions, even in the most scholarly and impressive one, *An Appeal to the World*, prepared by W. E. B. Du Bois for the NAACP. What was humiliating was the force with which they cut against the grain of U.S. foreign policy, exposing distinct racial limitations on the American campaign for freedom. *An Appeal to the World* appeared precisely in the midst of the administration's Cold War initiatives, with the president dividing the world into "free" and "slave" halves in the Truman Doctrine and bolstering the "free" half with the Marshall Plan and NATO. "It is not Russia that threatens the United States so much as Mississippi; not Stalin and Molotov but Bilbo and Rankin," Du Bois wrote. "This protest is a frank and earnest appeal to all the world" in hopes of inducing "the nations of the world to persuade this nation to be just to its own people."[151] It was hard to imagine a more pointed refutation of Truman's argument that the United States should now be going abroad to promote justice and freedom in other nations.

The Truman administration responded to the persistence of such critics who sought international assistance with two contrasting but complementary tactics. American policymakers deeply resented African Americans who encouraged foreigners to examine what they viewed as purely an internal matter. As the mercury of the Cold War thermometer plunged, the leftist associations of most of those critics made it easy for many government officials to assume they were trying to help the Soviet cause. The FBI, the State Department, the Justice Department, and Congress all contributed to the successful campaign to limit their audiences, publicly labeling them tools of the Soviet Union and canceling their passports.[152]

At the same time, the Truman administration sought to defuse these criticisms through a more positive approach. Most fundamentally, this meant emphasizing the president's own real commitments to racial reform and desegregation at home. Truman's advisers timed certain initiatives to counteract the negative publicity associated with the petitions: releasing the final report of the President's Committee on Civil Rights six days after the NAACP petition was published, and issuing the U.S. Information Agency's pamphlet "The Negro in American Life" partially in response to the Civil Rights Congress petition.[153] The boldest aspect of this approach was to send more conservative African Americans abroad to tell a different story than a Du Bois or a Robeson. Foreign critics of race relations in the United States could be most effectively rebuffed by the presence of visitors like Max Yergan and Edith

Sampson, who would not deny the continuing existence of racism at home but who would emphasize instead that the American system was steadily bringing greater equality to all.[154] The appointment of Edward R. Dudley as the first black American to the rank of ambassador (to Liberia) in 1949 represented a similar tactic.[155]

Beyond the ideologically charged arguments of Cold Warriors and leftist radicals, the bottom line for people of other nations evaluating American democracy along the color line was how the white American majority behaved. Actions in the South and around the country were important, while racial practices in Washington and New York and among Americans abroad often made the most dramatic impressions. Washington, the Justice Department argued before the Supreme Court, was "the window through which the world looks into our house."[156] What the world saw was no model of democracy: pervasive discrimination, widespread poverty, and the political disenfranchisement of all its citizens, an anomaly in the capital of a republic.[157] The District of Columbia was governed by congressional committees usually under the control of white supremacist Southerners like Theodore Bilbo, who vowed to preserve the nation's capital as a model of segregation. The city's *Evening Star* observed in 1946 that "it must be viewed as one of the ironies of history that the Confederacy, which was never able to capture Washington during the course of that war, now holds it as a helpless pawn."[158] Six weeks after Truman spoke to the nation and the world about the need to uphold freedom in distant capitals like Athens and Ankara, the executive secretary of his Committee on Civil Rights concluded: "As the nation's capital Washington is now a symbol of a failure in democracy."[159]

As the largest employer and landowner in the district, the U.S. government was not easily separable from the Jim Crow practices prevalent there. The State Department, despite its clear interest in improving domestic race relations, remained one of the most completely segregated federal agencies.[160] Foreign visitors to what was supposed to be the citadel of democracy could not avoid observing the prevalence of American racial discrimination, and visitors with dark skins often experienced the visceral dislike of many white Americans.[161] In contrast to the Dixie-dominated legislative branch, the president at least recognized the need to eradicate this highly visible stain on the nation's reputation. He could not control the political machinery of the district nor get his civil rights package through Congress, but he did order the elimination of racial discrimination in the federal civil service in July 1948.

Truman thus initiated the desegregation of Washington's public sphere, a process largely completed in the late 1950s.

The second major portal into the practice of American democracy was New York, the nation's largest city and the home of the new United Nations. Like Washington, New York received growing numbers of diplomats of color in the postwar years. But unlike the national capital, New York lay well north of the Mason-Dixon line, unencumbered by a firm legal tradition of segregation. Ruled not by segregationists in Congress, Gotham was instead part of one of the country's most liberal states and the heart of cosmopolitan America. None of these differences prevented dignified nonwhite foreigners from experiencing the humiliations of racial discrimination in the realm of public accommodations and real estate. The explicit commitment of all UN member states to racial equality, embodied in the 1945 UN Charter and the 1948 Universal Declaration of Human Rights, further highlighted the embarrassing gap between the host nation's words and practices.[162] U.S. officials had from the beginning feared potential UN involvement in human rights issues within the United States. Southerners and other conservatives in Congress refused to ratify the body's Convention on the Prevention and Punishment of the Crime of Genocide, owing to concern that American race relations might be put under an international microscope.[163] Diplomats from outside Europe varied widely in their overall views of American society and U.S. foreign policy, but they were nearly unanimous in their dislike of American racial practices.

The third demonstration of U.S. race relations was made by Americans traveling overseas. After 1945 they covered almost the entire globe, but certain areas were especially important. For all their colonial practices abroad, the more homogeneous Western Europeans were typically less racially discriminatory at home than the heterogeneous Americans, as the British had demonstrated in their welcome of African American soldiers during the war.[164] Violence by white Americans against their black countrymen, usually among military forces, troubled many Europeans, who resented such incidents happening in their countries.[165] Swedes, for example, were shocked in 1947 when white merchant marines with knives and razors assaulted black sailors at a dance hall in Malmoe City Park. The local paper *Expressen* regretted the damage to America's image by such "strange customs of intercourse among their citizens."[166] Less violent incidents were also troubling. Two officials in the U.S. embassy in Luxembourg refused to help or even be polite to the visiting Jubilee Singers of Fisk University in 1951, although the students were on precisely the kind of goodwill tour that the Truman administration was promoting. "We have been trained as Foreign Service Officers," one of the diplomats declared to the U.S. ambassador, "and we are not interested in negroes of any kind."[167]

Perhaps the most revealing demonstration of U.S. racial policies and practices overseas came in the Panama Canal Zone. Strict racial segregation there was the responsibility solely of the U.S. government, which ruled the canal zone without any local representation. Almost all whites in the zone were relatively wealthy U.S. citizens, while black workers came from Caribbean countries, had no rights of U.S. citizenship (nor could their children born there become American citizens), and enjoyed a much lower living standard. The dominant voices controlling the zone, as in the District of Columbia, were the Southern chairmen of the congressional oversight committees.[168] Unlike Washington, however, the canal zone was not surrounded by segregated Southern and border states. It sat squarely in the middle of another nation, whose darker-skinned majority resented American racial prejudices and the support lent to Panama's light-skinned elite.[169] Latin Americans throughout the region were concerned about U.S. sovereignty a thousand miles south of the Rio Grande, and Yankee racial discrimination was not reassuring. American rule along the Panama Canal seemed to represent a pure case of U.S. overseas imperial policy, directly analogous to European colonial policies abroad, and unlike in the Philippines, the Americans were not leaving anytime soon. Civil rights problems varied in other U.S. offshore holdings, from less pronounced tensions in the organized territories of Hawaii, Puerto Rico, Alaska, and the Virgin Islands to plainly exploitative relationships in Samoa and Guam.[170]

The largest multiracial group of Americans overseas in the Truman years was the armed forces. Racial policies in the U.S. military served as a lightning rod for domestic and foreign critics of inequality. Need rather than principle had created a temporary breach in the wall of military segregation during the winter of 1944–45, when the German counteroffensive at the Battle of the Bulge convinced General Dwight Eisenhower to allow volunteer black platoons to fill the gaps in larger white units. Their impressive performance under fire led to some minor postwar army reforms, but segregation—and rising African American resentment—remained the rule before the president's July 1948 executive order banning racial discrimination in the U.S. armed forces.[171] This was Truman's most significant action in favor of racial equality. The army and the marines, led by recalcitrant officer corps, resisted doggedly (in contrast to the navy and the air force), but the executive order provided the legal basis for what the war in Korea would make a military necessity two years later.[172] Desperate personnel shortages drove the process of desegregation in Korea, as American commanders seized on the efficiency gained by eliminating racial barriers. The president and his advisers embraced the political benefits of reducing discrimination in a multiracial force fighting in Asia. Both political and military leaders also recognized the controversy such policies tended to stir up at home, especially in Congress, and they tried to minimize domestic publicity about the process of Jim Crow's discharge from American ranks by discussing it with the press in only the broadest terms.[173]

The Korean War demonstrated the ways in which race continued to complicate U.S. foreign relations right through the final days of the Truman administration. Historic anti-Asian sentiments, which had reached the boiling point during the war in the Pacific a few years earlier, reemerged as North Koreans and Chinese began killing American soldiers on the battlefield.[174] U.S. commanders like General Douglas MacArthur radically underestimated the skill and tenacity of the Chinese People's Liberation Army, and young American men paid with their lives in the bloody retreat from the Yalu River. Disdain for China's military capacity stemmed in part from its lack of heavy artillery and air power and its primitive logistics, but it also reflected American racial thinking.[175] Captain Fred Ladd, an aide to MacArthur's chief of staff, recalled later how U.S. officers en route to the Korean peninsula "thought that they were going to go over there and 'stop the gooks'—just the same as in Vietnam. Just who 'the gooks' were, they didn't know, and they didn't want to know. You could have asked any American senior officer in Korea: 'Who commands the Korean 42nd division—ROK or Communist— and what's his background?' He wouldn't have known what you were talking about. A gook is a gook. But if the Germans had been the enemy, he'd have known."[176]

Establishing bases in a friendly, if occupied, Japan and having South Korean allies in the field required distinguishing "good Asians" from "bad Asians," but the fundamental inclination to view Asians in racial terms remained. Evidence of this attitude abounded at the UN's prisoner-of-war camp at Koje-do, where U.S. guards expressed their disdain for Chinese and

North Korean prisoners in explicitly racial language. Those prisoners also witnessed considerable conflict, sometimes armed, between white and black or Latino Americans guarding them.[177] The integration of U.S. forces and the presence of Asian allies did not eliminate overnight the entrenched habit among white Americans of seeing the world in racialized terms.

U.S. commanders court-martialed a disproportionate number of African American soldiers during the conflict and sentenced them more severely on average than their white counterparts.[178] An investigation by the NAACP's Thurgood Marshall suggested that white officials suspected black troops of "fragging"—murdering by stealth—their officers, who were typically white and often Southern.[179] One of the most basic fears of some white Americans in the early Cold War—that African Americans might serve more as a fifth column behind U.S. lines rather than as a loyal force—had evidently not been put completely to rest. This concern was amplified by stories of North Korean and Chinese "brainwashing" of American prisoners, which was blamed for the decision by some twenty-two U.S. prisoners of war, including three African-Americans, not to return home when offered their freedom after the war.[180]

One other complication along the color line showed up during the Korean War. The sexual relationships of American soldiers of all colors with Korean and Japanese women took various forms, and many veterans returned home with Asian wives. Accepted quietly in much of the United States, this development created serious problems when it involved white Southern men returning to their home region. The essential fabric of segregation appeared to be threatened, and state justice officials in Georgia and Mississippi refused to recognize the validity of the marriages.[181] This issue highlighted in acute form the ambiguity still surrounding the nexus of race and foreign relations in the United States at the conclusion of Truman's presidency. The Korean War had ended further efforts at domestic social reform, including civil rights legislation, and had revealed anew abundant white American prejudices against both Asians and blacks.[182] But the conflict had also hastened military desegregation, allowing white and black soldiers to live together in unprecedented closeness, and resulted in many white American men finding themselves undeterred by racial differences in the pursuit of their most intimate relationship. The experience of transcending distinctions of color, even in limited ways, helped lay some of the groundwork for the success of the coming civil rights movement. In the quintessential dilemma of postwar America, old habits of discrimination would continue to conflict with a growing disinclination to view the world though racial lenses.

BY DESTABILIZING THE social order, wars bring unexpected changes in their wake. The larger the conflict, the more extensive such alterations tend to be. As the greatest conflagration in human history, World War II left a swath of extraordinary destruction across much of the globe, while also creating the conditions for both reform and revolution. After 1945 the U.S. government faced a situation in which traditional hierarchies of power, especially racial ones, had been greatly disturbed in the United States and abroad. How these would be reconstructed was an open question.

Harry Truman presided over an administration riven by the tensions between older traditions and structures of discrimination and newer commitments to equality. On one side stood the segregated backgrounds of U.S. policymakers, the latent racial prejudice of most white Americans, the power of the white South in Congress, the political bent of the FBI and other segregated bureaucracies in Washington, and the colonialism of America's closest allies. On the other side were aligned the decline of scientific racism, the Holocaust's delegitimation of racial discrimination, African American demands for decent treatment, the rising tide of independence in the colonial world, and Soviet ideological and diplomatic competition. The tension between these two conflicting tendencies marked Truman's presidency from its first day to its last.

The gap between official U.S. rhetoric and actual American practices regarding human equality was vast in 1945 and remained marked in 1953. That it shrank in some significant ways between those years is a tribute, above all, to the persistence of people of darker hue who refused to accept discriminatory treatment; their resistance and pressure created the force to drive reforms. Truman himself bears a measure of credit as well for promoting certain changes, for a combination of reasons including political expediency, international pressure, and personal belief. It is perhaps worth remembering that the United States was hardly the only nation not living up to its international agreements regarding human rights after World War II. But because of its extraordinary position of power and its decision to take on the mantle of international leadership, what the United States government said and did mattered enormously for the course of post-1945 world history. Framing its foreign policy in terms of containing Soviet power and influence in order to promote and sustain a "free world," the Truman administration exposed itself to accusations of hypocrisy in an era of Jim Crow, colonialism, and apartheid.

Truman and his advisers strongly believed that the security of the nation and the world depended upon halting the expansion of Communist rule. The president and more liberal members of his administration also agreed that racial discrimination must end, but none of them viewed racism as a threat to American interests equal to Communism. Indeed, they had little patience with those who did—some leftists, many African Americans and other people of color, and most Third World revolutionaries—seeing them as naive at best and Soviet agents at worst. Like the white supremacists they opposed in the South and in southern Africa, such dissenters threatened the president's efforts to assemble a multiracial anti-Communist alliance at home and abroad. Antiracists and American policymakers shared the conviction that domestic happiness during the Cold War could no longer be separated from a sound and stable international order. Where they differed was in the designation of the foremost threat to that order.

Ultimately, much but not all changed along the color line in the Truman years. Progress at home included military desegregation, a Democratic Party increasingly supportive of racial equality, and federal judicial decisions against elements of discrimination. Abroad, it encompassed the independence of most colonies in Asia including the two most populous ones, India and Indonesia. Left still standing were a largely segregated America and an Africa yet under colonial control, with an expanding core of enthusiastic white supremacism at its southern end.[183]

3

THE LAST HURRAH OF THE

OLD COLOR LINE

STOPPING OVER IN FIJI on a return trip from Asia, U.S. Secretary of State John Foster Dulles donned swimming trunks for a dip at the beach. His bodyguard, Louis Jefferson, and a local black policeman watched. "Meestair Doolays, he is very white. Very, very white," the policeman commented quietly. "Yes, he is," agreed Jefferson. The policeman continued, "Some of you whites are whiter than others." The bodyguard acknowledged that was probably true. The policeman smiled: "You whites see us blacks as having many shades of black, brown, and so forth, in our color, but most of you do not know that *we* see *you* as having many shades of white. Many shades. It does help us to tell you apart."[1]

Dulles was indeed "very, very white," in his politics as well as his skin tone. He was representative of the administration of Dwight Eisenhower, which governed the United States from 1953 to 1961, a period of extraordinary change in race relations both in the domestic sphere and in the international system. It was the opportunity and the misfortune of Eisenhower to occupy the White House during the years when African Americans moved to the forefront of American politics and Africans initiated their final push for liberation from colonialism. Those twin struggles for racial equality, the American civil rights movement and the African quest for independence, caught most white Americans by surprise and challenged their simple assumptions about a world divided by Cold War tensions between East and West. America's new preeminence in global affairs in the 1950s meant that its traditional patterns of racial segregation and discrimination would not likely survive intact in a world of changing race relations. In the last week of the Eisenhower administration Chester Bowles, Connecticut congressman and soon-to-be undersecretary of state, emphasized in a *New York Times* article that in matters of race Americans "will appear [to the world] no better than we actually are. In this sense, as in so many others, the division between 'domestic' and 'foreign' policies no longer has meaning."[2]

Well prepared for dealing with most foreign policy issues and enormously popular at home and abroad, Eisenhower faltered in the face of challenges to the existing racial status quo. In the aftermath of *Brown v. Board of Education,* the epochal 1954 Supreme Court decision outlawing racial segregation in public schools, the former five-star general refused to exercise the leadership for which he was famous. Despite his administration's emphasis on America's proclaimed moral and spiritual superiority to Communist nations like the Soviet Union, the president and his advisers failed to recognize either the central moral issue involved in racial inequality or the significance of race relations in the modern world. The men of Washington tried likewise to sidestep the issue of African independence. They hoped above all to keep the inexorable processes of domestic desegregation and foreign decolonization quiet and gradual. They accepted only the slowest of changes toward greater equality, as they sought to bolster the international social order in which the United States had become the dominant force. Robert Burk, the most careful student of the administration's civil rights policies, has argued convincingly that Eisenhower was content with "the political containment of racial problems rather than their solution." And those racial problems were not merely domestic but also international.[3]

The Eisenhower Administration, Black Americans, and Civil Rights

Dwight Eisenhower was a child of the nineteenth century. Born of a Southern mother in Texas in 1890, the year of the massacre at Wounded Knee, he grew up in Abilene, Kansas, not far from what contemporaries considered the "Indian frontier." He was six years old when the Supreme Court established in *Plessy v. Ferguson* the "separate but equal" doctrine of racial segregation as the law of the land. No African Americans lived in his hometown, nor did any attend the U.S. Military Academy at West Point with him. His military career before World War II took him to army posts almost exclusively in the southern United States, or in the Panama Canal Zone and the Philippines, where racial discrimination was at least as rigid as below the Mason-Dixon line. The U.S. Army remained segregated throughout his career in it, and his primary experience of people of color was in the role of servants. Few were surprised by his 1948 testimony to the Senate Armed Services Committee in opposition to desegregating the American armed forces. Never an explicit bigot, Eisenhower occasionally showed a dislike of raw prejudice. But he had neither experience nor models of egalitarian multiracial cooperation. He was a full generation older than Martin Luther King, Jr., and other civil rights organizers who would challenge the American social order he so cherished, and he remained essentially a man of his times.[4]

Eisenhower valued white Southerners and felt at ease in the segregated South. He had met his wife, Mamie, in San Antonio, and they considered retiring there. He vacationed frequently at the southern Georgia plantation of his secretary of the treasury, George Humphrey, playing bridge and hunting wild turkey while riding in carriages driven by old black servants. Georgia and the rest of the Deep South gained an additional tie to the president when the group of six millionaires who were his closest friends built him a vacation home on the segregated Augusta National Golf Course. While in the White House, Eisenhower enjoyed elaborating on his extensive Southern ties when negotiating with recalcitrant Southern senators. He was troubled by the Supreme Court's attention in the *Brown* decision to the inner feelings of black schoolchildren rather than to those of white Southern youngsters. He also resented the way the Court's decision strained his personal ties with white Southerners. Even in his rejection of Southern white violence against African Americans and Jews in a 1958 news conference, Eisenhower posed as the *real* defender of supposedly genteel Southern traditions against a wave of terror bombings by a group calling itself "the Confederate underground": "From babyhood I was raised to respect the word 'Confederate'—very highly, I might add—and for hoodlums such as these to describe themselves as any part or relation to the Confederacy of the mid-nineteenth century is, to my mind, a complete insult." These were not the words of a man much troubled by regional racial traditions, like the slavery at the heart of that honored Confederacy.[5]

If white Southern racial practices did not usually bother Eisenhower, the presence of blacks in any but subordinate positions did. By the late 1940s he was long accustomed to abundant black or Filipino servants, but he remained uneasy in social gatherings with prominent people of color such as the Nobel Prize–winning diplomat Ralph Bunche.[6] As president he did recognize the importance of African American voters, and he sprinkled a small handful of lower-level black appointees throughout his administration. But the relatively easy access of prominent black Americans to the White House under Harry Truman evaporated after the 1952 election. Only once in his eight years as president—years of momentous changes in American and world race relations—did Eisenhower even meet with black political leaders, in June 1958 for forty-five minutes.[7]

Eisenhower's most influential advisers rarely encouraged him to advocate racial equality. The strongest supporter of civil rights in the administration, Attorney General Herbert Brownell, resigned early in the second term, leaving only Vice President Richard Nixon as an occasional voice opposing racial discrimination. The most important figures on foreign affairs, Foster Dulles and his brother Allen, the head of the Central Intelligence Agency—descendants of South Carolina slave owners—showed little interest in racial matters. FBI Director J. Edgar Hoover viewed the mere advocacy of racial equality as a subversive act.[8]

What little openness to combating racial injustice existed in the administration ebbed further in the last few years of the decade, replaced by anger at what officials perceived as black ingratitude toward the president. Offended as early as 1956 by what they saw as a new aggressiveness among black Americans—especially, in the words of White House staffer Maxwell Rabb, "an ugliness and surliness in manner"—the president and his advisers maintained little real interest in civil rights. Even the notably polite early phase of the civil rights movement was apparently threatening to them. In February 1959 segregationist and Alabaman Wilton Persons replaced Sherman Adams as the White House chief of staff. Persons told Frederic Morrow, the sole black member of the White House staff, to speak with him about all issues except those regarding race and civil rights; the latter should be excluded, said Persons, for emotional reasons. Here was a revealing window on the racial blinders of the Eisenhower administration: the man controlling access to the president of the United States in 1959, in the midst of African decolonization and American civil rights activity, would not even *talk* with a black assistant about anything having to do with race relations.[9]

Morrow's tenure as a White House assistant provided a catalog of racial insensitivity and discrimination at the highest level of the U.S. government. His colleagues on the president's staff regularly ignored or made fun of him, from baiting him with "nigger" jokes to refusing to acknowledge his presence in social situations. He was forced to live in dilapidated housing in the still segregated city of Washington. On the campaign trail and on other travels for the Republican Party, he was frequently mistaken for a hired servant and harassed by hotel employees, who assumed he was surreptitiously luring white women into his room. Morrow accepted such indignities calmly because he believed that the symbolism of his presence in the administration was an important first for African Americans and because he considered Eisenhower a decent and admirable man. But he grew increasingly unhappy with his superiors in the president's second term, as their lack of commitment to racial justice was highlighted by rising white violence against blacks in the South. Morrow ultimately came to suspect that other blacks were right when they accused the administration of using his presence to pacify African Americans.[10]

Eisenhower's lack of concern for civil rights stemmed in part from his ignorance of the circumstances in which most Americans of color lived. His extraordinarily broad travels to every state and almost every continent had not included visits with darker and poorer people. He knew almost nothing of the appalling housing options and limited employment possibilities facing his fellow citizens of darker hue—what even Hoover, an inveterate opponent of racial equality, admitted in 1960 were "very trying conditions."[11]

But the president also chose to pay little attention to evidence of mounting white violence toward African Americans during his years in office. That wave of racial fury rose most ominously in the Deep South, typified by the murder of Emmett Till in 1955 and the lynching of Mack Parker in 1959. In Birmingham in 1957 six Ku Klux Klansmen randomly seized and castrated black war veteran Edward Aaron, taunting him in the process with the name of the Supreme Court chief justice associated with the *Brown* decision: "Look here, nigger! You ever heard of a nigger-loving Communist named Earl Warren?" Assaults on blacks knew no regional boundaries, however, as crowds of thousands in the South Deering section of Chicago turned out in August 1953 to harass the first black tenants of Trumbull Park Homes, a formerly all-white public housing project. Physical intimidation continued in the neighborhood for years thereafter. One of the more revealing indications of the conditions facing black Americans in the Eisenhower years was the title of a 1956 travel guide published by the American Automobile Association, listing public accommodations that served guests of all colors: "Vacation without Humiliation."[12]

Eisenhower believed that neither personal nor political benefit would accrue from his becoming involved in civil rights. He felt so uncomfortable with the issue that he simply avoided it. He disliked the words "discrimination" and "racial" and kept them out of his speeches. Refusing to meet not only with black leaders but also with Southern white ones at critical points, Eisenhower worked to confine race relations to courtrooms, statehouses, and school boards. Roy Wilkins, the director of the NAACP in the 1950s, concluded in retrospect: "President Eisenhower was a fine general and a good, decent man, but if he had fought World War II the way he fought for civil rights, we would all be speaking German today."[13]

Personally comfortable with segregation, the president viewed it as primarily a political rather than a moral issue. This approach required balancing the claims of all contestants, including white Southern voters whom he strove to bring into the Republican fold in large numbers for the first time. So Eisenhower refused all opportunities to wield his considerable moral authority by speaking out against racial injustice and in favor of racial harmony. He could not agree with Martin Luther King, Jr., that race was "indeed America's greatest moral dilemma." In contrast to 1948, when it had been a central part of the presidential contest, civil rights in the 1952 campaign barely showed up on the major parties' political radar. Eisenhower ran hard below the Mason-Dixon line, winning four states and more Southern votes than any Republican presidential candidate since Reconstruction. With the Democrats determined to avoid the party split over civil rights of four years earlier, nominee Adlai Stevenson chose the segregationist Alabama senator John Sparkman as his running mate. Not until after its second defeat by Eisenhower in 1956 would the national Democratic Party begin to shift toward a stance clearly in favor of racial equality. During his years in the White House, Eisenhower felt little pressure from his political opponents to stand up for racial justice.[14]

As president in some of the chilliest years of the Cold War, Eisenhower did recognize the need for at least symbolic improvements in American race relations. He was acutely sensitive to the image of the United States abroad and did not enjoy seeing it tarnished by the more extreme brutalities of Jim Crow. His conservative view of the role of the federal government in American society precluded significant federal intervention in matters of segregation, which he saw as falling under local and state jurisdiction. But one area in which Eisenhower did support some greater federal involvement was voting rights. He believed in the Fifteenth Amendment and understood that the denial of voting rights to African Americans, especially in the South, undercut the nation's claim to being a democracy. He also recognized that granting the ballot to blacks would not necessarily end social segregation, with which he and most other white Americans were still comfortable.[15]

The Eisenhower administration's greatest achievements in civil rights came in two highly symbolic areas of federal authority: the District of Columbia and the nation's armed forces. Washington's history as a Southern city controlled by Southern-dominated congressional oversight committees made it a peculiar liability in the Cold War competition for the goodwill of newly independent countries. The prejudice and humiliation experienced by non-white visitors to the nation's capital—unable to use public facilities like restaurants, theaters, and hotels—seemed to trumpet to the world American hypocrisy about freedom. Eisenhower recognized this problem even before taking office, and he eliminated racial barriers in public places in the district between 1953 and 1955. After the *Brown* decision of 1954, he moved swiftly to include public schools.[16]

While successful in removing the most egregious symbols of inequality in what the president called the citadel of democracy, the administration hardly transformed the district into a showplace of American egalitarianism. There was minimal progress during the 1950s in reducing extensive discrimination in employment and housing in Washington. Federal agencies and city departments remained segregated when Eisenhower left office in 1961, and massive white flight to the suburbs during the decade undermined the effort to integrate either schools or neighborhoods. The patterns of urban decay, poverty, crime, and discrimination that marked the nation's capital received little attention from the Eisenhower administration.[17]

The American armed forces, stationed throughout the world since 1945, served as an even more visible symbol of the United States than its capital city. Truman had ordered their desegregation in 1948, and the process was well under way when Eisenhower took office. By 1955 the American military had become an integrated institution in an otherwise still segregated society. The armed forces thus served as a laboratory for social change in the years before the rise of the modern civil rights movement. "In effect," historian Bernard Nalty notes, "the Department of Defense was operating a racially integrated society sealed off from nearby civilian communities, with its own stores, its own sources of recreation, and in many instances its own family housing." Military desegregation was not perfected in the 1950s, as many National Guard and reserve units remained all-white, and incidents involving civilian employees and black service personnel continued to fester. The scale of the problem was, after all, enormous: the massive institution of the American military had to make a 180-degree turn, from enforcing segregation in every detail of its collective life (as it had for most of two hundred years) to banning it completely. But the symbolism of the change could not be missed by foreign observers, as Eisenhower—like Truman—intended.[18]

Beyond these circumscribed areas of clear federal authority, Eisenhower believed that changes in race relations could only come slowly in the United States. He assured his "great friend" James Byrnes, the segregationist governor of South Carolina, over lunch in the summer of 1953 of "my belief that improvement in race relations is one of those things that will be healthy and sound only if it starts locally. I do not believe that prejudices, even palpably unjustified prejudices, will succumb to compulsion." Six years later he wrote to Atlanta newspaper editor Ralph McGill that "coercive law is, by itself, powerless to bring about complete compliance with its own terms when in any extensive region the great mass of public opinion is in bitter opposition." To those who insisted on the elimination of racial discrimination, he counseled—to their dismay—only greater patience. "No one," he told a skeptical group of black newspaper publishers in May 1958, "is more anxious than I am to see Negroes receive first-class citizenship in this country . . . but you must be patient." Whites, he believed, deserved more time to adjust themselves.[19]

Eisenhower repeatedly condemned what he called extremists on both sides of the civil rights issue, a calculus equating whites engaged in violence with polite, peaceful protesters seeking the fulfillment of their constitutional rights. He insisted that those "who want to have the whole matter settled today" were no different than those "so filled with prejudice that they even resort to violence." He rejected suggestions that he meet with civil rights leaders, arguing that he would then also have to meet with officers of the Ku Klux Klan. The president failed to grasp the elemental moral distinction between people of enormous courage, willing to sacrifice their lives for the highest principles of American life, and people of marked cowardice, eager to wound and even kill others while showing no willingness to risk their own well-being.[20]

In contrast to such "extremists," Eisenhower hewed to what he called "the center line, which is the only path along which progress in great human affairs can be achieved." He was sure that the passage of time would bring change in the hearts of white segregationists, and he believed that more observant African Americans recognized this. Speaking with Republican congressional leaders in the summer of 1956 about his administration's pending civil rights bill, the president repeated a story he had heard recently in Augusta from golfing legend Bobby Jones, of a black field hand supposedly saying, "If someone doesn't shut up around here, particularly these Negroes from the North, they're going to get a lot of us niggers killed!" Like self-described white moderates of the South, Eisenhower saw himself as a protector of the majority of moderate blacks, who were imperiled by "outside agitators" likely to provoke a violent white response. Ultimately, Dwight Eisenhower believed that he was the right man to lead all Americans responsibly along "the center line," including all black Americans.[21]

Declarations of Liberty and Equality: Brown, Bandung, and Montgomery

The nineteen months from mid-1954 to late 1955 marked a sea change in American and international race relations. The U.S. Supreme Court declared racial segregation in public schools unconstitutional; leaders of color from around the world gathered for the first time in history, to declare their solidarity and to reject white supremacy; and the African American community of Montgomery began a boycott of segregated city buses that initiated the modern black civil rights movement. The Court's decision in *Brown v. Board of Education* in May 1954 was carefully limited: it addressed only schools, and it was to be enforced only "with all deliberate speed" rather than with any specific timetable. In fact, school desegregation took place so slowly that by the end of the Eisenhower administration almost seven years later, only 6.4 percent of black schoolchildren in the South were attending integrated classes. It would take ten years before Congress broadened the scope of that judicial decision by outlawing racial discrimination in all public places. But the symbolic significance of the case could not have been greater. By a unanimous decision that included three Southern justices, the nation's highest court reversed the political and moral context of race relations in the United States. Racial discrimination was no longer allowed by the laws of the land. What had previously been considered by most whites subversive—racial integration—acquired the full moral authority of the Constitution.[22]

The Court's logic in *Brown* was at least partly international. The Justice Department brief in the case, filed just weeks before Eisenhower took office, emphasized that "it is in the context of the present world struggle between freedom and tyranny that the problem of racial discrimination must be viewed." The Court's decision did not specifically cite the Cold War or the retreat of colonialism in the rest of the world, but the justices could not help being affected by the dominant political and social realities of their time. American racial segregation, so little questioned in legal circles even a decade earlier, carried much heavier baggage in a world now dominated by the United States but marked by the rapid decline of white supremacy.[23]

The Eisenhower administration certainly understood this reality. The Voice of America immediately beamed the news of the Court's decision around the world in thirty-four languages, along with background materials emphasizing the history of supposedly steady black advancement in the United States. Foster Dulles and the State Department were enthusiastic about the benefits of the decision for America's foreign relations. International reaction to the *Brown* decision was uniformly favorable, with the predictable exception of the South African government. Some African Americans resented the administration's emphasis on Cold War logic over genuine belief in the fundamental validity of racial equality. Sociologist Franklin Frazier, for example, argued that "the white man is scared down to his bowels, so it's be-kind-to-Negroes decade at last." But most blacks, including even some Communist Party members, were impressed by the Court's decision. Cold Warriors of all but the most racist stripe found cause for celebration. The black-owned *Pittsburgh Courier* agreed with the administration and with most other newspapers outside the South that "this announcement will . . . stun and silence America's Communist traducers behind the Iron Curtain."[24]

The administration as a whole, however, along with many other white Americans, seemed more sure of what they wanted the world to think of the *Brown* case than of what they thought themselves. Many local media outlets were silent, and Universal Newsreels declined to mention the case because of fears that it would prove too controversial. Eisenhower himself refused to endorse the Court's decision or to offer any leadership to the country—beyond a quiet agreement to desegregate Washington schools—as it grappled with the implementation of this monumental cultural change. The president also declined to speak of the case as one involving a moral issue. A few months before the Court's announcement, he had even gently lobbied Chief Justice Earl Warren on behalf of segregationists. "These are not bad people. All they are concerned about is to see that their sweet little girls are not required to sit in school alongside some big overgrown Negroes." Afterward, Eisenhower made clear in private his dislike for the decision, his fears of white Southern resistance, and his certainty that the Court had guaranteed a worsening of race relations in the South. Four years later he admitted publicly that he would prefer a slower version of school desegregation, to which NAACP chief Roy Wilkins could only respond that if the snail's pace of the previous four years represented swiftness, "then we certainly do not understand the word speed." Eisenhower's abdication of leadership on school desegregation would come back to haunt him in the 1957 crisis in Little Rock, before an international audience.[25]

Eleven months after the Supreme Court changed forever the shape of race relations in the United States, an event of comparable significance for international race relations unfolded in the Indonesian city of Bandung. There in April 1955 representatives from twenty-nine nations of primarily Asia, Africa, and the Middle East met in what host president Sukarno of Indonesia called "the first international conference of colored peoples in the history of mankind." The agenda was threefold: to promote cooperation among the nonaligned nations of the Third World; to deliberate about such common problems as colonialism and racism; and to advocate world peace. The gathering had an ecumenically ideological cast, mixing Communist, pro-Western, and neutralist states. What almost all the representatives shared, besides a non-European phenotype, was a common historical relationship of colonialism with the West. The conference therefore carried an implicit condemnation of the West that was bolstered by an explicit call for eliminating vestiges of colonialism and racial discrimination.[26]

Delegates at Bandung tended to hold a nuanced view of the United States. President Carlos Romulo of the Philippines noted the centrality of the doctrine of white supremacy in all versions of Western colonialism and underlined "the role played by this racism as a driving force in the development of the nationalist movements in our many lands." He urged Western nations to cut themselves free from the "albatross" of racial discrimination that still hung around many of their necks—an undisguised reference to Jim Crow practices in the United States and South African apartheid. But Romulo also noted the considerable amount of goodwill evident at Bandung toward the United States, stemming in part from such evidence of progress as the *Brown* decision. Sukarno observed in his opening speech on 18 April the anniversary marked by that date: Paul Revere's ride through the Massachusetts countryside, warning of the approach of British troops and helping precipitate the American Revolution, "the first successful anti-colonial war in history." Toward the United States, then, the new nations of color in the 1950s had mixed feelings.[27]

Few white Americans expected anything good to come out of the Bandung conference. The nonaligned position of the representatives regarding the Cold War made Dulles and his colleagues deeply suspicious. In contrast, the

Soviets and Chinese strongly supported the gathering, and the United States feared that their influence would subvert the neutralism of the new African and Asian nations. The Soviet Union was excluded from the actual meeting as a white, European nation, despite its claim to being part Asian. China, of course, was invited as an Asian nation, and the U.S. government worried about its impact on the others. As the National Security Council (NSC) noted a few years later, the People's Republic had "an important advantage" over the Soviet Union in the Third World: "They can put themselves forward as an under-developed country which has recently been freed from foreign domination." The Bandung conference was not likely to win much support in official Washington, a place dominated in 1955 by such figures as Texan Lyndon Johnson, the Senate majority leader, Georgia segregationist Walter George, the chair of the Senate Foreign Relations Committee, and segregationist sympathizer Dwight Eisenhower.[28]

The Eisenhower administration sent no greeting to the conference and did its best to ignore it or sabotage it. Foster and Allen Dulles agreed in January 1955 that Bandung likely portended an aggressive new Communist strategy in Africa. The secretary of state convened a meeting of the Southeast Asian Treaty Organization in Bangkok in February as a forum to condemn neutralism. Escalating U.S.-Chinese tensions in the Taiwan Straits that spring led to Eisenhower's public warning about American readiness to use nuclear weapons in the region; Navy Chief of Staff Robert Carney then leaked to the press American contingency plans for an all-out attack on China in case of war between the two nations. Administration officials hoped to discourage the consolidation of an African-Asian bloc in an effort to keep Africa more oriented toward the West, and they were careful not even to use the term "Afro-Asian group" without the adjective "supposed."[29]

Adam Clayton Powell, one of three black members of the U.S. House of Representatives, attended the conference in Indonesia as a journalist. With great fanfare he criticized U.S. foreign policy as insufficiently anticolonial, but he strongly defended American race relations. His presence in Bandung, he reported to Eisenhower afterward, gave "living proof to the fact that there is no truth in the Communist charge that the Negro is oppressed in America." Powell returned to the House to receive a standing ovation, a singular experience for a man generally resented by his colleagues, especially those from the South. Several months later Foster Dulles tacitly acknowledged the success of the gathering in Indonesia when he called privately for a "Bandung Conference in reverse" as a forum for colonial and ex-colonial powers to lay out an independence program "in a dramatic way": as a means, that is, for re-treating white powers to cut their ideological losses.[30]

African Americans in the cradle of the old Confederacy later that year fol-lowed the same logic as the Supreme Court and the Bandung delegates in de-claring their own opposition to the tradition of white supremacy. The *Brown* decision had set the Constitution squarely on the side of desegregation, and the black American press widely hailed the Bandung conference as, in the words of the *Baltimore Afro-American,* "a turning point in world history" and perhaps even the most important event of the twentieth century in its "clear challenge to white supremacy." The Montgomery bus boycott initiated the modern era of mass civil rights organizing in the United States. Lasting from December 1955 to its successful conclusion one year later, the boycott took advantage of the easing of the Red Scare that had dominated the American political landscape of the early 1950s and had made efforts at social change appear traitorous. The peaceful black protesters in Alabama, while framing their actions in patriotic language and while concerned primarily with local issues, also identified their cause with that of people of color abroad. King de-scribed the Montgomery movement as part of a global process in which "the oppressed people of the world are rising up" against colonialism, imperialism, and segregation. The retreat of Europeans from Asia and Africa seemed clearly to African Americans to be part of their struggle for racial justice.[31]

Despite its explicitly Christian style and commitment to nonviolence, the Montgomery bus boycott represented just the kind of social change "from below" that the Eisenhower administration hoped to avoid. The president himself simply sat it out, refusing to make any public comment on the issue. He ignored appeals from King and a variety of other Americans to support the boycotters in their quest for polite treatment on public transportation. Eisenhower confessed in his diary in November 1956 his disappointment in the Supreme Court's upholding of a lower court decision banning segrega-tion on intrastate buses: "In some of these things [I] was more of a 'States Righter' than the Supreme Court." Other Americans were pleased by the outcome of the boycott, finding in its peaceful, patriotic, and religious char-acter precisely the kind of affirmation of the American political system that the country needed in the Cold War. Even many white citizens of Montgom-ery seemed ready to accommodate themselves to some change in race rela-tions. On 21 December 1956, the first day of integration on the city's buses, one white man found himself—for the first time in his life—sitting behind a black man and piped up: "I see this sure isn't going to be a white Christmas."

His fellow passenger smiled back, "Yes, sir, that's right," and those within ear-shot laughed. Such good humor seemingly found little echo in the Oval Office that holiday season.[32]

Southern white resistance to the changes suggested by *Brown*, Bandung, and Montgomery tended to divide into two camps that differed in strategy and often in class origins but not in essential goals: the violent and, in the common phrase of the time, the "respectable." The first approach had an impressive pedigree in Southern history, renewed for a national and international audience in August 1955 by Roy Bryant and J. W. Milam of Money, Mississippi. Enraged at reports that a fourteen-year old black Chicago resident, visiting relatives nearby, had whistled at Bryant's wife, the two men kidnapped, tortured, and murdered Emmett Till. This was actually the fourth known interracial homicide of the year in Mississippi. The first three victims had been prominent black Mississippians engaged in voter registration campaigns. Those political assassinations had received some notice in the national press, but nothing like the enormous attention brought by Till's death. The difference had to do with his age, his being merely a visitor to the South, and his mother's insistence on an open casket at his funeral back in Chicago, which allowed press cameras to record the particular brutalities inflicted upon the young man. Bryant's and Milam's immediate acquittal by an all-white jury, despite eyewitness evidence against them (and, indeed, their later admission of guilt to a journalist paying them for their story), further inflamed opinion outside the white South and beyond the United States.[33]

Many white Southerners sympathized with the murderers of Emmett Till, seeing their action as a regrettable but understandable response to the *Brown* decision and to black Southern organizing, as well as to the supposedly ever present threat of black male lust for white women. One local white at the Bryant and Milam trial explained the realities of preserving the status quo to a Northern reporter with a nod toward the Tallahatchie River, where Till's body had been found: "That river's full of niggers." Some white Southerners dissented, like novelist William Faulkner of nearby Oxford, who argued that in 1955 to "be against equality because of race or color, is like living in Alaska and being against snow." Black Americans were disturbed, if not surprised, by what the NAACP called the "state of jungle fury" among white Mississippians. Foreign observers tended to share that estimation, in Europe as well as in the Third World, and the Eisenhower administration could not miss seeing how the Till case undercut the goodwill that the *Brown* decision had recently won for the United States abroad. The FBI, for its part, was certain that the only whites who cared about the Till case must have been Communists. Hoover could find no other explanation for Chicago mayor Richard Daley's protest against federal inaction in the case than that he had been duped by the Communist Party. Liberals used anti-Communist logic in a converse fashion, believing the jurors in the trial of Till's murderers deserved, in the words of NAACP chairman Channing Tobias, "a medal from the Kremlin for meritorious service in communism's war against democracy."[34]

"Respectable" defenders of the South's traditional racial hierarchy avoided the overt viciousness of Emmett Till's murderers but tapped into the same broad reservoir of resentment of black self-assertion in the second half of the 1950s. White moderates no less than white conservatives were uncomfortable with African Americans who did not defer to their wisdom and power. White anxieties escalated as the decade wore on and evidence piled up that Jim Crow was under siege: the founding of the Southern Christian Leadership Conference (SCLC) in 1957 as the political arm of the black church in the South; the rising alienation of some civil rights workers, like Monroe, North Carolina, NAACP chapter president Robert Williams, who spoke openly by 1959 of meeting "violence with violence"; and the growth of the Nation of Islam, which whites unanimously disliked for its unflinching rejection of white leadership and even white people. Like their president, most white Americans were not yet ready for social upheaval along the color line.[35]

White Southerners generally saw the *Brown* decision as a terrible mistake. Powerful U.S. senators like Strom Thurmond and Richard Russell believed that racial integration would subvert the fundamental social order of the United States. The civil rights movement was as great a threat to the national security of the country as the Soviet Union. They were determined that racial equality would not be, in the words of one South Carolina newspaperwoman, "rammed down our throats." Governor Herman Talmadge of Georgia argued that the Supreme Court had "violated the law" in the *Brown* decision and called for white Southerners as a "sovereign people" to now become "the Court of Last Resort."[36]

In the absence of support from Washington for the implementation of the *Brown* decision, "respectable" Southern resistance built up over the next two years, culminating in the Southern Manifesto of 12 March 1956. Almost every U.S. senator and representative from the South signed this Declaration of Constitutional Principles, as it was officially known, condemning the Court's decision as an unconstitutional "abuse of judicial power" that usurped states' control of public education. Several state legislatures in the former Confederacy had already passed resolutions declaring that *Brown* had no legal force within their borders. In the first week of March, the University of Alabama expelled Autherine Lucy, a black student whom federal courts had just ordered admitted. These developments were keenly watched abroad; American journalist Carl Rowan, traveling in India and Ghana, was struck by the intensely negative responses to these events that he encountered. The early spring of 1956 thus marked a high-water point of national division over whether or not the United States would join the rest of the world—outside southern Africa—in moving away from racial discrimination.[37]

In this period of crisis for the United States in determining its future as a society, the Eisenhower administration maintained a low profile. Having ducked the original Court decision, the president also avoided confronting subsequent Southern resistance until his hand was forced in Little Rock in the fall of 1957. Eisenhower responded publicly to the announcement of the Southern Manifesto by seeing the issue from the white Southern viewpoint: he emphasized that the Court had upheld segregation for the previous sixty years, and that white Southerners' "deep emotional reaction" would take time to moderate. Brownell's Justice Department was working quietly on some limited civil rights proposals regarding voting, but white Southerners kept a close eye on the White House. Governor Allan Shivers of Texas, a Democrat who had supported Eisenhower in 1952, defiantly tested the president in September 1956 by using state police to prevent court-ordered desegregation in two Texas cities. Eisenhower ducked again, evading reporters' questions and denying there was any federal role in the case. White Southerners understood the White House's message, and violence against blacks by both police and private citizens continued unabated. In January 1957 the Koinonia Community, a racially integrated Christian group of sixty people farming near Americus, Georgia, appealed directly to Eisenhower for protection from the barrage of physical harassment and destruction—shootings, burnings, and dynamitings—they faced daily. The administration replied that it had no jurisdiction in the situation. Two weeks later, the president flew to Thomasville—just one hundred miles south of Americus—to enjoy twelve days of white Georgian hospitality at the Humphrey plantation.[38]

The presidential election of 1956 demonstrated anew the tepidness of support for equality in the United States in the era of Dwight Eisenhower. African Americans who could vote split their ballots between the two candidates, finding little in either to arouse enthusiasm. In his continuing appeal to white Southerners, Eisenhower refused to let the Republican platform explicitly link his administration to the *Brown* decision. Unlike four years earlier, when the platform had called for the "liberation" of white "captive peoples" in Eastern Europe, the president's party avoided any inkling of support for the liberation of black Americans from the chains of Jim Crow. The Democrats again provided little pressure on their opponents to identify with the cause of equality. Presidential nominee Adlai Stevenson represented a party deeply divided along sectional lines by the issue of civil rights. He therefore hewed to the middle ground of advocating only gradual implementation of the Supreme Court's ruling on school desegregation, and he opposed the use of federal troops for that purpose. Leadership on the crucial issue of race relations did not come from within the traditional political leadership of the United States in the 1950s.[39]

With little enthusiasm, Eisenhower allowed his attorney general to continue developing proposed legislation bolstering black voting rights. Brownell was concerned by the increasing tension in the South between violent whites and organizing blacks; he hoped that the bill would appeal to moderates of all colors by focusing on voting rather than the more emotional issue of school desegregation. The purpose was to restrain further racial polarization in the South. Signed into law after considerable compromise with the Southern congressional leadership, the Civil Rights Act of September 1957 represented the first federal law of its kind in eighty years, but its course through Congress also revealed how little the president cared about it or lobbied for it. Two months earlier he had declared at a press conference that he did not understand parts of his own bill, the parts that were toughest on violators of black voting rights. Recognizing encouragement when they saw it, white Southerners seized the opportunity to amend further the legislation. Most Southern senators saw the weak final version as such a clear victory for their side that they were embarrassed by Strom Thurmond's unsuccessful last-minute filibuster. Civil rights advocates were equally dismayed at the law itself. King warned Nixon that any impact of such inadequate legislation depended on "a sustained mass movement on the part of Negroes"—precisely what the Eisenhower administration hoped to avoid.[40]

Like their domestic counterparts in the American South, defenders of the traditional international social order also resisted change in the mid-1950s. Just as white Southerners reluctantly gave up peonage, prison labor leasing, and frequent lynchings, so Western Europeans relinquished most of their direct colonial control of the Third World. But the imperialists abroad, like the segregationists at home, refused to retreat quietly. While white Southerners drew the line at school desegregation in 1956, the British and the French joined Israel in drawing their line at the Suez Canal that same year. Their co-ordinated assault on Egypt at the end of October after Gamal Abdel Nasser's nationalization of the canal confronted the Eisenhower administration with an international crisis that played out along racial lines. *Time* magazine noted that "the 'nonwhite' nations of the world lined up against Britain and France in a virtually solid front." Eisenhower told his advisers not to overemphasize the role of Nasser, as the British had, because the Egyptian president merely embodied the broader desire of Arab peoples for full independence and for "slapping the white man down."[41]

The assault on Suez represented precisely the kind of escalation of conflict between the First and Third Worlds that the Eisenhower administration sought to avoid. The president was furious with the precipitous action of the British, French, and Israelis, which he and Foster Dulles saw as such an egregious act of aggression that they had to condemn it. Eisenhower particularly resented his closest allies, the British, for forcing him to line up against them at the United Nations. His reaction would be the same a year later when white Southerners put him in a similar situation in Little Rock. The president and his advisers shared the dismay of one British Labour Party official who bemoaned the reversal of supposedly traditional racial roles: "Nasser's behaving like an Anglo-Saxon—we're behaving like Arabs." Henry Cabot Lodge, Jr., the U.S. ambassador to the UN and one of the most liberal members of the administration, reported to Eisenhower with relish "the enthusiastic reactions from Afro-Asian nations about your policy here in the Near East crisis"—forcing the British and French to withdraw from the canal—and other U.S. officials were also pleased by the positive responses from non-European countries. But the president discounted his novel, and brief, status as a hero in the Third World, criticizing instead what he saw as non-Europeans' hypocritical lack of concern with the contemporaneous, brutal Soviet occupation of Hungary. To Asians and other nonwhites, he complained to the British ambassador, "colonialism is not colonialism unless it is a matter of white domination over colored people." The president found few non-European governments that he could trust.[42]

White resistance to racial change in the Eisenhower years culminated in the desegregation crisis at Central High School in Little Rock in September 1957. There the violent and "respectable" streams of segregationism flowed together. Governor Orval Faubus and other Arkansas officials defied the federal courts first by using the Arkansas National Guard to prevent nine black students from entering the school and then, after the courts ordered them to halt such obstructionism, by openly encouraging the white mob that shut down Central High. Faced with blatant defiance of federal authority and apparent anarchy in the streets of Little Rock, a furious Dwight Eisenhower had no choice but to send in the U.S. Army. On the evening of 24 September, the joking young men of the Arkansas National Guard who had abetted the mob disappeared. They were replaced by a thousand unsmiling, steely-eyed paratroopers of the 101st Airborne Division, a unit whose experiences with the Nazi Wehrmacht and the Chinese People's Liberation Army left it unimpressed by civilian hooligans. They escorted the black students to school for the next two months. The president's years of sympathizing with segregationists had finally caught up with him, as white Southern extremism forced Eisenhower to become the first president since Reconstruction to use military might to protect black American citizens and their constitutional rights.[43]

Eisenhower responded to the situation in Little Rock much as he had to the crisis at Suez, with white Southerners in the role of the First World and black Southerners cast as the Third World. In each case he lost his patience with people with whom he usually agreed but who had finally gone too far by choosing violence rather than moderation. And in each case he seemed somewhat uncomfortable with the approval he received from anticolonialists and racial egalitarians, whom he saw as not much less extreme than the members of the hysterical crowd outside Central High. In a televised address to the nation on 24 September, Eisenhower explained his actions in Little Rock as a matter of preserving domestic order, not promoting racial justice. The issue was anarchy, not desegregation, he said. The president emphasized his "many warm friends" in the South and his own years of residence there, and declared that his personal opinion about the *Brown* decision had "no bearing on the matter of enforcement" of federal court decisions. Eisenhower felt personally betrayed by Faubus, who had backed out of an agreement with the White House not to obstruct the court order further. As David Halberstam observed, the former five-star general "did not look kindly on frontal challenges by junior officers." Insurrection, not integration, brought the crisis to a head for the man in the Oval Office.[44]

Little Rock displayed America's racial dilemmas to a fascinated international audience. Television cameras from around the world whirred as vicious white mobs taunted and assaulted the handful of orderly, well-dressed black schoolchildren behaving with great dignity. African American residents of the city were astonished and encouraged by the scope of world attention, including mail from well-wishers around the globe to the nine black students. The name "Little Rock" quickly became the foremost international symbol of American racism, with hostile crowds in Venezuela chanting it as they attacked Nixon's motorcade in Caracas in 1958. Rejecting American warnings about Communist influence in South Africa, Nelson Mandela declared that Africans "do not require any schooling from the U.S.A., which . . . should learn to put its own house in order before trying to teach everyone else." Moscow Radio mockingly included the Arkansas capital in its daily itineraries of cities passed over by Sputnik I, the world's first orbiting satellite launched just ten days after Eisenhower's dispatch of troops. Western Europeans emphasized the connection between American behavior at home and its leadership abroad. French commentators concluded, in the words of an administration summary, that "the United States could hardly hope to set world standards for morality in the face of such degrading spectacles." A white captain based at Little Rock Air Force Base asked the obvious question about U.S. foreign relations in a letter to the editor of the *Arkansas Gazette:* how could the peoples of Asia and the Middle East "have faith in our government when we so openly show hate for anyone outside the white race?"[45]

The Eisenhower administration again seemed more certain of how it wanted its racial policies portrayed abroad than at home. Once the president had dispatched U.S. troops, his advisers moved to ensure maximum propaganda benefit from the decision. His initial address to the nation was swiftly translated into forty-three languages, and the Voice of America broadcast generous details of the federal military intervention in an American state. The Eisenhower administration, as Moscow Radio enjoyed pointing out, showed more concern with the international repercussions of the events in Little Rock than with the actual violations of human rights and democratic practices there. The president himself refused to offer any hint of approval for the federal court decisions underlying the situation, emphasizing instead how "our enemies are gloating over this incident and using it everywhere to misrepresent the whole nation." Above all, he regretted how "it would be difficult to exaggerate the harm that is being done to the prestige and influence" of the United States abroad. For Eisenhower, white racial violence appeared to remain a strictly political rather than moral problem.[46]

Race as an International Issue for the United States

Despite significant progress in the military and the schools and in popular sports and entertainment, racial discrimination remained a fact of American life. Indeed, it was even exported, to the fierce resentment of people of color abroad. In the private sector, this meant the establishment of segregated workplaces by American companies abroad, such as Firestone in Liberia, of which Eisenhower himself was fully aware. In the public sector, this included the often disdainful attitudes of American diplomatic personnel posted in various Third World countries. Not until 1956 did the State Department seem to realize how offended Africans were by the American practice in South Africa and Rhodesia of inviting only whites to Fourth of July celebrations at the U.S. embassies and consulates. Members of the U.S. delegation to the UN lobbied the U.S. government to change this tradition, especially in light of "the recent disgraceful race riots in Alabama" over the possibility of integrating the state university in Tuscaloosa. Dulles agreed in principle but refused to issue any specific orders on the matter. Those American diplomats abroad who refused to join all-white social clubs, such as Henry Ramsey in Madras, India, often felt the stinging resentment of colleagues who believed they had "gone Asiatic." The White House chief of protocol complained about having to invite "these niggers" from Africa to receptions, and African diplomats experienced enormous difficulties in finding decent housing both in Washington and in New York. The new states of Africa, swiftly passing legislation to ban racial discrimination, could not avoid knowing the deep American ambivalence about racial equality.[47]

Racial thinking influenced Eisenhower's evaluation of the two states he most feared, the Soviet Union and the People's Republic of China. With his advisers and friends, the president referred to the Chinese as "hysterical" and "fanatical" and warned, "we are always wrong when we believe that Orientals think logically as we do." Long-standing anti-Asian feelings among white Americans had been exacerbated by three years of war against the People's Republic in Korea, as old images of the Chinese as cruel, inscrutable, robotlike, and ultimately inhuman resurfaced. One American officer recalled Chinese soldiers' bravery in the face of withering U.S. firepower: "It was like dealing with mass lunacy." Those Americans impressed by the Chinese military performance showed no less surprise, like the administration official who exclaimed: "I was brought up to think the Chinese couldn't handle a machine.

Now, suddenly, the Chinese are flying jets!" Like many Americans, Eisenhower was angered by what he saw as Chinese ingratitude for all that Americans had done for their country. American frustration with the Chinese apostasy in rejecting the pro-American Chiang Kai-shek in favor of the Communist Party remained enormous in the 1950s. The convergence of the Yellow Peril with the Red Peril in the People's Republic helped China to surpass Russia in the 1950s as the enemy whom many Americans most loathed.[48]

The racial and ideological constitution of the Soviet Union was more ambiguous to Americans in the 1950s than that of China. Fresh from their revolution and from holding their own in Korea against the vaunted U.S. military, the Chinese openly encouraged the spread of anticapitalist revolution throughout the Third World. By contrast, the Soviets shifted after Stalin's death to a less confrontational stance toward the West. The Eisenhower administration appreciated this change. In contrast to the "hysterical" and "irrational" Chinese who did not seem to fear American nuclear threats, Russian leaders, according to the president, despite "all differences in culture and tradition, values or language, . . . were human beings, and they wanted to remain alive." Foster Dulles called the Russians "calculating" in their actions, while arguing that "emotion played a large part in [China's] conduct." Nevertheless, the inclusion of a vast swath of Asia and its peoples within the Soviet Union made many Americans unsure of the country's racial dependability. In its coverage of the Soviet occupation of Hungary in 1956, *Time* described the "big new tanks . . . protected by trotting groups of Asian-Russian infantrymen." Obsessed by racial categorization, Americans were not completely sure what to make of the Soviets.[49]

Cold War competitiveness with the Soviet Union highlighted the hypocrisy of American racism amid U.S. claims to leadership of the "free world." Eisenhower campaigned in 1952 on a platform calling for liberation of the "captive peoples" of Eastern Europe, and Congress joined him in proclaiming a Captive Nations Week in 1959. Such a narrow vision of who was not free omitted vast numbers of people south of the Mason-Dixon line and the Sahara. An editorial in *The Crisis,* the monthly journal of the NAACP, pointed this out: "The time has come for the 'Free World' to devote as much attention to liberating the 'enslaved peoples' behind the various bamboo and liana curtains of Asia and Africa as behind the Iron Curtain of Europe." African Americans denounced the discrepancy between their secondary status in their own country and the generous refugee programs established by the U.S. government for white aliens fleeing the 1956 invasion of Hungary. "The advocacy of free elections in Europe by American officials is hypocrisy," Martin Luther King, Jr., declared, "when free elections are not held in great sections of America." Africans, Indians, and other nonwhite foreigners were equally skeptical of U.S. policies favoring those with fairer skins, as numerous observers pointed out and as Eisenhower and Dulles privately acknowledged.[50]

The standard American explanation of racial discrimination in the South during the 1950s involved the doctrine of federalism and states' rights: the division of authority among federal, state, and local governments precluded federal interference into such matters as education and criminal justice, which were the states' responsibility. The awesome reach of the U.S. government around the globe, however, convinced many foreigners that such reasoning was either absurd or disingenuous. John Hope Franklin discovered this opinion at a scholarly seminar he attended in Salzburg, Austria, in the summer of 1958. A French participant told him: "It appears to the outsider that federalism stands in the way of nothing that the national government actually wants to do; but it is always used as an excuse for the national government's not protecting the rights of Negroes." A Yugoslav at the same conference noted the contrast between Eisenhower's ready dispatch of troops to Lebanon in 1958 and his great hesitancy in sending paratroopers to Little Rock a year earlier. The Eisenhower administration likewise refused at first to become involved in the 1958 North Carolina "kissing case," in which two black boys under ten were arrested and sentenced to indefinite terms in the state reformatory for kissing a white female playmate. The administration changed strategy, encouraging a swift release of the young "subversives," only after local black protests spurred extensive international media coverage of the case, which led in turn to major demonstrations at U.S. embassies throughout Europe.[51]

As Americans wrestled with issues of racial equality and Communist influence, the two most contentious problems of the 1950s, a debate emerged about their precise relationship to each other. Liberals and racial egalitarians argued that the Cold War competition with the Communist bloc required the United States to live up fully to its proclaimed ideals of freedom and equality for all. Otherwise, the Soviets would win over the world's nonwhite majority and the future of the globe would be theirs. Black labor leader A. Philip Randolph encouraged Eisenhower in 1953 to stand up for African independence; since Africans were now determined to be free, he wrote, the United States had to help them see that the Soviets were not "the only champions of revolutions for nationalism and revolt against landlordism, poverty, disease, illiteracy and tyranny." Congressman Adam Clayton Powell of New York called the arrests of black bus boycotters in Montgomery in February 1956 "another ghastly victory for communism" and urged federal intervention. Eisenhower's refusal to render any public judgment about the events in Alabama was, other critics noted, "exactly what the communists desire" and directly counter to what the United States had fought for in World War II. African American journalist Rowan learned on a 1954 tour of India how offended Indians were by domestic American racial practices and how effectively local Communists could make political points out of them. *Blitz*, a Communist Party newspaper, connected the antiblack riots in Chicago's Trumbull Park with Rowan's simultaneous State Department–sponsored lecture tour under the headline "ROWANS BABBLE WHILE NEGROES BURN."[52]

Conservatives and segregationists disagreed fiercely with the liberal conjunction of domestic and foreign racial policies. Their equally anti-Communist language targeted support for racial equality and integration as subversive of American freedom. Even as red-baiting in general declined in American politics after the end of the Korean War in 1953 and the Senate censure of Joseph McCarthy in 1954, white Southerners increased their use of anti-Communist rhetoric after the *Brown* decision as part of a strategy to equate integration with Communism. Governor Talmadge explained this position in *You and Segregation,* a 1955 book immensely popular among white Southerners: "Too many things are being done in this country and by our country because we keep looking back over our shoulders at the Communists. Who cares what the Reds say? Who cares what *Pravda* prints?" Talmadge and other prominent Southerners like Georgia senator Richard Russell believed liberals were being suckered by Moscow's Cold War rhetoric. Rather than try to satisfy Soviet demands, the United States should stand up proudly for its own traditions. If Communists supported racial integration, could there be any clearer sign of its immorality?[53]

The Eisenhower administration agreed with elements of each side in this debate about Communism and racial equality. The president and his advisers recognized the logic of not driving African states toward the Soviet bloc for assistance, but they shared an underlying assumption about the potential subversiveness of alienated African Americans. Dissatisfaction breeds dissent, and J. Edgar Hoover argued that the civil rights movement in the South, being angry and reformist, was thoroughly penetrated by Communist Party operatives. No one in the administration disagreed with the nation's chief policeman. The FBI responded to school desegregation incidents by investigating them for disorder and subversion rather than for the local obstruction of federal court orders. Accordingly, bureau reports were treated above all as "domestic intelligence" and were sent first to the Internal Security Division of the Justice Department and only later to its Civil Rights Section. Foster Dulles revealed the depth of the administration's distrust of black people in a 1953 lament about "the problem of getting colored people cleared by the FBI." As recorded by a note-taking assistant, Dulles said, "there was practically no negro, even Ralph Bunch[e], who could come through an FBI check lily white, because all of their organizations had been infiltrated at one time or another." Despite the White House's insistence on "getting lily white clearances on everybody" working for the government, the secretary of state concluded that "it is impossible to do this with negroes."[54]

Even with their uncertainty about the loyalty of black Americans, Eisenhower and his advisers recognized that accelerating changes in race relations abroad in the late 1950s would inevitably impact racial practices at home. Nixon returned from Ghana in 1957 to tell the president bluntly: "We cannot talk equality to the peoples of Africa and Asia and practice inequality in the United States." In a discussion of the 1959 lynching of Mack Parker with Governor J. P. Coleman of Mississippi, Eisenhower did not indicate any concern for the victim, but he made very clear his unhappiness with the damage such incidents were doing to the U.S. image abroad. The end of white supremacy in the rest of the world, journalist Isaacs observed, "made its survival in the United States suddenly and painfully conspicuous . . . It was like being caught naked in a glaring spotlight alone on a great stage in a huge theater filled with people we had not known were there." The increasing appearance of African diplomats in Washington and at the United Nations in New York, and their respectful treatment by white American policymakers who had never previously dealt with people of color as peers, signaled the decline of rigid racial hierarchies.[55]

For Americans of African descent, the liberation of Africa from colonial rule carried particular meaning. As citizens of a country that had long associated Africa with savagery and darkness, few black Americans had been able to embrace their African heritage without ambivalence. Some, like John Lewis, grew up cheering from segregated balconies for the African warriors in Tarzan movies; more probably agreed with Roger Wilkins that "to identify with the black clods of Tarzan's Africa never entered my mind." Malcolm X recalled, "we did not realize that in hating Africa and Africans we were hating ourselves," including the very features and hues of an African phenotype.

The appearance of proud, capable statesmen from the emerging independent African nations in the halls of the UN in the late 1950s changed that perception of primitive Africa with dramatic suddenness. Black Americans across the political spectrum shared a newfound enthusiasm for African culture and an admiration for black Africa's political leaders. The new civil rights movement drew encouragement from events across the Atlantic. Martin Luther King, Jr., repeatedly equated colonialism with segregation—"they are both based on a contempt for life, and a tragic doctrine of white supremacy"—and emphasized African Americans' awareness that their "struggle for freedom is a part of a worldwide struggle." Antiracist organizers understood at least as clearly as American policymakers that racial hierarchies in the United States and abroad formed a seamless web, and they correctly foresaw that independence in Asia and Africa would hasten the demise of Jim Crow.[56]

Many black Americans traveled to the new nations of West Africa in the first years of independence, from 1957 onward, searching for their roots in their ancestral continent. They were inevitably impressed by the sight of black people occupying all positions in a society, from the most menial to the most elite. They thrilled to the absence of racial harassment and humiliation. "Two whole months," one told an American journalist, "for the first time in my life, two whole months without incidents or insults." For most African American visitors, however, the intense feeling of racial familiarity was balanced by the reality of alienation from a vastly different African culture. Like all peoples traveling abroad, they tended to become aware of their own strong personal and cultural attachments—in this case, to the United States. Faced with African pride in hard-won independence over white colonial rule, black Americans were sometimes made to feel disadvantaged because of their still segregated American homeland. Some African Americans in West Africa found themselves in the unexpected position of defending the United States against African criticism. For most black Americans, the precise character of their renewed relationship with the ancestral continent remained open to negotiation.[57]

Eisenhower, the Third World, and Africa

In contrast to his counsel of patience for people of color at home, Eisenhower seemed to understand that nonwhites abroad had gone too far down the road of change to wait any longer for gradual progress. Most of the peoples of Asia and the Middle East had achieved their independence from European rule before 1953, and Africa swiftly followed. As at home, the issue of race was crucial abroad. Colonialism had been grounded in racial inequality and discrimination, and the successful struggles for national liberation brought a new day of formal racial equality to the world's nonwhite majority. Eisenhower generally recognized the power of nationalist sentiment and commitment in the Third World. Within certain ideological boundaries, he tried to position the United States in support of the governments of newly independent nations. The U.S. government itself controlled a handful of largely nonwhite territories, most importantly Hawaii and Alaska. Unlike Southerners in Congress who opposed statehood for them because of the accompanying infusion of citizens of color, Eisenhower supported the two territories' bid to join the Union on an equal basis. He signed the bills designating them the forty-ninth and fiftieth states in 1959.[58]

Beyond a basic recognition of the inexorable character of anticolonial independence movements, however, Eisenhower and his secretary of state, Foster Dulles, knew little about the Third World. Like most white Americans concerned with foreign affairs in the 1950s, they were men of Europe. Eisenhower had forged close ties with the British government as commander of the European theater in World War II, links he further strengthened as NATO commander in the late 1940s. His most prominent biographer, Stephen Ambrose, notes that the Atlantic alliance was "the issue closest to his heart." With the exception of his work on the Japanese peace treaty, Dulles's extensive background in foreign affairs had focused primarily on Europe before 1953. The secretary of state's general opposition to European colonialism in the 1950s derived from his understanding of U.S. interests in the Cold War, not from sympathy with nonwhite liberation fighters. Even his differences with the Europeans over preserving formal colonialism sometimes blurred. In February 1956 he dismissed the Indian claim to the city of Goa—a tiny vestigial Portuguese colony surrounded by Indian territory—insisting instead that "all the world regards [Goa] as a Portuguese province." "The world" for the American secretary of state in the 1950s still meant Europe.[59]

The administration's thin veneer of anticolonialism masked not only a sturdy Eurocentrism but also a marked disdain for most non-European leaders, including even those with close ties to the West. White House assistant C. D. Jackson was plainly disgusted by what he called "the swirling mass of emotionally super-charged Africans and Asiatics and Arabs that outnumber us." Even Vice President Nixon, the administration's "liberal" on racial issues, who generally argued for greater U.S. attention to Africa, referred to the continent in a 1960 NSC meeting as "a horrible place." Foster Dulles's successor, Christian Herter, warned the president that Cuba's Fidel Castro was "very much like a child" who became especially "voluble" and "wild" when he spoke in Spanish; Eisenhower responded by recalling that India's Nehru "accumulated emotional frenzy" as he spoke. The image of Third World leaders as volatile children was especially enduring for the administration. Sounding remarkably like his white South African counterparts, Foster Dulles lamented in June 1958 the "tremendous surge in the direction of popular government by people who have practically no capacity for self-government and indeed are like children in facing this problem." Whether these "children" could mature remained uncertain to U.S. policymakers.[60]

Non-Europeans abroad were frequently skeptical of American foreign policy and occasionally demonstrated their displeasure in ways that further alienated U.S. officials. Crowds of angry protesters in Caracas, for example, assaulted and nearly killed Nixon on his would-be goodwill trip to Latin America in the spring of 1958. The Tokyo government canceled its invitation for Eisenhower to visit Japan in the summer of 1960, after enormous street protests broke out; a fierce attack by the crowds on the advance party of press secretary James Hagerty required U.S. marines to rescue him by helicopter. Even Third World peoples more admiring of the war hero president showed some ambivalence, like the sign visible in Rio de Janeiro during his stopover there in February 1960: "We like Ike; We like Fidel, too." Third World dubiousness about the administration's actions in Iran, Guatemala, Indochina, the Taiwan Straits, Lebanon, Indonesia, Cuba, and the Congo—often stridently expressed—exacerbated official American dislike for much of the world beyond Western Europe.[61]

The willingness of the Eisenhower administration to acknowledge the seriousness of nationalism in the Third World was undercut by its growing anxiety about Communist influence there. From 1955, with Stalin dead and postwar Europe mostly stabilized, the Soviet government under Nikita Khrushchev sought to expand its contacts with the new governments of Asia and the Middle East and the emerging nations of Africa. Mao Tse-tung's similar intentions provoked American fears that China's history as the largest nonwhite nation exploited by the West might position it to provide leadership to other similarly exploited countries in the Third World. Just as the FBI and other elements of the U.S. government assumed that African Americans in the South were naturally docile and could be organized only by subversive and usually white outsiders ("outside agitators"), Eisenhower and his advisers believed that dark-skinned, childlike leaders of non-European nations were vulnerable to the pernicious influences of subversive outsiders like the Soviets and the Chinese. In the steely light of the Cold War, American policymakers saw the dark hand of Moscow behind all challenges to a Western-dominated capitalist order.[62]

The Eisenhower administration found it difficult to convince Asians and Africans that the expansion of Soviet influence represented a new and more powerful form of colonialism. For most countries, European colonialism remained a far more clear and present danger than Communism. The close American relationship with the colonial powers of Western Europe and with South Africa, along with American racial practices at home, often brought U.S. rather than Soviet intentions into question. The National Security Council acknowledged this problem privately in March 1958: "African desires for self-government[,] as well as the racial policies of the Union of South Africa, make Soviet pretensions to be the champion of the oppressed colonies acceptable in some areas, so that Western attempts to picture Soviet Russia as a colonialist power itself have simply not been believed by African leaders." Later that year the U.S. Information Agency (USIA) warned the White House of the deteriorating image of the United States abroad. American influence in the world was being perceived increasingly as a form of cultural and military imperialism, it feared, with the creation and maintenance of states like South Vietnam, Taiwan, and South Korea—with their Christian, pro-Western leaders—as prime examples. The USIA paraphrased the Indian perspective: "We don't see the Russian fleet in Oriental waters. We see only the American fleet. We don't see the Russian Army in mainland China but we see a good deal of the American Army in Formosa, and Japan, and Korea, and Okinawa, and the Philippines." Worst of all, the USIA concluded that the United States was behaving exactly like one of the Third World nations that the administration disdained: "Our chief problem is to grow up psychologically." Boasting about "our richness, our bigness, and our strength, . . . we continue to act like adolescents."[63]

Fearful of Soviet influence and distrustful of the newly independent nations, Foster Dulles lashed out at those countries seeking to remain nonaligned in the Cold War. In a commencement address at Iowa State College in June 1956, the secretary of state declared that neutrality "has increasingly become an obsolete conception, and, except under very exceptional circumstances, it is an immoral and shortsighted conception." Even those in the

State Department more sympathetic to the Third World considered neutralism a product of "the strong irrational nationalist tendency" in those countries, ruled by people the NSC called "still immature and unsophisticated with respect to . . . the issues that divide the world today." An internal State Department study of the problem in 1955 did acknowledge, however, that neutralism was a complicated issue, as the United States had itself promoted its neutral position regarding other nations' conflicts for almost all of its history, most recently in the years 1914–1917 and 1939–1941. The study also emphasized that "color consciousness is closely linked to anti-colonialism," and that neutralism was often defended on the grounds that "the racial intolerance of the Western powers is just as bad as anything that might be practised by the Communist bloc." Despite Dulles's severe public words about neutralism, his department concluded privately that "this sensitivity about race is one of the chief obstacles to sympathy with the West."[64]

The success of independence movements in Asia and the Middle East in the Truman years meant that the problem of colonialism for the Eisenhower administration was confined increasingly to Africa. Africa's significance for the United States in the 1950s included economic, strategic, and political elements. This bright continent represented the world's last great frontier of natural resources, offering vast quantities of such critical minerals as gold, diamonds, columbite, cobalt, chromite, and uranium. "It is our Last Frontier," declared John Gunther in 1955 in his popular book *Inside Africa.* "Africa lies open like a vacuum, and is almost perfectly defenseless—the richest prize on earth." The American economic stake in the region, while smaller than in other parts of the world, still amounted to 3.5 percent of U.S. foreign trade. The continent's strategic importance for raw materials for modern weapons was amplified by its location astride major shipping lanes and its utility in wartime for military bases, as Eisenhower himself had learned in commanding forces along its northern shore in World War II. Africa's political significance to the United States lay mostly in its meaning to America's NATO allies. The Eisenhower administration acknowledged privately the pervasive African resentment of European economic exploitation and racial discrimination. "Should serious disorders develop in the area," the NSC warned the president in 1957, "there might be a further military and economic drain on some of the more important NATO powers." Worse, the current easy access to Africa's mineral wealth might be cut off. The NSC concluded that the United States therefore had "a very real interest in orderly political evolution in Africa."[65]

Uranium was the key to the American perspective on Africa at the start of the Eisenhower administration. The preeminent military might of the United States depended on its nuclear arsenal, and over 95 percent of the fuel for those weapons since the original Manhattan Project had come from the Shinkolobwe mine in the Belgian Congo. It would be difficult to overstate the strategic importance of this resource for the United States in 1953. The chairman of the U.S. Atomic Energy Commission, Gordon Dean, was careful to remind Foster Dulles of this in his first weeks as the administration's chief foreign policymaker. Just behind the Belgian Congo came the troubled land of apartheid, Dean told Dulles: "Within two to three years South Africa will be our most important uranium supplier." As the 1950s progressed, other sources of the crucial ore became available, from expanded operations in Canada and on the Colorado Plateau at home, while production at Shinkolobwe tailed off until the mine finally closed in 1960. But throughout the years of the Eisenhower administration, first the Belgian Congo and then South Africa held a unique place in American strategic planning.[66]

South Africa epitomized the problems that Eisenhower faced in Africa. Control of that nation's vast mineral wealth and strategic materials rested in the hands of its small white minority. By the mid-1950s the largest producer of uranium ore, South Africa also received the bulk of American investment and trade with the entire continent. His State Department superiors emphasized to Henry Byroade as he settled in as U.S. ambassador to Pretoria in 1956 that American relations with the apartheid state "are very friendly and harmonious. South Africa is strongly anti-Communist and pro-West" and looked increasingly to the United States rather than Britain as its closest ally.[67] But these same friendly white folks had already begun, as one State Department official reminded Foster Dulles, "to create a police state." The NSC privately agreed, regretting that South African "political development has been retrogressive." And the whites seemed immune to friendly persuasion. The next U.S. ambassador to Pretoria after Byroade, Philip Crowe, referred privately to Prime Minister Henrik Verwoerd as "a fanatic with a fanatic's absolute faith in his own righteousness," especially about racial issues. So while an earlier U.S. ambassador, Waldemar Gallman, had urged Dulles in 1953 to remember that "peace in the mines, in the homes and on the streets [of South Africa] is essential" to American security interests, other U.S. officials grew more worried as the decade progressed. Mason Sears of the State Department warned Dulles of the "contagious influence of racism" on neighboring African states, and many in the administration began to fear that the brutalities of anti-Communist apartheid might drive Africans into the arms of the Communist Party.[68]

The Eisenhower administration believed it could have little influence on the racial policies of the South African government. Following the ruling Nationalist Party's electoral victory in April 1953, Dulles directed Ambassador Gallman to suggest gently to the prime minister the virtues of some mild reforms. Gallman agreed with that approach, while warning Washington that the possibility of massive African organizing in South Africa was more imminent than he had previously believed. African National Congress (ANC) literature reflected increasing hostility toward the United States during the 1950s, as Washington preserved its close ties to Pretoria. One young ANC activist suggested African disillusionment with the United States: "I think America has lost African friendship. As far as I am concerned, I will henceforth look East where race discrimination is so taboo that it is made a crime by the state."[69]

South Africans, however, could not help noticing changes in American racial practices. In January 1955, the *U.S.S. Midway* docked in Cape Town for a shore leave; the American officers not only refused a South African request to keep sailors of color on board but also opened up the ship to South African visitors on a nonsegregated basis. Over 20,000 South Africans of all colors came aboard over two days to see the inside of an integrated American warship. In a quiet but revealing administrative shift, the State Department by 1956 moved its coverage of South Africa out of the Bureau of European Affairs and into the Bureau of Near Eastern, South Asian, and African Affairs. It was as though, for Washington, South Africa had disembarked from Europe and finally arrived in Africa. If this was true, then perhaps the African majority of South Africa would now show up more clearly on Washington's radar.[70]

The overall thrust of U.S. policy toward Africa in the 1950s was the same as the administration's policy toward civil rights at home: to avoid it as much as possible. The greater American interest in Africa by the end of the decade came only in response to decolonization, not in support for it. If top U.S. policymakers knew little about the world outside Europe in general, they knew almost nothing about Africa. Dulles devoted only two sentences to the continent in his initial speech as secretary of state. Blaming "Communists" rather than white authorities for any unrest in the area, he worried "*if* there should be trouble there"—indicating that the Defiance Campaign of Africans in South Africa against apartheid, the Mau Mau rebellion of Kenyans against British rule, the full-scale revolt in Madagascar in 1947, and the outbreaks of anti-French violence in Morocco and Tunisia had made little impression on the chief U.S. diplomat. Africa remained a future problem for the administration, one that it figured to avoid for now.[71]

At least until 1958, when accelerating African independence began to require the United States to establish direct political ties to most of the continent, Eisenhower and his advisers sought to leave African issues generally to the Western European colonial powers. Europe and Africa remained "fundamentally complementary areas," according to the NSC. Sounding much like a white American Northerner speaking of white Southerners and the need to preserve close ties to them, Henry Byroade—then an assistant secretary of state—emphasized in a 1953 speech the crucial importance of African interests to the Western European economy, "which we have contributed so much to support." The success of the Marshall Plan investment, in other words, depended to a great extent on Africa. Byroade warned that "a sudden break in economic relations might seriously injure the European economies upon which our Atlantic defense system depends." American policymakers believed that orderly progress toward African self-rule would develop best under European tutelage. The U.S. government itself depended in large part on European experts on Africa in the 1950s, owing to the dearth of such specialists in the United States.[72]

Official American policy toward colonialism in Africa changed little between 1953 and 1958. Independence was the eventual goal, but "premature independence" remained the great danger. "It is a hard, inescapable fact that premature independence can be dangerous, retrogressive, and destructive," Byroade declared in 1953 in the administration's first major statement on Africa. In a world populated by Soviet predators, there was no room for romanticizing about immediate liberation. "Unless we are willing to recognize that there is such a thing as premature independence," Byroade concluded, "we cannot think intelligently or constructively about the status of dependent peoples." Here was the key: while supporting the principle of "eventual self-determination," the United States implicitly claimed the right to determine when African peoples were "mature" enough for independence. The image of parent and child—and the reality of what Africans increasingly called neocolonialism—could not have been more clear. In the administration's second major statement on Africa, George Allen sounded the same notes three years later. On the very cusp of the great turn in African history toward independence, the U.S. assistant secretary of state could only manage to call the Europeans in Africa "the so-called colonial powers" and attribute antiracist organizing in South Africa to "Communists."[73]

The Eisenhower administration did not intend this approach as a recipe for simply preserving the status quo. Gradual but steady preparation for eventual independence was crucial to avoid a rerun of the previous decade's events in Asia. In his report to the president on his 1957 visit to Africa for Ghana's independence celebration, Nixon spoke of his "distinct impression that the communist leaders consider Africa today to be as important in their designs for world conquest as they considered China to be twenty-five years ago." Many Americans in the 1950s were still traumatized by "losing" China to Communist revolution and hoped to prevent a similar development in Africa. Realizing that "the grandchildren of Dr. Livingstone's porters [now] go to Oxford," as John Gunther put it, they sought to make allies rather than enemies of the new nationalists taking power south of the Sahara.[74]

Beginning with the Suez crisis in the fall of 1956, several developments converged to move the administration to a position of greater support for African independence and even a limited openness to African neutralism in the Cold War. As black African independence began to unfold in Ghana in 1957 and in Guinea in 1958, civil rights also emerged as a major political issue at home with the Little Rock school desegregation crisis of 1957–58, while Vice President Nixon's ambitions for the White House in 1960 led him to push for a more active U.S. policy regarding Africa. By 1960, the year in which seventeen new nations emerged on the continent, Eisenhower himself could be heard at the United Nations declaring America's acceptance of African nonalignment. U.S. officials shifted ground to increase American influence with the new nonwhite majority in the UN General Assembly and its prominent political role in international affairs. Race relations were an important element in producing this change. Americans both inside and outside the government acknowledged that racial discrimination was "in many ways the heart of the colonial question," as Chester Bowles noted. At the same time, the rise of the civil rights movement in the United States was changing the domestic calculation in African policy. George Allen advised Dulles of the importance of a "multi-racial approach" to African problems: "Any other course of action would, in the long run, meet with such domestic opposition within the United States that it would be next to impossible to carry out." A new day was coming in American relations with Africa.[75]

African Challenge and American Response

The white Southern challenge to desegregation that crystallized at Little Rock was soon eclipsed for the Eisenhower administration by the rising wave of African independence. That wave began to crest in Ghana in 1957 and buried most of the remaining European colonial ties to the continent in its thunderous crash from 1960 to 1963. Asian independence in the early postwar years had provided a crucial precedent, bolstered by the addition of three Arab north African states—Libya, Morocco, and Tunisia—by 1956. Most British and French efforts to prop up collaborative regimes in the Middle East failed during the decade; the 1958 revolution in Iraq was just one example. The outbreak of the lengthy Mau Mau uprising in Kenya in 1952 and the start of the protracted Algerian revolution two years later marked the spread of wars of national liberation to Africa. Severe British repression of the Kenyan rebellion highlighted the racial hostility present under colonial rule. Senator John Kennedy's July 1957 speech supporting Algerian independence signaled his new prominence in foreign affairs, while his future opponent in the 1960 presidential race, Vice President Nixon, worried instead about the actions of the minority of one million Frenchmen living in Algeria should the colony become independent. Racial polarization in much of Africa by the late 1950s was creating a similar problem for the Eisenhower administration to that represented by the crisis in Arkansas, as politically active black Americans were quick to note.[76]

Racial divisions in Africa stemmed above all from the resistance of white settlers to African liberation movements. Tensions rose in direct proportion to the number of white settlers in an area; South Africa, Southern Rhodesia, Kenya, and Algeria formed the strongest white redoubts, while West Africa moved most quickly to majority rule. Metropolitan administrations mattered, too: the British, moving steadily toward independence for their colonies, seemed more like moderate white Southerners, while the Portuguese and the South Africans, digging in for a last stand, appeared analogous to white "massive resisters" in the South. The administration was stumped by the question of how "to force the obdurate right wingers among the 'settlers' to become cooperative with the Africans," and feared that a mere "handful of . . . white die-hards" might spark an international racial conflict certain to damage the interests of the United States abroad, not to mention its domestic stability. The administration had precisely the same concern regarding violent white Southerners.[77]

In addition to apartheid in South Africa and insurrections in Kenya and Algeria, another manifestation of the white settler problem that coincided precisely with the tenure of the Eisenhower administration was the Central African Federation (CAF). Formed in 1953 out of the British territories of Nyasaland and the two Rhodesias, the federation was intended by Britain to tie the self-governing colony of Southern Rhodesia more closely to the rest of black Africa to the north, thereby discouraging the growing political orientation of Southern Rhodesia's white community toward South Africa. Apartheid might thus be contained at the Limpopo River. The British government framed the creation of the CAF as a step toward "racial partnership" in an eventually independent, multiracial state. But the actual effect of the federation was to increase white political and economic domination over Africans. Africans in the three territories had opposed the idea of the CAF from the beginning, and they resented the intensifying racial discrimination and harassment they experienced during the 1950s. The NSC acknowledged in March 1958 that "the hardening of European settlers' attitudes" was "undermining the principle of racial partnership" and thus destroying the "U.S. hope that the Federation would serve as an example for the development of multi-racial societies." African resistance to white rule in the CAF accelerated after 1957, leading to independence under majority rule for Nyasaland and Northern Rhodesia early in the 1960s. The more numerous and powerful whites of Southern Rhodesia, like their neighbors in South Africa and the Portuguese colonies, held out. As one boasted to an American visitor, "this country is like the American South at its best."[78]

The problem for the Eisenhower administration in dealing with Africa was that Africans across the continent saw white domination rather than Communism as their chief stumbling block. Ndabaningi Sithole of Southern Rhodesia spoke for most Africans in his widely read 1959 book, *African Nationalism*, which declared that "African nationalism is a struggle against white supremacy." Sithole and others argued that white rule in Africa was precisely analogous to Soviet control of Eastern Europe. Washington's anti-Communist prescription for dealing with unrest elsewhere in the world therefore was unlikely to work in Africa, as American policymakers reluctantly admitted. The CIA acknowledged that most African nationalists were political moderates "well-disposed by education (and sometimes religion) toward the West." In fact, the agency reported, "most of them have sought to avoid violence except as a last resort," and outside South Africa "none is a known Communist." The agency warned instead of increasing interracial tensions in areas of heavy white settlement. And an aide to Foster Dulles, returning from a tour of the continent in 1958, emphasized to the secretary of state the often savage character of white rule in African colonies. As a result, he said, "the black African's attitude toward the white man shades from universal envy through mistrust and fear to burning hate."[79]

At midnight on 5 March 1957, in a celebration rich in symbolic importance, Ghana became the first black African nation to declare its independence from European colonial rule. Africans, white settlers, and colonial officials alike knew this marked the beginning of the end of the Africa they had known. Events in Ghana encouraged nationalists elsewhere on the continent in their own struggles against white rule. African Americans also recognized the impact that events in West Africa could have on their movement for equality in the United States. Martin Luther King, Jr., had been unable to get an audience with Richard Nixon before the two met by chance in Accra at the independence celebrations; in that atmosphere, the astute vice president treated King with great respect and arranged to see him again at home.[80]

President Kwame Nkrumah of Ghana overnight became the leading political figure on the continent. Like Sithole, Nigerian journalist Nnamdi Azikiwe, and most other African leaders, Nkrumah had important personal and cultural ties to the West. He had spent ten years studying and working in the United States and recalled that period with affection in conversations with Nixon. But he also remembered his first experience south of the Mason-Dixon line, when the future world figure asked a white waiter in a bus terminal restaurant for a glass of water and was told with icy disdain: "The place for you, my man, is the spittoon outside." Nkrumah's determination to help lead all of Africa out of colonialism made him skeptical of Eisenhower's lukewarm approach to African independence. Events in the Congo in 1960 divided him further from the United States, and Ghana moved soon thereafter into a new and close relationship with the Communist bloc.[81]

After Ghana, independence came swiftly to most of the rest of the continent. Throughout 1958 African representatives gathered in a series of conferences to encourage the liberation of the remaining colonies. The All-African Peoples Conference that convened in Accra in December symbolized a reversal of the famous Berlin Conference seventy-four years earlier; where the delegates at Berlin had divided up the continent under European rule, those at Accra called for the departure of the remaining colonialists and the unification of the continent under African rule. Guinea declared its independence from France in October 1958, and by the end of 1960—the Year of Africa—

seventeen more nations had come into existence, with the rest outside southern Africa soon to follow. Prime Minister Harold Macmillan aligned the British government with these developments by traveling to the heart of white supremacy—Pretoria—and telling the South African parliament on 3 February 1960 that "the growth of national consciousness in Africa is a political fact and we must accept it as such." The leader of the greatest colonial power in Africa affirmed that "the wind of change" blowing through the continent was not to be denied.[82]

The U.S. government responded to the rise of neutralism in Asia and the Middle East and the growth of nationalism in Africa in part by emphasizing cultural diplomacy with the Third World. The State Department, for example, dispatched African American speakers to India and other nonwhite nations to explain the progress being made at home in eliminating racial discrimination. In 1956, the Voice of America began broadcasting jazz, an American music form wildly popular in much of the world. In the spring of that year, Louis Armstrong played before an enthusiastic audience of 100,000 in the Gold Coast, and trumpeter Dizzy Gillespie received State Department sponsorship to tour Asia and the Middle East with his racially mixed eight-piece jazz orchestra. Gillespie remembered afterward that he "wasn't going over to apologize for the racist policies of America," but he did enjoy dispelling some of his audiences' worst assumptions about the state of race relations in the United States. He recalled how being the black leader of a multiracial band surprised many foreign listeners: "They'd heard about blacks being lynched and burned, and here I come with half whites and blacks and a girl playing in the band. And everybody seemed to be getting along fine."[83]

The Eisenhower administration's strategy of using African Americans as cultural diplomats in the Third World undoubtedly helped create some goodwill abroad, but it also occasionally backfired. This was hardly surprising, given the ambivalence most black Americans felt about their country and its treatment of them. Gillespie and his band found that while they were shown off abroad by the U.S. government, they were often ignored or mistreated at home, as when they were harassed by police in the Atlanta airport for not waiting to board a plane until all white passengers were on board. The news of white treatment of black schoolchildren in Little Rock—and the initially hesitant response of the president—so angered Armstrong that he promptly canceled his much anticipated tour of the Soviet Union for the State Department. He was dismayed by Eisenhower's role in allowing the crisis to develop. When asked by the press about the Russian trip, Armstrong replied: "I'll do it on my own. The people over there ask me what's wrong with my country, what am I supposed to say?" Given the limits on the role black Americans could play in this administration's diplomacy, Henry Cabot Lodge, Jr., suggested to Eisenhower that high-ranking white officials "who are temperamentally so inclined" could be put to great use in Asian and African countries; they could demonstrate that white Americans, unlike Europeans, would not only work with people of color but also would "play with them and treat them as social equals." Lodge concluded that "a certain amount of sociable drinking, dining, dancing and laughing" by elite American policymakers visiting Third World nations would go a long way toward building goodwill for the United States.[84]

Lodge was one of a handful of influential Americans pushing the administration toward a more positive view of an independent Africa by 1956. Republican Congresswoman Frances Bolton of Ohio was another. She toured the continent at the end of 1955 and called for an immediate effort by Americans to overcome their "apathy or ignorance" regarding the continent. A third was the rising critic of the administration's African policy in the Senate, John Kennedy. The most important was probably the vice president, who returned from Ghana in 1957 to recommend paying much more attention to Africa and creating a bureau for African affairs in the State Department in recognition of the continent's new significance. By 1958 Nixon was arguing in the National Security Council for the United States to "take the initiative in encouraging neutralism, which the national independence movements favor, instead of assuming that a neutral is on the Soviet side." Nixon concluded that in place of military or political connections with the West for Africa, the United States should instead promote "educational and cultural ties." NSC note takers recorded the revealing argument of Eisenhower himself in August 1958 that "rather than slow down the independence movement," he would "like to be on the side of the natives for once."[85]

The president was articulating, in his own condescending fashion, the shift in American policy toward Africa that occurred in 1958. The new Bureau of African Affairs created that summer in the State Department initiated the idea of a unified U.S. Africa policy. No longer would the continent be understood primarily through its relationship with Europe. Career diplomat Joseph Satterthwaite filled the new post of assistant secretary of state for African affairs, and, despite his known sympathies for white South Africans and his subsequent ambassadorship to Pretoria, eschewed references to "premature independence" or the priority of America's NATO allies. Instead, the administration focused on countering the Soviet and Chinese diplomatic initiatives on the continent and preventing Communists from posing as the friendly alternative to colonialists. The NSC emphasized the need to avoid letting "thwarted nationalist and self-determinist aspirations" be turned to the Eastern bloc's advantage. The United States could no longer afford to remain behind the curve in its relations with Africa.[86]

On 30 October 1958 the administration signaled another change in its thinking about Africa by voting, for the first time, in favor of a UN resolution condemning South Africa's apartheid policies. Lodge had lobbied for a year in favor of this new position, arguing that the *Brown* decision and the Little Rock crisis had made it "even more important that we express ourselves positively on racial issues." Ambassador Byroade tried afterward to mollify an unhappy South African foreign minister, Eric Louw. When Louw suggested that the change in the U.S. position on apartheid resulted from the United States' anxiety about its own domestic racial conflicts, Byroade conceded that it could "hardly be denied that our problems at home had made more people aware of and think about racial problems than in years of the recent past." By 1959 the United States even voted for a resolution opposing the continuation of racial discrimination anywhere in the world—a remarkable stance in light of Jim Crow's abiding presence in the South at home.[87]

Africa's acquisition of a more positive policy from Washington after 1958 did not mean that Africans had won over the hearts and minds of the Eisenhower administration. Even as they established friendly relations with the new nations of the continent, white American policymakers could not hide their ambivalence about newly independent black people. A 1958 NSC report praised by Eisenhower captured the dilemma for many Americans: "The Spirit of 1776 was running wild," it said—showing American pride in being a model for other nations—but this phenomenon was "rather terrifying" in Africa. Other advisers to the president deplored the "juvenile delinquencies of these new nations," while admitting the need to "cheer for freedom." The NSC referred to "the contagion of the nationalist fever," a disease sure to spread even to the isolated and presumably healthy Portuguese colonies. Part of the problem, as one American official admitted, was the "serious and unnecessary handicap" that burdened the administration because it had "relatively little direct association with Africans, or knowledge of what they are thinking." That lack of information was a direct cost for U.S. foreign policy of the administration's willing participation in racial segregation, both at home and abroad: if one talks only with whites, one can know only what whites think. The extent of this ignorance of Africans among American policymakers as late as 1960 sometimes appeared unlimited: Nixon, the administration's strongest supporter of Africa, announced to a National Security Council meeting in March that "some of the peoples of Africa have been out of the trees for only about fifty years." Budget director Maurice Stans responded that his recent visit to the continent convinced him that "many Africans still belonged in the trees."[88]

While Britain, France, and even Belgium were easing themselves out of most of Africa by the end of Eisenhower's second term, the Portuguese dug in to stay. The U.S. government, as one American diplomat noted in 1955, was "well aware of the oppressive and medieval practices of the Portuguese" in Angola and Mozambique. The same NSC report that Eisenhower praised so highly emphasized how "badly governed and administered" the two colonies were, how common forced labor—that is, slavery—was, how rare educational opportunities of any kind were for Africans, and how "mercilessly" the Portuguese authorities exploited the resources of the area. None of this apparently troubled the president. Fresh from the disappointment of his canceled summit meeting with Khrushchev in Paris in May 1960, he stopped over in Lisbon to an enthusiastic welcome. There Eisenhower praised the "real progress" made by the dictatorship of Antonio Salazar, affirmed that officials from Washington and Lisbon "have worked together without a single difference of opinion," and called for even closer ties between Portugal and the United States in their mutual struggle against Communism. The *New York Times* report of the visit could not help noting the disappointment of Portugal's own persecuted democrats, who watched Eisenhower's appreciation for the Salazar regime provide cover for "the continuing lack of civil liberties, censorship, repression of political activity and police rule" at home in a member nation of NATO.[89]

The experiences of the growing number of African diplomats working and traveling in the United States indicated the tentativeness of the new U.S. warmth toward Africa in the last three years of the Eisenhower administration. High-profile visitors, like Nkrumah and Sekou Touré of Guinea, were carefully shielded by their American handlers from chance encounters with venomous white racism. Others were less fortunate, especially along the Maryland and Delaware sections of the corridor between New York and Washington. The eviction of Ghanaian finance minister H. A. Gbedemah from a Howard Johnson restaurant in Dover on 9 October 1957 received considerable publicity, coming as it did at the height of the Little Rock crisis.

Eisenhower acted swiftly to make amends by inviting Gbedemah to the White House for breakfast. But the ease with which any white American could embarrass the U.S. government in this way was striking. Whether in Baltimore, Charlottesville, Houston, or elsewhere, white American citizens frequently failed to hide the feelings they shared with their vice president and president about people of color and wound up insulting African visitors. The swiftness of the administration's response to such incidents mollified many black diplomats from abroad, but the contrast between Eisenhower's concern for the treatment of foreigners of color and his seeming indifference to the treatment of Americans of color raised doubts about his sincerity. African Americans were both amused and disturbed to see white officials suddenly fawning on dark-skinned foreigners while keeping mum on Jim Crow for the home folks.[90]

1960: The Challenge Intensifies

During Eisenhower's final year as president, three major challenges arose to the American hope for a multiracial, postcolonial Western alliance against the Communist bloc. The first was the Sharpeville massacre that completed the racial polarization of South Africa; the second was the crisis of national disintegration that followed the Congo's independence; and the third was the campaign of Cuban president Fidel Castro against U.S. leadership, waged partly along racial lines. All of these emerged against the domestic backdrop of the student sit-in movement that began on 1 February in Greensboro, North Carolina, and spread rapidly throughout the South. African American resistance to segregation at home thus increased hand-in-hand with the Third World struggle for racial equality. Phaon Goldman noted in the *Negro History Bulletin* in the fall of 1960 that much of the students' inspiration came from African nationalists "who recognize that the 'go slow' school of moderates didn't tell the Hungarians and Tibetans [in their rebellions against the Soviet Union and China] to 'go slow' and they don't want to hear it either." Four black residents of Haughton, Louisiana, distilled the problem for Eisenhower in a letter in January. Citing the denial of their right to vote and the campaign of terror against African Americans throughout much of the South, they told the president: "This country can no longer be a leader of the free world and speak for freedom when its citizens suffer as we do."[91]

On 21 March 1960 police in the small town of Sharpeville opened fire on a peaceful crowd protesting the apartheid pass laws for nonwhites, killing sixty-nine—many of them shot in the back—and wounding over two hundred. So began a major turn in South African history. The outraged black response led to the banning of political organizations opposed to apartheid. With all forms of protest against white supremacy now illegal, the ANC and other black groups went underground and took up weapons for the first time. U.S. ambassador Philip Crowe reported home that the radicalization of Africans in South Africa in the days after the massacre had created "a dangerous and explosive situation" in which African leaders "no longer are thinking of very gradual revolution some time in the future." CIA chief Allen Dulles shared with the National Security Council his discouragement about the increasingly remote prospects of a negotiated settlement of the South African racial conflict. He saw only trouble ahead: "Especially after the Congo becomes independent [in three months], there would be great opportunities for smuggling arms to the natives of South Africa."[92]

The confused response of the Eisenhower administration to the Sharpeville killings demonstrated its uncertainty about how to accommodate rising racial egalitarianism while protecting America's perceived interests abroad. Typically, the U.S. government avoided critical comment on internal events in other countries, especially friendly ones. But the slaughter at Sharpeville seemed too egregious for the United States to have "no comment." Without checking first with their superiors, State Department officials released a mild statement of "regret" at what had happened and of hope that Africans would be allowed to use peaceful means to "obtain redress for legitimate grievances" in South Africa. Many black South Africans and other opponents of apartheid were encouraged, believing that the world's greatest power was at last on their side. Lodge reported the "unanimous thanks" coming in at the United Nations from Asian and African delegates. But Eisenhower and Secretary of State Christian Herter, Dulles's successor, were furious. The president insisted on an apology to the South African ambassador—though "this action should be kept secret"—and told Herter that if it were his choice, he would fire the bureau chief responsible for the statement. Herter was especially troubled by such criticism "on a subject which not only has world-wide interest, but also involves domestic political factors." In a discussion with British prime minister Macmillan a few days later, Eisenhower emphasized his sympathy with white South Africans and their "difficult social and political problem"; it reminded the president of his good wishes for his "friends in Atlanta on some of their difficulties."[93]

In addition to having sympathetic feelings toward white South Africans and perceived vital economic and military interests, American policymakers believed pragmatically that the apartheid regime was not about to crumble in the face of rising dissent. When asked by an audience in Ann Arbor, Michigan, a few months earlier what he predicted for the future of South Africa, Kenyan labor leader Tom Mboya had responded simply: "Black." Washington's view would have been "black eventually, white for now." At the first NSC meeting after the news from Sharpeville arrived, Allen Dulles affirmed that South African authorities "would probably be able to maintain order for some time"—a period the CIA calculated as at least the next few years. An NSC analysis of Africa two months earlier had underlined the "resiliency" of the South African economy, with its vast gold reserves and industrial diversification. Confronted with this situation, and given U.S. strategic and economic interests in the region, including plans for a new missile tracking station on South African soil, the Eisenhower administration chose to maintain its close relationship with the masters of apartheid.[94]

The next crisis of 1960 with strong racial overtones unfolded three months later a little further north on the African continent. Belgium's unexpectedly swift withdrawal from the Congo left that vast new nation on the verge of disintegration after its 30 June independence ceremonies. The mineral heartland of Katanga province immediately seceded, under the leadership of anti-Communist Moise Tshombe and with considerable help from Belgian mining corporations, along with Belgian citizens and soldiers. Prime Minister Patrice Lumumba, a man of neutralist and socialist inclinations, stood firmly for national unity. He called for outside assistance—first from the UN, and later from the Soviets—in opposing the Belgian-backed effort to detach the wealthiest part of the brand new country. Eisenhower's goals in this crisis were twofold and somewhat contradictory: to preserve Western access to Katangan minerals (for which Tshombe was the best instrument), and to maintain a unified Congo with a central government oriented more to the West than the East (for which Lumumba was probably the best instrument). Belgium emphasized the first goal. The United States gave priority to the second, while keeping both options open. With the Soviets eager to intervene as a way to jump-start their influence on the continent, the Cold War thus came to Africa in the same year that the civil rights movement embarked on a newly mobilized campaign in the United States.[95]

The Eisenhower administration decided within two months of Congolese independence and the beginning of the political crisis there to remove

Lumumba from power. State Department officials were at first hesitant to adopt such a radical view of American needs in central Africa. Throughout July and well into August, U.S. diplomats in the Congo and their superiors at Foggy Bottom believed Lumumba was "an opportunist and not a Communist." They saw him as a neutral African nationalist. They were impressed with his leadership skills and believed there was "no better alternative on the horizon than a government built around him." But Lumumba was determined to drive out the Belgian forces defending secessionist Katanga, and his willingness to ask the Soviets for assistance after a slow initial response from the UN brought a sharp change in the United States' attitude. In late July Allen Dulles told Eisenhower that Lumumba was "a Castro or worse" who "has been bought by the Communists." The American ambassador in Leopoldville, Clare Timberlake, began referring to the "evil" Congolese prime minister as "Lumumbavitch." In the frigid Cold War climate of mid-1960, amid the Laos and Cuban crises, the downing of the American U-2 plane, and the cancellation of the Paris summit, the president was taking no chances. He apparently granted permission to the CIA in late August to try to eliminate Lumumba.[96]

The discussions within the Eisenhower administration that surrounded the decision to try to kill Lumumba revealed an array of American anxieties about this proud African nationalist. Foremost was white Americans' resentment of what they saw as the Congolese leader's ingratitude to the departing Belgians. The American press had reported favorably on Belgian rule in central Africa during most of the 1950s. "Nowhere in Africa is the Bantu so well fed and housed, so productive and so content as he is in the Belgian Congo," *Time* effused in 1955; thanks to Western technological improvements, "the sons of cannibals now mine the raw materials of the Atomic Age." Eisenhower's closeness to King Baudoin of Belgium made him unlikely to appreciate Lumumba's refusal to offer humble or even polite words at the new nation's independence ceremonies. "We are no longer your monkeys!" the prime minister declared. "We have known ironies, insults, and blows which we had to undergo morning, noon, and night because we were Negroes." This was precisely the kind of language that the American president was already hearing more than enough of across the American South in 1960. Lumumba concluded in that speech that the Congolese would not forget the deaths and imprisonment at Belgian hands of countrymen who resisted "a regime of injustice, suppression, and exploitation."[97]

This flagrant lack of appreciation for the benefits of Belgian colonial rule suggested to some in the Eisenhower administration that Lumumba might not even be mentally competent. This was a major question for American policymakers who dealt with the Congo. During Lumumba's visit to the UN on 26 July 1960, Lodge found him "a little flighty and erratic" but someone who "knows exactly what he is doing" and, overall, "not a bad man to deal with." Lumumba also made fairly positive impressions on several State Department officials in Washington and on Averill Harriman during the latter's tour of the Congo on behalf of presidential candidate John Kennedy.[98] But most of the administration viewed the Congolese prime minister differently. Undersecretary of State Douglas Dillon judged him "impossible to deal with" and "just not a rational being"; Herter agreed with Secretary General Dag Hammarskjold of the UN that Lumumba was "definitely a dope fiend"; Ralph Bunche declared him "crazy"; and Allen Dulles concluded that Lumumba's actions "indicate that he is insane." The leap from mental debility to witchcraft followed swiftly. Dillon agreed with the assessment of Pierre Wigny, the Belgian foreign minister, that Lumumba was "a sorcerer" who could entrance the Congolese parliament, and the U.S. ambassador to Belgium, William Burden, referred to an upcoming meeting with him as a seance. This was demonization in its most literal sense.[99]

Ingratitude and insanity combined with one other issue that stirred the men of the Eisenhower administration: sexual intercourse between black men and white women. During the first weeks of the Congo crisis, which was initiated by the mutiny of Congolese soldiers against their Belgian officers, reports began reaching the United States of Congolese troops raping white women, including nuns. Stimulated by press coverage of the reports and detailed accounts given to the UN Security Council by an outraged Belgian foreign minister, this issue became, in the words of a State Department memorandum, the "continuing preoccupation" of the White House and the State Department. The passion and fascination with which the subject was discussed within the administration finally drove one State Department official to remind his superior in writing that "the UN did not go into the Congo to save white women from being raped." Southern segregationists shared the horror of the administration and were explicit about where they thought desegregation would lead. Conflating Africans with African Americans, white supremacist organizer Leander Perez told a segregationist Citizens Council meeting in New Orleans that fall: "Don't wait for your daughter to be raped by these Congolese. Don't wait until the burr-heads are forced into your schools. Do something about it now."[100]

Suspected of being a dupe of the Soviets and nominally responsible for the apparent sexual atrocities of Congolese troops, Lumumba helped seal his fate with the Eisenhower administration during his one visit to Washington by asking his State Department liaison, Thomas Cassilly, to procure him a blonde companion for the night in Blair House. The request was turned over to the CIA, which satisfied it, and was duly reported to the White House. Eisenhower and his advisers were outraged. Sympathetic to the Soviets, antiwhite, ungrateful, probably insane, head of a nation of black rapists, and now lusting after blonde American women: Lumumba seemed to represent everything wrong about African independence. The extreme licentiousness of certain American political leaders, such as the Democratic presidential candidate that year, was not used as a standard for judging Lumumba—not when the issue was seen as race. The only historical analogy left for white Americans to call on in viewing the Congo leader was that of a slave revolt. Budget director Maurice Stans provided this comparison in an August NSC meeting when he argued that it was "the objective of Lumumba to drive the whites out and take over their property." The oldest and deepest fear of white Americans seemed to be coming true on the international stage in the midst of the Cold War: Patrice Lumumba was Nat Turner.[101]

Eisenhower liked Lumumba even less than Nasser, and he disapproved of the Belgians for creating a problem in the Congo similar to that of the British and French at Suez. The Congolese Force Publique mutiny had begun 6 July after Belgian officers made it clear that independence would have no effect on their control of the army. One general, for example, wrote on a blackboard in front of his troops: "After independence = before independence." Once Belgium sent in military reinforcements to protect Belgian lives and defend Katanga's secession, Eisenhower reluctantly supported UN military intervention in the Congo to reestablish order, hasten the evacuation of Belgian forces, and end Katanga's would-be independence. Belgian behavior had called attention to the racial coding of the conflict in the Congo—just as white Southerners had done with their massive resistance campaign at home—and the administration was eager to avoid further racial polarization that could only hurt the West and benefit the Soviets. Eisenhower insisted that the UN troops going to the Congo not be white, but preferably African or at least Asian or Middle Eastern. Ambassador Timberlake reported from Leopoldville that some Belgian soldiers were shooting civilians in the streets without provocation. The Belgians, he concluded with dismay, "have become completely irrational and in many instances have behaved worse than the worst Congolese." Supposedly traditional racial roles were once again not holding up. American policymakers feared that Belgians determined to remain in an independent Katanga were opening another front in the white settler problem of Africa, like the situations in Algeria, Kenya, Rhodesia, and South Africa.[102]

Tensions along racial lines between First and Third Worlds in 1960 were pointed up by one other international incident: Fidel Castro's ten-day stay at Harlem's Hotel Theresa while attending the September session of the United Nations. The Eisenhower administration's initial view of the revolutionary Cuban leader and his associates as needing to be treated "more or less like children," in Allen Dulles's words, had darkened considerably by this point, as Cuba's relationship with the Soviet Union warmed. Castro responded to U.S. hostility in part by using race as a weapon. Though phenotypically "white" himself, his racially egalitarian government was popular with Afro-Cubans, who constituted a substantial minority of the island's population. Castro equated and denounced the racial prejudices of many of Cuba's counterrevolutionary elites and of white American Southerners, abetted by the U.S. government's disinterest. When the somewhat bohemian Cuban entourage met with hostile treatment at the plush Shelbourne Hotel in New York, the traditional lodging for prerevolutionary Cuban diplomats, Castro sought a public relations coup against the Eisenhower administration. He briefly considered camping in Central Park and then seized on the idea of staying in Harlem to highlight American racism and Cuba's greater commitment to racial equality.[103]

Castro's sojourn in Harlem proved a triple success in racial terms: it improved his standing among Afro-Cubans, black Americans, and Africans. Harlem residents, regardless of their views of the Cuban revolution, appreciated the global attention their neighborhood received, as world leaders like Khrushchev, Nehru, and Nasser came up to 125th Street to meet with Castro. Black Americans applauded the Cuban leader's opposition to racial discrimination and savored his explicit challenge to white American authority. They enjoyed the prominent place in the Cuban delegation accorded to the black army chief of staff, Major Juan Almeida, in an era when Americans could hardly dream of General Colin Powell's comparable stature thirty years later. Castro also met with Malcolm X and discussed their common support for Third World independence from white control. In his five-hour speech to the UN General Assembly, the Cuban leader stressed his country's support for Lumumba in the Congo crisis and for African self-rule throughout the continent. He emphasized the parallels between Cuba and the Congo, as both emerged from rule by racially discriminatory elites, whether pro-American white Cubans or colonialist Belgians. He and Lumumba also shared the unenviable status of being targeted for elimination by the CIA. In this widely publicized visit, Castro challenged the idea of U.S. leadership in a predominantly nonwhite world. By putting race at the center of his message, he pinpointed America's greatest weakness in the Cold War.[104]

FOR EIGHT YEARS Dwight Eisenhower presided over a nation undergoing, like the larger world around it, fundamental change in its race relations. Both at home and abroad, people of color mobilized in the late 1950s to overcome racial discrimination and colonialism. Most whites resisted this challenge to their traditional authority. Whether in the American South, in the settler communities of Africa, or in the metropolitan governments in Western Europe, they worked to build bulwarks against the rising tide of equality, independence, and desegregation. It was the unique historical opportunity of the Eisenhower administration to wield power at a turning point in national and global relations. How the administration performed in these difficult circumstances says much about Eisenhower's leadership and wisdom, as issues of race pervaded the politics and diplomacy of the 1950s.

The president and his advisers were astute, sophisticated men, as two decades of "Eisenhower revisionist" scholarship have now made clear. Recognizing at least some of the political costs of Western white racism in the Cold War, they took a few symbolic steps to accommodate the new realities of a changing global racial order, like desegregating Washington's public spaces and trying to avoid close identification with segregationism or colonialism. A few members of the administration, like UN ambassador Lodge, openly supported desegregation and swift decolonization. The administration as a whole was quicker to acknowledge change abroad than at home, as the reality of African independence forced its hand in a way that the still developing civil rights movement at home could not yet do. Most notable, American policymakers showed their frustration with short-sighted white resistance in the crises at Suez, at Little Rock, and in the Congo, which required the president to align himself—briefly and somewhat uncomfortably—with Third World and African American causes. These three incidents marked the high points of the administration's efforts to contain racial hostilities and co-opt demands for racial equality.

But Eisenhower and those closest to him also remained in part blinded by their nostalgia for the stability of the white-ruled era now slipping away. They consistently refused to demonstrate almost any sympathy for black Americans or Africans in struggles for liberation. Put off by even the supremely polite, respectful behavior of early civil rights organizers like King, they could hardly imagine the perspective of those promoting armed self-defense by the late 1950s, like Malcolm X of the Nation of Islam or Robert Williams of the Monroe, North Carolina, chapter of the NAACP. Yet those voices only grew louder and more numerous after 1960, as did those in southern Africa calling for armed struggle to overcome the entrenched forces of white supremacy there.[105]

4

REVOLUTIONS IN THE AMERICAN SOUTH

AND SOUTHERN AFRICA

THE BRIEF PRESIDENCY of John F. Kennedy spanned a period of intensifying change and conflict in race relations around the globe. Hard-won African independence rolled south from the Mediterranean Sea in these years, halting only at the Zambezi River by the time of the young president's death in 1963, while segregation in the United States faced unprecedented challenges. The epicenters of racial tension lay in the American South and southern Africa, as the expanding civil rights movement encountered increasingly violent white resistance in the United States, and armed insurgencies replaced nonviolent protests against white rule in Angola and South Africa. Faced with the beginnings of the final chapter in the world struggle against legal white supremacy, Secretary of State Dean Rusk acknowledged in 1961 that the Kennedy administration confronted the "historical problem of coming into some reasonable relationship with [the] non-white world."[1]

Previous presidents had largely been able to avoid the escalating racial conflicts at home and abroad. Much as he might have wished to, Kennedy did not have this option. "The time comes in the life of any nation when there remain only two choices: submit or fight," the African National Congress announced on 16 December 1961, as it initiated its campaign of sabotage against government buildings. "That time has now come to South Africa." Angolans had made the same choice ten months earlier, taking up weapons against their Portuguese colonial masters just three weeks after Kennedy took the oath of office. In the form of Freedom Rides and peaceful mass mobilizations across the American South, that time had also come to the United States in 1961. Both Africans and African Americans were determined to make the U.S. government take sides in struggles that seemed to them utterly clear in their moral justice and world historical in their political significance. This was their "Cold War" against racial inequality: replete with the same full measure of righteousness and clarity that Kennedy and other Cold Warriors saw in their own conflict with Communism. Just as anti-Communists disdained or mistrusted neutralism regarding the Soviet Union, antiracists, the Central Intelligence Agency reported to the White House, saw "no room for nonalignment in this dispute."[2]

There was also little room for avoiding the dispute. Intensifying racial conflict threatened a crucial tenet of Kennedy's foreign policy: what the State Department called "the global U.S. strategy of fostering a cooperative community of free nations across the North-South dividing lines of race and wealth." That strategy of preserving a unified "free world" alliance of the First World and the Third World was integral, not peripheral, to the American policy of containing Soviet and Communist influence in the early 1960s, as Kennedy's unprecedented attention to the nations of the Third World indicated. Conversely, Rusk argued, Moscow would "make the most of its opportunity to develop divisions within the free world and to be the champion of the colored races—the same old game." Since the death of Stalin in 1953 and the start of European withdrawal from Africa a few years later, Soviet leaders had begun to show greater interest in expanding their influence south of the Mediterranean. The secretary of state concluded that any encouragement of racial strife "would enhance Sino-Soviet Bloc opportunities in Africa" and diminish American unity at home.[3]

Kennedy entered the White House after campaigning on a platform that supported domestic civil rights and African independence. After eight years of Dwight Eisenhower's coolness toward their cause, antiracists greeted the new administration with optimism. But Kennedy and his advisers had first to carry out their primary mission: winning the Cold War. For that purpose the president believed he needed the support of both black Americans and white Americans, both Africans and Europeans. To secure this support, the administration pursued greater racial equality, but it tried to do so without alienating either side. Finessing their differences was the key, and compromise and the diminution of conflict were the preferred tactics. While the new African nations sought to "make us choose between Portugal and South Africa, on the one hand, and the rest of Africa, on the other," White House aide Arthur Schlesinger, Jr., reminded Attorney General Robert Kennedy, "we wish to evade that choice." The strategy of evading the choice was precisely the same at home, where civil rights workers in the South pushed the administration to support either them or their opponents. The president needed white Southerners as well as African Americans and their liberal white supporters in the Democratic Party, and he was determined to avoid having to choose between them.[4]

Southern Africa and the American South were the two places where the global problem of racial conflict was intensifying most dangerously in the early 1960s. The nations of the Third World, newly prominent in the United Nations and a central concern of the Kennedy administration in its competition with the Soviet Union, watched these two "Souths" for signs of the true intentions of the U.S. government. Would the United States and its new president lead as a multiracial model in a world arching toward greater racial justice? Or would they instead continue as a defender of white privilege in the international system? These questions went to the core of American national identity in a global community.

The Kennedy Administration and Racial Change

Alone among Democratic presidents since 1945, John Kennedy entered the White House without strong personal feelings about civil rights and racial discrimination. The others—Harry Truman, Lyndon Johnson, Jimmy Carter, and Bill Clinton—all came from former slaveholding states, and in their different ways each wrestled with problems of race and took clear stands against Jim Crow before their ascension to the presidency. The lone Northern Democrat in the Oval Office during the Cold War, Kennedy came from a background of great privilege in nearly all-white Massachusetts. Like the rest of his family, he knew few African Americans and had little sense of the impact of the color bar on their lives. By 1961 the only black person he had spent significant time with was George Thomas, his personal valet.[5]

Inexperience does not necessarily imply prejudice, however, and Kennedy as president surprised many with his complete personal ease around people of color in social settings. The generational contrast with his predecessor in the White House seemed to show up here more clearly than anywhere else. While Dwight Eisenhower's attitudes had been formed earlier in the century, people of Kennedy's age were dramatically affected by World War II and the ways it delegitimated white supremacy. His own experience of prejudice against Roman Catholics may also have disinclined him to discriminate blindly against others. Civil rights activist James Meredith was not far from the mark in remembering later that, in personal terms, Kennedy was the first American president who was not a racist.[6]

Even so, he did not automatically take the side of antiracists in the political conflicts of the early 1960s. The president and his brother Robert shared a mistrust of moralistic idealism and of the liberals they associated with it; they saw themselves instead as cool-headed pragmatists. As he had done in the Senate, the president hoped to keep both black and white Southerners reasonably satisfied with him, while focusing on his higher priority of stopping Soviet expansionism around the world. This was a political strategy not much different, ironically, from that of Eisenhower. But the accelerating conflict in the South did not allow Kennedy to avoid the issue as much as his predecessor had, and by 1963 he was forced to take sides.[7]

The pattern was similar abroad. Kennedy had built a reputation in the Senate as a critic of European colonialism, even while staunchly defending America's alliance with Western Europe. As president he seemed to take a strong interest in the new nations and peoples of Africa, inviting twenty-eight heads of state from the continent to visit him in the White House. Africans appreciated his attentions—so much so that the National Security Council recognized that America's "one important asset" in dealing with Africa was "the President's status and personal relations" with leaders there, "on which we can draw heavily." Like the presence of unprecedented numbers of African Americans in his administration, Kennedy's ease with people of color abroad had great symbolic value. Thus he pursued close ties with both First and Third Worlds, working to mediate their differences just as he did those of white and black Southerners at home.[8]

What especially roused the president's ire about both segregation and colonialism was the hindrance they caused to the struggle against Communism. Such forms of racial discrimination were anachronistic and deeply irrational to a modernist like Kennedy: wastes of emotion, energy, and time. In a life-and-death struggle with the Soviet Union, and with the loyalties of the vast Third World in the balance, he had difficulty understanding the shortsightedness of white supremacists. Kennedy strongly hoped not to have to force desegregation on those who would resist it, but he considered such matters as sitting next to people of different skin colors a requirement of public civility in the modern world.[9]

The 1960 presidential campaign reflected these concerns. Kennedy had started out as the weakest Democratic candidate on civil rights, but he moved quickly to a more active stance. In doing so he represented the broader shift of the center of the Democratic Party away from an accommodation of its white segregationist Southern element, the latter having failed to put a dent in Eisenhower's two victorious campaigns. Republican candidate Richard Nixon took the opposite tack, moving to a less liberal civil rights stance with appeals to white Southern voters in a partial preview of his 1968 "Southern strategy."[10] Kennedy won a narrow electoral victory, in no small part by maintaining more effectively than Nixon a delicate balancing act about race. On one hand, he intervened to help get Martin Luther King, Jr., released from a Georgia prison and spoke repeatedly on the campaign stump about Africa, as a way to appeal to black and liberal voters in Northern cities without threatening white Southerners. On the other hand, he used the support of his running mate, Lyndon Johnson, and other white Southern officials to win back Louisiana and the key state of Texas, which the Republicans had taken in 1956.[11] The international context of this racially charged campaign was obvious to both camps, as Kennedy and Republican vice presidential candidate Henry Cabot Lodge, Jr., competed in calls for an end to racial discrimination to win the friendship of the new African nations. The awareness of the international arena had helped drive the effort to free King, whose sentence to months of hard labor for a minor traffic offense was, as Robert Kennedy put it, "making our country look ridiculous before the world."[12]

In his first press statement after the close election of 1960, Kennedy sought to pacify conservatives by announcing that the directors of the FBI and the CIA would continue in their positions. This gesture built into the bureaucratic structure of the new administration the staunch hostility of J. Edgar Hoover to the civil rights movement at home and of Allen Dulles to Third World nationalism abroad. Bureau agents in the American South and agency operatives in southern Africa proved extremely "soft" on racism in their common quest to root out Communist subversives. Their priorities made them insensitive to racism's corrupting influence on civil society. The CIA worked closely with South African security forces, and Hoover became obsessed by 1963 with discrediting the civil rights movement and especially King. The performance of these powerful security bureaucracies did not always satisfy Kennedy and his advisers, but their control of classified information about political activists and revolutionaries (and about the president himself) gave them considerable influence on a White House focused above all on the Cold War.[13]

The new administration respected Kennedy's long-standing ties to the white South and the power of its representatives in Congress, particularly the Senate. The political struggle in the South in the early 1960s over racial equality was essentially an intra–Democratic Party contest, and a Democratic president with a narrow margin of victory felt he had to work both sides of the street. From his Senate days Kennedy had personal links to white Southern leaders, like George Smathers of Florida, and was naturally inclined to see other prominent figures in his own party as reasonable people he could work with. Southern Democrats chaired the majority of congressional committees, exerting enormous influence on the legislative process; in combination with Republicans on many domestic issues, they functioned as the real majority party. Their control of federal funding created roadblocks even for such symbolic executive initiatives as desegregating the National Guard. From appointments of segregationist judges to the federal bench in the South to a refusal for more than two years to press for civil rights legislation, the president made real efforts to accommodate the rulers of Dixie.[14]

There were limits to that accommodation, of course, as his choice of a secretary of state showed. For an administration so concerned about foreign affairs and so vulnerable to criticism regarding American racial discrimination, there was no more symbolic appointment. Kennedy personally favored the chair of the Senate Foreign Relations Committee, J. William Fulbright, but the Arkansan's domestic segregationist voting record made him unacceptable—clear evidence of how much had changed in sixteen years, since South Carolina's more enthusiastically segregationist James Byrnes had been easily appointed Truman's secretary of state.[15] White Southerners with traditional racial attitudes were no longer qualified to represent the United States to an outside world that was mostly nonwhite.

The president instead pulled an end run around this dilemma, choosing a racial liberal from the Deep South whose grandfathers had both served in the Confederate Army, native Georgian Dean Rusk. Rusk wielded his geographical and cultural roots with aplomb, reminding Southern senators during a testy public exchange on racial discrimination's effects on American foreign relations that he was at least as "Southern" as they were. When Strom Thurmond asked, "Mr. Secretary, aren't you lending support to the communist line?" Rusk dissented indignantly, noting that he spoke as the American secretary of state. "Mr. Secretary," Thurmond persisted, "I'm not sure that you understood my question. I am from South Carolina." "Senator," replied Rusk, "I understood your question. I am from Georgia."[16] The secretary of state early on declared American racism to be "the biggest single burden that we carry on our backs in foreign relations," and he helped the president try to counteract it with symbolic gestures like sending the first black American ambassadors to European countries.[17]

While juggling competing loyalties to white Southerners and to civil rights supporters, the administration had to contend with a deep division between traditional, Eurocentric Cold Warriors, concerned with the fate of European colonial empires abroad, and advocates of greater attention to the emerging nations of the Third World. What the NSC called the "very bloody struggle" within the State Department between Africanists and Europeanists represented at an institutional level the president's own personal ambivalence about this problem.[18] Undersecretary of State Chester Bowles led the forces attuned to the Third World, and his replacement by Europeanist George Ball in November 1961 represented a turning point in the administration's orientation. Bowles had written the civil rights plank for Kennedy's campaign platform, and his supporters were dismayed at the signal that his removal would send both to nonwhite nations abroad and to racial liberals within the Democratic Party. White House civil rights adviser Harris Wofford warned Kennedy that this change would encourage the numerous racists in the State Department and undercut the president's own warmer policy toward Africa. "You need someone," Wofford concluded, "who, from the time he wakes up in the morning until he goes to sleep, knows that the Cuban invasion is wrong and that our new position on Angola is right."[19]

Instead, the administration began shifting its attentions away from the concerns of Third World nations, who had "names like typographical errors," in Ball's words, and whose leaders were "over-sensitive," according to the NSC's Robert Komer, or in the case of Arabs "rug merchants," as the State Department's William Bundy put it.[20] Removed from his influential position in the State Department and shunted off to the Third World as the president's special representative and adviser on African, Asian, and Latin American affairs, Bowles regretted that Kennedy's "principal advisers on foreign policy were heavily oriented towards Europe" and the Soviet Union.[21] This reassertion of Cold War thinking within policymaking circles encouraged the NSC in its argument that the continuing decolonization of Africa was now making Soviet-dominated Eastern Europe by far the largest bloc of unfree peoples in the world. Europe, not Africa, could now be the proper focus of world concern for national independence and self-determination.[22]

Kennedy, the Third World, and Africa

John Kennedy's foreign policy combined a fierce anti-Communism with an unusual interest in the emerging nations of the Third World. His commitment to West Berlin during the wall-building crisis of August 1961 indicated the centrality of Western Europe in his strategic thinking, while his recognition of the power of nationalism and the weakness of colonialism marked his presidency as a time of change in U.S. relations with the world outside Europe. He and his advisers were determined to be both more successful in containing the expansion of Soviet influence and more sensitive to the nonwhite world's concerns than Eisenhower had been. "The great battleground for the defense and expansion of freedom today," Kennedy told Congress in May 1961, "is the whole southern half of the globe—Asia, Latin America, Africa, and the Middle East, the lands of the rising peoples."[23] The administration created the Peace Corps as a way of recognizing and addressing the growing importance of the North-South axis of tension in world affairs. This was a racially coded axis, as Sargent Shriver, the Peace Corps's first director and the president's brother-in-law, emphasized: a division "between the white minority and the colored majority of the human race."[24] The choice of Shriver for the position reflected Kennedy's awareness of the importance of race in American relations with the Third World, as Shriver had a long history of involvement in the national Catholic interracial movement and was the most liberal member of the Kennedy clan—"the house Communist," as others liked to call him on occasion.[25]

Real Communists, usually of the Soviet and Chinese varieties, created or exacerbated the most pressing problems in the Third World, according to the Kennedy administration.[26] One reason for establishing the Peace Corps, the president explained, was to counteract the overseas work of Eastern bloc professionals—doctors, engineers, nurses, teachers—who served as "missionaries for international communism."[27] Khrushchev and Mao seemed to compete with each other in their support for "wars of national liberation" in remaining colonial areas, to Kennedy's chagrin. The State Department warned that "the Communist Bloc will continue to fish in troubled African waters" and feared that the Chinese "may be in a particularly strong position [there] on grounds of race."[28] Kennedy worried that Beijing's efforts at promoting its leadership of the non-European world would intensify further with its imminent acquisition of nuclear weapons.[29] By contrast, racial considerations would not likely be a source of strength for the United States. American military commanders trying to stem the Communist-led insurgency in South Vietnam were leery of using U.S. combat troops in part because it might "stir up [a] big fuss throughout Asia about [the] reintroduction of forces of white colonialism" into Indochina, as Admiral Harry Felt explained to General Maxwell Taylor in October 1961.[30]

Kennedy's stiffest competition for African friendships may in some ways have come from a less likely source, much closer to home. Fidel Castro, like Kennedy in his presidential campaign, began courting African sympathies in 1960. Castro's personal challenge to Kennedy as another new world leader of youth, vigor, virility, athleticism, and idealism, but also anticapitalism, has long been noted. It has been less well remarked that the Cuban revolutionary's words and actions also challenged and, in fact, surpassed those of the American president on matters of racial equality and African independence.[31] The revolution's commitment to ending racial discrimination had made it especially popular among the island's large Afro-Cuban population. Castro's increasing attention to Africa may have been in part a way of gratifying Afro-Cuban supporters while sidestepping the difficult task of fully eradicating traditional white Cuban racism—based on precisely the same logic as Richard Mahoney's argument about Kennedy's attention to Africa during the campaign as a way to please African Americans while not provoking white Southerners.[32] Regardless of Castro's personal motivations, however, his pursuit of diplomatic ties with African nations and his enthusiastic support for Patrice Lumumba in the Congo and for antiracist guerillas in the white redoubt of southern Africa were warmly appreciated south of the Mediterranean. In Washington, such initiatives and his increasingly close relationship with Moscow were instead deeply resented.[33]

Revolutionary Cuba's identification with the cause of international racial equality deepened the hostility of Third World reactions to the failed invasion at the Bay of Pigs in April 1961. Few African or Asian delegates at the UN shared the Kennedy administration's view of the Havana government as illegitimate. On the contrary, they were generally pleased when Robert Williams, a U.S. Army veteran and civil rights activist, sent a telegram to the UN denouncing the contrast between U.S. actions in the Caribbean and inaction in the American South.[34] When Williams fled North Carolina five months later to avoid being lynched by a white mob, he was given asylum in Havana. There he began broadcasting a twice weekly radio program across the Florida straits under the clever title *Radio Free Dixie*, encouraging armed African American rebellion. Williams appears to have been one of the few observers of the early 1960s to highlight explicitly the parallel between black Americans in the Jim Crow South and Eastern Europeans under Soviet-backed rule. While in Cuba he also established close contacts with other antiracist liberation fighters such as the South African ANC representative Ambrose Makiwane.[35]

Despite his radicalism, Williams was not alone among black Americans in his reading of the racial meanings of the contest between Havana and Washington. A wide range of African American intellectuals and political commentators, in contrast to the anti-Cuban hostility that pervaded the white media, associated the Bay of Pigs invasion with racist elements in the U.S. government, particularly the CIA.[36] Just as Southern segregationists were especially enthusiastic in their loathing for Castro, South Africa was the sole African nation to support the U.S.-backed invasion.[37] Few observers could miss the racial symbolism that Cuba and its chief antagonist seemed to be acting out, even if they did not know of the CIA's ongoing efforts to assassinate Castro and simultaneous collaboration with South African security forces.[38] Here we find one of the roots of Cuba's assistance to Angolan rebels against South African invaders in 1975.

The relationship of the United States to Africa would ultimately be determined not by the actions of America's rivals like Cuba or the Soviet Union but by the United States itself. Kennedy was the first American president to face fully the new reality of African independence. It has long been observed that his administration initially paid considerable attention to the continent and sought to accommodate radical nationalism there, but that subsequently a less sympathetic anti-Communism and neglect came to pervade Kennedy's Africa policies as the Cold War heated up in Berlin, Indochina, and Cuba.[39] Kennedy undoubtedly created much goodwill among Africans for the United States. This had begun with his support for Algerian independence in 1957; it had continued with his attentions to the continent during the 1960 campaign; it was boosted by his appointment of a strong cadre of Africanists to staff the African bureau in the State Department, led by Assistant Secretary of State G. Mennen ("Soapy") Williams; and it was sustained by the president's personal attention to African diplomats and visiting heads of state, and by his growing commitment to civil rights.[40] Several times Williams toured the continent, where his declaration that "Africa is for the Africans" was not well received by white settlers and colonial officials. When asked about the statement at a press conference, Kennedy's response—"I don't know who else it would be for"—underlined the sense among many Africans that an American president was finally on their side.[41]

Nevertheless, the Kennedy administration's interest in improving U.S. relations with Africa never approached the level of its commitment to American political, strategic, and economic priorities in Europe and East Asia. South of the Sahara, there were no equivalents to NATO or to U.S. trade and investment relationships with Europe or the Pacific rim. The discovery of alternative sources had reduced American dependence on African strategic minerals since the Korean War, and U.S. trade with the continent remained relatively minor. "From a global perspective," Pentagon planners concluded in 1963, "Africa is not an area of primary strategic importance to the US, and we therefore have a strong interest in restricting our involvement" there. In other words, there would be no strategic airlifts, à la Berlin, to sustain the besieged citizens of Soweto in their struggle with totalitarianism, nor would there be any counterinsurgent campaigns, as in Indochina, to contain the expanding influence of apartheid in Southern Rhodesia.[42]

The degree of intimacy that the Kennedy administration developed with the new nations of Africa was ultimately limited by the differences between African and U.S. perspectives and priorities. For Americans in the early 1960s, Africa was what the State Department called "probably the greatest open field of maneuver in the world-wide competition between the [Communist] Bloc and the non-Communist world."[43] Not surprisingly, Africans did not tend to see their countries as an "open field" for anything. Instead, for all their differences, they were fully united on the need to bring an end to the continuing brutalities of racial discrimination on the continent. As Soapy Williams reminded Rusk, "there are no African 'moderates'" on the twin issues of apartheid and colonialism.[44] A group of U.S. ambassadors in Africa warned their superiors in Washington in 1961 that "the most highly-charged issues in sub-Saharan Africa today are the war in Angola and racial discrimination in the U.S."[45] These did not match up well with the Soviet Union, Cuba, and Berlin, the priorities of the men in Washington. The CIA acknowledged that south of the Sahara, the United States, despite Kennedy's personal popularity, was seen as a nation "which does not fully understand Africa's problems or fully support African aspirations."[46]

Lingering elements of racism and segregation within the U.S. government reinforced this perception. Africans appreciated Kennedy's efforts to include darker faces in his administration, but they could hardly fail to notice how white the overall visage of American foreign policy remained. Below the ceremonial level of certain ambassadorial appointments, desk officers in the State Department and the other bureaucracies involved in foreign relations were still all white in the early 1960s.[47] The close personal ties of top U.S. policymakers with European and even South African officials contrasted sharply with their nearly complete ignorance of African officials and southern African nationalist leaders.[48] The explicit racial prejudice of many older foreign service officers was an obvious problem, but insensitivity to ingrained racism was perhaps more pervasive. When the U.S. embassy in Ghana asked the State Department for a short film to use at its Fourth of July reception in 1961, it received *The Life of Theodore Roosevelt*, filled with images of San Juan Hill certain to remind viewers of the Bay of Pigs invasion—and, even worse, of Roosevelt in a pith helmet leading overburdened African porters through the bush on his 1910 safari in east Africa.[49]

There was no clearer reminder to Africans of American racism at the highest levels than the spectacle of Senator Allen Ellender, Democrat of Louisiana, touring Africa in the fall of 1962 and declaring in a press conference in Salisbury, Southern Rhodesia, that Africans did not have the ability to govern themselves.[50] The State Department labored mightily to separate the Kennedy administration's policies from the senator's sentiments, and observers of all colors knew that Ellender was not an intimate of Kennedy.[51] But the declarations of a senior U.S. senator from the president's own party, made in southern Africa—along with his assurances that he had "never seen any subjugation" in South Africa—only served to bolster long-standing African skepticism about the intentions of American leaders.[52]

The acid test of the Kennedy administration's relationship with Africa came in the region south of the Zambezi River, the most racially polarized place on earth in the early 1960s. There, the State Department observed, the southward sweep of anticolonial liberation had "finally rolled up against the last remaining redoubts of control by European settlers, investors and administrators."[53] With no apparent room for compromise between those committed to permanent white minority rule and those dedicated to majority rule, the two sides appeared to be digging in for a showdown. Each side was organizing alliances throughout the region, revealing how integrated the region had become and how inseparable would be the fates of South Africa, Southern Rhodesia (Zimbabwe), South West Africa (Namibia), Angola, and Mozambique. The State Department worried about the increasing consolidation of "the Welensky-Verwoerd-Salazar axis," which prompted other Africans to organize military support for the armed guerillas in South Africa and Angola.[54] "Our enemies fight collectively," Nelson Mandela reminded Africans all over the continent in January 1962, as he recruited them to do the same.[55] American hopes that Southern Rhodesia, at least, might be encouraged to start down the path of racial liberalization—that it might "move north" rather than "move south"—were dashed by white elections in December 1962 that brought the white supremacist Rhodesian Front to power.[56] Southern Rhodesia, Williams told Rusk, "is [the] new African time bomb."[57] In sum, the State Department warned that if current trends continue, "we may be faced with a major racial war."[58]

The stakes in such a scenario for the United States were perilously high. Policymakers varied in their predictions of whether the conflict would unfold in the form of continued guerilla actions within the southern African states or with actual invasions across international boundaries from the north.[59] But either would have what the State Department called "a direct impact on two of the major axes of tension in the world: the latent division between North and South along racial lines and the active East-West competition for world influence." If the new African nations believed they had to chose between cooperating with the West and helping liberate Africans in the southern end of the continent from white supremacist rule, the results could undermine one of the most basic elements of American foreign policy in the Cold War. "It could fatally compromise our over-all strategy of fostering a cooperative community of free nations," the State Department concluded.[60] The racial policies of southern Africa's whites were a cancer threatening to destroy the "free world" alliance from within.

The Kennedy administration therefore followed a policy of containment in southern Africa—the containment of racial conflict. "To mitigate the polarization of the races into two warring factions" was the goal, according to the State Department.[61] A policy designed not to alienate either side inevitably made neither fully happy, as American officials criticized apartheid and colonialism while resisting substantive sanctions against South Africa and Portugal.[62] The parallels to Kennedy's policies toward racial conflict in the American South were striking: rhetorical support for racial equality, combined with calls for restraint on all sides and an abiding sensitivity to the political power of white authorities in the region. Even the complaints of officials in Pretoria and Lisbon sounded like those from Montgomery and Jackson. Like Southern segregationists talking about African Americans organizing against Jim Crow, Portuguese and white South Africans argued that whatever African unrest existed in their territories was the result of outside agitators, including the meddlesome U.S. government itself.[63] With substantial interests on both sides in southern Africa, "we have sailed an improvised, often erratic course between the antagonists," the National Security Council noted three weeks before Kennedy's death. "Since we can't now bet on a winner, we should [still] be hedging our bets and buying time."[64]

To many observers, it seemed that the racial conflagration threatening to engulf the region had already begun in the Congo.[65] The secession of the mineral-rich southernmost province of Katanga (Shaba), under the leadership of the anti-Communist, pro-Western Moise Tshombe, undermined the new nation's independence from its opening moments in June 1960. The execution of popular socialist prime minister Patrice Lumumba in January 1961 at the hands of Tshombe's soldiers then deepened the crisis. Though black himself, Tshombe had close ties to Belgium and hired white mercenary soldiers—including hated South Africans and Rhodesians—to preserve his control of an independent Katanga, framing the Congo crisis in racial terms.[66] The white governments to the south viewed Katanga as a buffer against black majority rule. Foreign Minister Franco Nogueira of Portugal explained this to the U.S. ambassador in December 1962, noting that the imminent assault by UN forces on Katanga was eagerly anticipated by Angolan rebels working out of Leopoldville (Kinshasa), who planned to use a "liberated" Katanga as another base of operations against Portuguese forces across the border in Angola.[67] The State Department recognized that "the struggle between Leopoldville and Elisabethville [the respective capitals of the Congo and Katanga] has represented in many minds the opening round in the battle over Southern Africa."[68]

Despite some obfuscation by conservative whites, the racial coding of the conflict in the Congo seemed clear to most observers. Africans, African Americans, and antiracists around the world condemned Tshombe as a stooge of Belgian mining interests and were outraged by Lumumba's assassination, which they likened to an international lynching.[69] Tshombe's white American defenders included an array of prominent conservatives in both parties who appreciated his anti-Communism, openness to foreign investors, nominal Christianity, and general orientation toward the West. On a continent filled with Africans resentful of European colonialism and skeptical of American racism, the Katangan leader seemed a refreshing alternative. In contrast to Lumumba, who appeared to listen to the wrong whites (those in the Eastern bloc), Tshombe was eager for the advice and support of whites from the "free world." He seemed to represent a solution to the Congo crisis, which had convinced so many whites of the dangers of "premature independence." But the American Committee for Aid to Katanga Freedom Fighters was also conspicuous for the enthusiastic support it received from leading segregationists, who understood precisely what Pretoria and Lisbon saw in Tshombe: Africa's most prominent black champion of white colonial interests.[70]

As with the rest of southern Africa, the Kennedy administration sought a compromise position between the two sides.[71] Despite considerable sympathy toward Tshombe among leading Europeanists in the State Department and prominent senators of both parties, the president and his closest advisers understood the unacceptable racial implications of Katangan independence. Fears of Soviet influence in a balkanized Congo bereft of its wealthiest province have usually been cited as determining Kennedy's response, but it was racial polarization that Kennedy feared would provide the opening to any Communist incursions. Recognizing Tshombe as the Congo's "most determined agitator of racism," Kennedy broke with Eisenhower by clearly opposing Katangan separatism and supporting the UN's mission to reunite the Congo.[72] He hoped the UN could do so without a war, and he avoided committing American troops, but after the international army's swift rout of Tshombe's forces in January 1963, the president was pleased by African praise for his position.[73] In denying Tshombe a visa to visit the United States a year earlier, Kennedy had provoked a storm of protest from conservatives, including prominent journalist and old Kennedy family friend Arthur Krock, a mainstay of the famously all-white Metropolitan Club in Washington. The president underlined the racial politics of U.S.-African relations in 1962 by dryly suggesting to aides an alternative arrangement: "I'll give Tshombe a visa and Arthur can give him dinner at the Metropolitan Club."[74]

If some in the Kennedy administration feared that the conflict in the Congo might light a fuse leading south, others focused on the racially charged war in southern Africa that had already begun: across the border in Angola. The rebellion against Portuguese rule began in Luanda in early February 1961 and escalated sharply the following month, spreading across the country to the north.[75] In some ways colonial Angola represented a clearer choice for U.S. policymakers than the independent Congo, where Tshombe was at least black, and South Africa, where whites tended not to be colonial officials but rather descendants of settlers who believed they had nowhere else to go. Indeed, Kennedy again broke sharply with Eisenhower's precedent by voting against Portugal in the UN and openly supporting Angolan self-determination, to the delight of Africans.[76] But U.S. policy was complicated by having substantive interests on both sides in the conflict: the goodwill of Africa and American standing in the Third World versus continued access to the strategic Portuguese-controlled air bases of the Azores.[77] The administration again tried to straddle the fence in an effort to avoid antagonizing either side. Hedging its bets in the most literal sense, the U.S. government provided aid both to the Portuguese armed forces, in the form of weaponry for a NATO ally, and to the most pro-Western elements of the Angolan rebellion under the leadership of Holden Roberto, who was put on retainer by the CIA. The results were sometimes bizarre, as when rebels and civilians wounded by American bombs and napalm were treated with American medical and food supplies in refugee camps across the border in the Congo.[78]

American assistance to the dictatorship of Antonio Salazar, the only European government to reject utterly the goal of independence for its African colonies, created a potential public relations disaster for the Kennedy administration. The Portuguese counterinsurgent campaign was unrelentingly savage in its slaughter of civilians and other atrocities. Rusk worried that "Portuguese repression in Angola is even bloodier than has come to light," and that the United States might be identified with it.[79] Images of white soldiers and settlers murdering and sometimes torturing thousands of unarmed black Angolans in their quest to root out nationalist guerillas were precisely the kind of spur to racial warfare that the U.S. government sought to avoid. When a London newspaper reported in August 1961 that part of a bomb dropped on an Angolan village carried the clearly visible insignia "Made in America," Kennedy was not pleased.[80] NBC's televised report a month later of American-made napalm being used against Angolan civilians further complicated the administration's dilemma.[81]

No other issue divided the Kennedy administration as clearly between Africanists and Europeanists as Angola. The latter emphasized the strategic importance of the Azores and the NATO alliance and worried about Communist influences among the rebellious Angolans. The former focused on the harsh treatment of Africans by Portuguese authorities, including vast immiseration, lack of education, and the modified form of slavery known as forced labor.[82] Assistant Secretary of State Williams made these points most strongly within the government, arguing that "while most of the world understands clearly the issues in Portuguese Africa, Berlin by comparison is primarily a white man's power struggle." For many observers in the Third

World, the Cold War seemed more like an intramural contest within the industrialized North than the ultimate conflict of two opposing worldviews, as was the case in southern Africa. Williams reminded Bundy of the fundamentally different priorities of the U.S. government and most people of color abroad: "If we expect Afro-Asian understanding and support in the great crisis facing us in Berlin, then we must have shown them by our actions on Angola that our opposition to oppression does not have geographic or racial boundaries."[83]

Regardless of their differences in emphasis, none of the elite white men who made policy decisions in the Kennedy administration was entirely immune to long-standing prejudices and assumptions about Europeans and Africans. While acts of savagery characterized both sides in the terrible conflict unfolding in Angola, U.S. government observers clearly reported the vastly greater force that the Portuguese brought to bear on their opponents and especially on civilians. It was "state terrorism," though they did not use this term. When American diplomats spoke frequently of "the terrorists in Angola," however, they were referring exclusively to African guerillas.[84] Even Williams accepted this terminology, while arguing that the United States should still deal with the insurgents, as "today's terrorist is tomorrow's statesman."[85] American policymakers used precisely the same linguistic distinction for the South African government and the ANC: Nelson Mandela, not Hendrik Verwoerd, was the terrorist. White people in the early 1960s, it seemed, could be oppressive and unjust, but they could not be terrorists.

The Portuguese government's perspective on the war in Angola, while opposite that of the insurgents, created comparable problems for the Kennedy administration in its broader Cold War strategy. Lisbon's anti-Communism and NATO membership did not prevent it from emphasizing the North-South axis over the East-West one, just as its African opponents did. "As far as Portugal is concerned," Foreign Minister Franco Nogueira reminded Washington, "Angola is much more important than Berlin."[86] The parallel to the situation in the American South was again clear: both white supremacists and antiracists showed far greater concern for race relations and colonial relations than they did for the struggle against Communism. Calling the rebel leaders in Angola "Congolese" and "Communists" to identify them as outsiders and blaming the United States for creating Angolan nationalism where it had supposedly not existed before, Portuguese officials made no secret of their contempt for what they considered American meddling in their country's internal affairs.[87] Salazar himself reminded Rusk that in authoritarian

Portugal the Americans were not dealing with a Britain or a France, constitutional democracies that had willingly relinquished their African territories: "I cannot give in Africa what I cannot grant to my own people."[88] Limitations on liberty in the "free world" were not restricted solely to its Third World precincts.

The real danger of the Angolan war for the Kennedy administration was its potential for spreading the virus of racial conflict. The immediate threat was to the surrounding southern African region, but problems in the Portuguese colony might also link up with racial antagonisms across the Atlantic. The NSC warned in January 1962 that the Angolan revolt meant that "Portugal will now be regarded as vulnerable" in Africa, encouraging Angolans to persevere and Mozambican nationalists to take up arms as well. The domino effect of African liberation would speed this process, as Tanganyika's independence next door provided a sympathetic neighbor and possible rebel bases, much as the Congo was providing for Angola.[89] Two weeks before Kennedy's death, the State Department reported that the fighting in Angola was certain to bring a further "sharpening of racial antagonisms" and the increased involvement of Africans from outside the country.[90]

The other direction in which the Angolan problem might expand was into domestic American politics. Portugal's aggressive public relations campaign in the United States threatened to tap into segregationist and other conservative sentiments, bolstering a loose international white resistance front.[91] Such an alliance was symbolized by white Rhodesian admiration for Republican senator and likely presidential candidate Barry Goldwater, including delight at Goldwater's belief that he had relatives in Southern Rhodesia.[92] Sympathy for Portugal and its actions in Angola were by no means restricted to Republicans and white Southerners, however. Dean Acheson, stalwart of the northeastern Democratic establishment and key Kennedy adviser on Portugal and other NATO issues, made no secret of his disdain for the Africanist position within the administration. He told Williams that just because the former Michigan governor was from "the dark city of Detroit," Williams should not think that the United States had "to pander to the dark and delirious continent of Africa."[93]

Acheson's fears were put to rest by Kennedy's decision during 1962 to shift attention from Angola to the Azores, as negotiations for renewing U.S. base rights there approached. Official American criticism of Portuguese colonialism diminished, in accord with Bundy's and Rusk's strategy of not letting liberals like Adlai Stevenson, ambassador to the UN, and Williams "make too much of a moral issue of what is essentially a political problem."[94] Having chosen Lisbon's camp, the Kennedy administration was still annoyed at the shortsightedness of its European ally. Presidential aide Ted Sorensen captured the White House's sentiment that the Azores issue was "not unlike other U.S.-European problems, in which the United States is expected to endure pain and criticism from our allies in order that we might maintain the right to protect them from the Soviet Union."[95] Both European colonialists and African revolutionaries, like Southern segregationists and African American civil rights workers, kept placing their own struggles above the anti-Soviet one. Managing this ongoing brawl within the "free world" remained an enduring problem for Kennedy and his advisers.

The Kennedy administration faced the same dilemma in South Africa that it did in Angola. Seceding from the British Commonwealth and declaring itself an independent republic in May 1961, South Africa hardened its apartheid system by cracking down on all forms of dissent. It was in the Kennedy years that South Africa—the heart of the white-ruled southern African region, with the most powerful economy and the largest white population by far—became an effectively authoritarian police state.[96] In response, the U.S. government faced, as it had with Portugal, what the State Department called "an embarrassing choice between security requirements and basic political principle."[97] American security requirements no longer centered on access to South African uranium, as they had in the first decade after World War II, but they still included a variety of strategic minerals.[98] The republic's fine harbors and location astride the east-west shipping route remained important to U.S. military planners, especially in light of the recent closure of the Suez Canal in 1956.[99] South African gold continued to stabilize the global capitalist monetary system, and American traders and investors were profitably involved in the republic's economy.[100] The one new American security interest in the land of apartheid in the Kennedy years was the establishment of a missile and satellite tracking station at Grootfontein, outside Pretoria. The U.S. desire to preserve access to that facility undercut its ability to distance itself politically from South Africa's apartheid policies and the enormous backlash they were creating at the UN and throughout the Third World.[101]

Apartheid's greatest problem for the Kennedy administration was its potential for undermining African and other Third World support for the West in the struggle against the Communist bloc. The president and his chief advisers did disapprove of the brutalities of apartheid on moral grounds, and they occasionally made public criticisms of the South African government's failure to live up to the human rights provisions of the UN charter.[102] The racial attitudes of Kennedy and Prime Minister Verwoerd were, after all, nearly opposites: the American, a social moderate whose trajectory was moving him into the camp of those supporting full racial equality, versus the South African, a reactionary white supremacist. But the administration seemed at least as anxious about how the apartheid policies of an anti-Communist government were actually helping the Soviet Union and its allies in their quest for greater influence in the Third World. Anti-Communism associated with white supremacy, in other words, encouraged Communism. Conversely, the prominence of South African Communists in the anti-apartheid struggle tended to identify Communism with freedom for many Africans. Williams explained to British officials that apartheid served as "a breeding ground for communism" that made "U.S. relations with the rest of the continent very difficult."[103]

Unfortunately for the Kennedy administration, South Africa was well on its way to pariah status in the international community for another reason besides its internal racial policies. It was also a colonial power in South West Africa, and a particularly illegitimate one in the eyes of most of the world because of what the State Department called its "persistent disregard of its obligations under the old League of Nations mandate" establishing its control there in 1919. Pretoria's determination to share the fruits of apartheid with Namibians further enraged anticolonialists and antiracists around the world, placing the territory "in the forefront of Afro-Asian attention."[104] The convergence in South West Africa of colonialism and apartheid thus concentrated the frustrations of most Third World observers. Simultaneously, the publication in 1962 of Allard Lowenstein's popular book *Brutal Mandate: A Journey to South West Africa* increased American public awareness of the colony and of the white authorities' treatment of Africans there.[105] Third World delegates made South West Africa a frequent subject of discussion and controversy at the UN throughout Kennedy's presidency; unlike apartheid, control of a separate territory prevented the Verwoerd regime from arguing that this was strictly an internal South African concern.[106] Supporters of Angolan liberation around the continent, like Prime Minister Ahmed Ben Bella of Algeria, watched the spread of apartheid northward toward the Angolan border and declared their commitment to Namibian liberation next.[107]

The U.S. government's response to these troubling developments was to try to separate matters involving race relations from all other elements of its relationship with South Africa. Kennedy took a rhetorically tough stand against apartheid, while seeking to disturb the fruitful aspects of its ties with the republic as little as possible. Thus his announcement of a unilateral partial arms embargo on South Africa in August 1963 symbolically disassociated the United States from the land of apartheid, while undermining the UN campaign for a more severe embargo and not materially damaging Verwoerd's military strength.[108] Francis Plimpton, a U.S. delegate to the UN, could assert boldly that "how and when the South African Government will abandon its hateful racial policies we cannot know, but abandon them it will," while Kennedy rejected economic sanctions to bring substantive pressure on Pretoria.[109] Bundy could advise the president to risk losing access to the missile tracking facility by agreeing to some kind of arms embargo, while Rusk urged his colleagues to remember that South Africa was ultimately an ally in America's "total confrontation [with the Soviet Union] affecting the life and death of our own nation."[110] There was some room in the administration for differences of opinion that could be kept under the roof of a single policy—a policy, ironically, of "apartheid," the Afrikaans word for "separateness," in this case the separateness of race relations from all other issues. Rusk emphasized this point with South African foreign minister Eric Louw, as recorded for telegraphic purposes: "While US and SA cannot agree on matter race relations," there was "no need why this disagreement should infect total range our relations."[111] Apartheid represented a kind of viral outbreak that had to be sealed off—contained—to limit its damage to more important interests within the anti-Communist alliance.

The Kennedy administration's unwillingness to go beyond rhetorical condemnation of South Africa's racial policies reflected its reluctance to encourage any destabilization of the Pretoria government in light of the political alternatives in the republic. With white support for Verwoerd overwhelming and dissent banned, the ANC was forced to go underground and began its sabotage campaign under Mandela's leadership in December 1961.[112] American diplomats watched with foreboding the growing pessimism of the ANC's Albert Luthuli, winner of the 1962 Nobel Peace Prize, whose commitment to Christian nonviolence paralleled that of Martin Luther King, Jr. Luthuli lamented to the U.S. ambassador Joseph Satterthwaite that the South African government seemed to be closing off all avenues of change for Africans besides violence.[113] It was as though the Birmingham police had used machine guns rather than water hoses on demonstrators, and the federal government had just watched with satisfaction. Political repression was so thorough that the Defense and State Departments agreed, "the U.S. does not wish to assume a stance against revolution, per se, as an historical means of change" in South Africa. It was instead only "communist inspired, supported or directed subversion or insurgency" that had to be totally opposed. This truism of American Cold War strategy had a peculiar meaning in a realm of white supremacy where the South African Communist Party was closely allied with the country's leading opposition group, the ANC. The ANC thus qualified as an organization of unacceptable revolutionaries, because it, U.S. officials believed, was "dominated by the Communist Party at the leadership level."[114]

By refusing to understand the Communist Party's legitimacy in the eyes of Africans in South Africa, the Kennedy administration identified Mandela and his colleagues as dangerous enemies of the United States. Mandela himself was neither "a Communist nor a member of the CP," as he told the court during his 1964 trial on charges of sabotage, but an admirer of the British and American political systems who sought some form of socialism in his country. The lawyer-turned-guerilla reminded listeners that Communists both abroad and at home had been stronger opponents of apartheid than had non-Communists, with the result that many Africans "today tend to equate freedom with Communism."[115] To the U.S. government, however, ANC members were not freedom fighters but "subversives." For all their racial liberalism and their distaste for racist brutality, the president and his advisers shared the fundamental assumption of their White House predecessors that the South African regime was a legitimate government. The changes in racial consciousness that were just beginning among white Americans in the early 1960s were not yet far enough advanced to induce elite white policymakers to question that elemental understanding. Similarly, African Americans had not established themselves as an influential voice on U.S. African policy, as they would by the 1980s.[116] Anti-Communist racial totalitarianism may have been problematic and even abhorrent to American leaders, but it was not yet illegitimate.[117]

Perhaps the most intriguing piece of evidence for the continuing close ties between Washington and Pretoria in the Kennedy years was the role that the CIA apparently played in helping South African security forces arrest Mandela outside Durban on 5 August 1962. The defense and intelligence establishments of the two countries worked closely together. This relationship was reconfirmed by the 15 June 1962 agreement on a tracking station, which included arrangements for South African purchases of U.S. weapons.[118] South Africa's departure a year earlier from the British Commonwealth had meant a decline in British intelligence support, and Pretoria's relatively small intelligence forces were happy to receive help from the well-funded and experienced CIA. Their mutual anti-Communism and suspicion of Third World nationalists inclined them to a similar view of the ANC, and the agency had at least one source within Umkhonto we Sizwe (the Spear of the Nation), the ANC's military wing, which Mandela headed.[119] Thus, reports that emerged in 1990 of the CIA tipping off South African authorities to Mandela's whereabouts in August 1962 fit well with established relationships of the time, even if those reports have yet to be confirmed with documentation.[120] The pattern also accounts for the evident sympathies of the FBI with segregationists in the U.S. South at the same time. Abroad as at home, American security agencies tended to array U.S. power on the white supremacist side in racial conflicts, especially when the president and his chief advisers generally paid little attention to the details of southern African policy, focusing instead on priorities north of the equator.[121]

Kennedy and the Civil Rights Movement in International Context

In John Kennedy's ideal world, there would have been neither Communists nor racists. Unfortunately for him, the real world had plenty of both. His presidency, he declared in his inaugural address, would be "a celebration of freedom," and indeed the early 1960s were a time of epic struggles to extend freedom. But there was a central conflict during the Kennedy years about what people most needed to be freed from or kept free from: Communism or racism. Anti-Communists looked mostly for freedom abroad, while antiracists focused on freedom at home. The president and his advisers believed firmly that the spread of Communism, by destroying free enterprise and the principle of political liberty, would remove the economic and political structures necessary for further social improvements.

Kennedy saw himself as the national and international leader with the clearest vision of the greatest threat to human liberty, and he tended to look at civil rights organizers as unruly followers inclined to jump out of line, run up to the front, and seize the spotlight for a less important cause. When confronted by Theodore Hesburgh, chair of the U.S. Civil Rights Commission and president of the University of Notre Dame, with problems like the absence of blacks in the Alabama National Guard, the president responded, "Look, Father, I may have to send the Alabama National Guard to Berlin tomorrow and I don't want to have to do it in the middle of a revolution at home."[122] Like Dwight Eisenhower, Kennedy grew frustrated with both sides in the conflict in the American South, as neither side seemed to him to be keeping its eyes on the ultimate prize of the Cold War and worldwide stability. Indeed, both civil rights organizers and those resisting them found race relations at home far more important in their daily lives than events in far-off Berlin or Vietnam. This was a perspective quite alien to the president.

The one way the civil rights movement could get the Kennedy administration's full attention was to clearly impact U.S. foreign relations. The president feared how easily another Little Rock crisis might happen, embarrassing the country before a global audience. Robert Kennedy explained the administration's international logic in his first speech on civil rights as attorney general, given at the University of Georgia. He argued that the graduation of the two African Americans just admitted there "will without question aid and assist the fight against Communist political infiltration and guerilla warfare."[123] John Kennedy disliked racial discrimination and understood the politics of symbolism, and he made highly visible gestures to diminish the nation's association with white supremacy: appointing dozens of blacks to important posts in the administration, desegregating the Coast Guard Academy, and hosting many more people of color at the White House than all of his predecessors combined. The contrast with Eisenhower, at the symbolic level, could not have been starker.[124]

Kennedy's primary concern about the civil rights movement was controlling it: moderating its tactics, channeling its demands, and limiting the social instability it stirred up in the South.[125] In this sense he fit King's famous description of "the white moderate" who agreed with the goals but not the direct action methods of the movement, "who paternalistically feels that he can set the time-table for another man's freedom," and whose greater devotion to order than justice, King believed, was blacks' "greatest stumbling block in the stride toward freedom."[126] Even in his symbolic actions Kennedy would go only so far. He refused for two and a half years the appeals from within and outside his administration to provide moral leadership to the nation on the issue of racial equality, despite widespread and extraordinary violence against American citizens throughout the South.[127] Indeed, the administration seemed determined to remain as neutral as possible in what increasingly looked like low-level warfare across the former Confederacy.[128]

The day before astronaut Alan Shephard lifted off on the first manned space flight on 5 May 1961, a group of young Americans set off to explore another new frontier of freedom, this one south of the Mason-Dixon line. The

Kennedy strategy of co-opting civil rights organizers to minimize racial polarization began with the Freedom Riders of the Congress on Racial Equality (CORE) and the Student Nonviolent Coordinating Committee (SNCC). While the Freedom Riders sought to force the Kennedy administration to uphold federal court rulings against segregated transportation, the White House desperately wanted to avoid such a confrontation with white Southern authorities. The timing was crucial: the president, embarrassed by the debacle of the Bay of Pigs invasion of Cuba a few weeks earlier, was about to meet Soviet chief Nikita Khrushchev for the first time at a summit conference in Vienna. The violence and terror that greeted the riders in Alabama, and the international attention they received, deepened the president's embarrassment and anger. The Freedom Rides, Assistant Attorney General Nicholas Katzenbach remembered, were seen in the White House "as a pain in the ass."[129]

By contrast, James Farmer and the other organizers of the rides believed they were on the front lines of the struggle for freedom. There was no more important business for the U.S. government, as they saw it, than the protection of the rights and safety of its own citizens, especially on the eve of any conference in which it would be posing again as the leader of the "free world." Focused on how the Soviets could use such violent incidents to demean the reputation of the United States abroad, Kennedy and his advisers did not see that for civil rights workers, the Deep South was like the Soviet Union in its negation of liberty and the rule of law and its determination to crush dissent. Here was a great irony: loudly anti-Communist Southern local authorities ruling a chunk of the "free world" in quasi-totalitarian fashion.[130]

The Kennedy administration was relieved to emerge from 1961 with no further major racial incidents, but the next year confronted the president with another version of Eisenhower's crisis at Little Rock. The vast, heavily armed throng of segregationists that tried to prevent James Meredith's enrollment at the University of Mississippi on 1 October 1962 was abetted by state officials all the way up to Governor Ross Barnett, and the riot that ensued brought precisely the kind of negative international attention Kennedy feared. Observers around the globe were stunned when members of the white mob killed a foreign reporter, Paul Guihard of the *London Daily Sketch,* along with another bystander, and wounded 166 federal marshals (28 of them by gunfire).[131] Like Eisenhower with Arkansas five years earlier, Kennedy was forced to send thousands of U.S. troops to Oxford to restrain white American savagery being displayed on television for a world audience. And like

Eisenhower, he was learning how treacherous the white Southern politicians of his party could be.[132]

The international context of the Ole Miss debacle was evident to many participants, though it was understood differently by different parties. The whites chanting "Go to Cuba, nigger lovers, go to Cuba!" at federal marshals captured the regional sentiment that the U.S. government should attend to the Cold War and leave Southern racial matters to the South, while the attorney general's half-joking question about photographs of Soviet missile installations in Cuba—"Can these things reach Oxford, Mississippi?"—suggested instead that the Cold War could no longer be fought effectively with people like Barnett in power. "The eyes of the nation and all the world are upon you and upon all of us," the president reminded Mississippians in a nationally televised address as the unrest escalated.[133]

Kennedy's tolerance for traditional Dixie mayhem across the color line ran out the following spring in Birmingham, "the Johannesburg of America." Brutality rained down with particular viciousness on the civil rights campaign that began there in early April and elicited for the first time a widespread and violent black response, across class lines.[134] Reporters and cameras from all over the country and the world poured into Alabama, capturing appalling scenes of police violence for a national and international audience. The international context was again crucial, as the founding conference of the Organization of African Unity (OAU) was meeting simultaneously in Addis Ababa, Ethiopia. Prime Minister Milton Obote of Uganda warned Kennedy that the "ears and eyes of the world are concentrated on events in Alabama," and a Nigerian journalist observed that the United States appeared to be becoming "the most barbarian state in the world."[135] The global attention encouraged civil rights activists, who were putting their bodies on the line in Birmingham. At the height of the crisis, King reminded antisegregation protesters that they were succeeding in forcing the U.S. government to take their side in the conflict because "the United States is concerned about its image. When things started happening down here, Mr. Kennedy got disturbed. For Mr. Kennedy . . . is battling for the minds and the hearts of men in Asia and Africa," who would no longer respect an America that allowed Jim Crow to survive.[136] One marcher turned the anti-Communist language of segregationists back against them with a sign reading, "Khrushchev can eat here—why can't we?"[137]

Full-scale street warfare and international opprobrium were too much, even for the cautious Kennedys. Disturbed by both the violence of local authorities and the willingness of so many blacks in Birmingham to fight back against the police, the president and his advisers recognized that the civil rights movement and American race relations stood poised on the brink of a more confrontational stage. Kennedy knew he must act forcefully to head off escalation; he had to provide leadership to the nation by framing the issue, for the first time, in moral terms.[138] He was not giving up on white Southern leaders, with whom he still had to work, but he was going to push them harder. He tried with Alabama governor George Wallace during a brief, shared helicopter ride from Muscle Shoals to Huntsville, but it was not a fruitful experience. Kennedy told the governor that for the good of Birmingham itself, the city must begin treating its black citizens more fairly if it was ever to be restored to social and economic health. Wallace responded with two of the oldest canards of white supremacists: the problem in Birmingham was "the influence of outside leadership," and King was "a faker" interested primarily in competing with Reverend Fred Shuttlesworth to see "who could go to bed with the most nigger women, and white and red women, too." Annoyed, Kennedy insisted on the seriousness of the black insurgency in Alabama and urged the governor to accommodate it.[139] Those present for the conversation could not have been surprised that it was the action of Wallace three weeks later, in symbolically attempting to prevent the desegregation of the last all-white state university in the nation, that finally moved Kennedy to seize the presidential bully pulpit on behalf of racial equality.

In his 11 June 1963 nationally televised address on the "moral crisis" facing the nation, Kennedy took sides as he had never done before. The atypical fervor with which he spoke reflected both personal revulsion at the violence inflicted on black Southerners and a determination to get new civil rights legislation passed by Congress in order—as Robert Kennedy put it—to get "people off the streets and the situation under control." The president made the international context for desegregation clear in his speech: "When Americans are sent to Vietnam or West Berlin, we do not ask for whites only." The struggle against Communism, he argued, now required the full racial integration of American public life. There would have to be more freedom at the heart of the "free world."[140] But the timing of the speech also suggested that the administration's view of the relationship between anti-Communism and antiracism might be changing, with the latter no longer merely an aid to the former. The day before, Kennedy had given a major address at American University calling for peace and a reexamination of American attitudes toward the Soviet Union. While perhaps coincidental, these two key speeches together pointed toward a future without either Jim Crow or the Cold War.[141]

Kennedy's call for full desegregation was intended for an international as well as national audience. Rusk instructed all U.S. ambassadors abroad to bear down on the task of countering the "extremely negative reactions" that racial incidents at home had elicited "from all parts of [the] world." The administration, he reminded them, was "keenly aware" of the impact of such incidents on the nation's image overseas, and the president's speech should be disseminated as widely as possible, as should the country's achievements to date in desegregation.[142] As hoped, Kennedy's civil rights bill was widely applauded in the Third World and especially in Africa, where newspaper pictures of police dogs attacking unarmed African Americans in the South—so reminiscent of apartheid and colonialism—had made a deep impact. Africans noted how different Kennedy's stand was from Eisenhower's, and some commentators even suggested comparisons to Abraham Lincoln.[143] But there was also a "strong note of skepticism" about how much progress was being made toward real equality in the United States, and State Department officials worried that the outlook for America's reputation in Africa was "likely to become worse" as reports of racial violence continued to mount.[144] The evidence along these lines was not encouraging: the most symbolic white Southern response to Kennedy's speech came a few hours later, when Mississippi NAACP leader Medgar Evers was gunned down in his front yard. The president himself had turned a corner, but whether he could pull segregationists around that corner with him remained an open question.[145]

African Americans, meanwhile, were moving rapidly down the road away from that corner, leaving the Kennedy administration to straddle a growing racial divide. The president's success in channeling the March on Washington in August into support for his civil rights bill marked the apex of the peaceful, integrated stage of the civil rights movement.[146] But antiracist organizers were increasingly disillusioned by the failure of the federal government to protect the lives and safety of American citizens in the South from the predations of local authorities. For a growing number of SNCC and CORE activists and their supporters by late 1963, the Kennedy administration was beginning to seem almost as much the enemy as the Ku Klux Klan and local law enforcement officers. One elderly black woman in Albany, Georgia, captured this rising sentiment in her response to a plea from activist attorney C. B. King not to lose patience with Washington: "You know what, lawyer, the Federal government ain't nothing but a white man."[147] The president, never enamored of civil rights organizers to begin with, reciprocated the feelings, privately referring to SNCC members as "sons of bitches" with an "investment in violence."[148]

White Southerners would have been pleased to know of such sentiments in the White House, for they deeply resented Kennedy's new support for civil rights legislation. The sharp increase in Republican voters in the South in the 1962 congressional elections, encouraged by the crisis at the University of Mississippi, had offered early evidence of the white backlash developing against greater national Democratic Party support for racial equality.[149] Dissenting Southerners vied with Kennedy to preempt the meaning of the language of the Cold War. While the president reprimanded segregationists' inadequate commitment to anti-Communism—"Why can a Communist eat at a lunch counter in Selma, Alabama, while a black American veteran cannot?"—George Wallace argued the opposite, insisting that the federal government that now sought "to persecute the international white minority" was filled with "the same people who told us that Castro was a 'good Democratic soul' [and] that Mao Tse Tung was only an 'Agrarian Reformer.'"[150] Segregationists noted accurately enough that racial egalitarianism in the South was a subversive idea, and one supported by those other subversives: Communists. "Some of these niggers down here would just as soon vote for Castro or Khrushchev," Sheriff Z. T. Mathews explained to reporters after intimidating a voter-registration rally in Terrell County, Georgia, in July 1962.[151] That Kennedy acted like a Soviet dictator seemed obvious to those white Mississippians who, after the Ole Miss crisis, put on their cars the license plate legend "Kennedy's Hungary" in reference to the Soviet Union's brutal repression of the revolution in Budapest in 1956.[152] The slogan that best captured segregationists' views of the connections between foreign affairs and the American South was the sign greeting Robert Kennedy along the road into Montgomery in April 1963: "No Kennedy Congo Here."[153] Many white Southerners believed they were not just like whites in Eastern Europe but even more like embattled white settlers in Africa, facing an onslaught of imminent black political participation and perhaps revenge with the support of the meddling U.S. government. Gus Noble, president of the Canton Citizens' Council, warned that equal rights in Mississippi would lead to the same horrors that had supposedly happened in former African colonies of Britain and France: "Those natives were not prepared for self-government. They were unstable. They ate each other."[154] By the end of his presidency, in sum, Kennedy found himself increasingly alienated from elements on both ends sides the growing racial divide in the South, even as he committed himself to ending racial dis-crimination in the United States.

Africans and Jim Crow

People of the Third World had long been interested in how people of color were treated in the United States, but it was only in the Kennedy years that significant numbers of Africans began viewing events on television or coming to America as diplomats and finding out some of the answers for themselves. Many African visitors were skeptical of the explanations offered for the con-tinued widespread racial discrimination in the leading nation of the "free world." U.S. officials most often cited the federal system of government, with its reserved rights for states and local communities, as restricting the national government's ability to ensure racial equality. Most Third World observers, along with many Europeans, found this argument less than convincing, espe-cially in light of the awesome power of the U.S. government to project mili-tary might around the globe.[155] If true, were white supremacists in the United States even more powerful than the extraordinary figure of the president him-self? Outspoken internationalist and civil rights advocate Robert Williams, who was engaged in an increasingly violent conflict with the Ku Klux Klan in Monroe, North Carolina, tweaked Kennedy's nose on precisely these grounds in 1961. In his famous telegram read aloud at the UN by the Cuban ambassa-dor during debate on the American role at the Bay of Pigs, Williams said that in light of U.S. aid to those fighting tyranny abroad, "oppressed Negroes in [the] South urgently request tanks, artillery, bombs, money, and the use of American airfields and white mercenaries to crush the racist tyrants who have betrayed the American Revolution and Civil War." There was considerable applause among the darker-hued delegates in the General Assembly, just as there was appreciation for the suggestion of Kwame Nkrumah of Ghana that Africans might offer a Reverse Peace Corps to teach respectful human rela-tions to Americans.[156]

Africans in the early 1960s continued to experience the United States pri-marily through New York and Washington. Since the spring of 1960, the State Department had been deeply concerned about how UN delegates from the new nations south of the Sahara would be treated by New Yorkers. Rusk and Adlai Stevenson were well aware of "the problem posed by representa-tives of African nations seeking suitable housing and office accommodations." That the presence of foreign diplomats in the nation's most cosmopolitan city should pose a "problem" indicated how clearly the Kennedy administration understood that racial discrimination remained a national rather than a Southern phenomenon. The U.S. mission to the UN went to great efforts to prepare the city's hotels, restaurants, and real estate agents for the novel experience of large numbers of nonwhite foreigners arriving with expectations of dignified treatment. The administration even arranged for private funding to hire a "trilingual specialist" to help African diplomats and their families find suitable housing arrangements while shielding them from white American racial prejudice—a kind of house hunting without humiliation. The contrast to the degree of federal government concern about dark-skinned Americans' experiences of discrimination in Manhattan and elsewhere was difficult to miss, especially for African Americans. The Cold War was the key to unlocking the Kennedy administration's concern about racism, and African independence had created a vast new source of potential embarrassment that the president and his advisers sought to preempt.[157]

Washington was a Southern town with a strong Jim Crow tradition, and diplomats of color were even more anxious about being posted there than New York. In a precise mirror image of how most American diplomats viewed positions in Africa, African ambassadors and their families and staffs found the American capital—the nerve center of the "free world"—to be a true "hardship post." Well aware of white racial violence in the South, Africans not only worried about finding decent housing but even feared being assaulted on the streets of Washington. Some African officials bound for the First World sought to be posted instead to Europe, which they considered much safer.[158] Eisenhower's desegregation of public accommodations in Washington, combined with the Kennedy administration's presence and concern, had in fact created a limited oasis there. But most African diplomats came to Washington by car from New York, and once they crossed the Delaware River on Route 40, they quickly discovered how unwelcome they were. In Delaware and Maryland they were regularly refused service in restaurants, humiliated, and harassed. Washington, surrounded by segregationist states, seemed to be for African visitors what West Berlin, surrounded by Communist East Germany, was for Western visitors: an isolated citadel deep in hostile territory, safely reachable only by air or by two restricted highways off which one dared not venture.

An important part of the logic of American Cold War strategy had been the belief that the more that new nations learned about the principal antagonists, the more they would prefer the United States over the Soviet Union. The first wave of Africans to test this theory found the results mixed. The Malian embassy in Washington passed along to Pedro Sanjuan, the assistant chief of protocol for the State Department, three letters it had received in 1961 from the American Nazi Party and included an analysis of the situation facing African diplomats who wished to travel in the United States. Malian officials found that "movements of African diplomats assigned to New York and Washington, are reduced, for these two cities, to a corridor constituted by Route 40 and the New Jersey Turnpike, which link the two cities. And their movements are free only if they do not get out of their automobiles." Excursions by airplane were generally safe, they continued, but trips from Washington of longer than a day "cannot be made if one chooses to travel by automobile unless one has a trailer." If one "wants to enjoy a rest on the beach, he is greeted on the beaches with hostile cries of 'No Negroes here!'" as happened to their ambassador "all along Chesapeake Bay" on an excursion on 4 September. The overriding sensation was not one of freedom but rather of hostility, which "causes African diplomats to consider the United States the most difficult post."[159]

The problems facing Africans in the United States were not entirely news to Kennedy, as the situation had been developing since 1957, and his own Senate staff had sometimes been involved in finding housing for insufficiently pale visitors from abroad.[160] But the frequency of such unhappy incidents increased during his first months in the Oval Office, along with media coverage of the confrontations, and the president quickly grew frustrated and sought ways to manage the problem. He continued Eisenhower's practice of mollifying insulted diplomats by inviting them to the White House for a personal audience, just as many governors and mayors across the country invited African visitors into their homes after similar occurrences in their own jurisdictions.[161] Kennedy also ordered the State Department to work on a solution, and the Special Protocol Service Section was set up under diplomat Pedro Sanjuan's enthusiastic leadership to buffer African diplomats from negative treatment. Sanjuan led a lobbying effort to desegregate public facilities along Route 40, first through partially successful persuasion and then through a change in Maryland state law. Sanjuan's zest for ending racial discrimination won him great affection among Africans but sometimes went further than his cautious superiors had expected. When he even encouraged CORE activists in their plans to help the desegregation campaign by picketing segregated restaurants along Route 40 in the fall of 1961, the president and his top advisers—whom CORE's role in the Freedom Rides a few months earlier still rankled—were not pleased.[162]

Still determined to move slowly on civil rights to avoid alienating white Southerners, Kennedy understood that the eradication of Jim Crow was the best strategy for improving U.S. relations with Africa.[163] He was pleased by the goodwill Sanjuan was creating for the administration among Africans. His civil rights legislation was widely applauded on the continent, as was his dispatch of federal troops to the University of Mississippi. In a typical conversation at the State Department two months after the latter incident, leaders of the South West Africa Peoples Organization pointed to Kennedy's actions at Ole Miss as evidence of the "great role" the U.S. government could play in ending racial totalitarianism in southern Africa as well as in the American South. Deputy Assistant Secretary of State Wayne Fredericks demurred, noting that Washington could act within U.S. borders in ways it could not abroad, but he did underline the American "hope that other countries will see this resolve [that we have shown]" on this issue.[164] Even Congress, while still unable to pass a civil rights bill over Southern opposition, indicated some willingness to accommodate the desire of people of color everywhere not to be gratuitously insulted; in 1963 it passed legislation purging "nigger" from all names applied to geographic features on U.S. maps.[165]

While the well-publicized appearance on American shores of a new wave of Africans in positions of authority created problems for many white Americans and for the U.S. government, it had a different impact on black Americans. Inspired by the respect accorded these dark-skinned diplomats, African Americans could not help comparing their own circumstances and struggles with the speed with which Africans had gained their independence and eliminated white supremacy in most of the continent. Students and other young civil rights activists were especially attentive to this contrast, which reduced their patience with white resistance. "Frequently I hear them say," King wrote in the *New York Times* in September 1961, "that if their African brothers can break the bonds of colonialism, surely the American Negro can break Jim Crow."[166] King himself vented the same frustration a year and a half later in his "Letter from a Birmingham Jail": "The nations of Asia and Africa are moving with jetlike speed toward gaining political independence, but we still creep at horse-and-buggy pace toward a cup of coffee at a lunch counter."[167] The growing presence of African students at black colleges in the United

States further heightened younger black Americans' awareness of their ancestral continent, as did the ties of some of their families to Africa, like that of SNCC's Julian Bond.[168]

The melee in the visitor gallery that interrupted Adlai Stevenson's first major address to the UN on 15 February 1961 revealed the rising international consciousness of black Americans. Some sixty African Americans struggled with guards while protesting the murder of Patrice Lumumba and the UN role in the Congo crisis. Removed from the building, they and hundreds of other supporters marched through midtown Manhattan chanting "Congo, Yes! Yankee, No!" James Reston of the *New York Times* observed these events, noting the intensity of black frustration, and concluded, "It is obvious that something is seriously wrong." Reston went on in a thoughtful analysis to suggest that "the young intelligent American Negro . . . has been identifying the battle for freedom in Africa with his own struggle for equality within the United States" far more than most white Americans realized. Here was "an event of momentous importance," Reston concluded: "a confluence of the world struggle for freedom in Black Africa and the struggle for equal rights in the Negro communities of America."[169] The Kennedy administration also acknowledged this confluence, as Wofford pointed out a month later in a public address: "Our pace [in civil rights] is in considerable part being determined by events abroad—by the emergence of Africa—by the establishment of a clear colored majority in the United Nations."[170] In their first appearance in the United States in significant numbers since the days of the slave trade, Africans were clearly bad news for Jim Crow.

There were limits on the extent of black American identification with Africa in the early 1960s. Cold Warriors had for the previous fifteen years wielded anti-Communism as an effective tool for repressing the anti-colonialist, internationalist perspective of the black Left, represented by such figures as W. E. B. Du Bois and Paul Robeson.[171] The domestic civil rights struggle that had spread like wildfire through the American South since the Greensboro sit-ins of February 1960 demanded the near total attention and energy of American antiracists during the Kennedy years—a point the president specifically remarked on with his advisers in discussions of U.S. policy toward southern Africa.[172] Such initiatives as the creation in mid-1962 of the American Negro Leadership Conference on Africa, representing most of the major civil rights organizations, demonstrated the real black interest in the continent, but the conference's limited influence and brief life span reflected its constituents' overwhelming concern with the cause of equality at home.[173] Even for those African Americans who identified strongly with the cause of African independence, their personal experiences with Africans often pointed up their "Americanness" more than their "Africanness." Black Peace Corps volunteers, for example, found that African responses to their presence tended to undermine any quest they may have been pursuing for a transnational racial identity as "black." Africans instead referred to them more frequently as "white," "European," "Westerner," or "foreigner," or at most "black white woman," "suntanned white woman," "black European," or "native foreigner."[174] The ties of African Americans to Africans were real and growing during the early 1960s, but they were also complicated and limited.

"CHRIST, YOU KNOW it's like they shoot this guy in Mississippi," John Kennedy complained to House Speaker Carl Albert on 12 June 1963, the day after Medgar Evers's murder. Racial tensions, he went on, had infected all politics: "I mean it's just in everything. I mean this has become everything."[175] By the end of his presidency, the civil rights movement and white resistance to it had indeed come to dominate the government's attention. The African American determination to end discrimination once and for all was captured by the speaker at a March 1963 mass meeting in Birmingham who declared, "There will be no nigger any more. It will be *Mister* nigger."[176] Simultaneously, the waves of African independence from the north crashed against the bedrock of apartheid and colonialism in southern Africa. A budding alliance among Pretoria, Salisbury, and Lisbon created a wall of white resistance, provoking the first efforts of the liberation campaigns that would ultimately wipe the last realm of white supremacist rule from the face of the earth.[177] In the southern parts of North America and Africa, racial polarization threatened to destroy the multiracial unity that the Kennedy administration believed crucial for its own political success at home and for American victory in the anti-Soviet struggle abroad.

Particularly troubling to the president and his advisers was the possibility that the struggles against racism on the two continents would get intertwined. Africans and black Americans recognized the two as parts of the same global problem, as did some Africanists within the U.S. government.[178] But for an administration determined to contain the political damage of racial conflict—and confident of its ability to manage crises—such connections were to be minimized. Only the Soviets would benefit from associations of Johannesburg with Birmingham, or Angola with Mississippi. International criticism of racial discrimination in the United States, in particular, was to be avoided. Thus Kennedy and Rusk succeeded in persuading King to cancel his scheduled UN testimony on apartheid in June 1963, to avoid "the danger that our domestic racial policies will be made the focus of attention."[179] African independence sent African diplomats in growing numbers to Washington and New York in the early 1960s, however, and there was no way to avoid black Americans' awareness of them or their experiences of American racial discrimination.[180]

Having substantial interests on both sides of the racial divides in the American South and southern Africa, the Kennedy administration sought to finesse the problem by avoiding choosing one side over the other. The president offered Africans and African Americans the extraordinary novelty of an administration studded at the top with believers in racial equality: Rusk, Bowles, Williams, and other Africanists in the State Department; Burke Marshall, John Siegenthaler, and Robert Kennedy in the Justice Department; Wofford in the White House; Shriver in the Peace Corps; and, above all, John Kennedy himself. The contrast with the Eisenhower administration was stark. And despite his initial resistance to leading publicly on the issue of civil rights, President Kennedy did ultimately come out strongly in favor of ending racial discrimination. The grief of many darker-skinned residents of both continents on 22 November 1963 was therefore definitely real, as was the satisfaction of many of their paler neighbors.[181] At the same time, the president bent over backward to avoid alienating the white authorities of Dixie and southern Africa. After much initial attention to African concerns, Kennedy and his chief advisers retreated to a generally less critical position regarding the staunchly anti-Communist white rulers of Angola and South Africa.

The actual policies of the Kennedy administration toward white supremacy in Africa and the United States thus moved in opposite directions between 1961 and 1963. Initially tough on racism in southern Africa and guardedly sympathetic to those seeking to end its sway, the president and his most important advisers shifted over time to a more neutral stance in which they worked to limit criticism of the white authorities there. Distracted by crises elsewhere, they revealed their traditional Cold War and European priorities and took positions that were, to paraphrase their own anti-Communist language, soft on racism. Conversely, Kennedy began his term aware of his political need for white Southern support in Congress and eager to avoid civil rights confrontations at home. By the end of 1963, however, disillusioning experiences with men like Alabama governors John Patterson and George

Wallace and Mississippi governor Ross Barnett had helped move him to a strong public stand in support of outlawing racial discrimination. Initially soft-pedaling racism at home, he became more outspoken over time. Ultimately, preserving social order within the fifty states was more immediately important to an American president than events at the far tip of Africa, and the Southern civil rights movement—aided by the viciousness of its opponents—managed to change Kennedy's political calculations and perhaps even his core attitudes in ways that the ANC and other insurgents south of the Zambezi River could not yet do. Their days of altering U.S. policy still stood more than two decades off in the future.

5

THE PERILOUS PATH TO EQUALITY

THE YEARS OF Lyndon Johnson's presidency marked both the apex of legislative racial equality in the United States and the culmination of racial violence in American cities and in Vietnam. From the great heights of the Civil Rights Act of 1964 and the Voting Rights Act of 1965, the political landscape fell away precipitously to terrifying uprisings in Los Angeles, Newark, Detroit, and more than a hundred other cities at home and a full-scale war between Americans and Asians abroad. For most American soldiers in Vietnam, the devastating combat took place increasingly along racial, rather than political, lines: the enemy was not just Communists but the Vietnamese. Moreover, rotating U.S. forces brought with them the sharpening racial tensions from home, so that black and white soldiers were often openly at odds with one another, even while fighting a single enemy. The Johnson administration confronted a downward spiral of racial antagonism and polarization rather than the era of racial progress and unity that it had worked so hard to promote. This pattern of racial crisis emerged between 1964 and 1968 not only in the United States and in Vietnam but also in southern Africa, forcing American policymakers to pursue a largely unsuccessful strategy of containing racial conflict both at home and abroad.

In the United States as well as in the Third World, Johnson tried to keep civil rights activists and anticolonial nationalists identified with the liberal, Western reformist model of his Great Society. He struggled to isolate them from the twin siren songs of Soviet Communism and nonwhite (black, African, or Asian) nationalism. As the most energetic proponent of racial equality to occupy the White House, Johnson hoped to become the *real* leader of the civil rights movement as well as of the non-Communist world. He also labored to blunt the impact of the white backlash in the United States against civil rights achievements and in southern Africa against the wave of independence ending colonial rule. Such vestiges of white supremacy, he knew, hindered his effort to promote a racially egalitarian Western alliance as a more attractive alternative to the Third World than Soviet or Chinese Communism. As a moderate man of the political center, Johnson believed that he occupied the middle ground between black and white extremism, and he sought to co-opt the energies of both.[1]

Lyndon Johnson, Civil Rights, and the Third World

Lyndon Johnson had experienced poverty as a child in the raw Texas hill country, but he entered the White House as one of the wealthiest presidents in U.S. history. His trajectory to political and material success represented the elemental American dream, and he was determined to embody national unity as the president of all Americans. His interest in the welfare of working people of all colors and his dislike of barriers to equal opportunity were legendary among those who knew him.[2] The liberalism of the Great Society involved helping those who were less fortunate. For Johnson such an approach was deeply genuine—and profoundly paternalistic. He craved love, appreciation, and recognition from those around him and from the public at large. As long as those receiving his assistance were sufficiently grateful to their patron and deferential to his power, his system worked.[3] But ingratitude and opposition from those he was trying to help dismayed Johnson and shriveled his generous instincts. By 1967 he would recite to his staff his civil rights achievements and lament: "I asked so little in return. Just a little thanks. Just a little appreciation. That's all. But look at what I got instead. Riots in 175 cities. Looting. Burning. Shooting. It ruined everything."[4] After 1965 the beleaguered president felt swamped by ingratitude from every direction: rioting African Americans, intransigent Vietnamese nationalists, and rebellious antiwar liberals who derailed his reelection. "How is it possible," he asked, "that all these people could be so ungrateful to me after I had given them so much?"[5]

Johnson knew little of other cultures. He had traveled abroad only rarely before becoming vice president. Like many of his generation, he assumed unquestioningly that American values were universal. He expected other peoples to admire the achievements of the United States and to aspire to similar affluence. On an official visit to Senegal's independence ceremonies in April

1961, he toured a small rural community and told the village chief: "I came to Kayar because I was a farm boy, too, in Texas. It's a long way from Texas to Kayar, but we both produce peanuts and both want the same thing: a higher standard of living for the people."[6] He let his enthusiasm fill in for cultural sensitivity when abroad, plunging into crowds to shake hands and kiss babies and ignoring advisers' warnings about dirt, disease, and diplomatic dignity. Although he had once announced that the trouble with foreigners "is that they're not like folks you were reared with," he treated them as if they could become so.[7] Johnson's vision of racial inclusiveness showed in his physical enthusiasm for African, Middle Eastern, and especially Asian crowds—"little brown people . . . packed as close as you could pack sardines"—whom he viewed as the salt of the earth.[8] By contrast, Scandinavians and other white northern Europeans often struck him as stuffy and unreceptive.[9] It was almost as though his dislike for northeastern elites and greater ease with Southerners and Westerners in the United States were mirrored abroad in an uneasiness with Europeans and a warmer feeling for people of the Third World.

As with people of color at home, however, Johnson expected his relationship with Africans and Asians to be that of a generous patron. In contrast to John Kennedy, who hated even to have to talk about agricultural policy, Johnson loved animals and the ranching and farming life. He believed that all farmers, whether in West Africa, Texas, or Vietnam, were essentially alike. He spoke frequently of building another Tennessee Valley Authority for the Mekong River, reflecting his desire to export the benefits of the New Deal and the Great Society for which he had worked so hard at home.[10] He was therefore mystified when the Vietnamese did not seem to appreciate his largesse. He once exclaimed to his aide Bill Moyers, "My God, I've offered Ho Chi Minh $100 million to build a Mekong Valley. If that'd been [American labor leader] George Meany he'd have snapped it [up]!" Robert Komer, Johnson's assistant to the national security adviser, remembered that Johnson "felt no particular need to delve into what made Vietnamese Vietnamese—as opposed to Americans or Greeks or Chinese."[11] In contrast to Ho, those South Vietnamese officials who emulated Americans made sense to the president. His compliment to the much shorter General Nguyen Cao Ky upon meeting him in Honolulu in February 1966 suggested his paternal delight: "Boy, you speak just like an American."[12]

Johnson's chauvinism had the peculiar flavor of his home state. His favorite historical analogy was that defining piece of Texan pride and nostalgia, the Alamo. "Vietnam," he asserted more than once, "is just like the Alamo. Hell, it's just like if you were down at that gate and you were surrounded and you damn well needed somebody. Well by God, I'm going to go—and I thank the Lord that I've got men who want to go with me, from McNamara right on down to the littlest private who's carrying a gun."[13] He was intensely proud of his two sons-in-law and their service in Vietnam. But the rising toll of American casualties there appalled and grieved him, even as he made decisions that further increased their numbers. He seemed less and less able to relax as the war to preserve a non-Communist South Vietnam continued. He gave up drinking his favorite Cutty Sark whiskey and, unable to sleep, appeared frequently in the White House operations room in the middle of the night to check on American casualties. Unlike many domestic opponents of the war who began to sympathize and even identify with the victims of the American war in Vietnam and with other Third World peoples, Johnson continued to focus on American lives. His attorney general, Ramsey Clark, recalled about the president: "I never sensed any concern for the other side. How many did the Vietnamese lose? How many were killed in the village? How many South Vietnamese, how many North Vietnamese, how many Vietcong? It was *our* lives, *our* country; and they didn't figure, those people."[14] As the war in Vietnam slipped away from him, Johnson seemed less and less engaged with those "little brown people" he had originally hoped to help.

In Africa there were no Alamos. Unlike most Americans in the early 1960s, Johnson did not derive his ideas of Africa from Tarzan movies and exotic animals in *National Geographic.* He had visited the continent once and at least thought of it in terms of people. In his sentimental but sincere fashion, he reflected on his trip to West Africa as vice president: "You know, in Senegal, when I looked into the eyes of the mothers there, they had the same look as the people in Texas, the mothers in Texas. All mothers want the best for their children, and the mothers in Senegal were no different from the white mothers in Texas."[15] If his great humanity allowed him to ignore skin color and cultural differences more than most Americans of his generation, Johnson's relentless political calculations ensured that he would avoid any new American engagement in an area of political turmoil and limited strategic significance to the United States. Domestic politics were his love, and Vietnam his distraction and eventual obsession. Crises in the Dominican Republic, the Middle East, and Europe took what little other time he and his advisers had, and Africa remained—as it had been for all his predecessors—the lowest priority in, literally, the world. Angered at a State Department spokesman's public acknowledgment of the new combat role of American personnel in Vietnam in the spring of 1965, Johnson made clear his version of diplomatic and political irrelevance: "He'll be giving his future briefings somewhere in Africa!"[16]

Both Africans and Vietnamese had reason to doubt Johnson's interest in them when he took the oath of office in November 1963, and most African Americans were also skeptical about their new president. Black Justice Department official Roger Wilkins remembered that "he was a Texan, and to me Texas was South, and he sounded South, and that's where my enemies were, more than in the Soviet Union, more than in North Korea." Other black leaders, including Martin Luther King, Jr., suspected that their strongest white allies might well be "converted" Southerners. As political strategist and Johnson adviser Louis Martin put it, "a reconstructed Southerner is really far more liberal than a liberal Yankee."[17] Freed from the segregationist white Texan constituency to which he had long bowed on racial matters, Johnson swiftly jettisoned his previous record of accommodating segregationism. "You know segregation is absolutely crazy," Attorney General Nicholas Katzenbach recalled the president telling him. "Eighty percent of the world is not white, and we have to live in this world as we have to live with everybody else in it."[18]

In his ascent from the South, Johnson cut himself loose from close ties with segregationist Southern Democrats in Congress. A product of their base in the American version of one-party states, Southern Democrats' disproportionate seniority and power in Congress had long mocked the idea of equality before the law in the United States. As vice president and then president, Johnson knew their power but grew to despise the virulent racism of such men as Mississippi senator James Eastland and that "runty little bastard," Alabama governor George Wallace. Johnson loved to mimic and caricature Eastland in private: "Jim Eastland could be standing right in the middle of the worst Mississippi flood ever known, and he'd say the niggers caused it, helped out some by the Communists."[19] Johnson understood that "if you can convince the lowest white man that he's better than the best colored man, he won't notice you picking his pocket. Hell, give him somebody to look down on, and he'll empty his pockets for you."[20] Johnson's anger about the dangerous demagoguery of Southern leaders boiled over after a frustrating December 1966 meeting at his Texas ranch with Southern Democratic governors who complained of his betrayal of his regional heritage: "Niggah! Niggah! Niggah! That's all they said to me all day. Hell, there's one thing they'd better know. If I don't achieve anything else while I'm President, I intend to wipe that word out of the English language and make it impossible for people to come here and shout 'Niggah! Niggah! Niggah!' to me and the American people."[21] To no small extent, he succeeded in that endeavor.

The Southerner who may have bothered Johnson the most was his former senatorial ally and chairman of the Senate Foreign Relations Committee, J. William Fulbright of Arkansas. Fulbright's anger at Johnson's misleading representation of events during the American occupation of the Dominican Republic in 1965 blossomed into an early public dissent from the administration's policies in Vietnam. The segregationist senator became identified with concern for nonwhite peoples in Southeast Asia, while the integrationist president called down a war of destruction on the Vietnamese people. Johnson resented this obvious irony and preferred a different interpretation of the two men's views on race and Vietnam: all Fulbright "thinks about is England, the Marshall Plan or Europe. He doesn't give a damn about Asians because they're brown."[22] The president believed that he understood the Alamo analogy and that Ho Chi Minh and Southeast Asian Communists—latter-day equivalents of Santa Anna and the Mexicans—represented a clear threat to American national security. "Now, Fulbright and Mansfield and Lippmann and RFK don't see this," he argued, "because they think of the South Vietnamese as yellow people not worth protecting."[23] The White House staff unsuccessfully sought ways to use Johnson's professed racial liberalism as a means of undercutting Fulbright's authority on foreign relations with the nonwhite Third World.[24]

Leaving behind most white Southern leaders, Johnson constructed close alliances with the heads of the major civil rights organizations. Roy Wilkins of the NAACP had the greatest access to the president, grounded in their shared belief in legislative lobbying, litigation, and voter registration as the paths to racial equality. A. Philip Randolph of the Brotherhood of Sleeping Car Porters and Whitney Young of the National Urban League were close seconds. African Americans like Young and Reverend Leon Sullivan of Philadelphia earned the appreciation of the president and his advisers for their determined efforts at improving the daily lives of urban black Americans combined with their refusal to criticize the war in Vietnam. The administration was delighted with the election of such mainstream black Democratic politicians as Carl Stokes as mayor of Cleveland. White House staffer John Roche referred to Stokes's victory as "the kind of 'black power' we need in the Democratic Party."[25]

Johnson put much less trust in the younger, grassroots activists of the Stu-

THE COLD WAR AND THE COLOR LINE

dent Nonviolent Coordinating Committee and the Congress on Racial
Equality, and he grew increasingly skeptical about—and hostile toward—
Martin Luther King, Jr., and the Southern Christian Leadership Conference.
The ultimate political insider, Johnson rejected mass public demonstrations
in favor of the careful application of interest group pressure on members of
Congress, the source, he believed, of real social change. "You cannot find one
minority group," he told a gathering of black candidates for elective office
in October 1966, "that wasn't able to ultimately overcome its oppression
by the ballots, by learning to use its power, P-O-W-E-R, capital at the polls,
P-O-L-L-S."[26] In the later years of his administration, as many younger civil
rights workers rejected liberalism in favor of black power and Marxism and as
impoverished black communities in the North erupted each summer in riots,
Johnson struggled to understand those developments. He listened carefully
to extensive FBI reports on black radicals, and he sent advisers to tour
burned-out areas and talk with residents. But Vietnam swallowed his attention, and for all his goodwill, the president could not fully comprehend the
changing mood of black America. White House aide Sherwin Markham reported back to him that staying in the black neighborhoods of Chicago "was
almost like visiting a foreign country—and the ghetto Negro tends to look
on us and our government as foreign."[27] After visiting Baltimore, Bill Moyers
agreed: "We didn't even speak the same language."[28] James Farmer of CORE
remembered how Johnson "was much better able to understand" the courte-
ous, middle-class representatives of the NAACP and the National Urban
League than he could "the angry young blacks who would tell it like it is, and
call him an MF [motherfucker]." Farmer emphasized the paternalism of
Johnson's older Southern manner, which did not "jive very well with angry
young black militants."[29]

Johnson was a most unusual Cold War era president. "I do not want to be
the President who built empires, or sought grandeur, or extended dominion,"
he told the nation in a televised address in March 1965. His heart was set instead on regenerating America itself, by eliminating poverty and racial discrimination at home in order for the United States to realize more fully its
promise as a model society for the world.[30] He took office in November 1963
on the heels of Kennedy's assassination, during a lull in international tensions
due to the recent resolution of the Cuban missile crisis and the nuclear test
ban treaty, which offered an opportunity to focus on domestic matters.
Warning aide Richard Goodwin, "Those civil rightsers are going to have to
wear sneakers to keep up with me," Johnson moved swiftly to make "that

moral commitment" to racial equality which he had urged John Kennedy to do in the summer of 1963. He had told Kennedy and his advisers that "the Negroes are tired of this patient stuff and tired of this piecemeal stuff." It was time, he had said, for the president to proclaim the immorality of racial discrimination, and it would be best to do it in the South while looking white Southerners right "in the face."[31] Johnson did exactly that in a rousing October 1964 campaign speech in New Orleans.[32]

Johnson knew that he was trying to ride the back of the tiger of the civil rights movement. The laborious, dangerous work of black Southerners had created the unusual opportunity he now seized to push civil rights legislation through a reluctant Congress. He believed it was time for a sympathetic administration to take over the leadership of the forces of change in the country, guiding them into responsible legislative reform. Johnson and his advisers also considered it crucial to minimize what they saw as extraneous mass demonstrations that provoked the fears of white conservatives and increased opposition to Great Society programs. In shepherding through the limited civil rights bills of 1957 and 1960 as Senate majority leader, he had sought to co-opt the energies and demands of the rising civil rights movement. So again, in 1964 and early 1965, he championed more substantial civil rights measures partly because he believed they were the right thing to do, but partly to siphon off the energies of the now much larger civil rights constituency. The president knew how much his support for racial equality would cost him and the Democratic Party in terms of white votes, especially in the South, and he was determined to contain further black political demands that would exacerbate that white backlash.[33]

Johnson's success in eliminating legal barriers to racial equality in the United States was embodied most dramatically in the Civil Rights Acts of 1964 and 1968 and the Voting Rights Act of 1965. The president wielded his particular skills—as a Southerner, as the preeminent legislator of his generation if not of the century, and as the recent master of the Senate—to make use of the opportunity provided by the national grief over the posthumously revered Kennedy.[34] He also appointed more African Americans to executive branch positions than the combined total from all previous administrations and pushed through comprehensive immigration reform in October 1965 to eliminate the racially discriminatory national origins system.[35] The international mood encouraged Johnson's reforms, as Africans and others paid close attention to American race relations and were disturbed by ongoing discrimination and violence against blacks, incipient race riots, and the developing white backlash represented by Republican nominee Barry Goldwater in the 1964 presidential campaign. Secretary of State Dean Rusk of Georgia emphasized in December 1963 the central importance for American foreign relations of a national commitment to racial equality: "We are looked upon as a leader . . . and when we fail to meet our commitments, this has a major impact on other countries."[36] The Johnson administration hoped that its leadership in civil rights would restrain criticisms of the U.S. government by people of color both at home and abroad. Black Americans, who ranged from conservative to radical, and the Third World might thus be maintained as parts of a large, liberal, "free world" alliance, which could outmaneuver threats from both the Communist left and the white supremacist right.[37]

Promoting Equality and Restraining Polarization

By the summer of 1964, just eight months after the new president took the oath of office, events in the United States and abroad began to strain his strategy of limiting and defusing racial conflict. Johnson signed the Civil Rights Act on 2 July, in the midst of a season of extraordinary violence by the Ku Klux Klan in Mississippi. A week earlier, three civil rights workers had disappeared in Neshoba County, Mississippi, victims of the forceful response of white residents of the Magnolia State to "the Invasion," as some called the Mississippi Summer Project to register black voters. Sixteen days after the Civil Rights Act became law, the first major race riot of the decade ignited in Harlem, signifying a rejection of white liberalism and its gradualist approach to solving problems of racial discrimination and poverty. Writer and black nationalist Julius Lester observed that the riveting news from Mississippi slipped from the front pages "when Harlem held its own summer project to protest the murder of a thirteen-year-old boy by a policeman. Summer projects, northern style, usually involve filling Coke bottles with gasoline, stuffing rags down the neck, and lighting them."[38] White resentment of such actions—dubbed "Goldwater rallies" by gloomy Democratic Party strategists—fueled the Arizona senator's nomination that same month.[39]

Trouble abroad quickly added to trouble at home. In early August Congress passed the Tonkin Gulf resolution, granting Johnson a free hand in dealing with the situation in Southeast Asia. Civil rights workers listening to the president's radio announcement of air strikes against North Vietnam worried that the administration was embarking on another war before the struggle to liberate Mississippi and the rest of the South was won.[40] The very next day, 5 August, rebels in the Congo captured the city of Stanleyville, provoking a crisis in central Africa that quickly led to the use of white mercenaries, American planes, and Belgian paratroopers to free white hostages held by African forces. The Johnson administration's aspiration to manage and control the forces of racial polarization represented an uncertain prospect, but the president believed, with reason, that his best days were still ahead.

The American response to the Congolese rebellion derived from the Johnson administration's determination to avoid entanglements in Africa. Outside South Africa, whose anti-Communist white government seemed firmly in control, the continent held little economic or strategic significance for Americans. Anti-American rhetoric from such radical nationalists as Ghana's Kwame Nkrumah made Johnson wary. Even the explicit efforts of the People's Republic of China and the Soviet Union to increase their influence there did not deeply trouble the administration, given the combination of Africans' poverty and strong nationalism. Unlike Southeast Asia, this was one part of the world to which American officials did not simplistically apply the domino theory. Africanists in the State Department generally lacked influence with Johnson, Rusk, and National Security Advisers McGeorge Bundy and Walt Rostow, while Defense, Commerce, and Treasury Department officials were far more concerned with South Africa than the rest of the continent. Key power broker George Ball, the undersecretary of state, had strong Europeanist inclinations that undergirded his determination to stay close to the British and avoid any other potential Third World morasses like Vietnam.[41]

In addition, by 1963 the initial romantic enthusiasm for African independence of many white American liberals, both in and out of government, had begun to ebb. The corruption, economic failings, and one-party rule of many new governments on the continent were helping drain away the interest in newly independent Africa evident in the Kennedy years.[42] Liberals in the Johnson administration no longer seemed so confident about making an analogy to the United States in 1776, in which Africans were seen as doing what white Americans had had to do two centuries earlier. Indeed, liberals worried more about the version of that analogy promoted by conservatives, who viewed Africans not as the white Americans but rather as the Native Americans, with white settlers in southern Africa seen instead as the true spiritual kinfolk of American pioneers.[43] This disillusionment with Africa coincided with rising militancy among elements of the civil rights movement and consequent white liberal dismay. Black folks everywhere, it seemed to whites like Lyndon Johnson, were getting harder to deal with.

African leaders were disappointed by the Johnson administration's lack of interest in them. They admired the president's commitment to civil rights legislation, but racial violence in Birmingham, Selma, Harlem, and elsewhere and the white backlash symbolized by Goldwater troubled them. They missed Kennedy and his greater interest in their continent, and they resented the American use of white mercenaries in the Congo and ongoing U.S. ties to the white minority regimes to the south.[44] The close American relationship with NATO partner Portugal did not improve the American image among Africans. In May 1964, for example, two deserters from the Portuguese army surfaced in the Congo to denounce the Lisbon government for its indiscriminate slaughter of civilians across the border in Angola, while the U.S. ambassador to Portugal toured Angola during the same month and publicly praised Portuguese colonial practices.[45] Robert Komer of the NSC optimistically summarized the problem for the president. The Azores base (which the U.S. leased from Portugal), he wrote, "makes it hard to be anti-Portuguese, while the UK's economic stake in Rhodesia and South Africa makes us reluctant to push them too hard." Since Africans could not ignore these concerns, he concluded, "to the extent that we can stay slightly ahead on these issues instead of being reluctantly dragged toward the inevitable, we can keep our African affairs in reasonably good repair." Staying only slightly ahead of white supremacy did not, however, represent the kind of leadership that Africans hoped the United States might provide.[46]

In the Congo in the summer of 1964, the American government quickly fell behind on an issue of great symbolic importance to much of Africa. After the withdrawal of UN peacekeepers from the country in June and the return of Moise Tshombe from exile in fascist Spain as the new prime minister on 5 July, much of the huge, unintegrated nation rose up in simultaneous but uncoordinated rebellions. Tshombe had made himself into perhaps the most hated black political figure in Africa through his leadership of the unsuccessful Katanga secession of 1960–1963, his ready use of white mercenary troops against fellow Congolese, his intimate ties with colonialist Belgian enterprises, and, above all, his responsibility for the murder of the Congolese independence leader and first prime minister Patrice Lumumba.[47] With a third of the country falling quickly into rebel hands, the Central Intelligence Agency warned Johnson that the pro-Western Tshombe's odds of political survival were only "about even." "Should Tshombe fall," the agency warned, "the prospects are dark. Extremists would be likely to gain increased influence in Leopoldville, secessionist regimes might break off and disorder would spread."[48] The Johnson administration's primary concern was to avoid the dissolution of this centrally located African nation, especially at the hands of leftist rebels. This was no time for "another Cuba." Johnson himself had long appreciated Tshombe as an African who openly admired the United States and disliked Communists, and the president failed to recognize the depths of African loathing for a man so closely identified with colonialist interests.[49]

The American response to the crisis in the Congo was to have the CIA organize a private air force to bolster Tshombe's government forces while working with him and the Belgians to hire a mercenary army of some seven hundred white South Africans, Rhodesians, and Europeans to quell the uprisings.[50] Having helped change the odds in favor of the government forces, the problem for the Johnson administration became one of managing the political fallout on the rest of the continent. George Ball either betrayed nearly inconceivable ignorance of American actions or simply deceived journalist Walter Lippmann in a conversation on 25 August, claiming that the United States was not involved at all in the Congo. Ball's metaphor aptly revealed the American view of the continent: the administration recognized, he said, that it "should not get bogged down in the African swamp." The undersecretary of state identified the ongoing problem for Washington officials regarding African affairs as the limits on "our ability to control" the Congolese and other governments there. Working in the "swamp," it seemed, was perilous, unpredictable, and not very rewarding. It should be strictly a last resort.[51]

The Congo in 1964 represented the first use of mercenaries as a direct instrument of U.S. policy in Africa. The white supremacist attitudes of South African and Rhodesian soldiers of fortune created obvious problems for an American government seeking to promote the appearance of a racially egalitarian alliance against leftist influences worldwide. British mercenary leader Mike Hoare explained to a white comrade that "we have a great mission here. The Africans have gotten used to the idea that they can do what they like to us whites, that they can trample on us and spit on us."[52] His attitude mirrored that of an increasing number of white Americans in late 1964 regarding African Americans. It seemed in one sense that the Johnson administration was colluding with the white backlash in Africa at the same time it was fending off the domestic version of that backlash in the United States. Journalist David Halberstam suggested the irony of the situation by noting that hiring

South African mercenaries to bring peace to the Congo was like the mayor of New York City bringing in the Mississippi Highway Patrol to halt riots in Harlem.[53] The Johnson administration's obtuseness about African sensitivities and realities surfaced again with the selection of Charles W. Englehard to represent the United States at Zambia's independence celebrations on 24 October. A major financial contributor to Johnson's campaigns, Englehard was well known in Africa as the largest American investor in South Africa and an intimate friend and enthusiastic supporter of the apartheid regime in Pretoria. The editors of the journal *Africa Today* declared him "no more fit to speak for the U.S. in Africa than is Governor George Wallace or the Grand Dragon of the Ku Klux Klan."[54]

In November Congolese rebels in Stanleyville seized two thousand white foreigners, mostly Belgians but also several Americans, as a hostage shield against an approaching force of mercenaries and Congolese government troops. On 24 November American planes flying from a British base on Ascension Island in the South Atlantic dropped Belgian paratroopers in a largely successful mission to rescue the hostages, although a few dozen were executed at the last minute by their captors. That same day the mercenaries also reached Stanleyville, and the combined mercenary and Belgian forces made swift and brutal work of the rebels and many civilians. As Bundy had warned Johnson a week earlier, an American role in the operation would carry with it "real political costs in the Congo" and in Africa generally, but the storm of criticism that followed seemed to dissipate quickly over the next few months.[55]

Africans objected to the rescue operation because they saw it as a reassertion of external white control over the sovereignty of an African state. They were appalled by Tshombe's use of white mercenaries against fellow Congolese, especially as they learned of what even the CIA admitted were the mercenaries' tactics of terrorizing civilian populations through robbery, rape, beatings, torture, and murder. While *Life* magazine described them as resembling "rough-hewn college boys," the mercenaries' attitudes were more accurately encapsulated in the answer Wally Harper, a South African, gave to a journalist's question regarding how he felt about fighting and killing people in the Congo: "Well, I've done a lot of cattle farming, you know, and killing a lot of beasts; it's just like, you know, cattle farming, and just seeing dead beasts all over the place. It didn't worry me at all."[56] Africans intensely resented both Western actions and news reports that valued white lives more highly than black lives. A Kenyan delegate at the UN Security Council rejected American claims that the hostage rescue was a humanitarian operation: "Where is this humanitarianism when the white mercenaries are allowed full license to murder innocent African men, women and children? . . . Where is this humanitarianism when American Negroes are brutally done to death in Mississippi and elsewhere?"[57] African American leaders across the political spectrum shared this skepticism, from Martin Luther King, Jr., with his mild but firm criticism to Malcolm X and his searing denunciation. Returning to New York from his second trip to Africa on the very day of the Stanleyville rescue, Malcolm called Tshombe "the worst African on earth" and the American policy of supporting him "insane." He warned that Africans were beginning to recognize the common ground in the U.S. government's apparent disregard for black lives in both the American South and the Congo.[58]

White Americans saw matters differently. The Johnson administration, the media, and the predominantly white public all overlooked the atrocities committed by the mercenaries against Congolese civilians. They focused exclusively on the white hostages and abuses by the rebels, and they were disgusted by African condemnations of the rescue effort as "imperialist aggression." The scenario of white hostages tormented by "savage" kidnappers touched a deep nerve in the American psyche, stirring emblematic memories of generations of violent European–Native American conflict, refreshed and distorted by the Hollywood westerns of the post–World War II era. This scenario also touched old fears of plantation slave revolts and potential black rapists, especially in the white South. Coming so soon after another summer of well-publicized white brutality against nonviolent blacks in the American South, the Congo involvement pushed many politically moderate Americans, whether reluctantly or eagerly, toward the more traditional color-coding of savagery in the American national narrative. The U.S. ambassador to the UN Adlai Stevenson accused African critics of "black racism," and *Time* magazine highlighted reports of rebel cannibalism and sexual mutilation of European nuns and priests. Citing the selfless medical work of American missionary physician Paul Carlson among the Congolese poor, *Time* put his picture on its cover and argued that his murder proved "that Black African civilization . . . is largely a pretense." The magazine blasted African criticisms of the rescue operation in revealing fashion: "The sane part of the world could only wonder whether Black Africa can be taken seriously at all, or whether, for the foreseeable future, it is beyond the reach of reason."[59]

Here, ultimately, was the problem: the West was "sane" and Africa was not, at least for now. Grounded in a fanciful depiction of the Congolese rebellion and its aftermath, one intended by its proponents to justify the CIA's actions, *Time's* conclusion about black Africans echoed claims of black immaturity long promoted by white Southerners to justify black subjugation. Bundy chose the paternalistic image of disobedient children in explaining the problem to Johnson in January 1965. Citing Africans' "willful misunderstanding of our paratroop rescue," the national security adviser recommended some careful, parental management of African feelings in the new year.[60]

As the political and racial crisis deepened in the Congo in the summer and fall of 1964, Johnson was confronted at home with the spreading dissatisfaction of African Americans with Democratic Party leadership.[61] A sharp split became evident at the party's national convention in Atlantic City in August, where simmering tensions between civil rights activists and the Democratic Party boiled over. Organizers in the South had for years faced violence and danger on a daily basis as they tried to register blacks to vote. They had operated in conditions often similar in many respects to those of a war zone. They perceived their opponents as the local Democratic Party elites, with the Ku Klux Klan as their enforcers.[62] The slow pace of change in the South and the persistence of white brutality were fraying the commitment of many organizers to nonviolence, especially among the less religiously motivated SNCC workers.[63] The violence of the white response to the Mississippi Summer Project further deepened their alienation, as did the ongoing failure of the federal government to protect civil rights workers.

When Johnson and the national Democratic Party then rejected the bid of the Mississippi Freedom Democratic Party (MFDP) to be seated in Atlantic City as the only democratically elected delegation from that state, accepting the all-white party regulars instead, many civil rights workers had had enough. With Washington's dominant party seemingly on the side of white Southern authorities, America itself appeared to many as no more than Mississippi writ large. They found it increasingly hard to disagree with Malcolm X's advice that "the best way to stop the Ku Klux Klan is to talk to the Ku Klux Klan in the only language it understands, for you can't talk French to someone who speaks German and communicate." The road to black power was opening up, and SNCC's Stokely Carmichael would soon conclude about the Democratic Party that it was "as ludicrous for Negroes to join as it would have been for Jews to join the Nazi Party in the 1930s."[64]

Like Africans abroad, African Americans at home were proving difficult for the Johnson administration to manage or control. During the fall presidential campaign, CORE and SNCC refused the administration's request through Roy Wilkins and civil rights activist Bayard Rustin for a moratorium on demonstrations to avoid encouraging further negative white reactions. The crux of Johnson's problem with the MFDP at Atlantic City had been its rejection of his compromise offer of two at-large delegates, as the president's forces refused to allow any possible dissenters to mar his renomination triumph.[65] Johnson believed he had to preserve the support of white Southerners for his reelection, and they had threatened to walk out of the convention if the black Mississippians were seated. Frustrated by televised pictures of MFDP pickets protesting the seating of the regular Mississippi delegates, the president seized his phone to order the top official of one of the national television networks, "Get your goddamn cameras off the niggers out front and back on the speaker's stand inside, goddamn it!"[66] Even his stalwart supporter Wilkins had to admit that Johnson's display of political power in defeating the MFDP, eased by access to FBI wiretaps of MFDP strategy sessions in Atlantic City, "wasn't one of his finest hours." "The lasting sense of grievance," Wilkins noted, caused "terrible damage to relations between white liberals and black organizers."[67]

Johnson's black support was beginning to slip away on the left at the same time as whites were starting to abandon him on the right. Much of the civil rights movement clearly had different perspectives and goals than the White House: social transformation of America from below versus gradualist leadership from above; distrust of rigid anti-Communism versus global leadership of the non-Communist world; and rejection of the legitimacy of Jim Crow versus respect for Southern officials. Perhaps the most crucial difference came in their views of the white Southern leadership. For civil rights workers, white supremacists like George Wallace and Jim Eastland were dire enemies in a struggle for control and direction of the American South. Black organizers were appalled that the U.S. government could send enormous armies halfway around the globe to defend the "free world," while refusing to order even one American soldier into the South to protect the legal right of American citizens to vote. For the White House, however, segregationist officials were an obnoxious but crucial part of the power structure of the federal and state governments and the Democratic Party—undoubtedly stubborn and backward-looking, but part of the respected leadership of the United States. Like the white authorities in southern Africa, they were in no way an enemy against whom the U.S. government should use force. For better or worse, Johnson believed, they were a part of "us."[68]

Despite these differences with the civil rights movement, the Johnson administration managed through 1964 to fend off most of the forces of racial polarization and keep its liberal coalition moving forward. Johnson's leadership in passing the landmark Civil Rights Act had been superb, and he succeeded in keeping the American role in Vietnam out of the fall presidential campaign. A landslide 61 percent of the voters—including almost all blacks casting ballots—elected him in November, sweeping into office as well a large Democratic majority in Congress more liberal than any in thirty years. Goldwater carried only his native Arizona and five states in the Deep South. While this portentous shift of white Southern voters away from a fellow Southerner and toward the Republican party suggested eventual trouble, for now Johnson savored his mandate for continued liberal reform, especially in the area of his starkest difference with Goldwater: civil rights. White supremacists in South Africa confirmed that reading of the election, registering their dismay that Goldwater did not symbolize what one representative of apartheid had anticipated as a "triumph of conservatism in the West," which would have allowed the United States to shake off its "sickly humanism."[69]

An explosion of racial violence in Alabama four months later, on 7 March 1965, shocked the American people and provided Lyndon Johnson with his greatest opportunity to demonstrate his leadership of the nation. Viewers of the Sunday evening ABC television movie *Judgment at Nuremburg* were treated to the irony of a news flash showing home-grown, state-supported storm trooper activity right in the United States. Heavily armed Alabama state troopers and Dallas County sheriffs viciously assaulted a line of peaceful, unarmed civil rights marchers at the Edmund Pettus Bridge outside Selma. Television cameras captured the ferocity of the beatings for a national and international audience. Having just been reassured a few months earlier that the events in Stanleyville showed racial savagery to be a problem of Africans (and perhaps their descendants), Americans and their Texan president were dismayed by the most brutal repression yet of any civil rights demonstration. No surreptitious act of night riders, this was a full-scale riot by the uniformed authorities of an American state.[70]

What did the events at Selma mean? For most white Alabamans, the police action saved them from the indignity of hundreds of black Americans marching symbolically down the Jefferson Davis Highway to Montgomery, the capital of the old Confederacy. The Alabama state legislature passed a resolution charging the marchers with conducting sexual orgies, leading several nuns in the group to joke about their lost reputations. SNCC chairman John

Lewis responded to the questions of the press: "These white segregationists always think about fornication. That's why you see so many shades of brown on this march."[71] For Vice President Hubert Humphrey and some others in the administration, the issue was not carnal but ideological: they believed Communist agents among the Selma demonstrators had successfully provoked a violent confrontation.[72] For FBI Director J. Edgar Hoover the problem was one of both sexual and political subversion, but he also understood that the strong national reaction against the police violence meant the ending of an era of quiet FBI alliance with Southern authorities in the defense of Jim Crow.[73] Perhaps the most illuminating evidence of the nature of the Alabama conflict appeared in the symbolism of flags: black Selma high schoolers in the march carried both the Stars and Stripes and the blue flag of the United Nations, while state patrol cars bore Confederate flags on their front license plates, and the capitol dome in Montgomery flew the Confederate flag above the Alabama state flag. Displaying the only American flags in sight and proclaiming their connection to a larger international community, young black Alabamans in their own state seemed to be walking through a foreign country.[74]

The events in Selma unfolded in a troubling international context. On the day after state troopers assaulted African American and white civil rights activists in Alabama, the first 3,500 U.S. marines waded ashore in South Vietnam to defend the Danang airbase. With his skull fractured by a police baton, John Lewis declared angrily, "I don't see how President Johnson can send troops to Vietnam [and] the Congo . . . and can't send troops to Selma, Alabama." Lewis and others could not believe that democracy deserved greater defense abroad than at home.[75] Harris Wofford, the associate director of the Peace Corps, flew to Montgomery on 8 March to join a second march because after "two years in Africa, I knew afresh that America's relationship with the world depended on how speedily and fully we ended racial discrimination at home."[76] That same day Bundy warned Johnson that "U.S. prestige and influence on the African Continent have never been lower," endangering American strategic, economic, and political interests there. The problem, Bundy said in language again implying a parental analogy, was that Africans were not convinced that anyone at the highest level of the U.S. government cared about them, while diplomats from China and the Soviet Union lavished public attention on them.[77] Events like those in Selma did little to improve African confidence in the United States and its leadership in a multiracial world. In a public meeting about American ties to South Africa, CORE's

James Farmer argued that "the fight against aggression and for equality is essentially the same struggle in South Africa and in Selma."[78] The Johnson administration would not have disagreed, but its primary concern was not swiftly achieving racial equality but avoiding friction and instability. In speech notes made three days after the Selma incident, Johnson defined America's mission in Africa as "[a]verting racial conflicts between [the white-ruled] parts of Southern Africa and the rest of the continent."[79]

Domestically, Johnson seized the opportunity arising from the national outrage about what came to be called Bloody Sunday in Selma to make a preemptive strike against further racial polarization. Aware that many younger African Americans agreed with SNCC's James Forman—"If we can't sit at the table of democracy, then we'll knock the fucking legs off"—the president delivered the most famous speech of his career to a joint session of Congress on 15 March.[80] "We have already waited a hundred years and more" since equality was promised, he pronounced in his most serious tone, "and the time for waiting is gone." With a determination unusual even for this forceful man, the president demanded that Congress pass a national voting rights bill to guarantee finally the most basic right of citizenship to all Americans. Selma marked a turning point in the nation's "unending search for freedom," he said, just as surely as Concord, Lexington, and Appomattox had done generations earlier. In a final twist that shocked even his most sympathetic listeners, Johnson seemed to identify himself, the U.S. government, and the American people for the first time fully with the cause of the civil rights movement. Pausing for emphasis, he declared: "And we . . . shall . . . overcome!" Stunned, the audience sat silent for an instant as one Southern congressman murmured, "Goddamn." Then the room erupted in a sustained standing ovation.[81]

Martin Luther King, Jr., watching the speech on television in a Selma living room, began to cry, something his aides had not seen before. Roy Wilkins, who was sitting in the audience on Capitol Hill that night, remembered that "at that moment, I confess, I loved L.B.J." At this high-water mark of the movement for racial justice in the United States, Lyndon Johnson was in the lead. He followed through by sending federal troops to protect the reconstituted marchers in Alabama and by pushing the Voting Rights Act through Congress with remarkable speed despite a Southern filibuster.[82] His assistants made certain that the international audience did not miss the significance of the stand he had taken, sending copies of his speech to the leader of each African nation.[83] In a graduation address at Howard University on 4 June, Johnson went further in his support for African Americans by laying out the logic of what would become known as affirmative action: "You do not take a person who, for years, has been hobbled by chains and liberate him, bring him to the starting line of a race, and then say you are free to compete with all the others, and still just believe that you have been completely fair." He "spelled out the meaning of full integration for Negroes," the noted black novelist Ralph Ellison recalled, in a way no other president ever had.[84]

Trouble in Vietnam, Los Angeles, and Rhodesia

The spring of 1965 marked the turning point in the Johnson administration's journey from great success to deep failure. From the moral heights of leading the campaign for national voting rights legislation at home, the president and his advisers tumbled toward the boggy ground of the Southeast Asian quagmire, the troubled waters of urban violence at home, and the swamp of white supremacy in southern Africa. The administration's growing international involvements, first in the brief occupation of the Dominican Republic in late April and then in the shift to a full-blown ground war in Vietnam in July, brought an end to most of its impressive domestic achievements in the areas of economic and racial justice.

The issue of race was never far below the surface for American policymakers dealing with Southeast Asia. General Maxwell Taylor, the U.S. ambassador to South Vietnam, laid out for the Joint Chiefs of Staff his strong reservations about committing American ground combat troops to the country: "The white-faced soldier, armed, equipped and trained as he is, is not a suitable guerrilla fighter for the Asian forests and jungles."[85] Administration officials on both sides of the debate about further involvement in Vietnam feared what dissenter George Ball called "the appearance of a white man's war."[86] Johnson's appointment in July of Major Hugh Robinson as the first black military aide to a U.S. president seemed no coincidence in this regard.[87] McGeorge Bundy advised the president to reject Ball's recommendation to withdraw from Indochina, but not before receiving "pretty tight and hard analyses of some disputed questions like the following . . . What are the chances of our getting into a white man's war with all the brown men against us or apathetic?"[88] The Johnson administration believed it could not afford to pull out of Vietnam, even as it acknowledged the likely racial character of a deeper involvement in the war there.

Bundy's "brown men" did seem particularly dubious about Johnson's new war. To preserve its precarious diplomatic balance in Africa, the administration had to resist the temptation to respond strongly to vocal critics like Ghana's Kwame Nkrumah.[89] Dissent grew quickly among black Americans, beginning with the speech of SNCC's Bob Moses at the first large antiwar demonstration on 17 April in Washington and accelerating among civil rights workers in the South that summer. A leaflet issued by a group of activists in McComb, Mississippi, in July urged resistance to the draft by connecting racial issues at home and abroad: "No Mississippi Negroes should be fighting in Viet Nam for the White Man's freedom, until all the Negro People are free in Mississippi . . . No one has a right to ask us to risk our lives and kill other Colored People in Santo Domingo and Viet Nam, so that the White American can get richer . . . [W]e don't know anything about Communism, Socialism, and all that, but we do know that Negroes have caught hell here under this *American Democracy.*"[90] Most Americans in 1965 still supported their government's anti-Communist stance abroad, but younger African Americans were sliding toward agnosticism on this issue. More moderate civil rights leaders opposed efforts at draft resistance, but that same month King also condemned Johnson's quest for a military victory in Vietnam and called for a negotiated settlement.[91]

The antiwar views of Moses and other SNCC staffers invited segregationists to identify them as traitors. U.S. Congressman Joe Waggoner of Louisiana seized the opportunity to condemn them as a "mob" with no students, "only radical, Communist-infiltrated gangs of agitators." Historian C. Vann Woodward tried to remind readers of the *New York Times* in August that the hearts of SNCC activists were "in Mississippi or Harlem, not in Moscow or Peking."[92] The FBI, however, was not buying any excuses for what it saw as subversive activity. With the concurrence of Johnson's Justice Department, on 15 June Hoover ordered a wiretap of the office phones of SNCC because it was "the principal target for Communist Party infiltration among the various civil rights organizations."[93] Going to war in Asia was clearly splitting the White House and civil rights workers further apart.

Thus far Johnson had been able to count on the commitment of most black civil rights activists to nonviolence as the key to limiting racial polarization in the United States, despite continuing white violence and a growing white political backlash. But the situation began to change in the summer of 1965 in ways that suggested the president's task as a keeper of national unity would soon get much harder. *Newsweek* and the *New York Times* both reported in August the growing popularity among Southern blacks of the Deacons for Defense and Justice, a group begun in Louisiana to provide armed security for civil rights workers and African American neighborhoods threatened by white violence and unprotected by local white police. The Deacons' effectiveness was clear in Bogalusa, Louisiana, where in April they had ended the traditional Saturday night entertainment for certain whites of shooting up the black section of town. Deacon leader A. Z. Young explained how: "Three rednecks [with rifles] got out of a car and began shouting in the middle of the night. About a dozen of us [with guns] come out from the bushes, from driveways, from dark houses. We ringed 'em around, and then, well, we talked to them a little. You know, they got mighty polite. They was all smiles. It was 'yes, sir' and 'no, sir' and so we let 'em go, and they ain't been back." Bogalusa had become "an armed camp" and a symbol of the Johnson administration's limited influence on the African American quest for equality and dignity.[94]

Small-town Louisiana provided merely a prelude, however, to the racial crisis that ignited in the Watts neighborhood of Los Angeles on the night of 11 August. Over the next six days, arson, looting, and gunfire left thirty-four people dead, over a thousand injured, and $40 million in property damage.[95] On a scale far larger than the disturbances of the previous summer, the Watts uprising was the first of the decade's insurrections in impoverished black urban neighborhoods to gain major national attention.[96] Most ironic and telling of all, the violence began just six days after Johnson signed into law the Voting Rights Act, probably the most effective single piece of legislation in promoting racial equality in the twentieth century. And it came only two weeks after he ordered over a hundred thousand American troops into Vietnam. As the United States entered into war abroad, national guardsmen patrolled the decimated streets of black Los Angeles in a scene that looked an awful lot like war at home.

Competing explanations of the violence quickly filled the national news, with responsibility assigned variably to black unemployment and poverty, white racism, black immorality and disrespect for the law, police brutality, and Johnson's coddling of African Americans.[97] The president himself reacted with shock, sorrow, and anger. The Voting Rights Act had been, he later told Doris Kearns, "a triumph for freedom as huge as any victory that has ever been won on any battlefield."[98] Expecting gratitude, he instead received the news of the Watts riot as a personal affront: "How is it possible after all we've accomplished? Is the world topsy-turvy?"[99] He literally did not know what to do at first, staying at his ranch in Texas where he had been vacationing and refusing for three days even to accept his assistants' calls from Washington—the only time he was ever so out of touch as president. After recovering his political instincts, he arranged for Republican John McCone to chair a state investigation of the riot, to defend his right flank: "An ex-CIA Director, conservative, if he says no communist conspiracy and describes the [awful] conditions in Watts, we'll be able to help those Negroes out there." Johnson feared that by increasing white resistance, Watts and any further disturbances might undermine future Great Society legislation and perhaps even threaten the liberal gains already made.[100]

In just five months, the dominant image of African Americans in the national media had changed from the hymn-singing, flag-bearing marchers of Selma willing to suffer for the cause of justice to uncontrolled hooligans reveling in random violence. Equating rioters with Klansmen, the president revealed his concern that Watts might be a prelude to a black return to savagery, perhaps like that of the African rebels in the Congolese jungle. "Negroes," he told his aide Joseph Califano, "will end up pissing in the aisles of the Senate" and making fools of themselves as they had after the Civil War a century earlier. The younger, Northern-born John Kennedy had not imbibed the Reconstruction myth of black depravity in its most racist form, and as president his experiences with white Southerners led him to doubt the traditional white view of post–Civil War history. His successor, raised in the segregated South, labored mightily to mold a nation that would not discriminate. But deep down, Johnson apparently still feared that beneath a thin veneer of civilization, primitive instincts lurked in the hearts of Americans of African descent.[101]

In the midst of these rising tensions, in October 1965, Congress passed Johnson's new Immigration Act, which abolished the long-standing national origins quotas and put all countries on an equal basis for the first time as potential providers of new Americans. The old system of ethnic and racial discrimination, with its strong preference for northern and western Europeans, had fit a segregated society. But the experience of World War II as well as the end of legal segregation made that vestigial structure an embarrassment to a multiracial nation at the head of the non-Communist world. By eliminating that system, the American government removed a major irritant to nonwhite nations in their dealings with the United States. The change would lead in subsequent decades to an unexpected new wave of immigration, especially from Asia, but in the fall of 1965 the new law's significance lay in its symbolism of a sharp break with the past: the United States would no longer legally seek to preserve an identity as a homogenous white society. There was little coincidence in this change being enacted in the wake of both the civil rights movement and the dispatch of American troops into war with an Asian enemy. The Johnson administration was determined to minimize racial conflict on all fronts.[102]

A few weeks later, the white backlash the Johnson administration had worked so hard to restrain at home erupted overseas. On 11 November Prime Minister Ian Smith of Southern Rhodesia issued a unilateral declaration of independence (UDI) from Britain. In an age of African states gaining their freedom from European colonial control, the UDI represented an exceptional form of decolonization: shifting power from the metropole not into the hands of representatives of the majority but into those of the tiny minority of white settlers and their descendants. The 95 percent of Rhodesia's population that was African held no political power and earned only 5 percent of the country's income, leaving white Rhodesians with one of the highest per capita incomes in the world. White Rhodesians had no intention of waiting for the continent's rising tide of racial equality and self-rule to reach their shore. They had observed recent events in the rest of Africa with dread, from the Mau Mau uprising in Kenya in the early 1950s to the Congo disorders since 1960. They had received thousands of British and Belgian refugees from those lands who bolstered their assumptions about Africans beyond white control as "savages." White Rhodesians were appalled that by 1964 Britain had granted full independence under African rule to Northern Rhodesia (Zambia) and Nyasaland (Malawi), the two other colonies of the former Central African Federation, while refusing to allow the same to their government in Salisbury, which had been self-governing since 1923. They had also been deeply impressed by Moise Tshombe's success in using a few hundred white mercenaries to gain control of huge portions of the Congo. "We have struck a blow for the preservation of justice, civilization and Christianity," Smith declared in his UDI speech, emphasizing his view of southern Africa as "the ultimate bastion against communism on the African continent." Equating the ending of racial hierarchies with Communism was part of a language familiar to American listeners: the language of the Dixiecrats, the Southern white segregationists. By setting its face against the tide of modern history, the Smith regime had chosen to follow the South African model, and in doing so it created a whole new front of racial polarization for the beleaguered Johnson administration.[103]

The British government refused to use force to restore its sovereignty, to the chagrin of Africans, who found themselves promoting British colonial control for the first and only time. London turned instead to the United Nations to promote diplomatic isolation and economic sanctions for Rhodesia as a way to pressure Smith.[104] African and Asian states reacted angrily to the UDI. Having just emerged from decades and centuries of European rule, they were appalled by the idea of a small number of whites trying to reverse the tide of anticolonial liberation. Africans condemned the unwillingness of Britain to use force against whites in Rhodesia, especially in light of its ready military response against rebellious nonwhites in other colonies like Malaya and Kenya. Even the governments of South Africa and Portugal responded to the Smith regime with some ambivalence, considering the renewed international condemnation it was sure to bring to all of the white-ruled states in the region.

African leaders spoke threateningly of possible race war. The Ivory Coast ambassador to the UN, Arsene Usher, warned the Security Council that if Belgians, British, and Americans had the right to rescue white hostages at Stanleyville, then Africans had the same right "to free their brothers taken as hostages by white rebels under a regime condemned by you and considered by world public opinion as illegal."[105] The CIA worried that in "exposing Africa's impotence before the world, the [Rhodesian] crisis has awakened bitter feelings of frustration, humiliation and racial inferiority which may lead to extreme, and sometimes irrational actions against Western interests."[106] A commentator in the *Christian Century* observed that Africans and other Third World peoples saw Rhodesia as "a touchstone of the moral integrity not only of Britain but also of the white world as a whole as they grapple with the injustice of racial discrimination."[107]

The Johnson administration in 1965 believed its record on racial discrimination was quite strong, and it had no interest in besmirching that reputation by a high-level engagement with the racially explosive situation in southern Africa. The timing of the UDI, just months into the new American involvement in a ground and air war in Vietnam, ensured that Washington officials would be distracted. Already battling Communists abroad and the forces of white conservative reaction at home, Johnson had no taste for fighting foreign racists in a region of low strategic priority. The American obsession with the war in Vietnam until 1973 and then with avoiding other such involvements for the rest of that decade would coincide precisely with the years of civil war in Rhodesia that ended in the establishment of Zimbabwe in 1980.

The Johnson administration followed Britain's lead in eschewing the use of arms and confrontation with either South Africa or the Portuguese rulers of neighboring Angola and Mozambique, while at the same time condemning the UDI and imposing sanctions on the Salisbury regime.[108]

The debates within the Johnson administration about how to respond to the UDI reveal the range of options policymakers could imagine. At one end stood White House aide Jack Valenti, who warned Bundy of the importance of the West resolving the Rhodesian problem rather than avoiding it: "A crisis faced now can return Rhodesia to legal rule, influence Portugal in a liberal sense, isolate South Africa and weaken its stand over S[outh] W[est] Africa, and keep alive in the rest of Africa distrust of Communism and belief in the West." Sidestepping the issue, he argued, would instead encourage Africa to turn away from the West and would produce "greater risks of bloody racial conflict" in Rhodesia and elsewhere in southern Africa in the long run. Valenti dismissed most supporters of the Smith regime as immigrants from Europe who "went to Rhodesia not to pioneer but to get sun, servants, no taxation and no welfare state"—not tough, self-reliant farmers supposedly comparable to white Americans' ancestors. He reminded Bundy sharply that there "*is* a contradiction between fighting to give South Vietnam the right to choose its own government and doing little or nothing about the rights of four million Rhodesian Africans. It perpetuates the Communists' cry that all rules are broken when white interests are at stake—a dangerous cry for a world in which the white man is outnumbered four to one." In the wake of the Watts riot, Valenti and others feared the impact of televised images of a potential Rhodesian race war on an American society increasingly marked by racial polarization and violence.[109]

At the other end of the spectrum of advice in the administration stood Dean Acheson, former secretary of state, original Cold War "wise man," and informal adviser to Johnson. Acheson strongly supported the Smith regime in Salisbury as well as the apartheid government in Pretoria, and he abhorred the UN sanctions against Rhodesia. He considered the sanctions a form of pandering to "the international juvenile delinquents" of the Third World and "the Black Mafia" of African ambassadors to the UN.[110]

Rusk and Bundy fell somewhere between Acheson and Valenti in their views of southern Africa. Unlike Acheson, neither the secretary of state nor the national security adviser had much personal sympathy for white supremacists. But Bundy seemed as concerned with avoiding American identification with "hot-headed Africans" as with the white government in Salisbury.[111] And Rusk declared flatly a year later, "The problem cannot be settled by force. The thought of using U.S. troops is appalling." For the vast majority of Rhodesians, of course, that same thought might instead have been encouraging, which raises questions about why, more precisely, the idea seemed so unthinkable to the secretary of state. Was Rusk simply reflecting the unavailability of U.S. forces already deployed in Southeast Asia? Was he also demonstrating the ongoing tendency of even racially liberal white Americans to identify with other whites abroad? From a key supporter of sending hundreds of thousands of American troops into Vietnam, such sensitivity about settling problems by force sounded false.[112]

While the Johnson administration struggled to avoid the issue of white Rhodesian independence as much as possible, the responses of other Americans demonstrated the chasm between opponents and defenders of the UDI. This gap reflected fundamental differences in definitions of "the West" and "Western values," as well as the widening split between mobilized African Americans and their segregationist countrymen. While the UDI did not generate much attention among the general public in the United States, civil rights workers and racial liberals strongly opposed the creation of what seemed to be another apartheid state. Martin Luther King, Jr., called on black Americans to monitor closely U.S. ties with the southern African states, and he condemned the considerable American investments in South Africa as undercutting Washington's verbal condemnations of apartheid and the UDI.[113] Even National Urban League director Whitney Young weighed in with the State Department to urge uncompromising opposition to Ian Smith. Young reminded George Ball that he had never discussed Vietnam publicly and had discouraged his colleagues from doing so, but that unlike Vietnam, Rhodesia was "purely a racial matter."[114] Opponents of Rhodesia argued that long-term American access to valuable Rhodesian minerals, especially chrome, would best be served by closer identification with that country's vast African majority than with its fleeting white rulers.[115]

The Rhodesian lobby, by contrast, represented a marriage of anti-Communism and white supremacy. White Southern senators and representatives like Strom Thurmond and James Eastland saw the Salisbury and Selma crises as closely related challenges to white authority that had to be resisted. Eastland complained bitterly in the Senate that, in contrast to U.S. nonrecognition of Rhodesia, "Every time a group of partially educated, half-savage tribes has constituted an alleged government and declared its country free and independent, we have been pressured by an increasing fear of world opinion into recognizing that government."[116] Anti-Communists like Barry Goldwater loathed the idea that the United States was expanding trade with the Soviet Union and China while cutting off economic ties with a government that was openly anti-Communist, pro-Western, Christian, and prosperous. After all, the Soviets and Chinese were aiding America's enemies in Vietnam while the Rhodesians spoke openly of sending troops to fight *with* the American forces there.[117]

For many white Southerners, the Rhodesian experiment represented a successful Western version of how the otherwise chaotic Third World ought to be run. U.S. Representative Joe Waggoner of Louisiana proclaimed that "Rhodesia has become the cornerstone of this nation's tenuous foothold in the entire Afro-Asian world."[118] If the Rhodesian government fell, the American group Friends of Rhodesian Independence argued, "both South Africa and Portuguese Africa will inevitably fall like dominoes, leaving the entire continent in anti-Western hands."[119] Here, then, was the heart of the conflict between Americans on opposite sides of the Rhodesian issue. Did Ian Smith have his finger in the dike, staving off a flood of savage African madness, or was he an unwelcome obstructionist, holding the door shut against simple representative government? Both defenders and opponents of the Salisbury regime believed they were promoting the values and security of the West against its worst enemies, but they differed dramatically in how they understood the West and their own country. Thus did the civil rights movement and the white backlash clash on international as well as domestic issues.

The international focus on Rhodesia brought renewed attention to the Republic of South Africa, the powerful hub of the white redoubt of southern Africa. In the first months of 1966, both SNCC and Senator Robert Kennedy increased their criticisms of American coziness with apartheid. SNCC members staged a sit-in at the South African embassy in Washington, while Kennedy traveled to the University of Cape Town to deliver the strongest anti-apartheid speech yet by a major American political figure. Within a year such pressures led to a new U.S. military policy of unofficially boycotting South African ports for refueling or shore leave.[120] Martin Luther King, Jr., accepted two invitations to speak in South Africa during 1966, but the South African government refused to grant him the necessary visas. Instead, he led a march in July by civil rights demonstrators seeking an end to residential segregation in Chicago, in which participants carried signs reading "End Apartheid in Real Estate." The response of white Chicagoans, gathering in crowds of thousands to throw bottles and rocks and shout obscenities at the peaceful marchers, only served to underline the parallels between racial discrimination in South Africa and that in the United States.[121]

Johnson faced an obvious dilemma in South Africa. A president so committed to ending racial discrimination could hardly have approved of apartheid, yet the United States continued to have substantial economic and strategic ties to South Africa. Along with Japan, South Africa had the fastest growing economy in the world in the 1960s. As one U.S. government study put it, "US private investment is continuing at a high rate in South Africa because of unusually favorable returns."[122] As the world's largest supplier of gold, the apartheid state had unique importance for the global economy and particularly for the economy of America's closest ally, Great Britain. The CIA emphasized that South Africa was "a major supplier of minerals to the free world" and that "consuming countries would have difficulty finding alternative sources."[123] Both the Defense Department and the National Aeronautics and Space Administration (NASA) operated space tracking stations within South Africa that they considered crucial for American space and missile programs.[124]

Nevertheless, South Africa was a growing political problem for the Johnson administration. International condemnation of the apartheid government was escalating, symbolized by the country's being banned from the Olympic Games in 1964. Undeterred, the Pretoria government by 1965 had crushed the initial black efforts at armed insurgency that had marked the first half of the decade, since the Sharpeville massacre of 1960, and had created a more ruthless, efficient version of a police state. "The rule of law has been set aside by arbitrary arrest and detention," observed the NSC, and "apartheid [is] now being drastically extended."[125] White liberal dissent virtually disappeared, with the opposition United Party even trying on occasion to "race-bait" the ruling Nationalists as insufficiently supportive of white rule in the region.[126] South Africa was, in fact, busy building additional facilities near its northern borders and in South West Africa better to project its military power into the surrounding area.[127] Meanwhile the Rivonia trial of 1964 resulted in life imprisonment terms for Nelson Mandela and eight other top leaders of the African National Congress who, as officials in the Johnson administration admitted privately, were "the leaders of the *moderate* African opposition in South Africa."[128] "Having been barred from all peaceful courses of action," the NSC noted, black South Africans were turning toward radicalism and "black consciousness," a movement that would parallel black power in the United States.[129] State Department emissary Charles Fahy warned Secretary of State Rusk upon returning from a fact-finding tour of South Africa that "for every Mandela there are 20 to take his place."[130]

South Africa's defiance of world opinion inevitably spilled over to roil its relations with the United States. A particular source of tension was South African treatment of nonwhite American officials. Rusk reacted with great anger to the apartheid state's efforts to prevent Ric Haynes, a young African American on the NSC staff, from visiting the neighboring British High Commission territories.[131] Haynes himself responded with humor to an invitation to a reception at the South African embassy in Washington two years later, telling his boss, Robert Komer, that it was probably a clerical error but that "nevertheless I think it's important I put in a brief appearance if only to prove the point that some of my best friends are South Africans, but I wouldn't want my daughter to . . ."[132] Administration officials in South Africa found their South African counterparts increasingly arrogant and even paranoid. The U.S. ambassador to South Africa, Joseph Satterthwaite, a man known for his sympathy for white South Africans, essentially gave up on talking with the dogmatic prime minister, Henrik Verwoerd. Lewis Douglas— prominent American businessman, former ambassador to England, investor in South African mining companies, and self-described "friend" of South Africa—spoke with the prime minister on a visit to the country in 1964 and, in Satterthwaite's words, "about reached [the] conclusion that Verwoerd is a psychopath."[133] If the South African leadership appeared insane to its friends, it looked much worse to those less at ease with apartheid.

While Johnson administration officials debated privately how best to respond to the deteriorating situation in the land of racial totalitarianism, the international community and civil rights leaders in the United States demonstrated what the NSC called "a growing dissatisfaction with our ambivalent policies toward South Africa."[134] Preoccupied with Vietnam and awash in conflicting advice about the need to alienate neither South Africa nor the independent black African nations, Johnson sought to muddle through and fend off domestic critics of his policies toward Africa. On 26 May 1966, two and a half years into his presidency, he gave his first speech about Africa and the importance of majority rule on the occasion of the third anniversary of the founding of the Organization of African Unity. White House aide Bill Moyers underlined the speech's importance not only for Africans but also as a way to trump two domestic critics: civil rights workers and Robert Kennedy. For the former, "it is a cheap way to keep them quiet on at least one issue." For the latter, "it will . . . make it difficult for Bobby to get far ahead of you" on the issue of political liberty for black Africans.[135] Anticipating that an imminent World Court decision would declare South Africa's control of South West Africa illegal, Johnson told the NSC on 14 July 1966 that perhaps a compromise statement by the United States would "relieve some of the pressure [on us] which will arise. We should try everything. Even a blind hog may find an acorn." The search for that acorn of compromise remained a Herculean task, however, as militant white supremacy in South Africa converged with that in neighboring Rhodesia, while both African Americans and Africans grew more skeptical of the motivations of the rooting American hog.[136]

The Escalation of Dissent

Racial polarization and antiwar dissent accelerated in the United States during 1966 and 1967, undercutting the Johnson administration's popularity and effectiveness. On 6 January 1966 SNCC became the first major civil rights organization to declare its opposition to the American war in Vietnam and to encourage resistance to the draft. One of its staff, Navy veteran Sammy Younge, was murdered for attempting to use a "white" restroom in Tuskegee, Alabama. Partly in response, SNCC's executive committee issued a statement condemning the U.S. government for being "deceptive in its claims of concern for the freedom of colored people in such countries as the Dominican Republic, the Congo, South Africa, Rhodesia, and in the United States itself." The war in Southeast Asia seemed to SNCC workers a demonstration of hypocrisy by liberal American politicians who declared their support for democracy around the world while tolerating undemocratic practices right under their noses in the American South. Strongly opposing violence as a means to liberation in the southern United States, those same political leaders were now unleashing extraordinary violence to prevent the apparent liberation of the South Vietnamese. Civil rights organizers were moving from a narrow critique of a double standard to a broader rejection of an American role in Southeast Asia. They increasingly connected the disenfranchisement of black rural Southerners with the disempowerment of Vietnamese peasants, and laid the blame for both at the U.S. government's doorstep.[137]

The predictable torrent of criticism that greeted SNCC's announcement emanated from whites of almost all political loyalties and from such mainstream civil rights organizations as the NAACP and the National Urban League. The Johnson administration cut off most of its contacts with SNCC, and the organization's white Southern opponents seized the opportunity to portray it as clearly subversive. SNCC's long-standing policy of not excluding Communists or anyone else willing to work for its goals made it vulnerable to red-baiting, especially now that it had moved beyond issues of racial equality to express opposition to an American war against Communists. SNCC worker Jimmy Garrett explained to a reporter that SNCC was actually "subverting" any Communists who might be working with it, as "we're more revolutionary than the Communists."[138] That insight did little to overcome the prevalent American assumption in the Cold War years that any organization not excluding Communists would soon come to be dominated by them. Even the Christian faith of many SNCC workers failed to bridge the growing gap between them and the anti-Communist mainstream in American life. When John Lewis, the SNCC chairman and a devout Baptist minister, was granted the first conscientious objector status of any black in Alabama history as a result of Justice Department intervention, his local draft board chairman complained, "But *we're* all Baptists and *we're* not C.O.'s."[139] Former SNCC staffer Julian Bond won election to the state legislature of Georgia in June 1965 but was temporarily prevented from occupying his seat in January 1966 because of his new would-be colleagues' outrage at SNCC's opposition to the draft. As if to emphasize the gap between America and the Third World on this issue, African diplomats at the UN hosted Bond with a luncheon in his honor.[140]

Despite their new pariah status in American politics, SNCC staffers were not alone in their skepticism about the expanding war in Vietnam. If Johnson could dismiss SNCC as a subversive organization, he faced a more formidable foe in William Fulbright. Beginning on 28 January 1966, the chair of the Senate Foreign Relations Committee held six televised hearings on the war in which emerged his doubts and those of other critics about American policies. Among civil rights groups, CORE moved in a similar trajectory to that of SNCC, declaring its opposition to both the war and the draft in July 1966. Southern Christian Leadership Conference staff shared most of that same perspective, and Martin Luther King, Jr., increased the mild but steady criticisms of American war policies he had begun in 1965. King refused to condemn Julian Bond or the SNCC statement against the war. As the cost of the war effort skyrocketed, he became increasingly disillusioned with how military spending was replacing funding for the poor at home. King hesitated for another year before giving up completely on the Johnson administration, but by the close of 1966 he had ended personal communication with the president. King, SNCC chair John Lewis noted, was no longer Johnson's "head nigger when they [civil rights leaders] come to the White House."[141]

SNCC remained at the cutting edge of African American disillusionment with the White House and with white Americans in general. Stokely Carmichael's defeat of John Lewis in the May 1966 election for SNCC chair marked a watershed for the organization, as it shifted to a less religious and integrationist, more rhetorically militant and nationalist position. SNCC's earlier denunciation of the Vietnam war blossomed into the angry rhetoric of "black power," beginning with Carmichael's effective use of that phrase during a 16 June march in Greenwood, Mississippi. Whites responded with dismay to this new language of confrontation and its rejection of the nonviolent style of the integrationist phase of the civil rights movement. These "angry children of Malcolm X," as Julius Lester called the new black radicals, had come to agree with Malcolm's assertion that "I believe in the brotherhood of all men, but I don't believe in wasting brotherhood on anyone who doesn't want to practice it with me."[142] King rejected the language and style of black power but, as civil rights struggles shifted from the rural South to the urban North, he acknowledged, "You can't communicate with the ghetto dweller and at the same time not frighten many whites to death."[143] The wave of white violence that met King's open-housing demonstrations in Chicago less than a month later left no doubt just how large the racial divide in America remained. Rising black anger and greater white hostility fed directly off each other.

SNCC had long been the least acceptable civil rights organization to the Johnson administration. By mid-summer 1966, SNCC members were calling Johnson a murderer of Vietnamese civilians and were arguing that white liberals were no better than Southern racists on matters of human equality. For Johnson, who had staked his reputation on the importance of precisely that distinction, this accusation was appalling. Carmichael's plan to protest the Vietnam war at the wedding of Johnson's daughter, Luci, on 6 August made their conflict more personal. When King, Wilkins, and other civil rights leaders appealed to the SNCC chairman not to go through with a demonstration "in [such] extremely poor taste," Carmichael responded with a withering blast at "your boss" Johnson's napalming of Vietnamese and scheduling of the wedding on the anniversary of the atomic bombing of Hiroshima.[144] Julius Lester registered his disgust with the NAACP and the National Urban League and their condemnations of black power, stating, "as sure as the cavalry always comes to rescue the settlers, here come Roy Wilkins and Whitney

Young riding across the plains, yelling, Here we is, white folks. Let's get them niggers. Even Tonto didn't *join* the damned cavalry."[145] Moreover, Carmichael and other black power advocates began broadening their discussion of racism and its causes to include capitalism as the fundamental problem in America. For many whites in Washington and elsewhere, this meant that J. Edgar Hoover's suspicions about the subversive nature of parts of the civil rights movement no longer seemed quite so preposterous.[146]

Lyndon Johnson's strategy of containing the direction and militancy of black political energies in the United States was faltering. The young radicals of SNCC by no means represented a majority of black America, but they seemed to be its most outspoken edge and thus perhaps its future. By the end of 1966 they had begun to switch allegiances: no longer Americans who believed in their government, they considered themselves domestic allies of Third World revolutionary forces like the National Liberation Front of Vietnam, fighting the American empire from within as a fifth column while the guerrillas did so from outside. Along with an increasing number of young white antiwar radicals, they rejected Johnson, insulting his family, even threatening to shoot his wife, and identifying with his enemies in Cuba, China, Algeria, and North Vietnam.[147] This shift was accelerated by travels to Africa and other parts of the Third World, which SNCC members had begun as early as 1964.[148] They knew the significance of Martin Luther King's journeys to Ghana in 1957 and India in 1959, and they realized the importance of Malcolm X's visits to Africa in 1959 and the Middle East in 1964 in shaping his understanding of, and identification with, the Third World.[149] Confronted with diminishing white support for substantive social change, ongoing black poverty and racial discrimination, and an expanding war in Vietnam, politicized young African Americans were trading white liberal allies at home for what they believed was membership in an international struggle for liberty by people of color everywhere.[150]

One of the most important ways in which black radicals identified their situation with that of Third World revolutionaries was, by 1966, through the growing use of the analogy of colonialism. "Black people in the United States have a colonial relationship to the larger society," Carmichael declared. Gone was the hope of a few years earlier that whites' personal prejudices could be changed through education; the problem had become a deeper one of social and economic structures that benefited whites over blacks.[151] By March SNCC staffers in Atlanta were speaking openly of white civil rights workers as analogous to "the white civil servants and missionaries in the colonial countries who have worked with the colonial people for a long period of time and have developed a paternalistic attitude toward them."[152] SNCC's James Forman pointed to the nation's capital to demonstrate the colonial relationship between black and white in the United States: a mostly black Washington citizenry was ruled directly by the white federal government without such democratic mechanisms as elected representation.[153]

The colonial analogy went hand-in-hand for many young African American radicals with a growing attraction to violence, or at least to the rhetoric of violence. Increasingly disdainful of King's nonviolent, religious, and integrationist vision, they turned to the writings of secular Algerian revolutionary Frantz Fanon, who saw violence as a "cleansing force" that liberates the colonized person from "his inferiority complex and from his despair and inaction; it makes him fearless and restores his self-respect."[154] The scale of American violence against people of color in Southeast Asia helped convince these radicals of the futility of continuing to try to convert white Americans through self-sacrificing love. King himself was moving toward a more radical critique of American society. By January 1966 he was referring to urban slums in the United States as "a system of internal colonialism" comparable to the traditional European relationship with the Third World.[155]

The turn to black power and the Third World by younger black American radicals and nationalists represented a defeat for the Johnson administration's efforts to create a unified and racially integrated America that would stand behind his war in Southeast Asia. "UHURU, LIBERTAD, HALAGUA AND HARAMBEE!" one black nationalist wrote to his local draft board in Detroit: "When the call is made to free South Africa . . . to liberate Latin America . . . to free the black delta areas of Mississippi . . . TO FREE 12TH STREET HERE IN DETROIT!: When these calls are made, send for me," but he would not fight against "my oppressed brothers" in Vietnam.[156] While King, Wilkins, and other civil rights leaders condemned the summer riots in black urban areas during the Johnson years, black power advocates cheered them on as signs of what they hoped might be a growing revolutionary consciousness among poor African Americans. After a riot in Cleveland, Carmichael announced: "When you talk about black power, you talk of building a movement that will smash everything Western civilization has created."[157]

If Johnson's attempts at channeling black political energies were failing, other white Americans stood ready with less friendly responses. White Southern segregationists had tapped into intense white resentment of the civil rights movement in the 1964 Democratic presidential primaries, where George Wallace had shown surprising popularity even in the North. A few months later, home in Alabama, Wallace had condemned the approaching Selma march as similar to the Communist "street warfare" that "ripped Cuba apart, that destroyed Diem in Vietnam, that raped China—that has torn civilization and established institutions of this world into bloody shreds."[158] The Alabama governor's description of the peaceful phase of the civil rights movement seemed, ironically, to match (albeit with different adjectives) Stokely Carmichael's version of what black Americans should now aspire to. The fantasies of the white right and the black left appeared to converge, far from the moderate middle ground of Lyndon Johnson and the rest of America, black and white. The election of race-baiter Lester Maddox as governor of Georgia in the fall of 1966 indicated that white racism was rising in step with black power.[159] Already angry about the decline of Jim Crow, Southern segregationists watched the Supreme Court strike down the last state antimiscegenation law in 1967 and then were dismayed in November 1968 to see Captain Kirk of *Star Trek* engaging in television's first interracial kiss with Lieutenant Uhuru. Senator Richard Russell of Georgia articulated their sexual anxieties by arguing that the new civil rights laws could be used to "compel [a black's] admittance into the living room or bedroom of any citizen."[160]

Southern segregationists in his own party siphoned off some of Johnson's potential support, but the real power of the white backlash showed up in the Republican Party. Goldwater's 1964 nomination and his support in the white South offered early evidence of this shift. The urban riots of the next three summers accelerated the trend: a growing number of white Americans saw these disorders as evidence of continuing black lawlessness and violence despite civil rights gains and federal government antipoverty programs. White demands for a reinstitution of order and severe punishment of rioters fed the Republican critique of Johnson as "soft" on crime, just as the GOP had condemned Harry Truman as "soft" on Communism twenty years earlier.[161] By 1966 some whites were taking to the streets in their own hostile response to civil rights protests and black riots. In Baltimore, a White House aide reminded Johnson, "Italian, Polish and other white minorities bellicosely marched into the Negro districts proclaiming 'white supremacy'" and other white racist groups held public rallies.[162] Another presidential aide thought "it would have been hard to pass the emancipation proclamation in the atmosphere prevailing this summer."[163] In November the Democrats lost forty-seven seats in the House of Representatives—more than they had added in the 1964 landslide. Johnson's liberal majority was gone.

While continuing to work for such legislative reforms as an open housing law, Johnson himself tacked closer to the prevailing political winds in his last two years in office. He spoke more about crime and less about civil rights.[164] The war in Southeast Asia commanded more and more of his attention. One other important force in the white backlash helped push him to the right: the FBI. Personal and bureaucratic racism encouraged Hoover and his agents to continue their long-standing vendetta against King ("the burrhead," as Hoover typically called him) and the civil rights movement as a subversive threat to the middle America they saw themselves defending.[165] Obsessed with the Communist Party, the FBI viewed the radicalization of the civil rights movement and the violence in Northern black ghettoes as further evidence of the essential subversiveness of African Americans and their vulnerability to Communist influences. "Communist," Hoover told his agents, "should be interpreted in its broadest sense."[166] Building on its already extensive spying on King and other black leaders, the FBI in 1967 launched a new counterintelligence program to disrupt and discredit black antiwar advocates and supporters of the black power movement.[167] Hoover fed Johnson a steady diet of information about King's sex life and friendships with leftists and supposed Communists. Already alienated from King, SNCC, and CORE, Johnson merely enjoyed the first kind of information but paid close attention to the second, which he found damning, especially in light of the war against Communism in which he was engaged in Vietnam. The president was skeptical of Hoover in some ways, but he also feared being identified with civil rights leaders should Hoover ever leak his information. With the cooperation of Johnson and his administration, the FBI thus helped further distance the White House from all but the most conservative civil rights leaders by the last years of Johnson's presidency.[168]

Lyndon Johnson's hopes for an America and a world undivided by race and united behind his leadership continued to spiral downward in 1967. Deepening American involvement in Vietnam, the largest race riots of the decade in Newark and Detroit, and the shift of Martin Luther King, Jr., to active public opposition to the war led the way. King's commitment to nonviolence had moved him to oppose the American war in Vietnam in a quiet way since its initial escalation in 1965. His dismay at military spending replacing antipoverty funding had deepened steadily since then. An encounter with photographs of napalmed Vietnamese children in the January 1967 issue of *Ramparts* magazine pushed him over the edge into active opposition and thus complete isolation from the Johnson administration. While the president remained tightly focused on American casualties, King's more internationalist perspective helped him count all the bodies in the war. On 25 February in Los Angeles he gave his first public speech against "a war that seeks to turn the clock of history back and perpetuate white colonialism" in Southeast Asia.[169] A month later in Chicago he helped lead his first antiwar march, and on 30 March he explained to a reporter that "the heart of the administration is in that war in Vietnam"—the opposite interpretation from Johnson's later claim that "that bitch of a war" had kept him from "the woman he loved," the Great Society. That same day King told the Southern Christian Leadership Conference board that in regard to Vietnam "the evils of capitalism are as real as the evils of militarism and evils of racism."[170] To Americans like J. Edgar Hoover—and Lyndon Johnson—this sounded awfully close to the language of Moscow or Hanoi.

On 4 April King gave his most significant address on the war at Riverside Church in New York City, assuring it much greater publicity than his previous antiwar statements. He condemned the U.S. government as "the greatest purveyor of violence in the world today" and its war in Vietnam as "an enemy of the poor." Just as the cost of the war was draining resources from impoverished Americans, the United States was siding with "the wealthy and the secure [the Saigon regime] while we create a hell for the poor" of Vietnam. King cited his Christian ministry and the commission to recipients of the Nobel Peace Prize (which he had been awarded in 1964) to work for "the brotherhood of man" as factors preventing him from viewing the war in Vietnam from an American nationalist perspective. Publicly opposing the war, he declared, was "the privilege and the burden of all of us who deem ourselves bound by allegiances and loyalties which are broader and deeper than nationalism and which go beyond our nation's self-defined goals and positions." Those allegiances and loyalties set King and others in the civil rights movement at odds with the militaristic nationalism of Lyndon Johnson.[171]

The Johnson administration now considered King beyond the pale. The president was particularly concerned with how King's new high-profile stance against the war and the draft might provoke further conservative backlash against Great Society programs already in place. Johnson suspected more strongly than ever that Communists were manipulating King. White House aide John Roche reported to him that King's speech was "quite an item" and indicated "that King—in desperate search of a constituency—has thrown in with the commies."[172] Infuriated at assertions by King that the Vietnam war was "a result of racist decision making" and by the general defection of a section of the civil rights movement from his liberal agenda, Johnson sought evidence from the FBI of a Communist conspiracy among radical black American leaders, a quest he renewed after the outbreak of violence in Newark and Detroit that summer.[173] Hoover was pleased to reply: "Based on King's recent activities and public utterances, it is clear that he is an instrument in the hands of subversive forces seeking to undermine our nation."[174]

The mainstream media responded to the Riverside Church address in similar fashion, rebuking King for his audacity in connecting the domestic issue of racial discrimination to foreign affairs. The *Washington Post* condemned him for doing "a grave injury" to "his natural allies," white liberals. *Life* was more direct in lambasting King's efforts at linking domestic and foreign developments: "He goes beyond his personal right to dissent when he connects progress in civil rights here with a proposal that amounts to abject surrender in Vietnam."[175] Such criticisms implicitly recognized the potential power of King's dissent in trying to link the civil rights and antiwar movements, and rejected such a rethinking of America's actions at home and abroad. King, prominent journalists said, was a civil rights leader and should remain just that. The NAACP board agreed, calling the Riverside Church speech "a serious tactical mistake."[176] Roy Wilkins argued that "you cannot serve the civil rights struggle at home by involving it in a struggle abroad." King had decided that the administration's determination to fight Communism abroad was squeezing out its interest in battling poverty at home, while the NAACP and the National Urban League continued to believe that the two wars could be conducted simultaneously.[177] During the early 1960s, King had commanded widespread, if often grudging, respect among the majority of Americans, white as well as black. But as he sailed into uncharted and unfriendly waters, responses to his voyage suggested the deepening split in American society. By 1968 younger black militants were dismissing him as an "Uncle Tom" while at the same time most whites feared him for his more militant and structural critique of American society and the war in Vietnam.

In the summer of 1967, black rage in Northern cities surged past dissent against the war as a challenge to the Johnson administration. Beginning on 12 July, Newark exploded in rioting, arson, and sniping that led to 26 deaths and over 1,500 injuries. Eleven days later Detroit erupted in the worst civil disorder of the century, leaving at least 40 dead and 1,300 buildings destroyed. Mayor Jerome Cavanaugh described the aftermath: "It looks like Berlin in 1945."[178] Touring the city at night for the Johnson administration, Justice Department official Roger Wilkins would not forget the impact of the military checkpoints: "There is no experience quite like being stopped at gunpoint by armed and uniformed men in the middle of an American city."[179] In the midst of increasing call-ups for the war in Vietnam, 4,000 paratroopers had to be diverted to Michigan to restore order in the city, the first such use of federal troops in a quarter of a century. Johnson was dismayed by the necessity of ordering mostly white National Guard and army units to occupy black ghettos, in part because he feared the increasingly frequent analogies between racial confrontations at home and abroad. He told Cyrus Vance, the top federal official in Detroit, "I'm concerned about the charge that we cannot kill enough people in Vietnam so we go out and shoot civilians in Detroit."[180]

Few Americans were pleased by the events in Newark and Detroit, but a small number of black radicals hoped they were witnessing the start of a revolution in America rather than merely ephemeral expressions of anger by the poor. H. "Rap" Brown, Carmichael's militant successor as SNCC chair, emphasized this distinction in a speech in Washington while Detroit was still burning: "Black folks try to loot when they should be shootin'. If you are gonna loot, loot you a gun store."[181] Brown called Johnson a "white honky cracker, an outlaw from Texas," and Carmichael, speaking at a press conference in Havana, where he was attending a conference of Latin American revolutionaries, urged African Americans to seek vengeance against Johnson, Rusk, and McNamara for their crimes in Vietnam: "we will kill first and we will aim for the head."[182]

The real violence in Newark and Detroit and the revolutionary rhetoric from some younger black leaders appalled white Americans as well as most African Americans. "Niggers with guns, niggers with guns" had been the muttered response of some white bystanders in Sacramento as thirty Black Panthers paraded through the California state capitol building with shotguns and rifles in May, and that thought was being murmured more widely by August.[183] When Carmichael announced on Havana Radio, on the second anniversary of the Watts riot, a call for African Americans to overthrow the "imperialist, capitalist and racialist structure of the United States," Senator Goldwater spoke for more than a few white Americans in urging Attorney General Ramsey Clark to arrest Carmichael and try him for treason, subjecting him to the death penalty.[184]

Johnson himself was pained and frustrated by the riots. He again sensed black ingratitude, for just a month earlier he had nominated the first African American ever for a seat on the Supreme Court, Thurgood Marshall. His leadership on behalf of racial equality seemed unappreciated either by younger, more militant black Americans or by white voters, who continued to shift toward the Republican Party. Public opinion polls after the riots showed sharp racial polarization. More than six times as many whites as blacks blamed the riots on outside agitators with Communist support. Conversely, two-thirds of blacks polled said police actions had contributed to the violence, while only one-sixth of whites agreed that police brutality even existed.[185] Congress reacted swiftly with harsh antiriot legislation and the defeat of such pending administration bills as the Rat Extermination Act for rodent eradication in urban slums, which opponents gleefully lampooned as a "civil rats bill." The riots led to a sharp drop in the president's public approval ratings, and Republicans emphasized how civil disorders had spread rapidly during his years in the White House.[186]

While opposition from white conservatives was intensifying on one side, liberal and black support for the administration was falling off on the other. Claiming that the riots were "caused by nice, gentle, timid white moderates who are more concerned about order than justice," King announced that he would go "all out" to prevent Johnson's reelection in 1968 and bring the war to an end. Senator Fulbright declared in the *New York Times* that the United States was fighting and losing "a two-front war" in Vietnam and in American cities. "The connection between Vietnam and Detroit," he argued, was in their conflicting demands on America for an "imperial role abroad" and "freedom and social justice at home."[187] Johnson created the National Advisory Commission on Civil Disorders (known as the Kerner Commission, after its chair, Illinois Governor Otto Kerner) to investigate the causes of the disorders, but he suspected that they had been instigated by a well-planned conspiracy of "Communist elements" and black nationalist agitators. He was unconvinced by the failure of the FBI and the intelligence services to find evidence of foreign funding of either the riots or black radical groups.[188]

Escalations of racial conflict abroad in the late summer of 1967 matched those at home. Despite some apparent victories on the ground in Vietnam, Johnson agreed in July to a significant escalation in the number of U.S. troops there. In southern Africa in August, African guerrillas crossed back into Rhodesia from neighboring Zambia, and suddenly white Rhodesian troops encountered their first pitched battles and suffered their first significant casualties. The war the UDI had invited had begun. The regional nature of the racial conflict also became evident that same month, as black insurgents of the South African ANC began working openly in Lusaka with their counterparts from Rhodesia. The Pretoria and Salisbury regimes responded in kind. South Africa declared in a "Monroe Doctrine for southern Africa" that its security frontier would now be the Zambezi River (the northern border of Rhodesia) rather than the Limpopo River (the northern border of South Africa).[189] The CIA's assurances to Johnson that Chinese and Soviet influence in the region was minimal and that the odds of a Communist government coming to power were "practically nil" did not alter the fundamental reality of a nascent race war developing on the southern end of the African continent.[190]

Toward Race War Abroad and Polarization at Home

In Lyndon Johnson's last year in the White House America descended to its worst levels of racial violence and conflict of the entire post–World War II era. From the My Lai massacre to the assassination of Martin Luther King, Jr., to the election of Richard Nixon, 1968 marked a grim season along the color line. On 8 February police gunfire killed three black students and wounded thirty more from South Carolina State College during a protest rally at an all-white bowling alley in Orangeburg. Such ready violence when Southern white officers found people of color in their cross hairs coincided with furious fighting in Vietnam, where the Tet Offensive had been launched nine days earlier by the North Vietnamese and the National Liberation Front.

The American forces in Southeast Asia were not all white, of course. In the Vietnam war for the first time the U.S. military was fully racially integrated in combat. Not surprisingly, racial tensions were common in rear areas, but more striking in some ways were the innumerable interracial bonds developed under fire on the front lines. *Time* observed accurately in 1967 that "black-white relations in a slit trench or a combat-bound Huey are years ahead of Denver and Darien, decades ahead of Birmingham and Biloxi."[191] Even some members of the Ku Klux Klan had such transformative experiences as being rescued and befriended by black comrades.[192] Whites accompanying fellow black GIs into off-base restaurants and bars at home in the South learned from a different perspective the meaning of "Hey! We don't serve niggers here!"[193] And African American soldiers likewise learned, most for the first time ever, to deal with whites as friends and peers.[194] The capacity of joint combat experience to erase color prejudice and even color consciousness emerged vividly in one black private's explanation to an African American reporter of why black soldiers tended to segregate themselves. Sitting next to a white friend with a pronounced Southern accent in a Saigon bar, he concluded: "White people are dull. They have no style and they don't know how to relax." When his friend objected, he cut him off: "Shut up. I'm not talking about you, nigger. I'm talking about white people."[195]

As the war ground on, however, and as draftees increasingly replaced career soldiers and overall morale declined, racial tensions within the American forces escalated. Blacks were at first disproportionately represented in combat units and thus in casualty figures, a result of lower socioeconomic status, fewer deferments or alternate service options, and higher reenlistment rates.[196] They resented the easier life available in the whiter rear areas like the huge base at Cam Ranh Bay, away from the front lines. To some it seemed there were really two different wars, one a multiracial affair in the combat zones and the other an almost all-white, country club scene in the support areas.[197] Johnson's Committee on Equal Opportunity in the Armed Forces had concluded in 1964 that one of the most powerful influences of white American troops on foreign countries was their racial prejudice, which forced local proprietors of businesses near American bases to adopt segregationist practices they had never used before: "The nationals of these countries themselves do not practice discrimination or segregation. At points distant from the bases Negro and white servicemen are treated alike. The local proprietors are mainly led to follow a pattern of segregation near our bases by what they understand to be the attitudes of our own service personnel." It seemed that white American soldiers often had to *teach* other peoples how to be sufficiently prejudiced. Racial discrimination, in other words, was apparently a standard but unfortunate American export.[198] In Vietnam, racial fights grew more common, especially in bars, over issues like interracial couples (still a troubling sight for many white American GIs) and musical choices (soul versus country-western). Japanese and other Asians hosting American troops on leave from Vietnam also witnessed the rising frequency of conflicts between black and white GIs.[199]

When not buffered by the distraction of incoming fire, race relations among American troops in Vietnam followed the same general pattern as those at home in the United States. As black power, urban riots, and the white backlash ascended, civility declined. Confederate flags, racist graffiti, and Ku Klux Klan organizing grew more common in U.S. barracks by 1968, while younger black draftees proved much less willing to accept such provocations quietly. Incidents of fragging, especially of white officers by black soldiers, increased sharply. King's assassination in April 1968 greatly exacerbated an already deteriorating situation, with hostile attitudes hardening on both sides of the color line. Even experienced African American officers with a professional commitment to the military were disturbed by the rise in open white racism among the ranks. Eddie Kitchen, a thirteen-year veteran, wrote to his mother in 1968 about the prevalence of Confederate flags around him: "We are fighting and dying in a war that is not very popular in the first place and we still have some stupid people who are still fighting the Civil War." Americans' efforts to change the world outside their borders were stumbling over their inability to bridge their own nation's deepest divide.[200]

Racial tensions and hatreds among Americans in Southeast Asia paled, however, in comparison to the actions of U.S. forces, both white and black, against the people of Vietnam. Massive, continual bombing and defoliation of the country made homelessness and death the largest products of the United States' efforts in Vietnam. Air Force General Curtis LeMay told Bundy and Johnson bluntly, "we should bomb them into the Stone Age."[201] The White House tended to conflate Vietnamese and Chinese Communists. When the president argued that the real enemy in Southeast Asia was "the deepening shadow of Communist China," LeMay urged that he go to the heart of the matter and solve the problem by "nuking the Chinks."[202] The war was supposed to be about ideology, about defending democracy or at least capitalism against Communism. But the difficulty of estimating potential enemies' ideology simply from their appearance was further complicated for American GIs by their tendency to see all Vietnamese as looking alike.

The result was what psychiatrist Robert Lifton has called "an atrocity-producing situation," specifically, "a counterinsurgency war undertaken by an advanced industrial society against a revolutionary movement of an underdeveloped country, in which the revolutionary guerrillas are inseparable from the rest of the population."[203] Defense Secretary Robert McNamara described the problem to Johnson in May 1967: "The picture of the world's greatest superpower killing or seriously injuring 1,000 noncombatants a week, while trying to pound a tiny backward nation into submission on an issue whose merits are hotly disputed, is not a pretty one."[204] But it was the strategy of attrition and the policy of the "body count" that most encouraged

American soldiers to kill Vietnamese regardless of their political beliefs. "[I don't] want anymore of your _____ prisoners. I want a body count," one colonel told soldier Robert Kruch in mid-1969. Whose bodies? "We were in a free fire zone," Kruch remembered being instructed, "and anybody we saw that was over 12 years of age that we thought was a male, was to be considered the enemy and engaged as such." Novelist and veteran Philip Caputo explained the impact of this policy: "Our mission was not to win terrain or seize positions, but simply to kill: to kill Communists [ideally, but] . . . 'If it's dead and Vietnamese, it's VC,' was a rule of thumb in the bush."[205]

Washington's preparation of young American men for this mission began in basic training, where Vietnamese were referred to as "gooks" or "dinks" and where women and children were understood to be potentially as dangerous as men in a guerrilla war.[206] Dehumanizing one's enemies in war is not unusual, but American troops in Vietnam dehumanized their allies as well. "Gooks" were not just Communists but all Vietnamese, and indeed all Asians. The other common meaning of "gook" as liquid slime suggests the esteem in which most Americans held Vietnamese. Images of Vietnamese as children, homosexuals, Indians, deer, or gophers abounded. GIs typically encountered the people of Vietnam as often resentful servants: cooks, launderers, bartenders, and prostitutes. The only Vietnamese most American soldiers did not see as weak and pitiful were, ironically, the soldiers of the National Liberation Front and the North Vietnamese army. The generally poor fighting record of the South Vietnamese army increased GIs' resentment, for they seemed to be fighting and dying for a people who would not fight for themselves. Hateful anti-Asian epithets tumbled readily from many American mouths, underlining the growing sense that the entire Vietnamese population rather than the National Liberation Foundation or the North Vietnamese was the enemy. The T-shirt favored by some American soldiers made this clear: "KILL THEM ALL! LET GOD SORT THEM OUT!"[207]

The Johnson administration's evident lack of respect for the South Vietnamese government helped contribute to the broad American confusion about whether Vietnamese Communists or perhaps all Vietnamese were the problem. Johnson, for example, visited American bases in Vietnam while president but did not bother to consult with his Saigon allies while in their country. By contrast, it is impossible to imagine Franklin Roosevelt avoiding Winston Churchill were he to have visited U.S. bases in England in 1944. Comments to Johnson like that of the racially liberal McGeorge Bundy about South Vietnamese head of state Nguyen Khanh as "very Asian and devious politically" reveal Washington officials' often unconscious disdain for those they were supposed to be fighting for.[208] Even opponents of the war sometimes contributed to the dehumanization of the people of Vietnam. Walter Lippmann, for example, explained in a September 1965 column that they "do not value their material possessions, which are few, nor even their lives, which are short and unhappy, as do the people of a country who have much to lose and much to live for."[209]

The administration's lack of respect for the Vietnamese, both allies and enemies, helps explain the ready use of the "quagmire" metaphor by commentators in the United States. Reimagining American forces as bogged down in some form of quicksand allowed the land of Vietnam itself to be seen as the treacherous aggressor and the Americans there as its victims. In this construction, not only the real American role in the war but also the actual Vietnamese people disappeared from the story. There was a parallel here to Africa, where the indigenous people were equally obscure to most Americans, and the landscape stood out as dark and perilous jungle—or, in the words of Undersecretary of State George Ball, "the African swamp" that the United States "should not get bogged down in." The invisibility of the enemy in Vietnam to most Americans was noteworthy: no television story during the war, for example, examined in depth the history, politics, or program of the North Vietnamese.[210]

American soldiers of color faced a painful dilemma about how to understand their role in a war that appeared increasingly racial in its essential conflict. Many, if not most, black, Latino, and Native American GIs used the same dehumanizing language as their white comrades about the Vietnamese. In fact, even some English-speaking Korean soldiers fighting on the U.S. side could be heard referring to Vietnamese as "gooks." Traditional American patriotism and the traumatic and unifying experience of being under fire from National Liberation Front and North Vietnamese forces help explain this apparent oddity.[211] These nonwhite GIs resisted National Liberation Front efforts to demoralize them through racial appeals, such as signs left in the jungle reading in precise English: "They call us gooks here and they call you niggers over there. You're the same as us. Get out, it's not your fight."[212] Similarly, some Vietnamese would point to American Indian soldiers' skin color and hair and then to their own, pronouncing them "same."[213]

Despite their patriotism and identification as Americans, many nonwhite GIs—like many white GIs—began to suspect that this really was not their war. Some even started to wonder if they might have as much in common with the "enemy" as with their white colleagues. They feared being used as pawns in an effort by their superiors to demonstrate that the war was not simply a conflict between white and nonwhite. Like black radicals back home, they began to speak of the Vietnamese as their "brown brothers."[214] Interviews indicated that black veterans, on average, held more positive views of the people of Vietnam than did white veterans.[215] This deepened the shame that some of them felt about their actions in the war. "We did to yellow people what whites do to us," one African American soldier explained in 1968.[216] A Native American veteran recalled resettling Vietnamese civilians as "just like what the whites did to us. I helped load up ville after ville and pack it off to the resettlement area. Just like when they moved us to the rez [reservation]. We shouldn't have done that."[217] Black soldiers were encouraged in their growing doubts about the war by rising opposition to the conflict among prominent African Americans at home, including such figures as boxer Muhammad Ali, comedian Dick Gregory, guitarist Jimi Hendrix, and singer Diana Ross.[218] Noting frequently that "no Vietnamese ever called me nigger," some even wondered if they should be fighting instead in southern Africa against "the white colonialist honkie."[219]

By 1968 the combination of the intensity of the fighting and the anti-Vietnamese fury of American commanders and soldiers led some U.S. forces down the dark path to full-blown racial savagery. On 16 March the 105 young American men of Charlie Company, burning with frustration and rage at the recent loss of several comrades, destroyed the entire village of My Lai, raping, torturing, and slaughtering some 500 unarmed women, children, and old men. Loose in what they repeatedly called "Indian country," American soldiers carried on as their predecessors had done on earlier frontiers in places like Sand Creek and Wounded Knee, as well as the Philippines at the turn of the century. One member of Charlie Company later recalled discussions just before the massacre of "wiping the whole place out" and "the Indian idea . . . the only good gook is a dead gook."[220] That nearly half of Charlie Company was black and its commander, Ernest Medina, was Mexican American did not lessen the murderous rage that led to the obliteration of the village.[221]

The tragedy at My Lai suggested that the Johnson administration had lost control of the racially focused furies of its soldiers abroad. The final report of the Kerner Commission published the same month brought more bad news for the White House. Concluding that "our nation is moving toward two societies, one black, one white—separate and unequal," the commission—made up of moderate national leaders chosen by Johnson—blamed white racism for "the explosive mixture which has been accumulating in our cities" and called for massive new federal spending on the nation's cities to avoid worse disorders in the future. The report implied that the Johnson administration's five years of work on race and poverty had fallen far short of solving America's urban problems, and it gave the president no explicit credit for his considerable efforts. Offended by what he saw as black and liberal ingratitude and not eager to provoke disgruntled whites further, Johnson coldly refused to receive the report in person and ignored most of its recommendations. In a nation of deepening divisions, the president was increasingly isolated.[222] Deeply disappointed by developments in America and Vietnam since the height of his popularity in early 1965, and facing serious opposition within the Democratic Party as manifested in antiwar candidate Eugene McCarthy's surprising strength in the New Hampshire primary on 12 March, Johnson decided he had had enough. On 31 March he announced to a stunned nation that he would not be a candidate for reelection in the fall.

The remainder of 1968 brought mostly bad news of a nation deeply divided by race and the war in Southeast Asia. King's assassination unleashed the greatest wave of urban violence of the decade. Looting, sniping, and arson in over a hundred American cities, combined with police gunfire, left 46 dead, over 3,000 wounded, 27,000 arrested, and $45 million in property destruction. More than 700 fires burned in Washington, D.C., alone, as army units in full combat gear took up battle positions around the Capitol and the White House. Leading antiwar presidential candidate Robert Kennedy was murdered in early June. The Poor People's Campaign planned by King before his death continued under the leadership of Ralph Abernathy, bringing several thousand impoverished Americans to camp on the grounds of the nation's capital in May and June to publicize their condition. At the climactic rally of 50,000 people on 19 June, Abernathy condemned the administration's record on race and poverty as a five-year string of "broken promises." Johnson, who had just pushed through national legislation banning racial discrimination in housing, watched bitterly as his support continued to fall away.[223] In August massive demonstrations and a police riot marred the Democratic Party's national convention in Chicago.

On 29 August a race riot at the Long Binh jail (nicknamed LBJ) in South Vietnam marked the worst prison riot in the modern history of the U.S. Army. Severe overcrowding, radically disproportional numbers of black inmates, and blatant racial discrimination by white military police led to a takeover of the stockade by 217 African American and 3 Puerto Rican inmates, resulting in 1 death, 34 injuries, and a month-long stalemate. The holdouts, as *Newsweek* described the scene, "proceeded to shed their uniforms and to don white kerchiefs and African-style robes that they made out of Army blankets," while "beating out jungle sounds on oil drums." Here in the jungles of Vietnam, Johnson's hopes for a racially integrated and progressive America seemed to have reached rock bottom, as black American soldiers rejected the white authority and culture they found so racist in favor of a separatist cultural style looking toward Africa.[224]

The road to Richard Nixon's election was open. The Republican candidate chose Maryland Governor Spiro Agnew for his vice presidential running mate as part of the "Southern strategy" of appealing to whites of that region. By October Agnew was declaring on the campaign trail that "the civil rights movement went too far the day it began."[225] The white backlash, already manifest in the independent candidacy of George Wallace, had found another voice. In the wake of King's death and the rise of black power, the civil rights leadership was divided and many of its constituents deeply alienated. Burke Marshall of the Justice Department noted that loss of unity in the fall of 1968, observing that the black community "wouldn't pay any more attention to Roy Wilkins than to George Wallace, maybe less."[226] In a contest between Nixon and Hubert Humphrey, Johnson's vice president and for two decades one of the strongest supporters of civil rights in the Democratic Party, Stokely Carmichael and Rap Brown suggested African Americans "go fishin'" on election day rather than vote for "either of the honkies."[227] Most black voters rejected that advice and supported Humphrey, but the halcyon days of liberal victory in 1964 seemed a distant memory as the Republicans regained the White House.

FOR ALL HIS IMPRESSIVE legislative achievements in the pursuit of ending racial discrimination, Lyndon Johnson could not hold back the tide of racial polarization that swept through the United States in the 1960s. He had "made the goddamnest commitment to the civil-rights cause I had ever heard" from a president, Roger Wilkins remembered, but Johnson was put on the defensive by the rise of black power, the onset of urban racial violence, and the growth of the white backlash.[228] Most African Americans still quietly supported the president, but younger radicals increasingly seized the spotlight. Already alienated from the Johnson administration by frustrating experiences in domestic civil rights organizing, they now linked their cause directly to that of revolutionaries abroad, from Vietnam to South Africa. One of the best remembered examples was the black power salute of champion sprinters Tommie Smith and John Carlos on the Olympic victory stand in 1968 in Mexico City, in protest of both racism in America and U.S. ties to South Africa and Rhodesia.[229] These younger radicals, often unknowingly, were reviving a critique of America's role in a white-dominated international system that had been articulated a generation earlier by African American leftists like W. E. B. Du Bois and Paul Robeson.

Like black radicals in the United States, nonwhite revolutionaries abroad such as those in the National Liberation Front and the African National Congress dismissed the paternalistic liberalism and the preference for gradual reform of the Johnson administration and much of the white West. Ho Chi Minh, like Martin Luther King, Jr., or Stokely Carmichael, remained well beyond the control of the American president. The freedom struggle in southern Africa also carried on beyond the circle of White House power, owing to the radical inappropriateness of gradual reform in the lands of apartheid and the UDI—the lands of the ultimate white backlash. The efforts of U.S. policymakers to contain the virus of domestic segregationism were more effective, although they had the ironic effect of helping to create a new conservative Republican majority in American politics. In southern Africa, the white backlash seemed to be consolidating its power, at least for the near future. Both abroad and at home, the Johnson administration's struggle to restrain strife along the color line and to channel the political aspirations of people of color in liberal, pro-Western directions broke apart on the shoals of racial polarization.

6

THE END OF THE COLD WAR AND
WHITE SUPREMACY

LYNDON JOHNSON'S PRESIDENCY marked the climax of the close, complicated relationship between changing race relations and American foreign policy during the Cold War. The convergence of civil rights reform, racial polarization, and an anti-Communist but increasingly racially charged war in Southeast Asia made the Johnson years unparalleled: so much hope, so much achievement, and so much disappointment. After 1968 the story of how these two themes intersected retreated from center stage in American politics.[1] Racial reform in the United States shifted from the moral high ground of desegregating the public sphere to the more ambiguous and less visible terrain of eliminating discrimination in hiring and housing. Violent racial conflict did not disappear after 1968, but it never again reemerged in the sustained form of the summer uprisings of the Johnson years. And the Cold War itself rapidly grew less intense after the election of Richard Nixon, with the new policy of détente pointing the way toward the eventual deescalation of the Soviet-American conflict by the late 1980s and the ultimate dissolution of the Soviet Union in 1991. The twin erosion of American segregation and international colonialism (with the exception of southernmost Africa) by 1968 relieved the U.S. government of much of its self-inflicted burden of trying to promote a "free world" abroad while maintaining racial hierarchy at home.

Richard Nixon and the Politics of Race at Home

At the Gridiron Club dinner of 14 March 1970 Washington's policymaking elite gathered for the annual bipartisan event. Hosted by prominent journalists, the occasion both honored and roasted the president. Democratic Senator Edmund Muskie provoked much laughter with a speech noting that the Republican Party's three problems were the war in Vietnam, inflation, and what to say on Lincoln's birthday. Indeed, the party of the Great Emancipator had a century later become the party of the white backlash against civil rights. Richard Nixon had boldly gone where no previous successful Republican presidential candidate had dared—explicitly wooing white Southern voters by playing to their antiblack sentiments—and he had won the White House. Just a few months later, Neil Armstrong and Edwin Aldrin fulfilled the age-old dream of reaching the moon and actually walking upon its surface, planting an American flag in the process. Even as American technological prowess seemed to portend an unlimited future, American racial politics appeared to be regressing into a too familiar past. The author of this Southern strategy surely did not anticipate Frank Wills, the African American watchman at the Watergate complex who would notice a suspiciously unlocked door one night in June 1972 and notify the police. But Nixon had his eyes wide open as he halted the momentum that the civil rights movement had acquired within the federal government since 1960, and ended Johnson's use of the symbolic and substantive power of the presidency to disavow racism as an acceptable part of American life.[2]

Ironically, Nixon's record before the 1968 campaign placed him among his party's most prominent supporters of greater racial equality. There is little evidence of notable prejudice in his upbringing in small-town southern California—for his family a difficult period of economic stress and the tragic early deaths of two of his brothers—or in his early political career. As vice president in the 1950s, he supported civil rights legislation and limiting the Senate filibuster used to defeat it. He traveled to West Africa in 1957 to represent the Eisenhower administration at Ghana's independence ceremony and came home to promote greater U.S. attention to that continent. Such positions earned him considerable black political support, including that of Martin Luther King, Jr., at the beginning of his unsuccessful 1960 bid for the presidency. He was even granted an honorary membership in a California chapter of the NAACP. If one can be known by one's enemies, it is worth noting that Nixon was regarded by Southern segregationists as the second worst figure in the Eisenhower administration, after Attorney General Herbert Brownell. Throughout the 1950s, he surpassed not only Eisenhower but also Kennedy and Johnson as a supporter of civil rights. In the 1960s, however, he moved in the opposite direction of his Democratic predecessors in the White House: while they emerged as stronger defenders of racial equality, Nixon tapped into the widening stream of white racial resentment.[3]

Nixon's electoral and governing strategies were rooted in a realistic appraisal of rising white resentment of African Americans. Traditional prejudice was exacerbated by several factors in the late 1960s: the summertime riots coming so swiftly on the heels of civil rights legislation, suggesting to many observers ingratitude; the antiwhite rhetoric of some black nationalists; the beginnings of school desegregation and the use of busing to achieve it; and growing anxiety about unemployment and inflation.[4] Anger about the failure of the American effort in Vietnam, and at nonwhite peoples more generally, also alienated many whites from an understanding of black political struggles. The leading newspaper in Greensboro, North Carolina, responded to a 1969 civil rights protest by Dudley High School students with the logic of U.S. officials blaming National Liberation Front successes on North Vietnam and China: the protesters, it editorialized, must have been "outsiders" engaged in "guerilla warfare" to manipulate a "terrorized majority."[5] In his inaugural address, Nixon spoke of overcoming these divisions, of bringing "black and white together, as one nation, not two."[6] But he understood better than most how resentment can serve as the engine of American politics. He positioned himself as the champion of aggrieved whites against what they saw as the special pleadings and privileges of African Americans.[7] By 1971, Spiro Agnew could declare publicly that "dividing the American people has been my main contribution to the national political scene since assuming the office of vice president . . . I not only plead guilty to this charge, but I am somewhat flattered by it." Some of that dividing was between young and old, left and right, protesters and war supporters. But a great deal of it was between black and white.[8]

The massive riots that had rolled through the nation's cities after King's assassination in April had set the tone for the 1968 presidential campaign. Riots dovetailed with increasing crime rates and antiwar protests to stimulate new fears of social instability among whites. Nixon had seized on these anxieties by running on a platform of "law and order." He generally avoided making explicit connections between race and social disorder, as white voters already associated the two from the urban uprisings of previous summers. But Nixon occasionally let down his guard, allowing code words to be replaced by plainer language. After taping a television advertisement attacking the decline of law and order, he said enthusiastically, "It's all about law and order and the damn Negro–Puerto Rican groups out there."[9]

Nixon won all but five Southern states to put him over the top in November's election. His key ally was Senator Strom Thurmond of South Carolina, the former 1948 Dixiecrat presidential candidate who had switched to the Republican Party in 1964. No one had stronger segregationist credentials, and no one could better convince white Southerners of Richard Nixon's commitment to their interests—especially slowing the process of school desegregation. It would have been hard to create a clearer contrast between Nixon's alliance with Thurmond and the Democratic candidate, Vice President Hubert Humphrey: Humphrey had led the successful civil rights campaign at the 1948 Democratic convention that had driven Thurmond and so many of his Southern colleagues out in rebellion. Nixon's ardent wooing of white Southern voters was a typical case of his savvy political calculations: "hunting where the ducks are," in Goldwater's phrase. There were at least twice as many whites as blacks in the South as a whole, and the white backlash was not limited to the region south of the Mason-Dixon line. "Remember, this isn't South Africa," one Republican strategist had remarked a few years earlier. "The white man outnumbers the Negro 9 to 1 in this country." No one understood that better than the Republican candidate.[10]

More dangerous to Nixon's Southern strategy than Humphrey was the independent candidacy of segregationist George Wallace. The former Alabama governor appealed explicitly to racial fears in showing how powerful a force white resentment could be. His running mate, retired Air Force General Curtis LeMay, was perhaps best known for his enthusiasm about the use of nuclear weapons. Wallace and his staff had, in fact, coached him extensively to stay away from the subject with the media during the campaign. But at the October press conference in Pittsburgh announcing LeMay's place on the ticket, the general could not resist a journalist's question about whether to use nuclear weapons in Vietnam: "I have to say that we have a phobia about nuclear weapons," he responded, and continued on at length, to the dismay of Wallace's staff.[11] Seizing victory abroad at whatever cost thus went with reasserting white authority at home. "On November 5, they're going to find out there are a lot of rednecks in this country," Wallace predicted: 9.9 million of them, in fact, constituting 13.5 percent of the total votes cast, and even 8 percent of votes outside the South.[12]

Nixon's choice of Agnew as his running mate marked the first payoff to the white South for its support of him in the primaries. The Maryland governor was best known by the late summer of 1968 for his tough response to the five-day Baltimore riot after King's death. At a well-publicized meeting with one hundred of the city's moderate African American leaders, most of whom had worked days and nights to calm angry black residents, Agnew stunned his listeners by denouncing them for not having condemned ahead of time young militants like Stokely Carmichael and Rap Brown—"the circuit-riding, Hanoi-visiting, caterwauling, riot-inciting, burn-America-down type of leader." Eighty of them got up and walked out of the meeting.[13] Together, Nixon and Agnew vied with Wallace during the fall to portray the Democrats as the party of failed liberal permissiveness—of welfare, crime, anti-American protesters, and favoring of blacks over whites. Just three years from the apex of Great Society legislative success, the Democrats now struggled against an image of not defending adequately the interests of most white Americans at home, while also failing to win the war in Vietnam and thus uphold the nation's interests in the Third World.[14] Nixon had replicated the Republicans' agreement of 1877 some ninety-one years later: winning the presidency by agreeing to end the "second Reconstruction" of civil rights, just as his predecessors had retained the White House by sloughing off their remaining commitment to equality and walking away from the first Reconstruction.[15]

As president, Nixon participated in the casual racism common among white Americans of his generation. He lacked the obsession with African Americans and their sexuality that characterized true race-baiters, but he did not spare the Oval Office references to "niggers," "jigaboos," "jigs," and "jungle bunnies."[16] News of a scholarship program for black students elicited this reply: "Well, it's a good thing. They're just down out of the trees."[17] That he had offered the same observation in an NSC meeting on Africa in the late 1950s suggested how little he had apparently learned from the ferment in race relations during the intervening decade. The president's instruction to White House chief of staff H. R. Haldeman that "we take a *very* conservative civil rights line" was grounded in what historian Hugh Davis Graham has called Nixon's unambiguously negative opinion of black Americans.[18] He informed aide John Ehrlichman more than once that blacks were genetically inferior to whites, while complaining several times that there were too many African American waiters in the White House.[19] His ease with the notion of white biological superiority was no doubt bolstered by similar theories from a handful of academics like Arthur Jensen of the University of California, reported to him by aide Patrick Moynihan as part of a "rather pronounced revival—in impeccably respectable circles—of the proposition that there is a difference in genetic potential" between the races.[20] The president's disdain for African Americans was no secret to them and was frequently reciprocated: Clifford Alexander noted in 1971, for example, that "blacks feel that it is hopeless to deal with Nixon. And they're right, because the man's a bigot."[21]

The significance of Nixon's attitudes toward people of darker hue can be clarified by noting his casual anti-Semitism. Like his views of blacks, the president's prejudices against Jews were more thoughtless than virulent. Indeed, several of his advisers were Jewish, including speechwriter William Safire and Henry Kissinger, who served both as national security adviser and as secretary of state. But the president shared the negative stereotypes of most gentiles of his generation, tending to see American Jews as a powerful, cohesive group committed to Israel's interests above all else. He enjoyed needling Kissinger to accentuate his vulnerability in this regard; after Kissinger had given an opinion on the Middle East in one Cabinet meeting, Nixon looked around the room and asked, "Now, can we get an American point of view?"[22] Nixon's attention to power above all else helps explain his greater respect for Jews than for blacks: while Jews had real economic and political clout, blacks generally did not. For Nixon the latter were weak and unorganized, able to get attention only by rioting.

For Nixon, African Americans were one of an array of domestic issues that he found ultimately boring and insignificant in comparison to the magisterial business of conducting international diplomacy. He often expressed the view that the Congress could run the country on the domestic side, while the president's job was to take charge of foreign affairs.[23] Like most white Americans, he associated blacks with bothersome domestic issues like welfare. At one point in the 1960 presidential campaign, he conceded to his staff that he would have to do a speech on "all that welfare crap." When aide Herb Klein gently suggested referring to the issue instead as "meeting human needs," Nixon replied, "I don't care. It *is* crap."[24] His interest in affairs within the United States picked up when they were likely to help his electoral chances; it was the politics rather than the substance of domestic policies that intrigued him. From 1968 on, black Americans were not part of his electoral strategy and therefore had little influence within his administration.[25]

Taking their lead from the president, Nixon's advisers demonstrated little concern for the interests of African Americans. Secretary of State William Rogers's support for civil rights since his days in the Justice Department under Eisenhower made him an exception, but one with little real power within the administration. Vice President Spiro Agnew's growing hostility to black militants earned him great popularity in the white South. Roger Morris covered African issues for the NSC staff and remembered ubiquitous racist humor about apes and smells; as Morris entered meetings, General Alexander Haig "would begin to beat his hands on the table, as if he were pounding a tom-tom." En route to a White House dinner for African ambassadors, Kissinger bumped into Arkansas senator J. William Fulbright, headed to the same event, and reportedly offered this greeting: "I wonder what the dining room is going to smell like?"[26] J. Edgar Hoover found fresh encouragement from Nixon's presence in the White House for his long-standing campaign against black militancy and black political organizations more broadly. The FBI director took the Black Panther Party at its word about planning "an armed black revolution" and viewed the Panthers as "the greatest threat to the internal security of this country." He sharply increased FBI surveillance of black communities nationwide during the Nixon years. Historian Kenneth O'Reilly has observed that Hoover had more in common with this president than with any of his six predecessors, under whom the bureau director also served.[27]

The racial inclinations of Nixon and the men around him matched those of many white Americans, but they did not create the racially polarized atmosphere of 1969. Four years of war in Vietnam and uprisings in black urban neighborhoods, plus the limits of civil rights gains and rising anger on both sides of the color line, had done that. Nixon sought advantage from a polarization between black power and the white backlash that he had inherited rather than manufactured. The United States' continuing destruction of much of southern Vietnam alienated more and more Americans, while younger African American militants increasingly identified with nonwhite revolutionaries in the Third World—including designated enemies of the U.S. government like Castro and Ho. Black nationalists began to call for Palestinian self-determination, a cause that a non-European majority supported in 1975 with a UN General Assembly resolution declaring Zionism "a form of racism and racial discrimination," even as white Americans were shedding some of their anti-Semitism and identifying with Israel as a bastion of "Western civilization" within the non-European world.[28] Moynihan warned Nixon not to underestimate the determination of African Americans to establish what he called their own "self-identity," the same issue motivating so many revolutionaries in the Third World: "Consider what those half starved, malaria ridden, pajama clad Viet Cong have done to the armies of the American Republic."[29]

Regardless of the accuracy of Moynihan's description of the National Liberation Front forces, there was no doubt by 1969 that those armies of the American Republic shared in the broader racial polarization of American society. This stemmed partly from the broader loss of morale among U.S. troops still involved in a losing struggle. Drug abuse began to reach epidemic proportions, and fragging escalated sharply. Racial conflict among American soldiers tended to disappear under fire, where the imminent threat of death and dismemberment bonded them tightly together. But away from the front, commanders could not prevent the suspicions and hostilities that pervaded American society at home from seeping into military life as well. Beyond dealing with discrimination from many of their white colleagues, African American GIs particularly resented the absence of blacks from positions of authority. In 1971, while blacks made up 11 percent of the American population and roughly 20 percent of military personnel, they accounted for only 2.3 percent of officers. A sometimes explosive tension developed between younger black soldiers, with a strong race consciousness and no longer willing to accept discrimination, and some white soldiers resentful of increasing demands for black recognition and parity. At Camp Lejeune in North Carolina, for example, a series of racial attacks and muggings in the spring and summer of 1969 culminated in a deadly brawl between black and white marines on 20 July.[30]

These conflagrations along the color line within the American forces were exceeded only by the racially coded violence of Americans against their perceived Vietnamese enemies, as had already become evident during Johnson's presidency. The slaughter of Vietnamese civilians forced its way into the consciousness of Americans at home in the fall of 1969. The initial cover-up of the destruction of the village of My Lai a year and a half earlier at last leaked to the press, and military authorities decided Lieutenant William J. ("Rusty") Calley was to blame. The very language of the indictment revealed the racial thinking of U.S. commanders: Calley was charged with the premeditated murder of at least seventy "Oriental human beings," a phrase eventually dropped by the prosecutors after repeated criticism but still used by the judge in his final instructions to the jury a year and a half later.[31] Calley was the only person found guilty of some of the hundreds of deaths at My Lai, as U.S. officials emphasized what they called the aberration of American behavior there. Some American soldiers took a different view. One of the hundred honorably discharged Vietnam Veterans against the War noted simply, "We all belong to the unit Lieutenant Calley belonged to."[32] Another GI in the same division as Calley told a reporter in Vietnam, "The people back in the world don't understand this war. We are here to kill dinks. How can they convict Calley for killing dinks? That's our job."[33]

The public response in the United States to the news about My Lai included outrage at the atrocity but focused mostly on how the tragedy affected the American soldiers involved. The mother of one member of Calley's Charlie Company spoke for many: "I gave them a good boy, and they made him a murderer."[34] Calley, seen by many as a scapegoat for higher officials, received an outpouring of public support after his conviction and sentencing at Fort Benning, Georgia. Governor Jimmy Carter of Georgia organized an American Fighting Man's Day, calling on Georgians to drive with their headlights on to "honor the flag" in solidarity with Calley.[35] Nixon added his own peculiar ethnic slant to the mix of responses to My Lai. Unhappy with the media attention being paid to the massacre and with the growing public opposition to his Vietnam policies, Nixon privately blamed the editors of moderate to liberal newspapers like the *New York Times:* "It's those dirty rotten Jews from New York who are behind it."[36]

Race remained to the very end a powerful yet ambiguous category for American thinking and behavior in Vietnam. Growing numbers of Americans were neither white nor black, but of Asian descent, and some of them fought in the war. One Japanese American GI, Scott Shimabakuro, remembered how his sergeant had told him not to marry his Vietnamese girlfriend "because she was a gook, which struck me as kind of funny because I was a gook also."[37] There was increasing slippage between "Americans" and "gooks": the former mostly killed, but sometimes married and sometimes even *were,* the latter. The final hours of the evacuation of the last U.S. personnel from Saigon on 29 April 1975 nevertheless suggested the continuing racial classification of "Americans." After checking the U.S. embassy parking lot one last time, administrative counselor Henry Boudreau reported with relief, "Didn't see any white faces out there."[38]

As president, Nixon implemented policies on racial matters in line with his campaign promises. He had confided to a supporter before the election, "If I am president, I am not going to owe anything to the black community."[39] "You received (probably) the lowest proportion of Negro votes of any President in history," Moynihan reminded him in his first months in office.[40]

Sensing the likely course of the president's policies, Jackie Robinson—former Nixon supporter, former baseball superstar, and conservative African American Republican—wrote to him in January 1969, "I sincerely hope you will forget the fact that the Black community . . . hardly had a role to play in your election."[41] For a man as focused on the maintenance and exercise of political power as Richard Nixon, this was unlikely. The advice of White House aide Patrick Buchanan made more sense: "The second era of Reconstruction is over; the ship of integration is going down; it is not our ship . . . and we cannot salvage it; and we ought not to be aboard."[42] The Nixon administration instead spoke for the white South: it urged Congress to slow court-ordered school busing and to weaken fair-housing and voting rights enforcement; it nominated two Southerners who supported a strict constructionist position on the Constitution for the Supreme Court; it hampered the civil rights offices of the Justice Department and the Department of Health, Education, and Welfare; and it fired officials seen as too eager to implement integration guidelines. Complaints about responsible enforcement of existing integration directives elicited this presidential command to John Ehrlichman: "tell them to *knock off this crap.*"[43] Nixon found his occasional meetings with black leaders useless and tiresome; to aides after one he sent a memo declaring, *"No More of This!"*[44] With Agnew as his point man, Nixon's popularity in the South continued to rise. George Wallace told NBC's *Meet the Press* in the fall of 1969, "I wish I had copyrighted my speeches. I would be drawing immense royalties from Mr. Nixon and especially Mr. Agnew."[45]

Unlike previous presidents who had opposed racial reform, Nixon faced novel limits on his ability to roll back gains for black equality. His administration had the sworn duty of executing federal law, including the 1964–65 civil rights bills and the Supreme Court's decisions requiring specific school desegregation plans. His solution to this dilemma was a clear instruction to aides to "do what the law requires and not *one bit* more."[46] Even his occasional initiatives at promoting black capitalism revealed ulterior political motives. The Philadelphia Plan, originally a Johnson-era Labor Department proposal to require minority preferences in hiring by federal government contractors—an early version of affirmative action—also served to drive a wedge between key Democratic constituencies: organized labor and African Americans.[47]

Nixon used Supreme Court appointments and school busing as his highest-profile demonstrations of sympathy with white Southerners. After the

Senate rejected his first choice, Clement Haynsworth of South Carolina, in part owing to evidence of segregationist attitudes, the president instructed his aides to look "farther South and further right" for another candidate.[48] In Florida they found G. Harrold Carswell, whose undistinguished legal career was matched by his prominence as a promoter of white supremacy. The administration's own private disdain for Carswell's qualifications suggested that his nomination was a calculated insult to Nixon's political opponents. Nixon turned a second Senate rejection to immediate political advantage in Dixie, declaring after the Senate's negative vote on Carswell: "I understand the bitter feeling of millions of Americans who live in the South about the act of regional discrimination that took place in the Senate yesterday."[49] Nixon's years in the White House were a time of considerable school desegregation in the South, effected through federal court rulings. But the administration made sure the courts, not the president, would be identified with that process. Nixon consistently opposed busing as a means of integrating schools and even had the Justice Department side with the state of Mississippi in its unsuccessful efforts to delay submitting desegregation plans. Few doubted where the president's sympathies lay.[50]

Nixon's position on civil rights issues helped lay the groundwork for his impressive reelection in 1972. "Go for Poles, Italians, Irish . . . don't go for Jews and blacks," he told Haldeman.[51] Winning forty-nine states to crush liberal Democratic candidate George McGovern, the president continued the Southern strategy of four years earlier. For the first time ever, the Democrats—the historic party of the South—won not a single Southern state. Nixon's real competition there would have come from George Wallace, who won a string of primary victories across Dixie before being partially paralyzed in an assassination attempt. With Wallace not in the running this time as a third-party candidate, Nixon sailed to victory. The Committee to Reelect the President had little need to engage in such dirty tricks as the Watergate break-in or its covert effort to weaken the early leading Democratic candidate, Senator Edmund Muskie of Maine, in the critical, nearly all-white New Hampshire primary. The latter took the form of late-night calls to white voters, with the person on the line claiming to be from a "Harlem for Muskie Committee" and informing them that Muskie had promised to do his best to help African Americans if elected.[52] The extent of Nixon's connection to black America in 1972 seemed to be summed up at the Republican convention in Miami, where conservative black performer Sammy Davis, Jr., hugged the president on stage after singing the anthem of the white South, "Dixie."[53]

The Nixon Administration and Race Relations Abroad

The Nixon administration's foreign policy focused on power, not race. The president and Henry Kissinger prided themselves on their realpolitik perspective: discounting ideology and pursuing American interests abroad based on the actions rather than the rhetoric of other nations. Hemmed in by a disastrous commitment in Vietnam, declining relative economic power, and new Soviet nuclear parity, Nixon and Kissinger renegotiated relations with the two greatest enemies of the United States. Opening links with the People's Republic of China led swiftly to détente with the Soviet Union. This new respect shown to China by a president long known for his disdain for the Communist government in Beijing indicated the priority of power over ideology. China had become a nuclear power in 1964, and its economic and political potential was vast. For a practitioner of realpolitik like Nixon, old anti-Asian and anti-Communist sentiments were not as important.

China was an exception by dint of its size and power, real and potential. The Nixon administration had little interest in the rest of the nonwhite Third World. There an absence of power dovetailed with darker skins and different cultures to put off the men of Washington. Kissinger was sometimes quite blunt about this lack of interest. He told Foreign Minister Gabriel Valdés of Chile, "Nothing important can come from the South. History has never been produced in the South. The axis of history starts in Moscow, goes to Bonn, crosses over to Washington, and then goes to Tokyo." Valdés interjected, "Mr. Kissinger, you know nothing about the South." The national security adviser answered, "No, and I don't care."[54] From the White House's perspective, no continent wielded less military and economic power than Africa, and, of course, no continent was more clearly marked by skin color as nonwhite.

For Nixon, Africans in the international system were similar to African Americans at home: a minority with little power or influence. Haldeman recalled the president telling him that "there has never been an adequate black nation, and they are the only race of which this is true."[55] Early in his first term, Nixon instructed his staff to reduce the number of visiting African statesmen he had to host.[56] The president's lack of interest in Africans was echoed by Americans living in white-ruled southern Africa, where only one out of ten U.S. businessmen surveyed in 1969 objected to apartheid. "I didn't mix with them in the States," said one manager at a Ford plant, "and I don't mix with them here."[57] "Them" were, apparently, the same: black Americans and Africans. Several hundred white Americans fought as mercenaries with the white government against black nationalist guerillas in the civil war in Rhodesia. The Nixon and subsequent Gerald Ford administrations were untroubled by Rhodesian recruitment efforts using American agents, despite such events as Rhodesia's widely publicized slaughter of black civilians at the Nyadzonia refugee camp in August 1976. Historian William Minter explains the significance: "If a black group in the United States or Britain had been actively recruiting for an African government that had just massacred over five hundred white civilians, a similar lack of reaction would have been unimaginable."[58] Nixon summed up his perspective on the importance of Africans in a phone call he made to Kissinger, intended to soothe the latter's jealousy at positive press coverage Secretary of State Rogers had received for a recent trip to Africa: "Henry, let's leave the niggers to Bill and we'll take care of the rest of the world."[59]

Nixon and Kissinger presided over a shift in American policy toward southern Africa that aligned the United States more closely with its white rulers. Their focus on powerful nations and their lack of interest in weaker ones meant that Africa would always rank low in their priorities, but it also meant that South Africa—the only industrialized state on the continent with a modern military force—would be given special consideration. Their geopolitics and their racial prejudices fit neatly together. The Nixon administration could take this position for its first five years, in part because of relative stability in the region. Until the coup in Portugal in 1974 that led to independence for Angola and Mozambique, the white authorities seemed in control. African nationalists had finally taken up arms in each territory in the 1960s in response to the refusal of the white regimes to move toward democracy, but they could not yet effectively challenge white power. Similarly, American supporters of color-blind democracy in southern Africa were not yet organized and powerful in the way they would become by the mid-1980s. The Congressional Black Caucus was not formally founded until 1971 and focused primarily—like most American race reformers—on domestic policies. Private organizations like the American Committee on Africa remained small. Younger black nationalists paid increasing attention to Africa, but their opinions were hardly welcome in the White House. Not until the presidency of Jimmy Carter did the voices of supporters of African liberation become more influential.[60]

The Nixon administration's tilt toward whites in southern Africa also reflected ignorance of African affairs and politics at the highest levels of the

U.S. government. In a December 1969 NSC meeting, for example, Agnew spoke at length about "South African independence" in 1965 and its similarity to American independence in 1776, in regard to blacks not having the vote. Nixon finally had to interject, "You mean Rhodesia, don't you?" The vice president paused, checking his notes, and agreed that he did.[61] This shift in policy took institutional form in the National Security Study Memorandum 39 (NSSM 39) of 1969, which claimed famously that "the whites are here to stay and the only way that constructive change can come about is through them."[62] Instead, within five years the Portuguese were withdrawing from Angola and Mozambique, as the white-ruled redoubt at the tip of the continent began to crumble. NSC staffer Roger Morris was involved in assembling NSSM 39 and recalled later how the process of putting it together "unearthed in the bureaucracy an often appalling ignorance of the history and politics of southern Africa."[63]

This tilt toward the white regimes reflected the administration's strategy of supporting the anti-Communist, pro-capitalist status quo in the region, as well as the lobbying of key agencies, allies, and individuals. The CIA wished to preserve its close links to the regimes in Salisbury, Pretoria, and Lisbon; the Conservative Party came back into power in Britain in June 1970, with a more sympathetic view of whites in the region; and influential former secretary of state Dean Acheson pressured Nixon and Kissinger to support the white governments. Acheson warned against what he called—in peculiarly revealing language for an American statesman—the "subjugation of Rhodesia to majority rule." He argued that southern Africa was "the most stable portion of an otherwise disturbed, indeed turbulent, continent," and that only the Soviets would benefit from changes there.[64] Few in the American business community or the Commerce Department wished to disrupt the benefits of links with South Africa: its vast gold production, the favorable U.S. trade balance there, and what NSSM 39 called the "highly profitable return" on private U.S investments in the apartheid state.[65]

In Rhodesia, the Nixon administration did not take the lead in drawing closer to the white minority regime. Instead, it allowed Rhodesia's enthusiasts in the American South, the U.S. Congress, and the business community to shift the terms of the bilateral relationship. Nixon did reluctantly follow Britain's lead by closing the U.S. consulate in Salisbury in March 1970, after Rhodesia cut its last ties with London. This action disappointed those Americans—particularly white Southern segregationists—who approved of the Ian Smith regime, including more than one out of every six members of Congress. Most of them emphasized the common American and Rhodesian commitments to anti-Communism and free enterprise and the shared history of declaring independence from England. Some questioned Nixon's simultaneous efforts to open relations with what they considered a more clearly "illegal" regime in Beijing, and some demanded that Britain close its consulate in Hanoi. Others were more blunt about the racial contours of the problem. Newspaper publisher Charles Jacobs of Brookhaven, Mississippi, informed the president that white Rhodesians "have no intention of turning their country over to the semi-civilized black Africans," who would only return it to "the savage conditions" prevailing on the continent wherever whites were not in control. Jacobs went on to note with some accuracy the extensive parallels in the two nations' development. For Rhodesia to engage in color-blind democracy—that is, democracy—"would be very similar [to] if in a fit of moral righteousness we turned America back to the Indians." Other than the reversed demographics of the two situations, this was true—and precisely what such Native Americans as those involved in the American Indian Movement were beginning to advocate.[66]

Other Americans supported Nixon's decision to close the Salisbury consulate. They framed their arguments around the simple fact that Rhodesia was retreating into greater racial discrimination and hierarchy at precisely the same time that the United States and the rest of the world were moving in the opposite direction. The State Department emphasized the global unanimity on this issue: not even Portugal or South Africa had granted diplomatic recognition to the Smith regime.[67] The Congressional Black Caucus urged the administration not to slide back into any kind of formal relationship with the Rhodesian segregationists, and to abide by UN mandatory sanctions on trade with the territory.[68] Forty-three Americans working with the Peace Corps next door in independent Malawi wrote to Nixon of the enormous symbolism involved. "We can assure you from our first-hand observations of Rhodesia and Rhodesians," they concluded, that current Rhodesian society "stands diametrically opposed to the history-shaping demands of equal dignity and equal opportunity for all men."[69]

The real turn in Rhodesian policy in the Nixon years came with the passage of the Byrd Amendment in 1971. Named for its sponsor, segregationist Senator Harry Byrd of Virginia, this measure opened up an exception for chrome ore in the previous U.S. policy of adhering to all UN economic sanctions on the illegal government in Salisbury. While providing critical cash for Rhodesia, the measure was defended as essential to avoiding American dependence on the Soviet Union—the other leading exporter—for chrome ore for steelmaking. A law grounded apparently in anti-Communism thus bolstered the forces of white supremacy, to the delight of several white Southerners in Congress. Until the Byrd Amendment's repeal in early 1977 under the Carter administration, the United States was the only nation in explicit legislative defiance of its UN obligations regarding sanctions—and a de facto partner with South Africa and Portugal in trading with Rhodesia.[70] The White House largely avoided the Byrd Amendment by adopting a neutral stand. Supporters of Rhodesia already knew the president's leanings, and there was no political advantage for the president in alienating Salisbury's opponents at home or in Africa. Nixon and Kissinger were happy to let Congress lead, with white Southern members providing a disproportionate amount of the political muscle to support their segregationist brethren across the South Atlantic—just as they had succored Moise Tshombe in Katanga ten years earlier during the Congo independence crisis.[71]

Portugal's withdrawal from Angola and Mozambique in 1975 led Kissinger to one last shift in policy toward Rhodesia. Nixon had resigned the presidency over the Watergate scandal, but his successor, Gerald Ford, preserved the essence of his antibusing position at home and his tilt toward the white authorities in southern Africa. Still secretary of state, Kissinger recognized that Portugal's retreat altered the entire political calculus in the region. The anti-Communist white-ruled states were now down to two—South Africa (including Namibia, which it controlled) and Rhodesia—and the fresh victories over white supremacy in Angola and Mozambique encouraged the liberation forces fighting in Rhodesia. Aid from the Soviets and Cubans had been crucial for the final outcome in Angola, and Kissinger feared the creation of a whole string of radical, pro-Soviet regimes, with Salisbury being the next domino to fall. "What is required, then," he agreed with his aides, "is a preemptive strategy" to take the momentum away from the Soviets and Cubans.[72] With this aim Kissinger toured Africa in April 1976 and declared in Lusaka that America supported majority rule in neighboring Rhodesia, touting "racial justice" as "a dominant issue of our age" and "an imperative of our own moral heritage."[73]

Africans were surprised at such words from the originator of NSSM 39 and dubious about the extent of his moral conversion, but they were pleased with the novelty of the U.S. government's support. Southern white Americans were a good deal less happy and used the concurrent Republican primaries to express their opinion. Standing in front of the Alamo in San Antonio, presidential candidate Ronald Reagan told a campaign rally that Kissinger's Lusaka speech could lead to a "massacre" of whites in Rhodesia and had "undercut the possibility of a just and orderly settlement there." White Texans responded to the symbolism of massacres of brave, embattled white folks and gave Reagan a resounding victory over Ford in their presidential primary a few days later. Kissinger's staff reacted by serenading him, bleakly, with "The Eyes of Texas Are upon You." Nevertheless, over the next several months, Kissinger's shuttle diplomacy in the region helped lay some of the groundwork for the eventual settlement of the civil war in 1979 and the creation of the new state of Zimbabwe out of the shell of Rhodesia.[74] The task of preventing further expansion of Soviet influence in southern Africa continued to conflict with the sympathies of some white Americans for the increasingly embattled rulers of the region. As Zambian prime minister Kenneth Kaunda told an American journalist in 1968, Zimbabweans were like other oppressed peoples such as the Czechoslovaks: they would eventually fight for their freedom. If the West would not help them, they would accept aid from the East. By the brutality of its actions and the accompanying rhetoric of anti-Communism, Kaunda concluded, Salisbury was actually inviting Communist influence.[75]

The key change in Africa during the Nixon-Ford years was the Portuguese withdrawal, following years of armed nationalist struggle and the 1974 overthrow of the authoritarian regime of Marcello Caetano in Lisbon. The departure of the last European colonial power on the continent intensified the pressure on the remaining white settler states, especially neighboring Rhodesia.[76] Nixon had supported Portuguese authority in Africa to the end of his presidency; earlier, in 1963, he had even visited Lisbon as a private citizen and told the foreign minister that independence was "not necessarily the best thing for Africa or the Africans."[77] Kissinger's and Ford's decision to support one of the three competing Angolan factions in the civil war that broke out in 1975 emerged from the particular political circumstances of the spring of that year: the final North Vietnamese defeat of South Vietnam, with the last Americans fleeing for their lives, and the ongoing congressional investigations of CIA covert operations. Kissinger's response was typically global in its conception: he was determined to show the Soviets that despite events in Vietnam and the debacle of Watergate, the United States would still effectively oppose leftist revolutions around the Third World. He and the State Department considered it "essential to the credibility of our policies throughout the world" to prevent the victory of the Soviet-backed faction in Angola.[78] After U.S. policy failed and the Marxist Movimento Popular de Libertação de Angola (MPLA) took power at the new nation's independence late in 1975, Kissinger was asked by reporters if it might not have been better to leave the Angolan issue to Africans to sort out. "You may be right in African terms," he answered, "but I think globally."[79]

Congress responded in an almost opposite manner to events in Angola. The critical context for many senators and representatives was not just the loss in Vietnam, but also the ongoing congressional investigations of CIA covert operations. Disturbed by tales of agency assassination attempts abroad, leading Democrats were determined to restrict further foreign adventures that might turn out like Vietnam. They were upset to discover in the fall of 1975 how poorly the White House had kept them informed of American assistance to the Frente Nacional de Libertação de Angola (FNLA) and the União National para a Independência Total de Angola (UNITA), and how little clear ideological difference there was between those two factions and the MPLA. Newspaper reports of South African forces fighting with UNITA deepened their disillusionment, as it became clear that few Africans would support the American position. In December the Senate Foreign Relations Committee approved the Clark Amendment, which ended U.S. aid to any combatants in the Angolan civil war—the first time Congress had specifically banned a covert operation.[80]

Once Portuguese forces had withdrawn, the struggle throughout 1975 in Angola seemed at first to have little racial significance. All sides were Angolan. But the South African intervention on behalf of UNITA and the FNLA radically altered the equation, as did the Cuban troops flown in at the MPLA's request. What had initially seemed a conflict among Angolans involved Cubans and South Africans killing each other by November, with the support of their respective superpower allies: the Soviets and the Americans. For many Americans including President Ford and his advisers, the South African endeavor was legitimate because it opposed the expansion of Soviet influence; the troops of the hated Fidel Castro were seen not as liberators but as mere tools of Moscow and spreaders of Communism in Africa. Their presence seemed to call for containment of the Soviets and the Cubans, and the South Africans were the only folks in the neighborhood for the job.[81]

For almost all Africans, however, the meaning of the Angolan conflict was the opposite. Attitudes varied about Communism and about the Soviet Union, but not about South Africa. The apartheid state had for Africans the same meaning that the USSR held for Americans: the ultimate institutionalized form of evil. For people newly freed from centuries of white domination, the sight of white South African troops pouring across the border from Namibia into Angola was analogous to Americans viewing Soviet troops invading West Germany. The same Cuban state that appalled most Americans impressed citizens of African nations, where Castro had long been an ardent supporter of struggles for national independence. The contrast with South Africa could not have been more obvious, as the Cuban troops defending the MPLA were overwhelmingly black. The State Department privately admitted the racial coding the conflict had taken on: "The South African involvement on the side of FNLA/UNITA creates a dilemma for us. It . . . risks bringing discredit by association upon the US." Indeed, many African states that had originally been inclined to support the FNLA or UNITA reversed themselves and recognized the MPLA government instead. The Americans and the Soviets seemed to have their proxies engaged in a war about Communism but also about race relations, with the United States represented by the forces of white supremacy.[82]

As NSSM 39 made clear, the willingness of the Nixon and Ford administrations to be associated with South Africa reflected geopolitical calculations about the near-term future of the region. The whites were apparently "here to stay"; South Africa's economy had grown impressively during the previous decade; and "Nixon's personal sympathies were with the South Africans," as one of his NSC staffers put it.[83] It also reflected a lack of close attention to South African domestic conditions inherited from the Johnson administration, along with sympathy among Southerners in Congress for the Pretoria regime.[84] But the U.S. government lagged behind the broader international community, whose views of South Africa shifted sharply between 1969 and 1976. Condemnation of apartheid and the continued occupation of Namibia led to concerted efforts to expel South Africa from the UN, which was prevented only by the vetoes of the United States, Britain, and France. The UN's 1975 resolution condemning Zionism as racism reflected similar international anger at what was seen as another unjust "white settler" state in the Third World (and the United States defended Israel much as it did South Africa). South African prime minister John Vorster noted the parallel, referring sympathetically in a 1971 interview to Israel's "apartheid problem" of governing Palestinians in Gaza and the West Bank.[85] Black schoolchildren in Soweto highlighted the problems of the U.S. government's tilt toward Pretoria on 16 June 1976, when they began vast street protests against the South African government's orders that half of their classes be taught in Afrikaans rather than English. The police opened fire as they had at Sharpeville fifteen years earlier, but the massacre of dozens this time did not end the protests, which rolled on and set the stage for the final phase of the struggle against apartheid. Just as Angolans and Mozambicans had proven American policymakers wrong about the strength of Portuguese rule, black South Africans now demonstrated, at terrible cost, that the internal stability of even the region's sturdiest white state was no longer guaranteed.[86]

Nixon's Southern strategy had turned out to have an international reach, incorporating whites in southern Africa as well as the American South. The NSC acknowledged how racial issues crossed national borders. "Because of the multiracial character of our society and our own racial problems," NSSM 39 concluded, "other countries tend to see our relationships with southern Africa as reflections of domestic attitudes on race."[87] Lyndon Johnson's mantle of inclusive paternalism regarding racial equality did not fit well on Richard Nixon, who seemed to have little real feeling for the experience of racial discrimination, but he occasionally tried it on. As Kissinger drafted Nixon's first foreign policy address to Congress, the president instructed him, "Make sure there's something in it for the jigs, Henry"; a few minutes later in the conversation, he circled back to the same point: "*Is* there something in it for the jigs?"[88] Similarly revealing was Nixon's performance with Agnew in a supposedly self-mocking skit at the Gridiron Club dinner on 14 March 1970. Seated at adjoining pianos and bantering (Agnew in mock Southern accent) about their Southern strategy, Nixon would start playing various tunes only to have Agnew drown him out each time with a manic version of "Dixie." The elite, bipartisan, nearly all-white crowd roared with laughter.[89] "There is no documentary evidence that this [kind of] racism was the decisive influence in Kissinger-Nixon policies in Africa, Vietnam, or elsewhere, policies for which there were other arguments and reasons, however questionable," Roger Morris concluded later. "But it is impossible to pretend that the cast of mind that harbors such casual bigotry did not have some effect on American foreign policy toward the overwhelming majority of the world which is nonwhite."[90]

Ultimately, Nixon and his closest advisers were neither ideologues nor people of deep commitments about race.[91] The president sought political power above all, which could mean playing different sides of racial issues in different circumstances. "Hunting where the ducks are" at home translated into courting Southern whites, while abroad it required seizing the initiative by courting the very heart of the "yellow peril," the Chinese. Indeed, Nixon enjoyed playing the racial egalitarian in contrast to Soviet leaders, whose grow-ing hostility toward Beijing in these years included occasional displays of anti-Asian racism—in remarks like "those squint-eyed bastards!" and "those yellow sons of bitches!"[92] The White House was delighted, similarly, with news from the FBI of a Black Panther Party member angered by experiencing racial discrimination on a visit to Cuba: more fodder for dividing antiracist opponents at home and abroad.[93] The administration's pursuit of political and diplomatic opportunities led to a focus on the most powerful nations, which could translate into political success at election time. Africa, by con-trast, seemed to offer few rewards, and the NSC recommended simply keep-ing issues there "manageable."[94] Both at home and abroad, Nixon officially opposed "systems based on racial discrimination" while refusing to put more than a token element of Washington's power behind eliminating them.[95]

Jimmy Carter and the New American South

The election of Jimmy Carter in 1976 offered a potentially sharp reversal of the Nixon-Kissinger obsession with power politics and disregard for matters of race relations. The first U.S. president from the Deep South and one dra-matically affected by the civil rights movement, Carter entered the White House in the aftermath of the nation's bitter experience in Vietnam. His fa-miliarity with the lives of African Americans was unprecedented in the Oval Office, and his commitment to racial equality was unquestioned by those who knew him. He promised to match this commitment with a reorientation of the nation's foreign relations, away from the anti-Communist focus of his predecessors and toward a novel attention to what he called "the newly influential countries in Latin America, Africa, and Asia."[96] The Carter ad-ministration sought to drop Washington's long-standing "zero-sum" view of international relations, by which political changes abroad had to be calcu-lated as favoring either the Soviet Union or the United States. The budget of the covert operations arm of the CIA—long a bulwark against change in the Third World—was reduced to less than a tenth of its size a decade earlier, while congressional oversight of the agency was strengthened.[97] This shift to-ward a restrained diplomacy more favorable to social change in the Third World eventually faltered in the renewed Cold War that emerged in the ad-ministration's last two years, 1979 and 1980. But for at least their first two years, Carter and his advisers brought racial justice to the forefront of U.S. foreign relations, with a particular concern for southern Africa.

No other occupant of the Oval Office in the Cold War era grew up with Carter's intimacy with African Americans. His father, a successful farmer and businessman in Plains, Georgia, employed as many as two hundred and sixty black workers at a time. His mother worked as a nurse, and he and his siblings were cared for by a black nanny who worked for the Carters for twenty-one years. When Jimmy's parents left Plains for adult vacations, he stayed with an African American couple employed by the family. His first playmates were the children of his father's farmhands, and they remained his best friends until the segregating effect of the school system slowly pulled them apart. This was, of course, a limited, Jim Crow kind of intimacy, one embedded in the larger hierarchy of color of the 1920s and 1930s. Carter's wife, Rosalynn, recalled how, on their honeymoon in the North Carolina mountains in 1946, she had glimpsed a white domestic servant on a wealthy estate: "It was shocking to me. In Plains, Georgia, I had never seen a white woman doing yard labor or housework for a white person."[98]

Carter's racial attitudes as an adult and in his early political career reflected, in part, the contrasting influences of his parents. Earl Carter's perspective on African Americans was typical of his time and place: he was an old-fashioned racist and segregationist, though one capable of generosity to individual blacks in his employment. Lillian Carter held far more egalitarian views than did the bulk of her peers, especially in southern Georgia. Indeed, she even served in the Peace Corps in rural India from 1966 to 1968, when she was in her sixties. Jimmy's personal attitudes were close to his mother's. As an aspiring businessman and politician in the 1950s and 1960s in Plains after his return from the Navy, however, he chose not to rock the boat about race relations. He largely accepted the reality of segregation, even during a term as chair of the Sumter County school board in the years immediately following the *Brown v. Board of Education* decision. By character Carter was a conciliator rather than a confrontational person. He sought to get along with everyone, as he built both his peanut farming and warehousing business and his reputation as a man of integrity and industry. No aspiring politician willing to work within the Southern system in the 1950s could afford to denounce segregation. For white Georgians, one old friend remembered forty years later, "integration back then was about like child molestation is today."[99]

The overthrow of the traditional racial order of Dixie that unfolded in the two decades after 1954 left no Southerner untouched. With his spiritually driven seriousness about moral issues and his pragmatic sensitivity to changing political winds, Carter stood out for his determination to be a leader in the "New South." Other whites in Plains knew that he had different beliefs from most of them: he refused to join the archsegregationist White Citizens Council in 1954, and he tried unsuccessfully to integrate his own Baptist church in 1962. Three factors pushed him further down the path to racial egalitarianism: his deepening Christian faith, his service in the military, and his witnessing of the civil rights movement in Georgia. The extraordinary integrated Christian farming community of Koinonia, just seven miles from Plains, had been established in 1942. While unwilling to identify too closely with Koinonia's rejection of Jim Crow as it faced terrorist bombings and shootings by the Ku Klux Klan in the late 1950s, Jimmy and Rosalynn clearly respected the community's ideals and worked discreetly to mitigate some of the hostility it experienced. Carter served as an officer on an integrated Navy submarine a few years earlier, and he had supported his crew's unanimous rejection of an invitation to a party for only its white members by British officials in the port of Nassau. Finally, he was appalled in 1963 by the intensity of white hatred and violence against peaceful African American protesters in Americus, just nine miles from Plains. Few who knew him well were surprised that by the late 1960s, Carter had deepened not only his Christian identity also his commitment to a broader emphasis on social justice for the least powerful, particularly the poor and those with darker skins.[100]

Carter's term as governor of Georgia from 1970 to 1974 demonstrated both his ambitious political pragmatism and his commitment to a new era in Southern race relations. Believing it necessary for success, he campaigned to the right during 1970, appealing specifically to "the people who voted for George Wallace" for president in 1968 and "the ones who voted for Lester Maddox," the archsegregationist current governor of Georgia who could not by law succeed himself in office. Carter and his staff cast his major opponent in the Democratic primary, former governor Carl Sanders, as a liberal who was by implication too close to the integrationist national Democratic Party. Carter's campaign was careful: he avoided explicitly racial language in its messages to white voters, and he also appealed discreetly to black voters. This strategy led to a surprising upset of the better-known Sanders, who won 93 percent of the black vote in the primary. In the general election in November, African Americans decided to stick with the Democratic candidate, helping sweep Carter into office.[101]

He then surprised most Georgians with a clarion call for whites to accept a new racial order. In his inaugural address in January 1971 Carter declared, "I say to you quite frankly that the time for racial discrimination is over. Our people have already made this major and difficult decision." The *New York Times* covered the speech on its front page, as the national media took note of Carter as a spokesman for the New South. He governed Georgia carefully in regard to race, trying to win over conservative whites while turning the state toward an integrated future. Thus he remained opposed to school busing, while doubling the number of blacks in the state government. His symbolic gestures had particular impact, especially integrating the highway patrol and the walls of the state capitol itself by hanging a portrait of Martin Luther King, Jr., there. In a state where the feelings of most whites toward the recently martyred civil rights leader still ranged from doubt to hostility, this was not an easy step to take. Black Georgians rewarded the governor with strong support.[102]

African Americans across the nation then provided the crucial votes for Carter's extremely narrow victory in the presidential election of 1976. His record as governor won him the enthusiastic endorsement during the Democratic primaries of black leaders like Martin Luther King, Sr., and U.S. Congressman Andrew Young, who had been a close assistant to Martin Luther King, Jr. In the general election against Gerald Ford, Carter won more than 90 percent of the black votes, which, as Carter acknowledged, provided the critical margin in several of the states that put him over the top in the electoral college.[103] This was precisely the opposite situation of Nixon in 1968. The *New York Times* noted the apparent irony of a "south Georgia white man with a mint julep drawl" being sent to the White House by the "grandchildren of slaves."[104] But black Southerners, like white Southerners, tended to appreciate Carter's rural Southern background and his evangelical Protestant faith.[105] Indeed, in the election of 1976 many white and black Southerners voted for the same presidential candidate for the first time.

Carter ran for president by positioning himself as an outsider to the federal government. He campaigned not so much against Ford as against Nixon and Kissinger and the shrewd and seemingly amoral politics and diplomacy they represented. Untainted by associations with Washington, Carter promised to promote the decency of the American people and restore the national government's moral standing after the debacles of the Vietnam war and the Watergate scandal. He did not emphasize specific issues. Rather, he asked voters to have faith in him as a person and as a leader. Many people of influence, especially within the Democratic Party, lacked such faith; they were deeply skeptical of a man they knew little about, whose identity in rural Southern culture seemed foreign to them. Many African Americans and white liberals from outside the South distrusted his anti-Washington stance, which they associated with conservatism and segregationism—the forces that would be sympathetic to the next anti-Washington president, Ronald Reagan.[106]

But the man who won the Oval Office in the bicentennial year did not fit easily into existing political categories. His concerns for efficiency and budget balancing in the federal government, reflecting his own background in business and the military, were matched by an often stubborn determination to follow the more compassionate elements of the Christian gospel as he understood them. "I owe the special interests nothing," he declared repeatedly on the campaign trail, and indeed traditionally powerful political and business elites were anxious about the new Georgian candidate.[107] In an impromptu speech to a group of wealthy Hollywood liberals at actor Warren Beatty's elegant apartment in August 1976, Carter made his populist inclinations clear: "If we make a mistake, the chances are we won't actually go to prison, and if we don't like the public school system, we put our kids in private schools. But the overwhelming majority of the American people" had no such options. Suggesting that his allegiances had shifted away from people of privilege, he concluded—with a glance at the California governor, a primary opponent who had just entered the room—that "public servants like me, and Jerry Brown, and others have a responsibility to bypass the big shots, including you and people like you and like I was, and make a concerted effort to understand the people who are poor, black, speak a foreign language, who are not well educated, who are inarticulate," and to run the government in a way that delivers the services they desperately needed.[108] In cultural terms, Carter felt closer to Southern populist George Wallace than to any of the other Democratic presidential candidates of 1976. Only on matters of race did the two Deep South neighbors differ sharply.[109] This was crucial: Carter was a president whose concern for people with fewer economic assets and darker skins might make him unusually well suited for dealing with a mostly nonwhite world.

Carter sought to place human rights at the center of his new diplomacy. Rejecting the amoral, realpolitik calculations of Kissinger and Nixon, the new Democratic administration believed that the United States needed to reclaim its own ideals by reemphasizing moral decency in its relations with the rest of the world.[110] In place of CIA operatives undermining Third World nationalism Carter envisioned evenhanded support for justice for all people. The State Department created in 1977 an independent Bureau of Human Rights and Humanitarian Affairs, headed by a new assistant secretary of state, Patricia Derian, whose political reputation had been forged in civil rights work in Mississippi.[111] This shift reflected a national desire in the mid-1970s for a more humane diplomacy, evident particularly among younger members of Congress.[112] Demanding human rights within Communist countries was not novel in Washington, of course. What Carter added was an equal imperative for the protection of human rights among non-Communist nations. He believed such an approach was necessary for the building of what he called the global community—a world less scarred by the brutalities of political repression.[113] Because Carter also remained committed to restraining the growth of Communist influence abroad—he had no illusions about political liberty within the Soviet bloc—he was taking on a most difficult task. His willingness to preserve close ties with such noted anti-Communist tyrants as the shah of Iran earned him pointed rebukes about the true meaning of "human rights." Despite such exceptions and limitations on the administration's humanitarian priorities, including the return to a fierce anti-Communism in 1979, Carter's concern for political prisoners and social justice around the globe—particularly in Africa and Latin America—did contrast sharply with the explicitly political calculations of his Republican predecessors.[114]

Carter's identity as a Southerner sharpened his emphasis on human rights. The civil rights struggle of the decades leading up to 1976 had transformed Southern race relations. White people of conscience like Carter were deeply affected, as they had learned with sometimes terrible clarity the necessity of protecting the lives and liberties of all human beings. Carter spoke of how as governor of Georgia he had cringed at the firestorm of international criticism aimed at the United States for the violence and discrimination of whites against African Americans in the South, and for American support of "racist regimes" and "dictatorships" in southern Africa. "I used to shrink up and dread the day each fall when the United Nations General Assembly began its deliberations," he told the Congressional Black Caucus, "because I knew that my country, which I love, would be the target of every attack, and the butt of every joke among two-thirds of the nations on earth."[115] The election of a white president from south Georgia with overwhelming black support symbolized the political reintegration of the South with the rest of the nation. It marked the end of the nation's acute vulnerability regarding the treatment of its nonwhite citizens.[116] What Carter called "the cancer of racial injustice" would no longer weaken the country's body politic nor pollute its relations with the rest of the world.[117]

The Carter Administration, Race, and Africa

Jimmy Carter believed that his mandate from the American people was to help heal the nation's wounds from Watergate, the war in Southeast Asia, and the social divisions that had emerged in the late 1960s. A conciliator, his personality seemed to fit the nation's need. His egalitarian inclinations showed immediately in his choice to walk rather than ride in his own inaugural parade. Gone were most of the limousines that marked the White House fleet, and gone was Nixon's love of the pomp and ceremony of the Oval Office. Carter was particularly determined to lead the nation symbolically toward greater racial and gender equality. He appointed unprecedented numbers of blacks and women to high federal offices; he and Rosalynn enrolled their daughter, Amy, in a desegregated public school in the District of Columbia; he spoke of racial discrimination as a disease comparable to polio; and his Justice Department worked quietly but diligently to enforce desegregation laws. Coretta Scott King praised him personally for "the new spirit of reconciliation which you and your Administration are causing in this nation."[118]

By the mid-1970s the battle for legal racial equality had been won. Struggles against discrimination now moved to the trickier ground of practice rather than law. Structures and habits of preferring whites over nonwhites in areas like employment, housing, and education represented less visible but more difficult barriers to overcome in many ways. One response to those deeply ingrained patterns of discrimination was affirmative action, articulated initially by Lyndon Johnson as a way of helping people of color overcome structural preferences against them. Affirmative action reached the Carter administration in the form of the suit filed by Allan Bakke against the University of California at Davis, charging that preferential treatment of less qualified minority applicants had resulted in his rejection for admission to the medical school there. The case went all the way to the Supreme Court, with Bakke's lawyers arguing that racial preferences of any kind were unconstitutional—that they represented "reverse discrimination" against white people.

The Bakke case became a lightning rod for contention about American race relations in the late 1970s. For many white Americans, Bakke's situation represented their growing frustration with what they believed was a new kind of racial discrimination against them. Roman Catholics and Jews remembered too clearly how a system of quotas had been used to limit their access to opportunities a few decades earlier. But for many African Americans and their white supporters, the Bakke case instead served as a litmus test of the nation's determination to remedy, rather than just talk about, the ongoing effects of past discrimination. For a seeker of national reconciliation like Carter, the case created a no-win situation. After extensive internal arguments, the Justice Department in September 1977 filed an amicus curiae brief with the Supreme Court that strongly endorsed the principle of affirmative action. Troubled by the idea of racial preferences in any form, the Carter administration nonetheless took the position that some kind of preferential consideration of race was necessary, owing to the damaging legacy of past discrimination and the continuing reality of covert discrimination. The Court agreed in its mixed decision of June 1978: actual racial quotas for admission were not constitutional, but taking race into less specific consideration as one of several admissions criteria was permissible. Carter himself did not take the opportunity of the Bakke case to make a ringing declaration of support for the principle of affirmative action, as his critics noted. His support was clear but quiet, as he sought to avoid alienating white voters.[119]

The Bakke case symbolized the broader post–civil rights dilemma of how to deal with persistent but unofficial racial discrimination. This was not a small problem for Americans of color. Discrimination shaped how the death penalty was administered, and it showed up in explicitly racial sentiments against Japanese expressed by white auto workers and even a Michigan congressman, frustrated by competition with Japanese car companies.[120] It included police brutality against Latino Americans in southwestern cities and at least one Klan kidnapping of a Latino outside San Diego, who was then delivered to the Border Patrol for deportation to Mexico—a vivid demonstration of the ongoing tendency of some white Americans to identify citizenship with phenotype.[121] It even encompassed the highest law enforcement officer in the Carter administration, Attorney General Griffin Bell. Bell's reluctance to resign from two whites-only country clubs and his conservative record on desegregation cases as a federal judge moved one early hero of the civil rights movement, Aaron Henry of Mississippi, to call him "just not a good man."[122] Discrimination even followed Americans of color abroad. Carter appointed Ric Haynes, an African American foreign service officer and former NSC staffer under Johnson, as U.S. ambassador to Algeria. Ambassador Haynes arrived in time for the 4 July 1977 party at the embassy in Algiers, only to overhear the spouses of two U.S. diplomats whispering about him, "Have you seen what Washington sent?"[123]

African Americans generally supported Carter throughout his presidency, but many of them grew disappointed as his term in office progressed. Most blacks remained confident that the president's heart was in the right place regarding matters of equal treatment and discrimination. What troubled them were his priorities, which increasingly focused on defense spending, balancing the budget, and fighting inflation. These left little room for promoting or even defending federal social programs, to the disappointment of members of the Congressional Black Caucus.[124] Regardless of his strong personal convictions about racial equality and his vocal defense of human rights abroad, the president chose not to use the bully pulpit of the White House to call the nation, as Lyndon Johnson had, to a greater commitment to civil rights at home.[125] Hemmed in by economic problems and political limitations at home and distracted by crises abroad, especially in Iran and Afghanistan, Carter did not provide as much leadership on civil rights as some wished to see.[126]

Carter's most significant step regarding race and foreign relations was his appointment of Andrew Young as ambassador to the UN. The sending of an African American for the first time to represent the United States to the world community at the United Nations carried considerable symbolic weight. Young's background in the Southern freedom movement and his strong support for the ongoing liberation struggles in southern Africa indicated that the new administration would part ways with its predecessors. Young served as a leader of the "regionalist" forces regarding Africa: those in the administration, including Secretary of State Cyrus Vance, who were inclined to understand African problems as essentially separate from the Cold War. While the UN ambassador did not hold the authority of either Vance or the national security adviser Zbigniew Brzezinski, Young's close relationship with the president helped focus American attention on the problem of human rights abuses in Africa to an unprecedented extent. Africans and other people of color around the world appreciated the sharp contrast between the politics of Andrew Young and those of Henry Kissinger, as human rights moved from the periphery to center stage. Brzezinski credited Young with "transforming American isolation in the Third World into American friendship."[127]

Part of Young's success in warming relations between Africa and the United States stemmed from his informal but effective style of diplomacy. Like his boss in the Oval Office, who arranged the Camp David accords between Egypt and Israel, Young was a conciliator by both nature and commitment—a man eager to bridge the gaps between conflicting positions and bring opponents together. In the civil rights movement his role had often been one of negotiating with white officials rather than receiving blows from police batons on the front lines. While unstinting in his pursuit of racial equality, he was no economic radical. He actively pursued the interests of American businesses in building trade with Africa, and he viewed American corporations operating in South Africa as a potentially liberalizing influence on apartheid. Even as the high-profile ambassador to the UN, Young had no immunity from enduring white attitudes about African Americans: headed into the garage at the Waldorf Towers in New York one day while dressed in a jogging suit, he encountered a white tourist who asked him to fetch his car. "There I was riding high being the ambassador and all," he recalled, "and this guy assumed I was the car hop. To him I was just another nigger." But such encounters did not deter Young's commitment to seeking racial reconciliation in American diplomacy.[128]

In contrast to most diplomats, Young was famously frank about his opinions. Depending on their politics, others found this alternatively refreshing or troubling. At various points, the ambassador argued publicly that Cuban troops brought "a certain stability and order to Angola"; identified the South African government as "illegitimate"; labeled the British government "a little chicken" on racial issues; called Nixon and Ford "racists"; and declared that there were "hundreds, perhaps thousands, of political prisoners [in jail] in the United States."[129] This was more than a simple breeze of fresh air in the U.S. government—it was a veritable gale of alternative perception. Carter was embarrassed by the barrage of criticism he received after each of these controversial assertions by his ambassador and friend, and more conservative Americans called for Young's removal from his prominent position at the UN.[130] Others, particularly Africans and their supporters in the United States, found Young's outspokenness bracingly accurate and encouraging. They considered it unarguable that apartheid rendered the South African regime illegitimate. They recognized Nixon's casual dislike for people of color and Britain's unwillingness to use force against rebellious whites in Rhodesia as it had against rebellious people of color in other colonies. They believed South Africa to be the real destabilizing force in Angola. And for people familiar with how the justice system in the American South had long served to incarcerate black organizers, including the ongoing case of the Wilmington Ten in North Carolina, Young's calculation about political prisoners seemed eminently plausible.[131]

Young was forced to resign from his position at the UN in August 1979 after meeting secretly with a representative of the Palestinian Liberation Organization (PLO), defying an administration promise to Israel not to do so. He was not the first administration official to meet with a PLO figure, and Carter himself had compared the PLO cause to that of the American civil rights movement.[132] Brzezinski had rejected Israel's proposed partial "home rule" for Palestinians in the West Bank and Gaza, comparing it to South Africa's "homelands" strategy for preserving white rule.[133] Few in American public life, however, were willing yet to negotiate with the PLO. And Young's firing ultimately reflected the feathers he had long ruffled with his frank diplomacy, the influence of American supporters of Israel, and the broader shift of the Carter administration to a more militarized and confrontational foreign policy. Young's refusal to think like a Cold Warrior conflicted with rising anti-Communism in the administration. The persistence of Cuban troops in Angola and their introduction into Ethiopia, in combination with the Nicaraguan revolution and insurgencies in other Central American states, raised concerns about leftist links between Africa and Latin America and about the expansion of Soviet influence on both continents.[134] Young's removal angered African Americans who considered him one of the primary symbols of Carter's commitment to black interests. It also exacerbated black-Jewish tensions in American politics, as blacks resented the influence of Israel's supporters and increasingly identified with Palestinians as struggling for self-determination against a more powerful "white" nation that discriminated against them.[135]

Before it tapered off in the renewed Cold War of 1979–80, the Carter administration's unprecedented interest in Africa arose from the confluence of several factors. The president himself was genuinely concerned about developments on the continent, on the basis of his early links to African Americans, his seriousness about social justice and human rights, and his interest in missionary work among the rural poor. He was, in fact, the first sitting American president to visit Africa, traveling to Nigeria in 1978. Carter's considerable involvement with development and health care issues in Africa in his postpresidential years has testified to these concerns.[136] The momentous political events in southern Africa in the 1970s had focused world attention on the region, especially the independence of Angola and Mozambique in 1975, the Soweto uprising of 1976, and the escalation of the civil war in Rhodesia/ Zimbabwe. The withdrawal of U.S. forces from Vietnam allowed American liberals and leftists to shift their attention from Southeast Asia to the problem of white supremacy south of Zambezi River. At the same time the success of the civil rights movement in ending legal discrimination in the United States cast apartheid in a worse light than ever before for most Americans. Finally, more traditional security interests also increased U.S. attention to Africa, as Nigeria became the second largest provider of imported oil to the United States in these years, and Cuban troops in Angola and Ethiopia suggested a growing Soviet or at least Communist influence on the continent.[137]

Washington's renewed attention to Africa came only in the wake of changing race relations in the United States. By the mid-1970s the process of desegregation left little room in the mainstream of American politics for defending racial inequality and discrimination. Conservatives still vigorously defended authoritarian anti-Communist governments in places like Iran and Nicaragua, but fewer and fewer voices spoke up for the white supremacist rulers of South Africa, Namibia, and Rhodesia. "I believe very strongly in majority rule," Carter declared on his third day in the Oval Office." There was no clearer violation of the human rights that the new president touted than the forms of control exercised by the rulers of those three states. Indeed, the administration feared that continued white rule might so alienate the African majority in the region as to lead to eventual Communist revolutions there. Helping negotiate independence for Namibia and a settlement of the civil war in Rhodesia, Vance argued, provided the only "workable strategy for improving [U.S.] relations with black Africa and blocking the spread of Soviet and Cuban influence in southern Africa."[138]

The Carter administration's support for majority rule in southern Africa reflected as well the growing influence of African American organizations on U.S. policy toward the continent. "The politicians were frank. They used to call us niggers," Young liked to tell white audiences when he was a member of Congress in the early 1970s. "But when blacks in a community got 30 percent of the vote, we became nigras. Later on, at about 40 percent, we were Nee-grows or colored people. And when we got the majority, we became their black brothers."[139] As African Americans gained political power in the 1970s, their interests in Africa received greater attention. The newly founded Congressional Black Caucus became a major voice on African issues, as did the lobbying organization TransAfrica, created in 1976 in response to the

Soweto uprising.[140] The NAACP and the Reverend Jesse Jackson also lobbied the administration to take stronger stands against the white governments of southern Africa.[141]

Carter's Africa policy could never fully escape Cold War and defense considerations, and the movement toward renewed conflict with the Soviet Union inevitably reduced American attention to the issue of majority rule. The Soviet airlift of twelve thousand Cuban troops into Ethiopia in 1977 sharply increased U.S. anxieties. The invasion of southern Zaire by Cuban-trained Katangans from Angola in May 1978 exacerbated those concerns.[142] For its part, the Castro regime was determined to support leftist revolutionaries in the Third World, and the large percentage of Cubans who were of African descent shared a concern for developments on the continent. Ethiopia had long been a symbol of African independence from white rule, and Cuba's assistance helped it to stave off both Eritrean separatists and Somali efforts to seize the Ogaden region. Many Africans appreciated Cuba's support for the revolutionary regimes in Addis Ababa and Angola, especially in light of South African opposition to the Luanda government.[143] The slippage of U.S. policy back toward containing Communism reflected the rising influence within the administration of Brzezinski, who had earlier edited a book with a title suggestive of his primary concern regarding the region: *Africa and the Communist World.*[144]

The Carter administration's attention to South Africa followed the same trajectory from a strong initial stance to a tapering off by 1979 as crises elsewhere seized Washington's attention. In contrast to the aftermath of the Sharpeville massacre sixteen years earlier, black protest continued to build and roil South African society after the Soweto uprising of 1976. The recent Portuguese withdrawal from neighboring Mozambique and Angola made the apartheid state seem newly vulnerable to its opponents. Vice President Walter Mondale met in May 1977 in Vienna with Prime Minister John Vorster to warn that relations between the two countries would deteriorate sharply if South Africa did not move away from apartheid. Mondale even used the phrase "one man, one vote" to explain the new administration's goal for South Africa—an unprecedented public declaration by a U.S. official in support of democracy there.[145] When the apartheid regime escalated its repressive tactics in the fall of 1977, murdering black leader Steven Biko during a prison interrogation, banning opposition groups, and detaining an array of other anti-apartheid activists, the U.S. government responded swiftly. The House of Representatives, led by the Black Caucus, passed the first congressional resolution critical of apartheid. In a symbolic gesture revealing how much had changed since the Nixon era, Carter sent a senior State Department official to Biko's funeral.[146] Young voted for a mandatory arms embargo against South Africa by the UN Security Council, the first resolution ever requiring punitive sanctions be imposed on a member state.[147] White South Africans resented U.S. actions and placed the blame squarely on Carter and his convictions about racial equality.[148]

The Carter administration was not yet willing to go beyond symbolic disassociations with apartheid to the level of structural disengagement from South Africa. Economic sanctions, the ultimate weapon for pressuring Pretoria, were discussed but put off.[149] This approach reflected in part the pro-business orientation of men like Carter and Young, who were well aware of the highly profitable character of extensive U.S. trade and investment relations with South Africa.[150] They tended to agree with American business leaders who argued that U.S. companies in South Africa provided important models of decent treatment of black workers.[151] They were also reconcilers, who sought a negotiated solution to South Africa's civil strife to avoid a full-blown revolution and race war. Carter told his Cabinet early on that "South Africa is like the South fifteen years ago," implying that its race relations could be reformed from within.[152] The president and his advisers feared that economic sanctions might have the negative effect of making white South Africans more defiant and isolated, while leaving them powerful enough to rule the country.[153] The American desire for Pretoria's assistance in finding a negotiated solution to the Rhodesian civil war next door also inclined the administration to preserve channels of cooperation with the apartheid regime.[154] White South Africans and their American supporters argued that South Africa was "an integral and indispensable part of the West" and a critical bulwark against the "Soviet assault on southern Africa."[155] The Carter administration was less inclined to agree with these arguments than its predecessors had been, but it still tended to view white South Africans as part of the Western world. As American attention shifted northward to the crises in Iran and Afghanistan and a renewed confrontation with the Soviet Union by 1980, the administration's interest in challenging Pretoria waned. The resignations of Young (August 1979) and Vance (April 1980) punctuated the turn away from Africa. Public attention returned to the Cold War, and one poll in 1979 found that only 18 percent of the American public had even heard of apartheid.[156]

The other holdout of white supremacy reached its demise in the Carter years, as Rhodesia became Zimbabwe in April 1980. Similar in many ways to neighboring South Africa, Rhodesia previewed how a combination of internal struggle and external sanctions would finally end white rule in South Africa fourteen years later. The Carter administration inherited a rather different set of policies regarding Rhodesia than those for South Africa, grounded in the distinctive histories of the two states. South Africa had gained its independence from Britain in 1910, when white supremacy was still the norm in Europe and North America, and the Pretoria government was therefore considered unquestionably legitimate by most other governments. The white settlers of Rhodesia, a much smaller percentage of their state's population, instead declared their independence from Britain in 1965 to preserve their dominance in an era when white rule was receding from most of the African continent. Changes in global race relations meant that Rhodesia was received differently than South Africa had been. No state recognized the Ian Smith regime as legitimate. The United States followed Britain's lead in imposing economic sanctions on Rhodesia, where there was much less American investment than in South Africa.

Carter thus inherited a tougher policy toward Rhodesia than South Africa. He quickly patched the major hole in the sanctions system, through which Rhodesian chrome had been flowing for six years, by successfully pushing Congress to repeal the Byrd Amendment.[157] Young lobbied his former congressional colleagues for this measure, describing it as a "kind of referendum on American racism"—powerful rhetoric in post–civil rights America.[158] Carter also came into office with Smith having already conceded in October 1976 that white rule was coming to an end, another sharp contrast to South Africa. In the meantime, however, the ongoing civil war between Smith's troops and the guerrilla forces of the Patriotic Front under Robert Mugabe and Joshua Nkomo threatened to destabilize the entire region, and Britain and the United States were eager to promote a negotiated solution that would minimize further racial polarization.[159] Then the wily Smith complicated the situation by promoting an "internal settlement" that would bring moderate blacks into the government while excluding the more radical Patriotic Front, as a way of preserving as much white political influence as possible. His American supporters, led by twenty-six Southern and conservative U.S. senators like Jesse Helms, invited him to visit the United States in October 1978, where he and they lobbied the public and the rest of the government for an end to economic sanctions against Rhodesia, based on his agreement to hold multiracial elections in April 1979.[160]

Carter resisted the considerable pressure that built up for ending sanctions after the election of Methodist bishop Abel Muzorewa as the first black prime minister of Rhodesia. The president and his advisers recognized that however much they might admire Muzorewa's moderate politics and religious background, the real power among Africans in the country lay with the Patriotic Front, which had rejected the Smith-controlled elections. Congressional Republicans had long argued that the Patriotic Front was supported by the Soviet Union and that its politics were explicitly Marxist. They found the insistence of Carter and the British government on including the Patriotic Front in any final settlement incomprehensible: weren't these precisely the kind of revolutionaries Americans were supposed to oppose? Senator Robert Dole even compared the front to the PLO, arguing that since the United States supported Israel and refused to talk to Palestinian terrorists, it should similarly promote the Rhodesian "internal settlement" and exclude Mugabe and Nkomo.[161] Andrew Young, of course, agreed with the logic of this comparison, though he drew the opposite conclusion: if we can talk to the Patriotic Front, we should talk to the PLO. Strongly supported by Africans and African Americans, Carter rode out the storm of criticism and preserved U.S. sanctions until after the Lancaster House agreements in London in September 1979, which led to more inclusive elections and the creation of Zimbabwe with Mugabe as prime minister.[162] Averill Harriman then reported to Brzezinski and Carter from Zimbabwe's independence ceremonies in April 1980 that Mugabe was turning out to be quite pragmatic: friendly toward the West, inclusive of whites in his government, eager for Western investment, and hostile to the Soviet Union.[163] In a month when the administration was awash in criticism for its failed military effort to rescue American hostages in Iran, this was welcome news of a success story partly attributable to the principled actions of the president.

The Carter administration ultimately allowed its social justice priorities to be eclipsed in its last two years by renewed anxiety about political revolutions and instability, which it feared the Soviet Union was either instigating or benefiting from. A growing Cuban military presence in Angola and Ethiopia, revolutions in Iran and Nicaragua, and especially the Soviet invasion of Afghanistan in December 1979 convinced many Americans, including the president, that Communist forces were once again on the march, as they had been in the late 1940s.[164] Attention to human rights and to building a "global community" waned relative to this renewed concern with traditional national security issues.[165] The resignation of Secretary of State Cyrus Vance in April

1980 and the rising influence of national security adviser Zbigniew Brzezinski symbolized this reorientation. Domestic racial politics developed in parallel fashion. While most African Americans remained favorably inclined toward the president throughout his four years in office, Carter's emphasis on budgetary restraint and fighting inflation placed him increasingly at odds with black political leaders and the social programs they supported. Young's resignation tied together these shifts to the right in both foreign and domestic policy. Young was the leading advocate of closer ties with the Third World, including leftist governments there, and he was the most visible and influential black official ever in the U.S. government; his departure bitterly disappointed many African Americans.[166] Carter's personal commitment to racial and social justice remained firm, but his administration's determination to contain the expansion of Soviet influence took precedence. Relations with the Third World slid down the priority scale, occasionally revealing surprising ignorance in Washington. "What the hell is an 'Ayatollah' anyway?" Vice President Walter Mondale asked CIA director Stansfield Turner at an NSC meeting on the Iranian revolution in September 1979. Turner replied, "I'm not sure I know."[167]

Jimmy Carter governed the United States in a period of transition, both in race relations and in foreign relations. In response to the civil rights victories of the 1960s and to growing black political participation and power, the nation was moving toward the more conservative racial politics of Ronald Reagan in the 1980s. From the reconsiderations of the Cold War in the early 1970s—the antiwar movement, détente, and distrust of imperial behavior abroad—Americans were now edging back toward the anti-Communist enthusiasms that would break out anew in 1980. Constrained by these political trends and by the overwhelming reality of double-digit inflation at the end of his term, Carter struggled to pull Americans free of narrow anti-Communism and to enshrine human rights as the country's highest priority abroad. "Foreign policy no longer has a single or simple focus," he explained, urging Americans to try to understand instead "the complexity of interrelated and sometimes disturbing events and circumstances" abroad.[168] With Andrew Young as a key spokesman and with robust African American support, this peanut farmer and businessman from the Deep South made the pursuit of color-blind democracy in southern Africa a higher priority for the U.S. government than had any other president. Carter's return to the Cold War fold by his last year in office blurred that focus, but not before he had warmed considerably the relations of the U.S. government with both the Third World and African Americans.

To the End of the Cold War and Beyond

After 1979 Americans seemed increasingly to agree about the evils of Communism and the Soviet Union and therefore the essential necessity of anti-Communism. Evidence of this abounds: from the popularity of Ronald Reagan for eight years in the White House, to the liberation of the Soviet empire and the implosion of the USSR itself, to the attention paid to recently declassified Soviet documents suggestive of extensive spying in the West during the Cold War. There were exceptions, particularly in the form of dissent against Reagan's support of right-wing governments and rebels in Central America. But overall, the transcendence of capitalist values and the delegitimation of socialist alternatives in the United States and indeed the world marked the last two decades of the twentieth century.

In the same period, the legitimacy in American public life of racial prejudice and discrimination continued to ebb. The epithet "racist" became a powerful rhetorical tool in politics, to an extent unimaginable a generation earlier. No demonstration of this trend could be clearer than the near universal condemnation of South African apartheid by the mid-1980s, including comprehensive economic sanctions against Pretoria passed by a Republican-controlled Senate. Again there were exceptions. Reagan vetoed those sanctions, requiring a congressional override, and George Bush played the race card successfully in the 1988 presidential campaign, while a broad white backlash dating back to the late 1960s helped end the federal welfare system and limit affirmative action. And prejudice and racial violence, of course, hardly disappeared. But explicit white supremacy as an avenue to mainstream political success was—like Soviet Communism—swept into the dustbin of history. A rising consensus in American life around racial equality and anti-Communism marked the final years of the Cold War.

Reagan rode to victory in 1980 on a wave of unhappiness with inflation, unemployment, and the seizure of American hostages in Teheran. His defeat of Carter carried racial significance. From his early years in Illinois and California, he had little experience with African Americans, like Kennedy, and he evidently demonstrated neither personal prejudice nor much interest in racial issues.[169] But his political ascendancy built from the Watts riots of 1965, as he was elected governor of California a year later as a proponent of law and order. His opposition to civil rights bills in the 1960s was well known.[170] In 1980 Reagan continued the Republican strategy of appealing to disaffected white Southern voters, in particular by rejecting affirmative action.[171] In implementing a Southern strategy Reagan even went so far as to open his campaign in Philadelphia, Mississippi, known primarily as the site of the infamous murder of three civil rights workers in June 1964. He made no mention of James Chaney, Michael Schwerner, and Andrew Goodman, declaring instead, "I believe in states' rights"—the precise language of those who had done the killing.[172] The Ku Klux Klan endorsed Reagan's candidacy, announcing that the Republican platform "reads as if it were written by a Klansman."[173] Twelve years earlier Martin Luther King, Jr., had too easily dismissed the California governor and aspiring presidential candidate as a "Hollywood performer lacking distinction even as an actor."[174] In 1980 and again in 1984, Reagan received the lowest percentage of black votes of any major party candidate in the twentieth century, but his brand of conservatism dominated American politics for a decade.[175]

The Reagan administration swiftly positioned itself as less sympathetic to the interests of African Americans than any administration since the 1950s. It appointed few blacks to top positions. Frustrated black foreign service officers eventually brought a discrimination suit against the State Department in 1986.[176] The Justice Department opposed affirmative action, narrowed the definition of racial discrimination, and supported federal tax exemptions for racially discriminatory schools. The administration's economic policies of favoring wealthier Americans and cutting social spending disproportionally hurt nonwhite citizens. Reagan appointed a raft of archconservative federal judges, including William Rehnquist as chief justice of the Supreme Court—despite Rehnquist's earlier record of sympathy for segregationist positions and his opposition to public accommodation laws outlawing racial discrimination.[177] The president's joke about differences within his administration regarding how tough to be with the Soviets— "Sometimes my right hand doesn't know what my far right hand is doing"— might have referred just as well to his civil rights policies.[178]

The Reagan administration's seeming indifference to African American conditions applied equally to the black majority in South Africa. "All Reagan knows about southern Africa is that he's on the side of the whites," Chester Crocker told a South African reporter a month before being named the administration's point man for African affairs.[179] American investors and foreign traders continued to benefit handsomely from their unparalleled position in South Africa's economy, but the president was not motivated primarily by such economic links.[180] He believed that instability under an anti-Communist government anywhere was caused by Communists. "Let us not delude ourselves," he explained in 1982. "The Soviet Union underlies all the unrest that is going on. If they weren't involved in this game of dominoes, there wouldn't be any hotspots in the world."[181] This simplistic view meant that, in South Africa, the banned African National Congress, which was allied with the banned Communist Party, was the source of trouble. As late as July 1986 Reagan blamed the ANC for a program of "calculated terror . . . designed to bring about further repression, the imposition of martial law, and eventually creating the conditions for racial war." The increasingly brutal repression of dissent against apartheid by the Pretoria government in the 1980s was, from the president's perspective, caused by its opponents.[182]

White South Africans were, of course, delighted at the good fortune dealt to them by American politics. They had been encouraged by the resurgence of anti-Communism in the late Carter administration, which gave them greater leverage against Western anti-apartheid pressures. Carter's assistant secretary of state for African affairs, Richard Moose, had noted in 1980 that "the apparent trend toward conservatism in the West also reinforces their world view."[183] But the contrast of Reagan's views on race relations and apartheid with those of Carter was stark and even more encouraging for them. The Pretoria government was almost euphoric at hearing the new U.S. ambassador to the UN, Jeane Kirkpatrick (unlike her predecessor, Andrew Young) defend apartheid against its critics by arguing that "racial dictatorship is not as onerous as Marxist dictatorship"—a rather bold opinion for someone who had lived under neither.[184] South African opponents of apartheid were dismayed. Reagan's policy of "constructive engagement" with Pretoria, Archbishop Desmond Tutu declared, was "immoral, evil, and totally unChristian."[185]

South Africa's severe crackdown on internal dissent in the early 1980s was matched by a series of military invasions to destabilize the neighboring "frontline" states of Mozambique, Zimbabwe, Namibia, and Angola, and thus weaken their ability and inclination to support ANC guerrillas.[186] A British Council of Churches delegation to Namibia in 1981 described the South African military occupation of that territory as a "reign of terror."[187] Congressional repeal of the Clark Amendment in 1985 allowed U.S. funds and weapons to flow again to the UNITA guerrillas in Angola, who were supported by South African forces in intense fighting against the Angolan government and its Cuban allies.[188] The CIA funneled military aid to UNITA through the Kamina air base in neighboring Zaire; the agency tried to keep this operation secret partly by using African Americans as ground personnel, in hopes that they would not be as conspicuous as whites.[189] Fidel Castro, for one, was not fooled. He reminded two visiting black U.S. officials in 1985 of the racial coding of the war in Angola: "We are the only country which has actually fought the South African racists and fascists."[190] Indeed, three years later Cuban forces finally halted the South African army's advance at a major battle at Cuito Cuanavale, setting the stage at last for a negotiated settlement of the political futures of Angola and Namibia. South Africa's regional offensive ultimately proved to be the final outward push of a cancer-ridden political system on the verge of internal collapse.

Apartheid was overcome primarily by the sustained resistance, at enormous cost, of black South Africans. The actions of Americans and other foreigners were secondary to developments inside the country. The dramatic upsurge in black protest in South Africa that began in 1984 did, however, affect American politics more directly than any previous African issue had been able to do. The South African government's declaration of a state of emergency in 1985 and its inability to halt massive protests made front-page news across the Atlantic. Best-selling books, popular films, and hit music albums in the United States focused on South Africa.[191] Archbishop Tutu's charismatic nonviolence and his winning of the Nobel Peace Prize reminded many Americans of Martin Luther King, Jr., a Nobel laureate two decades before. Indeed, the anti-apartheid movement that grew rapidly in the United States in the mid-1980s seemed to many of its participants to be a renewal of the American civil rights movement of the early 1960s. Under the leadership of TransAfrica and the Congressional Black Caucus, the movement focused with considerable success on two goals: getting American institutions to withdraw their investments from South Africa and from any companies engaging in business with South Africa, and moving Congress to enact broad economic sanctions on the apartheid state. Five years of "constructive engagement" yielded no evidence by 1986 of reform within the Pretoria regime, which instead continued to dig in its heels against change. The self-evident failure of the Reagan administration's policy opened the door to a new consensus in American politics on disengagement from South Africa.[192]

The comprehensive economic sanctions passed by Congress, including the Republican-controlled Senate, over Reagan's veto in October 1986, reflected the consolidation of a broad national agreement on the immorality of legal racial discrimination. The televised pictures and reports of police brutality in South Africa, based solely on the color of one's skin, dismayed American audiences. Despite the opposition of the president and of hard-line sympathizers with white rule like Jesse Helms, even most conservative Republicans decided to distance themselves from virulent white supremacy. Such long-time friends of the apartheid regime as Representative Newt Gingrich, soon to become Speaker of the House, began to put distance between themselves and apartheid's rulers.[193] For senators like Nancy Kassebaum of Kansas, the demands of black South Africans for the right to own property where they wished and to live free of government restriction and harassment seemed quintessentially American, even conservative. Political calculations were also part of the equation: Kassebaum and other Republicans feared losing congressional seats to the Democrats in the November 1986 elections if they were associated with human rights abuses in South Africa.[194] But even Reagan's own national security adviser, Robert McFarlane, was disgusted by the "hypocrisy" of Prime Minister P. W. Botha, whom he finally concluded was a "mean-spirited son of a bitch."[195] Secretary of State George Schulz decided that the domestic political costs of trying to work with Pretoria were increasingly not worth paying: the "flak factor" had gotten "completely out of hand."[196] The consensus against apartheid reflected, in part, the rising political power of African Americans and their votes, which also helped change the behavior at home of some formerly staunch segregationists. James Eastland gave the NAACP $500 shortly before his death; Strom Thurmond sent his son to an integrated school; and George Wallace was granted an honorary degree from the Tuskegee Institute.[197] The Reagan administration's hostility to the causes of domestic civil rights and southern African liberation, while important, was partially mitigated by the rise of the anti-apartheid movement and the tenacity of African American political organizing.

The election of George Bush in 1988 signaled, for the most part, a continuation of Reagan's policies regarding racial issues and foreign affairs. In contrast to Reagan, Bush came out of the more moderate, eastern tradition within the Republican Party, represented by his father, Prescott Bush, a U.S. senator from Connecticut. But like Reagan, he moved to the Sunbelt—Houston—early in his adult life and shifted politically to the right as his party did.[198] At heart a more cautious and less visceral conservative than Reagan, Bush adapted himself to the prevailing political winds of the 1980s and served as a loyal vice president for both of his predecessor's terms.

Like his early political patron, Richard Nixon, Bush took a much greater interest in international matters than in the intricacies of domestic governance. Indeed, his unwillingness to take actions to moderate a prolonged recession in the early 1990s contributed mightily to his failure to be reelected in 1992. With few African American supporters to sensitize him to the problems of discrimination, Bush viewed racial issues as only a minor aspect of the least interesting part of his job: domestic politics. Like Nixon, however, he proved adept at playing to whites' racial fears for political advantage. He allowed campaign manager Lee Atwater in 1988 to exploit fear of black rapists as a centerpiece of the campaign against Massachusetts governor Michael Dukakis. Republican advertising that fall focused on "Willie" Horton (a nickname manufactured to sound more "black"), a convicted criminal who committed rape and murder against a white couple while on parole from a Massachusetts prison. If the explicit message was that Dukakis was supposedly soft on crime, it required no great insight to see a calculated titillation of white racial and sexual fears.[199]

Once in the Oval Office, Bush continued Reagan's opposition to affirmative action policies. He responded to the May 1992 race riots that broke out in Los Angeles by blaming liberal social welfare policies that, he argued, had fostered dependency on the state and irresponsible personal behavior.[200] In a move seemingly almost as cynical as the "Willie" Horton campaign tactic, the president replaced retiring Supreme Court Justice Thurgood Marshall—a jurist of enormous reputation and dignity—with Clarence Thomas, a young nominee distinguished for such a position primarily by having conservative views and by being black.[201] The other African American to rise to prominence with the patronage of the Reagan and Bush administrations was General Colin Powell. Having served as Reagan's national security adviser in 1987–88, Powell became a household name as a result of his performance as chairman of the Joint Chiefs of Staff—the first black person to serve as the nation's foremost military commander—during the Persian Gulf war of 1991. Powell was a nonpartisan figure of widely admired strategic acuity and political skills. By the mid-1990s, in fact, he had become probably the most uniformly admired African American in the nation's history and was considered a possible presidential candidate in the mold of Dwight Eisenhower. He was no enthusiast for Republican racial politics and policies, however. "Even though Reagan and Bush are two of the closest people in my life," Powell recalled in 1995, "I've got to say that this was an area where I found them wanting."[202]

Two great epochs in international affairs came to an end while Bush was in the White House. One was the joint collapse of Communist rule in Eastern Europe in 1989 and of the Soviet Union itself in 1991. The other was the 1990 agreement of the last remaining white minority regime in the world first to grant independence to Namibia, arguably the world's last colony, and then to release political prisoners including Nelson Mandela and allow a free election in 1994 that would end apartheid. The speech by South African prime minister F. W. DeKlerk on 2 February 1990 announcing these decisions meant for legalized white supremacy what the fall of the Berlin Wall two and a half months earlier had signified for Soviet Communism: the final surrender.[203]

The Bush administration itself had little to do with these developments, which were driven primarily by indigenous actors in each region. The administration indicated little concern for events in South Africa, and the president's great caution regarding change in the Soviet bloc left the U.S. government mostly in the position of an observer.[204] But two more historic changes away from authoritarian rule of the many by the few would be difficult to imagine. The liberation of the last piece of Africa from white rule and the freeing of Eastern Europe and central Asia from Communist rule combined to close the curtain on the protracted process of decolonization, which had colored so much of twentieth-century world history. As the Cold War receded into history, contention over race relations was unlikely to exert as much influence again on international relations as it had in the era of African decolonization and America's civil rights struggle.

EPILOGUE

AFTER THE FALL of the Berlin Wall issues of race and foreign relations continued to intersect, just as they had before 1945, but without the prominence that the Cold War had given them. Former Arkansas governor Bill Clinton won the presidency in 1992 and again in 1996 as a New Democrat, supportive of the interests of American corporations and less committed to his party's traditional backing of federal social programs. He stole much of the Republican Party's thunder by presiding over a period of dramatic economic growth, eliminating federal budget deficits, and ending the federal welfare system in 1996. He nevertheless built an exceptionally close personal and political bond with African Americans. They stood with him more reliably than any other group of Americans, even during his 1998 impeachment trial stemming from an affair with a White House intern. Clinton defended affirmative action ("mend it, don't end it"), appointed far more people of color and women to the Cabinet and the federal judiciary than had his predecessors, visited African American churches with relish ("He can sing the songs without even looking at the hymnal," black Congressman John Lewis noted), and had a black best friend (Washington lawyer and former National Urban League president Vernon Jordan). His roots in Arkansas help explain his connection to African Americans. As an eleven-year old fifty miles away in Hot Springs, he had been riveted and horrified by the 1957 school desegregation crisis in the state capital. "It was Little Rock that made racial equality a driving obsession of my life."[1]

Clinton similarly paid more attention to Africa than had previous presidents. He visited the continent twice to enthusiastic receptions, as he urged negotiated solutions to armed conflicts and promoted American trade and investment there. His administration's initial enthusiasm for peacekeeping missions involving U.S. personnel diminished, however, after eighteen American soldiers and more than five hundred Somalis died in an October 1993 firefight in war-torn Mogadishu, where George Bush had first sent marines a year earlier to establish safe routes for famine relief workers. When the genocidal slaughter of half a million Tutsis by Hutus in Rwanda broke out a few months later in 1994, the U.S. government and the other members of the UN Security Council failed to intervene to stop the killing. Later that same year, the Clinton administration did intervene in the much nearer black Caribbean nation of Haiti. Roman Catholic priest and social activist Jean-Bertrand Aristide had been elected president in 1991, only to be overthrown in a right-wing military coup. As the regime's repression sent thousands of Haitians fleeing toward Florida, Clinton used the threat of an American invasion to force the island's military rulers, despite long-standing links to the CIA, to abdicate and return Aristide to office.

Racial cleavages at home still occasionally wounded the reputation of the United States in the 1990s. Race riots engulfed south central Los Angeles in 1992 after a jury found police officers innocent of brutality in the filmed beating of black motorist Rodney King. A UN special investigator in 1998 found that a criminal defendant's race continued to be a crucial factor in determining who received the death penalty in the United States.[2] Los Alamos nuclear physicist Wen Ho Lee was fired, arrested, and kept largely in solitary confinement for seven months in 1999–2000 on suspicions of passing atomic secrets to China. Well before the Justice Department abandoned almost all of its case against Lee (bringing an extraordinary rebuke from the presiding federal judge, who believed he had been misled by prosecutors), a backlash developed against what many saw as racial profiling of the Taiwanese-born U.S. citizen. Even the head of counterintelligence at the Los Alamos laboratory agreed, accusing federal investigators of racial bias in assuming that Lee's ethnicity might incline him toward espionage on behalf of another country.[3]

America's multiracial demographic profile could also be wielded as a positive model for a world suffering fierce ethnic and racial conflicts after the end of the Cold War. President Clinton called the 1999 NATO air war against Yugoslavia, which aimed to end the slaughter of Muslims by Serbs in the province of Kosovo, a victory of ethnic diversity over the euphemistically named "ethnic cleansing." Praising a racially diverse crowd of three thousand service personnel at Whiteman Air Force Base in Missouri, Clinton declared: "I invite the people of this world today who say that people cannot get along across racial and ethnic and religious lines to have a good look at the United

States military, to have a good look at the members of the United States Air Force in this hangar today."[4] With segregation, colonialism, and the Cold War behind them, Americans claimed anew their diversity as an important feature of leadership in the contemporary world.

FROM THE atrocities of World War II to the emancipation of the majority of the world's people from colonial rule, race formed a prominent contour of world history during the second half of the twentieth century. Race relations particularly colored the development of the United States, as the most powerful nation in international affairs and as a society struggling with its own multiracial character. There was no greater weakness for the United States in waging the Cold War than inequality and discrimination. Other problems hindered American leadership abroad after 1945—U.S. economic needs, self-righteousness, parochialism, expansive definitions of national security—but none served to deny so blatantly America's Cold War promises of liberty and equality. To plumb the depths of American difficulties in providing international leadership after the World War II, the changing state of race relations inside and outside the nation's borders must be considered.

Confronted with the erosion of European colonial control of Asia and Africa, U.S. policymakers sought to contain the expansion of Soviet influence by building a multiracial, anti-Communist alliance of the West and the Third World. In pursuit of their goal of a capitalist and preferably democratic world order, American leaders worked to convince newly independent nonwhite nations of the superiority of American institutions and ideals to those of the Soviet Union. Both ends of the North-South axis of colonial tensions might thus be won to the Western side of the East-West axis of the Cold War. The strategy of wooing Asians and Africans bolstered the rising movement in the United States for desegregation and full racial equality. A key tactic was to limit racial polarization, both in the American South and southern Africa, by containing the forces of white racism and channeling the energies of race reformers along moderate lines.

What impact did racial considerations have on U.S. strategic priorities and decisions after 1945? With the question framed in this narrow fashion, the answer is relatively little. Race did not determine the main outlines of American policy toward the Soviet Union or U.S. plans for constructing a global capitalist order, although it did often affect how those priorities were pursued.

But if the question is cast more broadly to ask what the relationship was between the waging of the Cold War and the historic dissolution of global white supremacy, the answer is more complicated and more significant.

In the Third World and particularly southern Africa, U.S. Cold War policies served primarily to slow down the process of ending white rule over people of color. This was especially true from 1945 to 1960. American determination to reconstruct a vibrant, anti-Communist Western Europe translated into tacit support for often bloody efforts to preserve colonial rule abroad by the governments of France, Belgium, Portugal, and Britain. U.S. funding through the Marshall Plan and the NATO alliance contributed to the ability of European powers to wage counterinsurgent wars from Vietnam to Kenya to Angola. The precise overlap of apartheid (1948–1990) with the Cold War (1947–1989) reflected white South Africans' effective use of anti-Communism to preserve the U.S. government's support for their undemocratic rule. Such U.S. policies allowed the USSR, China, and Cuba to appear more supportive of the cause of national independence and racial equality. "The Soviets are probably just as racist as other white Europeans," one black Namibian pointed out, "but at least they support the right cause against South Africa."[5] Decolonization provided independence and formal equality for the nations of the Third World, much as emancipation had done for enslaved individuals a century before. In this largest of post-1945 struggles for liberty, the weight of the U.S. government initially came down on the side of the white powers.

This pattern began to shift after 1960. Colonialism had fallen in Asia and was collapsing in Africa, and African Americans across the South were launching the final phase of the civil rights movement at home. "Problems of race prejudice and discrimination are worldwide," observed the editors of the NAACP journal *The Crisis* in May 1960, as sit-ins spread throughout Dixie and as Africans marched for freedom in South Africa and fought for it in Algeria. "And so is the rebellion against these twin evils . . . We do not think anyone can deny the fact that the present global turbulence is a precursor of the worldwide collapse of white supremacy."[6] Within five years, all but southern Africa was free from white rule and civil rights legislation had erased Jim Crow's legal sanction in the United States. The Johnson administration responded with hostility to the formation of a new white-ruled nation of Rhodesia in November 1965. The contrasting support from Washington and other Western seats of power in 1910 for the creation of a white-ruled South Africa provided a measure of how much had changed in world race relations in the intervening decades. U.S. policy toward South Africa subsequently cooled under the pressure of a growing anti-apartheid lobby, leading eventually to the 1986 sanctions.

Even as it succeeded in Asia and progressed in Africa, the struggle for racial equality was buffeted in the United States in the immediate postwar years by the powerful downdraft of anti-Communist fervor. Truman's initial steps to desegregate the military and the civil service came at the cost of repressing a left-leaning anticolonialist movement in the United States. Only after 1954 did the easing of the Red Scare, the coalescing of a self-proclaimed Third World, and the rising visibility of the Southern freedom movement combine to bring race relations closer to the center of Cold Warriors' attention. "With the Communists reaching out to the uncommitted people of the Middle East and Africa and Southeast Asia," Senator Paul Douglas argued in 1957, "each housing riot in Illinois, each school riot in Kentucky, and each bombing of a pastor's home or intimidation of a would-be Negro voter in Alabama or Mississippi becomes not only an affront to human dignity here in this country, but a defeat for freedom in its tough world struggle for survival."[7] Despite stiff opposition from Southern segregationists, this liberal anti-Communist perspective eventually won out in American politics in the years of the Kennedy and Johnson administrations.

The far-reaching changes that swept through the American South in the second half of the twentieth century cannot be understood apart from the international context of the Cold War. The evolving civil rights movement fit into the larger story of decolonization and the Cold War struggle over world leadership and the meaning of "freedom." The scale of change in the former Confederacy was occasionally startling. Fifty years after his presidential candidacy as an anti-Communist, archsegregationist Dixiecrat in 1948, South Carolina senator Strom Thurmond could be found not only seeking and winning black votes at home but also smiling broadly arm-in-arm before the world's news media with South African president Nelson Mandela, the former anti-apartheid guerilla and Communist Party ally.[8] One need not deny the continuing reality of discrimination and de facto segregation in much of the South to acknowledge how different the region was at the end of the Cold War compared to the beginning. Out of the fires of those changes emerged Southern and border-state presidents who became the strongest supporters of racial equality in the postwar White House.

Among post-1945 presidents, regional background seemed the best predictor—even better than party affiliation—of their degree of enthusiasm for ending racial discrimination once in the White House. Non-Southerners displayed reticence: Dwight Eisenhower of Kansas, Richard Nixon of California, Gerald Ford of Michigan, Ronald Reagan of Illinois, and George Bush originally of Connecticut. Southerners, by contrast, led the way: Harry Truman of Missouri, Lyndon Johnson of Texas, Jimmy Carter of Georgia, and Bill Clinton of Arkansas.[9] The only Democratic president of this period not from below the Mason-Dixon line, John Kennedy, entered office with less clearly defined commitments on race than the others of his party; his shift to full support for civil rights legislation came only as a result of events during his presidency. Presidents who had grown up in the segregated South had been forced to confront racial issues and resolve their own racial uncertainties and commitments more clearly than their peers from outside the region. They were personally familiar with African Americans and the injustices that limited the possibilities of their lives. For Southerners who rose to the Oval Office, racial prejudice and discrimination were not distant problems that could mostly be ignored but personal issues that affected people they knew and the reputation of their own beloved native region.

The Cold War provided the context for the final resolution in American public life and in international affairs of the historic conflict between formal racial hierarchy and racial equality. That resolution frequently involved arms, bloodshed, and high personal costs, paid almost always by people of color. The destruction of white supremacy represented, in world historical terms, a more momentous achievement even than the conquest of Soviet Communism. Prejudice and discrimination and the violence they often spawned did not disappear entirely from American life, but they now had to survive without the support of law or majority opinion. While rarely standing at the front of the march for civil rights and decolonization, American leaders after 1945 accommodated themselves with varying degrees of enthusiasm to the new era of racial equality. By the end of the Cold War the United States had emerged as the multiracial leader of a multiracial world.

NOTES

Prologue

1. "Big Step Ahead on a High Road," *Life*, 8 December 1961, 32–33.
2. "Department Urges Maryland to Pass Public Accommodations Bill," *Department of State Bulletin*, 2 October 1961, 551–552.
3. Recent scholarship examining aspects of the relationship between race and U.S. foreign affairs includes Paul Gordon Lauren, *Power and Prejudice: The Politics and Diplomacy of Racial Discrimination* (Boulder: Westview, 1988); Richard

Drinnon, *Facing West: The Metaphysics of Indian-Hating and Empire-Building* (1980; New York: Schocken, 1990); John W. Dower, *War without Mercy: Race and Power in the Pacific War* (New York: Pantheon, 1986); Michael H. Hunt, *Ideology and U.S. Foreign Policy* (New Haven: Yale University Press, 1987); Brenda Gayle Plummer, *Rising Wind: Black Americans and U.S. Foreign Affairs, 1935–1960* (Chapel Hill: University of North Carolina Press, 1996); Penny M. Von Eschen, *Race against Empire: Black Americans and Anticolonialism, 1937–1957* (Ithaca: Cornell University Press, 1997); Gerald Horne, "Race from Power: U.S. Foreign Policy and the General Crisis of 'White Supremacy,'" *Diplomatic History* 23 (Summer 1999): 437–461; Renee Romano, "No Diplomatic Immunity: African Diplomats, the State Department, and Civil Rights, 1961–1964," *Journal of American History* 87 (September 2000): 546–579; Mary L. Dudziak, "Desegregation as a Cold War Imperative," *Stanford Law Review* 41 (November 1988): 61–120; Alexander DeConde, *Ethnicity, Race, and American Foreign Policy: A History* (Boston: Northeastern University Press, 1992); Thomas Borstelmann, *Apartheid's Reluctant Uncle: The United States and Southern Africa in the Early Cold War* (New York: Oxford University Press, 1993).

4. The phrase "Third World" emerged out of a 1955 conference in Bandung, Indonesia, to refer to those nations unaligned in the Cold War, almost all of which were peopled by non-Europeans (the capitalist West constituting the "First World" and the Communist nations the "Second World"). First used by a French journalist, it quickly became a term of self-designation for nationalists in Africa, Asia, and the Middle East, as well as revolutionaries in Latin America.

5. Lauren, *Power and Prejudice,* 155–156.

6. Roy Preiswerk, "Race and Colour in International Relations," in *The Year Book of World Affairs, 1970* (London: Stevens, 1970), 76, n. 65. On the North and West, see Steven Lawson, *Running for Freedom: Civil Rights and Black Politics in America since 1941* (Philadelphia: Temple University Press, 1991), 172.

7. Herbert Shapiro, *White Violence and Black Response: From Reconstruction to Montgomery* (Amherst: University of Massachusetts Press, 1988), 395–396 (emphasis added).

8. William Minter, *King Solomon's Mines Revisited: Western Interests and the Burdened History of Southern Africa* (New York: Basic Books, 1986); Robert Kinloch Massie, *Loosing the Bonds: The United States and South Africa in the Apartheid Years* (New York: Doubleday, 1997).

9. Henry F. Jackson, *From the Congo to Soweto: U.S. Foreign Policy toward Africa since 1960* (New York: Morrow, 1982), chap. 4.

10. Aldon Morris, "Centuries of Black Protest: Its Significance for America and the World," in *Race in America: The Struggle for Equality,* ed. Herbert Hill and

James E. Jones, Jr. (Madison: University of Wisconsin Press, 1993), 62–66; Harold R. Isaacs, *The New World of Negro Americans* (New York: John Day, 1963), 52; George M. Houser, "Freedom's Struggle Crosses Oceans and Mountains: Martin Luther King, Jr., and the Liberation Struggles in Africa and America," in *We Shall Overcome: Martin Luther King, Jr., and the Black Freedom Struggle,* ed. Peter J. Albert and Ronald Hoffman (New York: Pantheon, 1990), 179.

11. For extensive discussion of these themes, see Von Eschen, *Race against Empire,* and Plummer, *Rising Wind.*

12. Lauren, *Power and Prejudice,* 3; Dower, *War without Mercy,* 145–146.

13. Thomas McCormick, "Drift or Mastery? A Corporatist Synthesis for American Diplomatic History," *Reviews in American History* 10 (December 1982): 326, cited in Elizabeth McKillen, "Historical Contingency and the Peace Progressives," *Diplomatic History* 20 (Winter 1986): 124.

14. Van Gosse, "Consensus and Contradiction in Textbook Treatments of the Sixties," *Journal of American History* 82 (September 1995): 665.

15. The extensive bibliographies of two recent books on the movement suggest how little the international sphere has figured in the writing of civil rights history: Harvard Sitkoff, *The Struggle for Black Equality, 1954–1992,* rev. ed. (New York: Hill and Wang, 1993), and Thomas R. West and James W. Mooney, eds., *To Redeem a Nation: A History and Anthology of the Civil Rights Movement* (St. James, N.Y.: Brandywine, 1993).

16. See, for example, Walter A. McDougall, "President Fails as National Shrink: One Lesson of *Sputnik?*" *Reviews in American History* 21 (December 1993): 701–702; James Chace, *Acheson: The Secretary of State Who Created the American World* (New York: Simon and Schuster, 1998), chapter 17.

17. Isaacs, *The New World of Negro Americans,* xiv.

18. See Michael Omi and Howard Winant, *Racial Formation in the United States: From the 1960s to the 1990s,* 2d ed. (New York: Routledge, 1994), especially 60; Jonathan Marks, *Human Biodiversity: Genes, Race, and History* (New York: Aldine de Gruyter, 1995).

19. Lauren, *Power and Prejudice,* 6 (Plato); Ronald Segal, *The Race War: The World-Wide Conflict of Races* (Harmondsworth, Eng.: Penguin, 1967), 301 (Cicero).

20. Elazar Barkan, *The Retreat of Scientific Racism: Changing Concepts of Race in Britain and the United States between the World Wars* (Cambridge: Cambridge University Press, 1992).

21. David A. Hollinger, "National Solidarity at the End of the Twentieth Century: Reflections on the United States and Liberal Nationalism," *Journal of American History* 84 (September 1997): 567.

22. Harold R. Isaacs, *Scratches on Our Minds: American Images of China and India*

(1958; Westport, Conn.: Greenwood, 1973), 283–284; Ronald Takaki, *Strangers from a Different Shore: A History of Asian Americans* (Boston: Little, Brown, 1989), 298–299.

23. Beth Bailey and David Farber, *The First Strange Place: The Alchemy of Race and Sex in World War II Hawaii* (New York: Free Press, 1992), 139.

24. See, for example, R. Gordon Wasson, "Popular Fallacies about Russia," *U.S. News and World Report,* 16 March 1951, 30–33. For a wide-ranging discussion of the problems with such continental designations, see Martin W. Lewis and Kären E. Wigen, *The Myth of Continents: A Critique of Metageography* (Berkeley: University of California Press, 1997).

25. Numan V. Bartley, *The Rise of Massive Resistance: Race and Politics in the South during the 1950s* (Baton Rouge: Louisiana State University Press, 1969), viii.

26. F. James Davis, *Who Is Black? One Nation's Definition* (University Park: Pennsylvania State University Press, 1991), especially 12–13.

27. Davis, *Who Is Black?* 7.

28. Madeleine G. Kalb, *The Congo Cables: The Cold War in Africa—from Eisenhower to Kennedy* (New York: Macmillan, 1982), 40.

29. William Attwood, *The Reds and the Blacks: A Personal Adventure* (New York: Harper and Row, 1967), 188.

1. Race and Foreign Relations before 1945

1. Edmund S. Morgan, *American Slavery, American Freedom: The Ordeal of Colonial Virginia* (New York: Norton, 1975), 6–18 (quotation at 13).

2. Gary B. Nash, *Red, White, and Black: The Peoples of Early America* (Englewood Cliffs, N.J.: Prentice-Hall, 1974, 1982); Drinnon, *Facing West;* Ronald Dale Kerr, "'Why Should You Be So Furious?' The Violence of the Pequot War," *Journal of American History* 85 (December 1998): 876–909.

3. Reginald Horsman, *Race and Manifest Destiny: The Origins of American Anglo-Saxonism* (Cambridge, Mass.: Harvard University Press, 1981), 1–6, 43–61, 208–229, 297; Thomas R. Hietala, *Manifest Design: Anxious Aggrandizement in Late Jacksonian America* (Ithaca: Cornell University Press, 1985), 10–54, 132–172.

4. Horne, "Race from Power," 443 (Stoddard); Lauren, *Power and Prejudice,* 54, 63–64; Isaacs, *The New World of Negro Americans,* 32–33; Hugh Tinker, *Race, Conflict, and the International Order: From Empire to United Nations* (New York: St. Martin's, 1977), 21; Gary Gerstle, "Liberty, Coercion, and the Making of Americans," *Journal of American History* 84 (September 1997): 549; Eric Tyrone Lowery Love, "Race over Empire: Racism and United States Imperialism, 1865–1890" (Ph.D. diss., Princeton University, 1997), 15.

5. Hunt, *Ideology and U.S. Foreign Policy,* 46–91, 116–117.

6. Peter Novick, *That Noble Dream: The "Objectivity Question" and the American Historical Profession* (New York: Cambridge University Press, 1988), 80–81.

7. James O. Gump, *The Dust Rose Like Smoke: The Subjugation of the Zulu and the Sioux* (Lincoln: University of Nebraska Press, 1994), 56.

8. Hunt, *Ideology and U.S. Foreign Policy,* 55.

9. Dower, *War without Mercy,* 151.

10. Walter L. Williams, "United States Indian Policy and the Debate over Philippine Annexation: Implications for the Origins of American Imperialism," *Journal of American History* 66 (March 1980): 810–831.

11. Dennis Hickey and Kenneth C. Wylie, *An Enchanting Darkness: The American Vision of Africa in the Twentieth Century* (East Lansing: Michigan State University Press, 1993), 19.

12. Alan M. Tigay, "The Deepest South," *American Heritage,* April 1998, 84–95.

13. Hazel M. McFerson, *The Racial Dimension of American Overseas Colonial Policy* (Westport, Conn.: Greenwood, 1997), 3. Over a century later, the one other possible exception is Puerto Rico, but it has not yet signaled a clear desire for statehood.

14. Thomas Pakenham, *The Scramble for Africa, 1876–1912* (New York: Random House, 1991), 28; Hickey and Wylie, *An Enchanting Darkness,* 8–13. On Stanley, see John Bierman, *Dark Safari: The Life behind the Legend of Henry Morton Stanley* (New York: Knopf, 1990). On the Congo Free State, see Adam Hochschild, *King Leopold's Ghost: A Story of Greed, Terror, and Heroism in Colonial Africa* (Boston: Houghton Mifflin, 1998).

15. Thomas G. Dyer, *Theodore Roosevelt and the Idea of Race* (Baton Rouge: Louisiana State University Press, 1980), especially 2, 19–20, 69–88, 114–115, 129–139; Hickey and Wylie, *An Enchanting Darkness,* 12–13. See also Michael McCarthy, *Dark Continent: Africa as Seen by Americans* (Westport, Conn.: Greenwood, 1983).

16. John Taliaferro, *Tarzan Forever: The Life of Edgar Rice Burroughs, Creator of Tarzan* (New York: Scribner, 1999); Hickey and Wylie, *An Enchanting Darkness,* 180–185.

17. Christopher Lasch, "The Anti-Imperialists, the Philippines, and the Inequality of Man," *Journal of Southern History* 24 (August 1954): 319–331.

18. Love, "Race over Empire," 83–84, 157–158, 162–164, 198; McFerson, *The Racial Dimension,* 2–3; Williams, "United States Indian Policy," 822.

19. Williams, "United States Indian Policy," quotations at 819, 825.

20. Ibid., 827–229; Dower, *War without Mercy,* 152.

21. Vicente L. Rafael, "White Love: Surveillance and Nationalist Resistance in the Colonization of the Philippines," in *Cultures of United States Imperialism,* ed.

Amy Kaplan and Donald E. Pease (Durham: Duke University Press, 1993), 185–186.

22. Drinnon, *Facing West,* 327–328 (quotation); Willard B. Gatewood, Jr., *Black Americans and the White Man's Burden, 1898–1903* (Urbana: University of Illinois Press, 1975), 261–279; Dower, *War without Mercy,* 152.

23. Drinnon, *Facing West,* 314.

24. Ibid., 309.

25. Lauren, *Power and Prejudice,* 61. For recent discussions of the German and Belgian cases and some of their continuing impacts, see Adam Hochschild, "Mr. Kurtz, I Presume," *New Yorker,* 14 April 1997, 40–47, and Donald G. McNeil, Jr., "Its Past on Its Sleeve, Tribe Seeks Bonn's Apology," *New York Times,* 31 May 1998, 3.

26. Anders Stephanson, *Manifest Destiny: American Expansionism and the Empire of Right* (New York: Hill and Wang, 1995), 104.

27. Gatewood, *Black Americans and the White Man's Burden,* 114–116, 129–153, 197–204, 241–245, 322–323; David S. Cecelski and Timothy B. Tyson, eds., *Democracy Betrayed: The Wilmington Race Riot of 1898 and Its Legacy* (Chapel Hill: University of North Carolina Press, 1998).

28. Bernard C. Nalty, *Strength for the Fight: A History of Black Americans in the Military* (New York: Free Press, 1986), 74–77; Willard B. Gatewood, Jr., *"Smoked Yankees" and the Struggle for Empire: Letters from Negro Soldiers, 1898–1902* (Urbana: University of Illinois Press, 1971), especially 3–6, 243–45; Kevin Gaines, "Black Americans' Racial Uplift Ideology as 'Civilizing Mission': Pauline E. Hopkins on Race and Imperialism," in *Cultures of United States Imperialism,* ed. Amy Kaplan and Donald E. Pease (Durham: Duke University Press, 1993), 441; Gatewood, *Black Americans and the White Man's Burden,* 183–186; Piero Gleijeses, "African Americans and the War against Spain," *North Carolina Historical Review* 23 (April 1996): 184–214.

29. Gatewood, *"Smoked Yankees,"* 199–200.

30. Stuart Anderson, "Racial Anglo-Saxonism and the American Response to the Boer War," *Diplomatic History* 2 (Summer 1978): 219–236 (quotation at 225); Thomas J. Noer, *Briton, Boer, and Yankee: The United States and South Africa, 1870–1914* (Kent, Ohio: Kent State University Press, 1978).

31. Anderson, "Racial Anglo-Saxonism," 229.

32. Isaacs, *The New World of Negro Americans,* 34–35; Pakenham, *The Scramble for Africa,* 470–486; Lauren, *Power and Prejudice,* 65.

33. Prasenjit Duara, "Transnationalism and the Predicament of Sovereignty: China, 1900–1945," *American Historical Review* 102 (October 1997): 1038.

34. Reginald Kearney, "Afro-American Views of the Japanese, 1900–1945" (Ph.D. diss., Kent State University, 1991), 34–64.

35. Lauren, *Power and Prejudice*, 67.

36. Walter LaFeber, *The Clash: U.S.-Japanese Relations throughout History* (New York: Norton, 1997), 80.

37. Roy Preiswerk, "Race and Colour in International Relations," in *The Year Book of World Affairs, 1970* (London: Stevens, 1970), 58.

38. Nell Irvin Painter, *Standing at Armageddon: The United States, 1877–1919* (New York: Norton, 1987), 364 (quotation); R. J. Vincent, "Racial Equality," in *The Expansion of International Society*, ed. Hedley Bull and Adam Watson (Oxford: Clarendon, 1984), 240–241; Lauren, *Power and Prejudice*, 69–71.

39. DeConde, *Ethnicity, Race, and American Foreign Policy*, 86.

40. Dimitri D. Lazo, "Lansing, Wilson, and the Jenkins Incident," *Diplomatic History* 22 (Spring 1998): 189 (quotation); David S. Foglesong, *America's Secret War against Bolshevism: U.S. Intervention in the Russian Civil War, 1917–1920* (Chapel Hill: University of North Carolina Press, 1995), 16; David S. Foglesong, letter to the editor, *The Society for Historians of American Foreign Relations Newsletter* 30 (June 1990): 17–25.

41. Hunt, *Ideology and U.S. Foreign Policy*, 115.

42. Herbert Shapiro, *White Violence and Black Response: From Reconstruction to Montgomery* (Amherst: University of Massachusetts Press, 1988), 145–157; Painter, *Standing at Armageddon*, 337–338, 363–365.

43. Isaacs, *The New World of Negro Americans*, 39 (quotation); Painter, *Standing at Armageddon*, 338.

44. Barkan, *The Retreat of Scientific Racism*, 57–59; Graham Smith, *When Jim Crow Met John Bull: Black American Soldiers in World War II Britain* (New York: St. Martin's, 1988), 17; Tyler Stovall, "The Color Line behind the Lines: Racial Violence in France during the Great War," *American Historical Review* 103 (June 1998): 737–769.

45. Lauren, *Power and Prejudice*, 106.

46. Steven A. Reich, "Soldiers of Democracy: Black Texans and the Fight for Citizenship, 1917–1921," *Journal of American History* 82 (March 1996): 1485–1486; C. Eric Lincoln, "The Race Problem and International Relations," *New South* 21 (Fall 1966): 6; Lauren, *Power and Prejudice*, 71; Smith, *When Jim Crow Met John Bull*, 7–17; Horne, "Race from Power," 450 (Wilson). On the generally respectful French treatment of colonial soldiers and its contrast to French hostility toward colonial workers in France during the war, see Stovall, "The Color Line behind the Lines."

47. Thomas A. Johnson, "Negro Veteran Is Confused and Bitter," *New York Times*, 29 July 1968, 14.

48. Wayne Addison Clark, "An Analysis of the Relationship between Anti-Communism and Segregationist Thought in the Deep South, 1948–1964" (Ph.D.

diss., University of North Carolina, 1976), 9–10 (quotation); Alexander Cockburn, "Beat the Devil," *Nation,* 3 May 1993, 582; George Frederickson, *Black Liberation: A Comparative History of Black Ideologies in the United States and South Africa* (New York: Oxford University Press, 1995), 140. Wilson worried that the "American Negro would be our greatest medium in conveying bolshevism to America." Melvyn P. Leffler, *The Spectre of Communism: The United States and the Origins of the Cold War, 1917–1953* (New York: Hill and Wang, 1994), 14.

49. Paul Gordon Lauren, "Human Rights in History: Diplomacy and Racial Equality at the Paris Peace Conference," *Diplomatic History* 2 (Summer 1978): 257–278; Lauren, *Power and Prejudice,* 82–98 (quotation at 93).

50. F. M. Carroll, ed., *The American Commission on Irish Independence, 1919: The Diary, Correspondence, and Report* (Dublin: Mount Salus, 1985), 37–38. Thanks to Rebecca Berens-Matzke for bringing this source to my attention.

51. Lauren, *Power and Prejudice,* 100–101 (quotation); LaFeber, *The Clash,* 123–124.

52. Kearney, "Afro-American Views of the Japanese," 87–116; Lauren, *Power and Prejudice,* 77–104.

53. Lauren, *Power and Prejudice,* 113.

54. Scott Ellsworth, *Death in a Promised Land: The Tulsa Race Riot of 1921* (Baton Rouge: Louisiana State University Press, 1982), 23 (quotation); Shapiro, *White Violence and Black Response,* 180–185.

55. Patricia Sullivan, *Days of Hope: Race and Democracy in the New Deal Era* (Chapel Hill: University of North Carolina Press, 1996), 27.

56. Gilbert C. Fite, *Richard B. Russell, Jr., Senator from Georgia* (Chapel Hill: University of North Carolina Press, 1991), 37–39, 166–168. See also Robert L. Zangrando, *The NAACP Crusade against Lynching* (Philadelphia: Temple University Press, 1980).

57. Kenneth O'Reilly, *"Racial Matters": The FBI's Secret File on Black America, 1960–1972* (New York: Free Press, 1989), 13–18; Theodore Kornweibel, Jr., *"Seeing Red": Federal Campaigns against Black Militancy, 1919–1925* (Bloomington: Indiana University Press, 1998).

58. Mae M. Ngai, "The Architecture of Race in American Immigration Law: A Reexamination of the Immigration Act of 1924," *Journal of American History* 86 (June 1999): 67–92; David M. Reimers, *Still the Golden Door: The Third World Comes to America,* 2d ed. (New York: Columbia University Press, 1992), 6–7; LaFeber, *The Clash,* 144–146; Lauren, *Power and Prejudice,* 110–111.

59. Alfred E. Opubur and Adebayo Ogunbi, "Ooga Booga: The African Image in American Films," in *Other Voices, Other Views: An International Collection of Essays from the Bicentennial,* ed. Robin W. Winks (Westport, Conn.: Greenwood, 1978), 350–357; Taylor Branch, *Parting the Waters: America in the King*

Years, 1954–1963 (New York: Simon and Schuster, 1988), 54. See also Edward H. McKinley, *The Lure of Africa: American Interests in Tropical Africa, 1919–1939* (Indianapolis: Bobbs-Merrill, 1974).

60. Barkan, *The Retreat of Scientific Racism*, xi–xii, 2, 76–90; Frederickson, *Black Liberation*, 213–214; Vincent, "Racial Equality," 246; Ashley Montagu, *Man's Most Dangerous Myth: The Fallacy of Race* (New York: Columbia University Press, 1942); Lee D. Baker, *From Savage to Negro: Anthropology and the Construction of Race, 1896–1954* (Berkeley: University of California Press, 1999).

61. Lauren, *Power and Prejudice*, 123–128 (quotation at 128).

62. Ibid., 119.

63. John Hope Franklin, *From Slavery to Freedom*, 5th ed. (New York: Knopf, 1980), 422, cited in Kearney, "Afro-American Views of the Japanese," 119. See also Joseph E. Harris, *African-American Reactions to War in Ethiopia, 1936–1941* (Baton Rouge: Louisiana State University Press, 1994); William R. Scott, *The Sons of Sheba's Race: African-Americans and the Italo-Ethiopian War, 1935–1941* (Bloomington: Indiana University Press, 1993); and the nicely subtitled *African Americans in the Spanish Civil War: "It Ain't Ethiopia, But It'll Do,"* ed. Danny Duncan Collum (New York: G. K. Hall, 1992).

64. Frederickson, *Black Liberation*, 151, 155.

65. Text of Atlantic Charter, 14 August 1941, in *FRUS, 1941*, 1:367–369. See also *The Atlantic Charter*, ed. Douglas Brinkley and David R. Facey-Crowther (New York: St. Martin's, 1994).

66. Lauren, *Power and Prejudice*, 154.

67. Piero Gleijeses, *Shattered Hope: The Guatemalan Revolution and the United States, 1944–1954* (Princeton: Princeton University Press, 1991), 22–23.

68. C. Vann Woodward, *The Strange Career of Jim Crow*, 2d ed. (London: Oxford University Press, 1966), 131.

69. Fred Pollock and Warren F. Kimball, "'In Search of Monsters to Destroy': Roosevelt and Colonialism," in *The Juggler: Franklin Roosevelt as Wartime Statesman*, ed. Warren F. Kimball (Princeton: Princeton University Press, 1991), 127–157; Lloyd C. Gardner, "The Atlantic Charter: Idea and Reality, 1942–1945," in Brinkley and Facey-Crowther, *The Atlantic Charter*, 45–81; Warren F. Kimball, "The Atlantic Charter: 'With All Deliberate Speed,'" in Brinkley and Facey-Crowther, *The Atlantic Charter*, 83–114; Christopher Thorne, *Allies of a Kind: The United States, Britain, and the War against Japan, 1941–1945* (New York: Oxford University Press, 1978), 158–159.

70. Bailey and Farber, *The First Strange Place*, 20–25, 134–135, 152–157, 165 (quotation at 157).

71. Dower, *War without Mercy*, 176–177 (quotation); Thorne, *Allies of a Kind*, 6; Kearney, "Afro-American Views of the Japanese," 181–182; LaFeber, *The Clash*,

217. When asked by a U.S. official about the effectiveness of Japanese propaganda among African Americans, one Howard University professor responded by telling him about a white officer who had used the term "yellow bastards" in a pep talk to the Howard ROTC unit. This did not encourage patriotic feelings in the black students. "You know," he told the official levelly, "many of us are yellow bastards, second and third generation." Isaacs, *The New World of Negro Americans,* 30.

72. Leonard Dinnerstein, *Antisemitism in America* (New York: Oxford University Press, 1994), 105–149; Ronald Steel, *Walter Lippmann and the American Century* (1980; New York: Vintage, 1981), 372–373; David S. Wyman, *The Abandonment of the Jews: America and the Holocaust, 1941–1945* (New York: Pantheon, 1984), 6–15, 42–58, 103, 189–192, 305–340.

73. Barkan, *The Retreat of Scientific Racism,* 1, 334; Preiswerk, "Race and Colour in International Relations," 59; Sherie Mershon and Steven Schlossman, *Foxholes and Color Lines: Desegregating the U.S. Armed Forces* (Baltimore: Johns Hopkins University Press, 1998), 105–114; Novick, *That Noble Dream,* 348.

74. Morton Sosna, "Introduction," in *Remaking Dixie: The Impact of World War II on the American South,* ed. Neil R. McMillen (Jackson: University Press of Mississippi, 1997), xvi–xvii; Richard M. Dalfiume, *Desegregation of the U.S. Armed Forces: Fighting on Two Fronts, 1939–1953* (Columbia: University of Missouri Press, 1969), 49; Mershon and Schlossman, *Foxholes and Color Lines,* 77–78; Wallace Terry, *Bloods: An Oral History of the Vietnam War by Black Veterans* (New York: Ballantine, 1984), 145–147.

75. Richard Polenberg, "The Good War? A Reappraisal of How World War II Affected American Society," *Virginia Magazine of History and Biography* 100 (July 1992): 321; Dower, *War without Mercy,* 348; Spencie Love, *One Blood: The Death and Resurrection of Charles R. Drew* (Chapel Hill: University of North Carolina Press, 1996).

76. Carr to President's Committee on Civil Rights, "Puerto Ricans in the United States," 10 June 1947, President's Committee on Civil Rights—Committee Documents, box 17, HSTL; Phillip McGuire, *Taps for a Jim Crow Army: Letters from Black Soldiers in World War II* (Santa Barbara: ABC-Clio, 1983), 227–241; Mershon and Schlossman, *Foxholes and Color Lines,* 67.

77. "Because Their Skins Are Brown," *London News and Chronicle,* 8 May 1943, clipping in Philleo Nash Papers, box 17, HSTL; David Reynolds, *Rich Relations: The American Occupation of Britain, 1942–1945* (New York: Random House, 1995), 216–237, 302–324.

78. Reynolds, *Rich Relations,* 220, 223.

79. Ibid., 218.

80. Walter F. White, *A Rising Wind* (Garden City, N.Y.: Doubleday, 1945), 146;

Smith, *When Jim Crow Met John Bull*, 31–32, 85–87, 120–129; Reynolds, *Rich Relations*, 302–306 (quotation at 302).

81. Nalty, *Strength for the Fight*, 154–155; Smith, *When Jim Crow Met John Bull*, 138–151; Reynolds, *Rich Relations*, 319–320, 324 (quotation).

82. Harvard Sitkoff, "Racial Militancy and Interracial Violence in the Second World War," *Journal of American History* 58 (December 1971): 667–681; James Albert Burran, "Racial Violence in the South during World War II" (Ph.D. diss., University of Tennessee, 1977); Pete Daniel, "Going Among Strangers: Southern Reactions to World War II," *Journal of American History* 77 (December 1990): 886–911; Neil A. Wynn, *The Afro-American and the Second World War*, rev. ed. (New York: Holmes and Meier, 1993), 60–78; Neil R. McMillen, "Fighting for What We Didn't Have: How Mississippi's Black Veterans Remember World War II," in *Remaking Dixie: The Impact of World War II on the American South*, ed. Neil R. McMillen (Jackson: University Press of Mississippi, 1997), 93–110; Shapiro, *White Violence and Black Response*, 310–341; John Dittmer, *Local People: The Struggle for Civil Rights in Mississippi* (Urbana: University of Illinois Press, 1994), 16–17; Sullivan, *Days of Hope*, 136–137; McGuire, *Taps for a Jim Crow Army*, 201–203. Most racial conflicts involved whites and blacks, but the zoot suit riots in Los Angeles in June 1943, in which white sailors assaulted Mexican American youths, demonstrated that Latinos were also vulnerable to outbreaks of virulent white racism. See Edward J. Escobar, "Zoot-Suiters and Cops: Chicano Youth and the Los Angeles Police Department during World War II," in *The War in American Culture: Society and Consciousness during World War II*, ed. Lewis A. Erenberg and Susan E. Hirsch (Chicago: University of Chicago Press, 1996), 284–309.

83. Richard Polenberg, *War and Society: The United States, 1941–1945* (Philadelphia: Lippincott, 1972), 108, 117–118, 124, 128; George H. Roeder, Jr., "Censoring Disorder: American Visual Imagery of World War II," in *The War in American Culture: Society and Consciousness during World War II*, ed. Lewis A. Erenberg and Susan E. Hirsch (Chicago: University of Chicago Press, 1996), 46, 56–58, 64; W. E. B. Du Bois, *Color and Democracy: Colonies and Peace* (New York: Harcourt, Brace, 1945), 85–92; Mershon and Schlossman, *Foxholes and Color Lines*, 124–127; *Freedom to Serve—Equality of Treatment and Opportunity in the Armed Services: A Report by the President's Committee* (Washington: U.S. Government Printing Office, 1950), 34–35; Borstelmann, *Apartheid's Reluctant Uncle*, 20.

84. Lauren, *Power and Prejudice*, 140–141.

85. "Chinese Exclusion Act Is Keynote of Japanese Propaganda to Asia," 1943, Nash Papers, box 10, HSTL.

86. White, *A Rising Wind,* 152–154 (quotation); Tinker, *Race, Conflict, and the International Order,* 44.

87. FBI, "Survey of Racial Conditions in the United States" [1943–44], WHCF, box 21, HSTL; Gary Gerstle, "The Working Class Goes to War," in *The War in American Culture: Society and Consciousness during World War II,* ed. Lewis A. Erenberg and Susan E. Hirsch (Chicago: University of Chicago Press, 1996), 118–119 (quotation); Kearney, "Afro-American Views of the Japanese," vi; Dalfiume, *Desegregation of the U.S. Armed Forces,* 110–111; Isaacs, *The New World of Negro Americans,* 30; Polenberg, *War and Society,* 101; Lauren, *Power and Prejudice,* 131.

88. Roger Daniels, *Prisoners without Trial: Japanese Americans in World War II* (New York: Hill and Wang, 1993); Ronald Takaki, *Strangers from a Different Shore: A History of Asian Americans* (Boston: Little, Brown, 1989), 379–392; Polenberg, "The Good War?" 319; Tinker, *Race, Conflict, and the International Order,* 45.

89. Polenberg, *War and Society,* 62.

90. David Campbell, *Writing Security: United States Foreign Policy and the Politics of Identity* (Minneapolis: University of Minnesota Press, 1992), 191, n. 85.

91. Gary Y. Okihiro, *Margins and Mainstream: Asians in American History and Culture* (Seattle: University of Washington Press, 1994), 169. In what must have been the supreme irony of their service, Japanese Americans in the segregated 442d Regimented Combat Team—which became the most decorated unit in the entire U.S. Army—even helped to liberate the Nazi death camp at Dachau, while their own families were still incarcerated in American concentration camps. Daniels, *Prisoners without Trial,* 64.

92. Takaki, *Strangers from a Different Shore,* 396–403.

93. Dower, *War without Mercy,* 36, 104–105. On the absence of enemy corpse mutilation by American soldiers in the European theater in comparison with its presence in the Pacific, see Roeder, "Censoring Disorder," 52.

94. Polenberg, "The Good War?" 319.

95. Isaacs, *The New World of Negro Americans,* 4. See also Jonathan G. Utley, *Going to War with Japan, 1937–1941* (Knoxville: University of Tennessee Press, 1985), 103.

96. Tom Englehardt, *The End of Victory Culture: Cold War America and the Disillusioning of a Generation* (New York: Basic Books, 1995), 98; Dower, *War without Mercy,* 164–170 (quotation at 170); DeConde, *Ethnicity, Race, and American Foreign Policy,* 121; Takaki, *Strangers from a Different Shore,* 367–378; Isaacs, *The New World of Negro Americans,* 29.

97. Aimé Césaire, *Discourse on Colonialism,* trans. Joan Pinkam (1955; New York: Monthly Review Press, 1972), 14; Ndabaningi Sithole, *African Nationalism*

(Cape Town: Oxford University Press, 1959), 19–23; Pakenham, *The Scramble for Africa,* 673–674; Lauren, *Power and Prejudice,* 161; Borstelmann, *Apartheid's Reluctant Uncle,* 23–25.

98. Isaacs, *The New World of Negro Americans,* 30–31; Borstelmann, *Apartheid's Reluctant Uncle,* 38–52.

99. Du Bois, *Color and Democracy;* White, *A Rising Wind;* Rayford W. Logan, *What the Negro Wants* (Chapel Hill: University of North Carolina Press, 1944). See also Plummer, *Rising Wind,* and Von Eschen, *Race against Empire.*

100. Minutes of meeting of U.S. delegation to UN, 26 May 1945, *FRUS, 1945,* 1:894 (quotation); Robert L. Harris, "Racial Equality and the United Nations Charter," in *New Directions in Civil Rights Studies,* ed. Armstead L. Robinson and Patricia Sullivan (Charlottesville: University Press of Virginia, 1991), 126–148; Lauren, *Power and Prejudice,* 146–157.

101. FBI, "Survey of Racial Conditions in the US" [1943–44], WHCF, box 21, HSTL; McMillen, "Fighting for What We Didn't Have," 101 (quotation); Charles M. Payne, *I've Got the Light of Freedom: The Organizing Tradition and the Mississippi Freedom Struggle* (Berkeley: University of California Press, 1995), 30–31; Harvard Sitkoff, "African American Militancy in the World War II South: Another Perspective," in *Remaking Dixie: The Impact of World War II on the American South,* ed. Neil R. McMillen (Jackson: University Press of Mississippi, 1997), 70–92; Polenberg, "The Good War?" 321; Wynn, *The Afro-American and the Second World War,* 134–135; Lawson, *Running for Freedom,* 9.

102. Thomas J. Schoenbaum, *Waging Peace and War: Dean Rusk in the Truman, Kennedy, and Johnson Years* (New York: Simon and Schuster, 1988), 78; White, *A Rising Wind,* 15, 35–38, 123ff. African American historian John Hope Franklin discovered a similar sense of shifting attitudes in the Louisiana state archivist who sneaked him into the officially segregated archives in Baton Rouge on V-J day in 1945, letting him work while the rest of the town celebrated the end of the war. Peter Applebome, "John Hope Franklin," *New York Times Magazine,* 23 April 1995, 37.

103. McMillen, "Fighting for What We Didn't Have," 96, 103.

104. David W. Blight, "In Retrospect: Nathan Irvin Huggins, the Art of History, and the Irony of the American Dream, *Reviews in American History* 22 (March 1994): 184; Michael Wreszin, "Radical in Wisconsin," *Nation,* 1 January 1996, 31; Mershon and Schlossman, *Foxholes and Color Lines,* 82; McMillen, "Fighting for What We Didn't Have," 102; White, *A Rising Wind,* 36; Applebome, "John Hope Franklin," 36.

105. Michael Cassity, "History and the Public Purpose," *Journal of American History* 81 (December 1994): 972–973.

106. Wynn, *The Afro-American and the Second World War,* 103–106, 133, 135 (Dixon); Sullivan, *Days of Hope,* 212 (Talmadge); Isaacs, *The New World of Negro Americans,* 114; Polenberg, *War and Society,* 50; Patrick S. Washburn, *A Question of Sedition: The Federal Government's Investigation of the Black Press during World War II* (New York: Oxford University Press, 1986).

2. Jim Crow's Coming Out

1. Harold R. Isaacs, "The Political and Psychological Context of Point Four," *Annals of the American Academy of Political and Social Science* 270 (July 1950): 56.
2. Lauren, *Power and Prejudice,* 187.
3. Article 1 of the United Nations Charter, quoted in Fahy to Carr, 17 June 1947, PCCR, box 6, HSTL.
4. Wallace Stegner, *One Nation* (Boston: Houghton Mifflin, 1945).
5. Isaacs, "The Political and Psychological Context," 51.
6. Edmund Davison Soper, *Racism: A World Issue* (New York: Abingdon-Cokesbury, [1947]), 272. See also Lauren, *Power and Prejudice;* Plummer, *Rising Wind;* Von Eschen, *Race against Empire.*
7. Melvyn Leffler, *A Preponderance of Power: National Security, the Truman Administration, and the Cold War* (Stanford: Stanford University Press, 1992).
8. NSC 68, "United States Objectives and Programs for National Security," 14 April 1950, *FRUS, 1950,* 1:234–292.
9. For the U.S. government's awareness of the potential significance of more inclusive Soviet racial policies, see CIA, ORE 25–48, "The Break-Up of the Colonial Empires and Its Implications for U.S. Security," 3 September 1948, Harry S Truman Papers, President's Secretary's File, box 255, HSTL.
10. David McCullough, *Truman* (New York: Simon and Schuster, 1992), 53 (quotation), 83; Alonzo L. Hamby, *Man of the People: The Life of Harry Truman* (New York: Oxford University Press, 1995), 4–5; William E. Leuchtenburg, "The Conversion of Harry Truman," *American Heritage* 42 (November 1991): 56–57.
11. Hamby, *Man of the People,* 238–239, 272, 364–365, 631; McCullough, *Truman,* 234, 247, 971–972; Donald R. McCoy and Richard T. Ruetten, *Quest and Response: Minority Rights and the Truman Administration* (Lawrence: University Press of Kansas, 1973), 14–16; Leuchtenburg, "The Conversion of Harry Truman," 68; Robert H. Ferrell, *Harry S Truman: A Life* (Columbia: University of Missouri Press, 1994), 292–299. For a very critical view that still acknowledges personal growth, see Kenneth O'Reilly, *Nixon's Piano: Presidents and Racial Politics from Washington to Clinton* (New York: Free Press, 1995), 145–165.
12. Dalfiume, *Desegregation of the U.S. Armed Forces,* 140–141; McCullough, *Tru-*

man, 323; Hamby, *Man of the People,* 271–273; Barton J. Bernstein, "The Ambiguous Legacy: The Truman Administration and Civil Rights," in Bernstein, ed., *Politics and Policies of the Truman Administration* (Chicago: Quadrangle, 1970), 271–272.

13. See, for example, Clifford's "Memorandum for the President," 19 November 1947, Clark Clifford Papers, box 23, HSTL.

14. David A. Mayers, *George Kennan and the Dilemmas of U.S. Foreign Policy* (New York: Oxford University Press, 1988), 56.

15. Kennan diary, 1933–1938, "Fair Day, Adieu!" George F. Kennan Papers, box 25, Seeley G. Mudd Library, Princeton University, Princeton, New Jersey. See also Mayers, *George Kennan and Dilemmas of U.S. Policy,* 49–53; Walter Isaacson and Evan Thomas, *The Wise Men—Six Friends and the World They Made: Acheson, Bohlen, Harrison, Kennan, Lovett, McCloy* (New York: Simon and Schuster, 1986), 171–172.

16. George F. Kennan, *Memoirs: 1925–1950* (1967; New York: Pantheon, 1983), 181.

17. Kennan, *Memoirs: 1925–1950,* 182–185; Michael H. Hunt, *Ideology and U.S. Foreign Policy* (New Haven: Yale University Press, 1987), 162–163; Mayers, *George Kennan and Dilemmas of U.S. Policy,* 271–272; Anders Stephanson, *Kennan and the Art of Foreign Policy* (Cambridge, Mass.: Harvard University Press, 1989), 157–175.

18. Kennan, *Memoirs: 1925–1950,* 476–482; Kennan memo to Acheson, 29 March 1950, *FRUS, 1950,* 2:601–602; Roger R. Trask, "George F. Kennan's Report on Latin America (1950)," *Diplomatic History* 2 (Summer 1978): 307–311.

19. Kennan, "Tasks Ahead in U.S. Foreign Policy," lecture at the National War College, Washington, D.C., 18 December 1952, box 18, Kennan Papers, Mudd Library.

20. Kennan, *Memoirs: 1925–1950,* 74 (quoting from a paper he wrote in 1938).

21. Ibid., 551. Other Americans in Moscow shared Kennan's views: Leslie C. Stevens, the naval attaché in the U.S. embassy from 1947 to 1949, referred to Russians' "half-Asiatic minds." Stevens, *Russian Assignment* (Boston: Little, Brown, 1953), 187.

22. Kennan, lecture 2 at Bad Nauheim, 1942, box 16, Kennan Papers, Mudd Library, cited in n. 73 of David S. Foglesong, "Liberating Russia? American Thinking about the Russian Future and the Psychological Warfare of the Early Cold War, 1948–1953," unpublished manuscript, 1997 (in author's possession). For further discussion of these issues, see David S. Foglesong, "Roots of 'Liberation': American Images of the Future of Russia in the Early Cold War, 1948–1953," *International History Review* 21 (March 1999): 57–79.

23. See, for example, Kennan, *Memoirs: 1925–1950,* 483–484.

24. Dower, *War without Mercy,* 310.

25. John Egerton, *Speak Now against the Day: The Generation before the Civil Rights Movement in the South* (New York: Knopf, 1994), 576–577. Byrnes's career after he resigned from the State Department in 1947 reflected his primary commitment to racial segregation. The former Supreme Court justice, World War II administrator, and secretary of state went home to South Carolina, where he served as governor from 1950 to 1954 and energetically opposed all movement toward desegregation. As the national Democratic Party and the Supreme Court shifted away from supporting Jim Crow, Byrnes grew increasingly alienated from those institutions that had been so central to his life and career. His trajectory from the apex of national life to a besieged regional redoubt reflected the manner in which changing race relations altered American political life in the decades after World War II.

26. George Streator, "Negro Youth Told Future Is in South," *New York Times,* 21 October 1946, 31.

27. Carr to Marshall, 23 May 1947, and Carr to Rusk, 11 August 1947, both in PCCR, box 6, HSTL; Marshall to Diplomatic and Consular Offices, 20 July 1948, *FRUS, 1948,* 1:595; William T. Bowers, William M. Hammond, and George L. MacGarrigle, *Black Soldier, White Army: The 24th Infantry Regiment in Korea* (Washington: Center of Military History, United States Army, 1996), 23. Marshall's initial response to the 1941 Atlantic Charter (when he was army chief of staff) revealed something of his negative view of self-government among nonwhite peoples abroad. Was this declaration of American war aims, he asked privately, "an effort to keep the people entertained?" Kimball, "The Atlantic Charter: 'With All Deliberate Speed,'" in Brinkley and Facey-Crowther, *The Atlantic Charter,* 86.

28. Dudziak, "Desegregation as a Cold War Imperative," 101.

29. Hunt, *Ideology and U.S. Foreign Policy,* 180.

30. "Negroes Ask Jobs on Acheson Staff," *New York Times,* 14 April 1951, 19; Plummer, *Rising Wind,* 272–273; Douglas Brinkley, *Dean Acheson: The Cold War Years, 1953–71* (New Haven: Yale University Press, 1992), 305; Stephen J. Rabe, *Eisenhower and Latin America: The Foreign Policy of Anticommunism* (Chapel Hill: University of North Carolina Press, 1988), 20.

31. Gordon H. Chang, *Friends and Enemies: The United States, China, and the Soviet Union, 1948–1972* (Stanford: Stanford University Press, 1990), 15; Brinkley, *Dean Acheson,* 303–304.

32. Dean Acheson, *Present at the Creation: My Years in the State Department* (New York: Norton, 1969), 112; Godfrey Hodgson, "Enemies and Empires" (review of Douglas Brinkley, *Dean Acheson: The Cold War Years*), *New Republic,* 28 December 1992, 36–40.

33. Nicol C. Rae, *Southern Democrats* (New York: Oxford University Press, 1994), 36–40; Fite, *Richard B. Russell,* 225.

34. Dittmer, *Local People,* 2.

35. Fite, *Richard B. Russell,* 199, 221, 232–236, 244–245, 265–266, 280 (quotations at 264 and 274).

36. Quoted in Rae, *Southern Democrats,* 38.

37. Douglas Little, "Cold War and Colonialism in Africa: The United States, France, and the Madagascar Revolt of 1947," *Pacific Historical Review* 59 (November 1990): 527–552.

38. For American awareness of this reality, see U.S. Office of Strategic Services, R & A No. 2852, "Regulations and Practices Concerning Native Troops in British, French, Spanish, Belgian, Portuguese, and Ex-Italian Africa," 27 February 1945, *OSS/State Department Intelligence and Research Reports* (Washington: University Publications of America, 1980), part 13, *Africa, 1941–1961* (microfilm, 11 reels with printed guide), reel 3.

39. Dalfiume, *Desegregation of the U.S. Armed Forces,* 105–106, 132–133. See also Logan, *What the Negro Wants;* Egerton, *Speak Now against the Day;* Sullivan, *Days of Hope;* Richard M. Dalfiume, "The 'Forgotten Years' of the Negro Revolution," *Journal of American History* 55 (June 1968): 90–106; Doug McAdam, *Political Process and the Development of Black Insurgency, 1930–1970* (Chicago: University of Chicago Press, 1982); Aldon D. Morris, *The Origins of the Civil Rights Movement: Black Communities Organizing for Change* (New York: Free Press, 1984).

40. Dittmer, *Local People,* 20; Michael J. Hogan, *A Cross of Iron: Harry S. Truman and the Origins of the National Security State* (New York: Cambridge University Press, 1998), 431.

41. See, for example, Patricia Sullivan, "Southern Reformers, the New Deal, and the Movement's Foundation," in *New Directions in Civil Rights Studies,* ed. Armstead L. Robinson and Patricia Sullivan (Charlottesville: University Press of Virginia, 1991), 96.

42. Perhaps the most well-known contemporary exploration and condemnation of this phenomenon was William Faulkner's novel of northern Mississippi, *Intruder in the Dust* (New York: Random House, 1948).

43. "Killing of Negroes Is Protested Here," *New York Times,* 2 March 1946, clipping in Philleo Nash Files, box 6, HSTL; "NAACP Swings into High as Lynch Wave Hits South," *NAACP Bulletin* 5 (August–September 1946): 1, clipping in President's Secretary's Files, box 131, HSTL; Gail Williams O'Brien, *The Color of the Law: Race, Violence, and Justice in the Post–World War II South* (Chapel Hill: University of North Carolina Press, 1999); Dorothy Beeler, "Race Riot in

Columbia, Tennessee, February 25–27, 1946," *Tennessee Historical Quarterly* 39 (1980): 49–61; Burran, "Racial Violence in the South," 229–256; Mary L. Dudziak, "Civil Rights and the Cold War: The Impact of Foreign Relations on Civil Rights Reform in the Truman Administration," paper presented at the annual meeting of the Society for Historians of American Foreign Relations, June 1993, Charlottesville, Virginia.

44. Carr to members of President's Committee on Civil Rights, 16 April 1947, PCCR, box 16, HSTL; "Massacre," editorial, *Crisis* 53 (August 1947): 233; *To Secure These Rights: The Report of the President's Committee on Civil Rights* (Washington: U.S. Government Printing Office, 1947), 20–32; Shapiro, *White Violence and Black Response,* 365–370; Burran, "Racial Violence in the South," 226, 257–290; McCoy and Ruetten, *Quest and Response,* 44–45. Woodward had been discharged from the army just three hours before the assault, making this one of the swiftest demonstrations of white Southern resolve to intimidate returning black veterans.

45. Harold Preece, "Klan 'Murder, Inc.,' in Dixie," *Crisis* 53 (October 1946): 299–301; Shapiro, *White Violence and Black Response,* 353–357, 369–370.

46. "South Carolina," *Time,* 2 June 1947, 27.

47. Theron Lamar Caudle, "Civil Rights and Federal Law," address to North Carolina Bar Association, Winston-Salem, N.C., 30 August 1946, Nash Files, box 26, HSTL.

48. Fite, *Richard B. Russell,* 230.

49. Carl T. Rowan, *Go South to Sorrow* (New York: Random House, 1957), 187–191; Dittmer, *Local People,* 21; Shapiro, *White Violence and Black Response,* 395; Eric W. Rise, "Race, Rape, and Radicalism: The Case of the Martinsville Seven, 1949–1951," *Journal of Southern History* 58 (August 1992): 461–490, especially 465–466.

50. *To Secure These Rights,* 79–80.

51. "One Killed, Seven Injured in Melee at Leavenworth Disciplinary Barracks," *Kansas City Call,* 9 May 1947, clipping in Nash Papers, box 27, HSTL.

52. Shapiro, *White Violence and Black Response,* 419. The recent literature emphasizing the limits on Northern racial liberalism after World War II is well represented by Thomas J. Sugrue, *The Origins of the Urban Crisis: Race and Inequality in Postwar Detroit* (Princeton: Princeton University Press, 1996).

53. Elie Abel, "Cicero Riot Scored as Defeat for U.S.," *New York Times,* 14 November 1951, clipping in Nash Files, box 5, HSTL.

54. Leo J. Margolin, letter to the editor, *New York Times,* 17 July 1951, clipping in Nash Files, box 5, HSTL.

55. Thomas J. Sugrue, "Crabgrass-Roots Politics: Race, Rights, and the Reaction

against Liberalism in the Urban North, 1940–1964," *Journal of American History* 82 (September 1995): 551–578, especially 556.

56. *To Secure These Rights,* 29.

57. Ibid., 24, 124–125. On Hoover's racial inclinations, see O'Reilly, *"Racial Matters."*

58. Johnson and Weston to Truman, 27 July 1947, WHCF, Official File, box 588, HSTL.

59. "Tennessee Race Riot Ends, Ten Injured; Recall '33 Lynching," *New York Evening Post,* 26 February 1946, clipping in Nash Files, box 6, HSTL.

60. Rowan, *Go South to Sorrow,* 42–43.

61. Jules Tygiel, *Baseball's Great Experiment: Jackie Robinson and His Legacy* (New York: Oxford University Press, 1983). Less well known, but just as revealing of the spirit of the times, was the response of the first black baseball player whom Brooklyn Dodgers president Branch Rickey approached about playing in the major leagues. Rickey asked Cuban shortstop Silvio García how he would react if a racist white American were to slap him. García snarled: "I kill him." Rickey went with Robinson instead. Thomas G. Paterson, *Contesting Castro: The United States and the Triumph of the Cuban Revolution* (New York: Oxford University Press, 1994), 49.

62. Dudziak, "Civil Rights and the Cold War," 26; Dudziak, "Desegregation as a Cold War Imperative," 84. On the miscegenation case, see Peggy Pascoe, "Miscegenation Law, Court Cases, and Ideologies of 'Race' in Twentieth-Century America," *Journal of American History* 83 (June 1996): 44–69, especially 61–67.

63. Dudziak, "Desegregation as a Cold War Imperative," 61–65, 103, 111, 118; Dudziak, "Civil Rights and the Cold War," 16–20; William C. Berman, *The Politics of Civil Rights in the Truman Administration* (Columbus: Ohio State University Press, 1970), 232. See also Bert Lockwood, Jr., "The UN Charter and U.S. Civil Rights Litigation: 1946–1955," *Iowa Law Review* 5 (1984): 901–956.

64. John D. Hargreaves, "Toward the Transfer of Power in British West Africa," in *The Transfer of Power in Africa: Decolonization, 1940–1960,* ed. Prosser Gifford and William Roger Louis (New Haven: Yale University Press, 1982), 135–136; Pakenham, *The Scramble for Africa,* 674.

65. For further discussion of these issues, see Borstelmann, *Apartheid's Reluctant Uncle,* and Von Eschen, *Race against Empire.* On Randolph's campaign, see Berman, *The Politics of Civil Rights,* 97–100, and McCoy and Ruetten, *Quest and Response,* 106–108. On Truman's commitment to Democratic Party unity, see Bernstein, "The Ambiguous Legacy," 284, and Berman, *The Politics of Civil Rights,* 238–239.

66. *To Secure These Rights,* 146.

67. "Truman's Address" [text of 29 June 1947 speech], *Crisis,* July 1949, 200. See also Walter White address (introducing Truman), 29 June 1947, Clark Clifford Files, box 3, HSTL; McCoy and Ruetten, *Quest and Response,* 73–74. The precise mix of motives behind Truman's actions in support of civil rights has been the source of some scholarly contention over the past three decades. His biographers (Hamby, McCullough) and some other scholars (McCoy and Ruetten, Donovan) have tended to emphasize his moral convictions about human equality, while dissenting historians (O'Reilly, Berman, Bernstein) have noted his calculations of political advantage. Bernstein and Berman have also been careful to credit Truman's egalitarian inclinations and his commitment to preserving social order and individual rights, as Hamby has been meticulous in acknowledging his political motives. Hamby, *Man of the People;* McCullough, *Truman;* McCoy and Ruetten, *Quest and Response;* Robert J. Donovan, *Conflict and Crisis: The Presidency of Harry S Truman, 1945–1948* (New York: Norton, 1977), and *Tumultuous Years: The Presidency of Harry S Truman, 1949–1953* (New York: Norton, 1982); O'Reilly, *Nixon's Piano;* Berman, *The Politics of Civil Rights;* Bernstein, "The Ambiguous Legacy."

68. Clark Clifford, memorandum for the president, 19 November 1947, Clifford Papers, Political File, box 23, HSTL.

69. Leuchtenburg, "The Conversion of Harry Truman," 66. On political calculations, see Hamby, *Man of the People,* 445–452, and Bernstein, "The Ambiguous Legacy," 284.

70. Berman, *The Politics of Civil Rights,* 83–85 (quotation); McCoy and Ruetten, *Quest and Response,* 99–100.

71. Channing Tobias et al. to Truman, 19 September 1946, President's Secretary's Files, General File, box 131, HSTL.

72. On African American votes, see Philleo Nash, memorandum for the president, 6 November 1948, President's Secretary's Files, Political File, box 55, HSTL. Truman had hoped for only a mild plank on civil rights at the Democratic Convention in order to avoid further alienating the white South. He resented the success of liberal forces led by Hubert Humphrey in writing instead a plank that incorporated the recommendations of the president's own Committee on Civil Rights, although that stand contributed importantly to Truman's victory in November. McCullough, *Truman,* 638–640. On the Harlem speech (and Truman's disinclination to talk about civil rights during other campaign speeches), see Ferrell, *Harry S Truman,* 296. For examples of how Truman's election encouraged racial liberals abroad, see Raymond A. Valliere, "South African Political and Economic Developments," 16 November 1948, enclosed in Winship to secretary of state, 16 November 1948, 848A.00/11–1648,

Confidential U.S. State Department Central Files, South Africa: Internal Affairs and Foreign Affairs, 1945–1949 (Frederick, Md.: University Publications of America, 1985 [microfilm, 14 reels with printed guide]), reel 4, and Frenise A. Logan, "Racism and Indian-U.S. Relations, 1947–1953: Views in the Indian Press," *Pacific Historical Review* 54 (February 1985): 77–78.

73. McCoy and Ruetten, *Quest and Response,* 171–204; Fite, *Richard B. Russell,* 245–246; Berman, *The Politics of Civil Rights,* 153–157; Hamby, *Man of the People,* 495.

74. Bernstein, "The Ambiguous Legacy," 292–296, 299; Hamby, *Man of the People,* 563; McCoy and Ruetten, *Quest and Response,* 283–288.

75. Barkan, *The Retreat of Scientific Racism.*

76. *To Secure These Rights,* 146–148; *Segregation in Washington: A Report of the National Committee on Segregation in the Nation's Capital* (Chicago: National Committee on Segregation in the Nation's Capital, 1948), 63–65; Harold R. Isaacs, *Two-Thirds of the World: Problems of a New Approach to the Peoples of Asia, Africa, and Latin America* (Washington: Public Affairs Institute, 1950), 43–44.

77. As noted in the Prologue, quotes around "race" are generally avoided throughout the text, but they are used here to remind the reader of the thoroughly socially constructed meaning of the term—a particularly important purpose of this paragraph.

78. Rowan, *Go South to Sorrow,* 150.

79. *Segregation in Washington,* 14; Logan, "Racism and Indian-U.S. Relations," 75–76.

80. *Segregation in Washington,* 6. For a similar incident in the restaurant of the House of Representatives, see Joseph D. Lohman and Edwin R. Embree, "The Nation's Capital," *Survey Graphic,* January 1947, 34.

81. Harris Wofford, *Of Kennedys and Kings: Making Sense of the Sixties* (New York: Farrar, Straus, Giroux, 1980), 110; Wofford, *Lohia and America Meet* (1951; Bombay: Sindhu, 1987), 43.

82. John Howard Griffin, *Black Like Me* (Boston: Houghton Mifflin, 1961).

83. Rowan, *Go South to Sorrow,* 52.

84. David M. Reimers, *Still the Golden Door: The Third World Comes to America,* 2nd ed. (New York: Columbia University Press, 1992), 61; DeConde, *Ethnicity, Race, and American Foreign Policy,* 141.

85. Memorandum of conversation of Acheson with Senator Benton, 21 June 1952, Dean Acheson Papers, Memoranda of Conversation, box 79, HSTL.

86. Takaki, *Strangers from a Different Shore,* 413.

87. Michelle Mart, "Tough Guys and American Cold War Policy: Images of Israel, 1948–1960," *Diplomatic History* 20 (Summer 1996): 357–380; Dinnerstein,

Antisemitism in America, 128–174; David S. Wyman, *The Abandonment of the Jews: America and the Holocaust, 1941–1945* (New York: Pantheon, 1984), 325; David A. Hollinger, "National Solidarity at the End of the Twentieth Century: Reflections on the United States and Liberal Nationalism," *Journal of American History* 84 (September 1997): 567.

88. John W. Dower, "Occupied Japan and the Cold War in Asia," in *The Truman Presidency,* ed. Michael J. Lacey (New York: Cambridge University Press, 1989), 368; Dower, *War without Mercy,* 301–311.

89. Max Hastings, *The Korean War* (New York: Simon and Schuster, 1987), 29.

90. Reimers, *Still the Golden Door,* 31. For an interpretation that emphasizes the unspoken threat of such an internment (through the McCarran Internal Security Act of 1950 and its provisions for incarcerating Communists in a national emergency), but on political rather than simply racial grounds, see Takaki, *Strangers from a Different Shore,* 415.

91. O'Reilly, *"Racial Matters,"* 28, 39; Mary L. Dudziak, "Josephine Baker, Racial Protest, and the Cold War," *Journal of American History* 81 (September 1994): 543–570.

92. Clark, "An Analysis of the Relationship between Anti-Communism and Segregationist Thought," esp. 14–15, 24–25.

93. Julian M. Pleasants and Augustus Burns III, *Frank Porter Graham and the 1950 Senate Race in North Carolina* (Chapel Hill: University of North Carolina Press, 1990); David R. Goldfield, *Black, White, and Southern: Race Relations and Southern Culture, 1940 to the Present* (Baton Rouge: Louisiana State University Press, 1990), 67–70.

94. Sullivan, *Days of Hope,* 180–181; McCoy and Ruetten, *Quest and Response,* 98–99.

95. Graham White and John Maze, *Henry A. Wallace: His Search for a New World Order* (Chapel Hill: University of North Carolina Press, 1995); Richard J. Walton, *Henry Wallace, Harry Truman, and the Cold War* (New York: Viking, 1976).

96. Sullivan, *Days of Hope,* 249–275; White and Maze, *Henry A. Wallace,* 276–282. Taylor was even arrested in Birmingham for breaking the city's segregation ordinance while trying to join a black youth group's meeting. The news that a U.S. senator could not speak freely with some Americans because of the color of their skin provoked comment around the world. "Taylor under Bail in Alabama in Racial Case, Will Make Test," *New York Times,* 3 May 1948, 1; Dudziak, "Desegregation as a Cold War Imperative," 83.

97. Sullivan, *Days of Hope,* 249.

98. In this regard, the president chose the advice of Clark Clifford and his political aides (to isolate Wallace by associating him solely with his Communist

Party supporters) over that of his Committee on Civil Rights, which warned that "public excitement about 'Communists' has gone far beyond the dictates of 'good judgment' and 'calmness' . . . A state of near-hysteria now threatens to inhibit the freedom of genuine democrats." Clifford, memorandum for the president, 19 November 1947, Clifford Papers, box 23, HSTL; *To Secure These Rights,* 49.

99. Clifford, memorandum for the president, 19 November 1947.

100. Von Eschen, *Race against Empire,* provides the best discussion to date of these issues. See also Charles I. Glickberg, "Negro Americans and the African Dream," *Phylon* 8 (1947): 323–330; Charles H. Wesley, "International Aspects of the Negro's Status in the United States," *Negro History Bulletin,* February 1948, 108, 113–118; Plummer, *Rising Wind;* Gerald Horne, *Black and Red: W. E. B. Du Bois and the Afro-American Response to the Cold War, 1944–1963* (Albany: State University of New York Press, 1986).

101. James L. Roark, "American Black Leaders: The Response to Colonialism and the Cold War, 1943–1953," *African Historical Studies* 4 (1971): 253–270; Mark Solomon, "Black Critics of Colonialism and the Cold War," in *Cold War Critics: Alternatives to American Foreign Policy in the Truman Years,* ed. Thomas G. Paterson (Chicago: Quadrangle, 1971), 205–239; Duberman, *Paul Robeson,* 388–390; Plummer, *Rising Wind,* 167–216; Von Eschen, *Race against Empire,* 107–144.

102. CIA, ORE 25–48, "The Break-Up of the Colonial Empires."

103. Lauren, *Power and Prejudice,* 201–202.

104. William Roger Louis, "American Anti-Colonialism and the Dissolution of the British Empire," *International Affairs* 61 (1985): 402–403, 411.

105. Plummer, *Rising Wind,* 127–128.

106. Robert J. McMahon, *Colonialism and Cold War: The United States and the Struggle for Indonesian Independence, 1945–1949* (Ithaca: Cornell University Press, 1981), 292–295.

107. Borstelmann, *Apartheid's Reluctant Uncle,* 111–114; J. K. Sweeney, "The Unwanted Alliance: Portugal and the United States," in *The Romance of History: Essays in Honor of Lawrence S. Kaplan,* ed. Scott L. Bills and Timothy Smith (Kent, Ohio: Kent State University Press, 1997), 214–228; Alfred E. Eckes, Jr., *The United States and the Global Struggle for Minerals* (Austin: University of Texas Press, 1979), 157–161; Scott L. Bills, *Empire and Cold War: The Roots of U.S.-Third World Antagonism, 1945–1947* (New York: St. Martin's, 1990), 210–212.

108. George W. Malone, "Colonial Slavery," *Congressional Record,* 28 June 1950, 9339; Plummer, *Rising Wind,* 189.

109. Truman, address to joint session of Congress, 12 March 1947, reprinted in

Major Problems in American Foreign Relations, 4th ed., 2 vols., ed. Thomas G. Paterson and Dennis Merrill (Lexington, Mass.: D.C. Heath, 1995), 2:259–261; Logan, "Racism and Indian-U.S. Relations," 76.

110. Assistant Secretary of State George McGhee recognized this by privately acknowledging that a 1950 U.S. diplomatic conference held in Tangiers on the subject of colonialism would have been comparable to the French setting up a conference in Mobile, Alabama, to discuss the status of African Americans in the South. See Plummer, *Rising Wind,* 237–238.

111. White (Central Intelligence Group) to Stewart, 5 August 1947, enclosing "Foreign Radio Comment on American Civil Rights," PCCR, Correspondence, box 6, HSTL.

112. Isaacs, *Two-Thirds of the World,* 42–43; Richard L.-G. Deverall, "The Struggle for Asia," *Crisis,* January 1951, 28–29. See also William Stueck, *The Korean War: An International History* (Princeton: Princeton University Press, 1995), 81.

113. Dudziak, "Civil Rights and the Cold War," 11–12.

114. Henry B. Ryan, *The Vision of Anglo-America: The US-UK Alliance and Emerging Cold War, 1943–1946* (Cambridge: Cambridge University Press, 1987), 170–171; Plummer, *Rising Wind,* 169; Fite, *Richard B. Russell,* 217. Churchill himself, as the son of an English father and American mother, literally embodied such an alliance.

115. Acheson, speech to the Delta Council, 8 May 1947, reprinted in Joseph M. Jones, *The Fifteen Weeks (February 21–June 5, 1947)* (New York: Viking, 1955), 247–281, quotation at 276; Dittmer, *Local People,* 26–27. Less than a month earlier, Du Bois had asked in a speech to the Progressive Citizens of America if it was possible to "export democracy to Greece and not practice it in Mississippi." Jonathan Rosenberg, "'Struggling to Save America': World Affairs and the U.S. Civil Rights Movement, 1945–1960," paper presented at the annual conference of the Organization of American Historians, April 1999, Toronto, Canada, 22.

116. Memorandum for the file, "Under Secretary's Meeting of Friday, April 7, 1950," Records of the Office of the Executive Secretariat, box 1, Record Group 59, National Archives, Washington, D.C. (McGhee quotation); I thank Frank Costigliola for bringing this document to my attention; McGregor memorandum, 3 February 1954, *FRUS, 1952–54,* 11:416–417; editorial footnote, *FRUS, 1952–54,* 414; Robert W. Moore, "White Magic in the Belgian Congo," *National Geographic* 101 (March 1952): 321–362. For more extensive discussions of these issues, see Borstelmann, *Apartheid's Reluctant Uncle.*

117. See, for example, Gallman to State Department, 14 March 1952, *FRUS, 1952–54,* 11:5; Sims to Feld, 16 June 1952, ibid., 21–22; memorandum of conversa-

tion, 21 August 1952, ibid., 25; George McGhee, *Envoy to the Middle World: Adventures in Diplomacy* (New York: Harper and Row, 1983), 131. One partial exception to this was Liberia, whose settlement by African Americans in the nineteenth century had given it a unique relationship to the United States. On the growing American economic involvement in Liberia after 1945, see Earl P. Hanson, "The United States Invades Africa," *Harper's Magazine*, February 1947, 170–177.

118. William Attwood, *The Reds and the Blacks: A Personal Adventure* (New York: Harper and Row, 1967), 36. On the common tendency of European officials in Africa to mistrust American motives there, see memorandum of conversation, 20 February 1952, *FRUS, 1952–54*, 11:1.

119. Dudley to State Department, 3 June 1952, *FRUS, 1952–54*, 11:19–20.

120. Borstelmann, *Apartheid's Reluctant Uncle*, 8–12.

121. Sims to Bourgerie, 18 March 1952, *FRUS, 1952–54*, 11:11 (quotation); minutes of U.S.-U.K. colonial policy discussions, 25 September 1952, ibid., 311.

122. Prosser Gifford, "Misconceived Dominion: The Creation and Disintegration of Federation in British Central Africa," in *The Transfer of Power in Africa: Decolonization, 1940–1960,* ed. Prosser Gifford and Wm. Roger Louis (New Haven: Yale University Press, 1982), 387–416.

123. Dorsz to State Department, 10 October 1952, *FRUS, 1952–54*, 11:347–348. See also Minter, *King Solomon's Mines Revisited,* 118–123; Robert B. Edgerton, *Mau Mau: An African Crucible* (New York: Free Press, 1989).

124. The best sources on the U.S. perspective on South Africa at the dawn of apartheid include Thomas J. Noer, *Cold War and Black Liberation: The United States and White Rule in Africa, 1948–1968* (Columbia: University of Missouri Press, 1985); Minter, *King Solomon's Mines Revisited;* Borstelmann, *Apartheid's Reluctant Uncle.* For an example of white South Africans' concern with spreading their influence northward, see memorandum of conversation, 29 March 1949, Acheson Papers, box 73, HSTL.

125. Borstelmann, *Apartheid's Reluctant Uncle,* 197–199. On American appreciation of South Africa's role in Korea, see memorandum of conversation, 20 April 1951, Acheson Papers, box 77, HSTL.

126. Kilcoin to Brown, 17 April 1952, *FRUS, 1952–54*, 11:909; Perkins to secretary of state, 11 June 1952, ibid., 920; memorandum by Lee and Thoreson, 16 September 1952, ibid., 932.

127. Memorandum of conversation, 23 October 1952, ibid., 946.

128. Department of State, Office of Intelligence Research, "The South African Constitutional Crisis," 2 July 1952, *OSS/State Department Intelligence and Research Reports* (Washington: University Publications of America, 1980), part 13 *Africa, 1941–1961* (microfilm, 11 reels with printed guide), reel 10.

129. McGhee, *Envoy to the Middle World,* 146, 131.

130. This difference had the capacity to shock the occasional African American visitor to South Africa, such as actor Sidney Poitier, who arrived there in 1952 to film *Cry the Beloved Country.* See Poitier, *This Life* (New York: Knopf, 1980), 147–150.

131. George M. Frederickson, *Black Liberation: A Comparative History of Black Ideologies in the United States and South Africa* (New York: Oxford University Press, 1995), chaps. 5–7.

132. Logan, "Racism and Indian-U.S. Relations," 76; Borstelmann, *Apartheid's Reluctant Uncle,* 102–103.

133. Memorandum of conversation, 14 October 1952, Acheson Papers, box 80, HSTL.

134. See, for example, Colon to Truman, 29 July 1946, WHCF, Official File, box 588, HSTL; "Greek Communists Use of Violations of Civil Rights in the U.S." [November 1946], PCCR, box 6, HSTL; Danciger to Wilson, 23 December 1946, PCCR, box 2, HSTL; Logan, "Racism and Indian-U.S. Relations," 73–74.

135. *To Secure These Rights,* 100–101.

136. White (Central Intelligence Group) to Krould, 18 June 1947, PCCR, box 6, HSTL.

137. Stueck, *The Korean War,* 80.

138. "Voice of America Listener Hits Lynching," 27 May 1947, PCCR, box 6, HSTL.

139. Bowles to Carr, 19 March 1947, PCCR, box 8, HSTL; Dudziak, "Desegregation as a Cold War Imperative," 89.

140. James W. Ivy, "American Negro Problem in the European Press," *Crisis,* July 1950, 413–418, 468–472; Dudziak, "Desegregation as a Cold War Imperative," 81–83.

141. Isaacs, "The Political and Psychological Context of Point Four," 56.

142. *To Secure These Rights,* 147.

143. Ibid., 146–147.

144. Lauren, *Power and Prejudice,* 194.

145. Minutes of meeting of U.S. delegation to UN, 10 November 1950, *FRUS, 1950,* 2:564–569; "United States Policy toward Dependent Territories," *FRUS, 1952–54,* 3:1097.

146. Dudziak, "Desegregation as a Cold War Imperative," 85. The staff of the President's Committee on Civil Rights was dismayed to learn the limits of Marshall's interest in domestic racial reform as a way of dealing with this problem. Committee staffers requested a letter from the secretary of state to bolster

their report's emphasis on the need for reducing racial discrimination, but Marshall instead argued that reports of racial problems in the United States were largely the work of Communist Party exaggerations. The committee decided Marshall's letter would actually hurt rather than help their cause, and instead used Acheson's older letter to the Fair Employment Practices Commission on the negative effects of discrimination on U.S. diplomacy. Carr to Marshall, 23 May 1947, and Carr to Rusk, 11 August 1947, both in PCCR, box 6, HSTL.

147. Dudziak, "Civil Rights and the Cold War," 12.

148. Dudziak, "Desegregation as a Cold War Imperative," 95.

149. *A Petition to the United Nations on Behalf of 13 Million Oppressed Negro Citizens of the United States of America* (New York: National Negro Congress, 1946); *An Appeal to the World: A Statement on the Denial of Human Rights to Minorities in the Case of Citizens of Negro Descent in the United States and an Appeal to the United Nations for Redress* (New York: National Association for the Advancement of Colored People, 1947); *We Charge Genocide: The Historical Petition to the United Nations for Relief from a Crime of the United States Government against the Negro People* (New York: Civil Rights Congress, 1951). See also Harris, "Racial Equality and the United Nations Charter."

150. Berman, *The Politics of Civil Rights,* 66.

151. *An Appeal to the World,* 20.

152. See Ellen Schrecker, *Many Crimes: McCarthyism in America* (New York: Little, Brown, 1998). Further insightful discussion of this repressive campaign can be found in Von Eschen, *Race against Empire;* Plummer, *Rising Wind;* Horne, *W. E. B. Du Bois;* and Duberman, *Paul Robeson.*

153. Shapiro, *White Violence and Black Response,* 517, n. 65.

154. J. D. Ratcliff, "Edith Sampson—Thorn in Russia's Side," *Reader's Digest,* April 1951, 107–109; Dudziak, "Desegregation as a Cold War Imperative," 99.

155. Michael L. Krenn, "'Outstanding Negroes' and 'Appropriate Countries': Some Facts, Figures, and Thoughts on Black U.S. Ambassadors, 1949–1988," *Diplomatic History* 14 (Winter 1990): 131–141. A handful of African Americans had previously served as heads of mission in places like Haiti and Liberia at a rank below ambassador.

156. Dudziak, "Desegregation as a Cold War Imperative," 109.

157. *To Secure These Rights,* 87–95; *Segregation in Washington,* 21–53. Residents of the District of Columbia did not have a voting representative in Congress.

158. *Segregation in Washington,* 88.

159. Carr to President's Committee on Civil Rights, 24 April 1947, PCCR, box 1 (copy also in box 15), HSTL.

160. Ibid.; "Meeting of Negro Leaders with Secretary Acheson on April 13, 1951," and Randolph et al., statement to Acheson, 13 April 1951, both in Acheson Papers, box 77, HSTL.

161. *Segregation in Washington,* 9–10.

162. Lauren, *Power and Prejudice,* 180, 192.

163. DeConde, *Ethnicity, Race, and American Foreign Policy,* 128–129; Shapiro, *White Violence and Black Response,* 421; Lauren, *Power and Prejudice,* 225; Natalie Hevener Kaufman, *Human Rights Treaties and the Senate: A History of Opposition* (Chapel Hill: University of North Carolina Press, 1990).

164. White, *A Rising Wind.*

165. Plummer, *Rising Wind,* 207.

166. "Swedish Press Scores American 'Racists,'" 6 June 1947, PCCR, box 6, HSTL.

167. Truman to Acheson, 11 August 1952, enclosing "Report on the Foreign Service Personnel of the American Legation in Luxembourg," Acheson Papers, box 80, HSTL.

168. *To Secure These Rights,* 80; Truman address to Congress, 2 February 1948, Nash Files, box 24, HSTL; George W. Westerman, "Canal Zone Discrimination," *Crisis,* April 1951, 235–237.

169. Davis to Secretary of State, *FRUS, 1949,* 2:706–707.

170. Collier statement, "Civil Rights Problems in Dependent Areas," PCCR, box 15, HSTL.

171. War Department, "The Utilization of Negro Infantry Platoons in White Companies," June 1945, PCCR, box 6, HSTL; War Department, "Army Talk," 12 April 1947, Alvan C. Gillem, Jr., Papers, box 1, DDEL; Mershon and Schlossman, *Foxholes and Color Lines,* 145–152. On African American organizing to pressure Truman to desegregate the military, see Dalfiume, *Desegregation of the U.S. Armed Forces,* 155–169.

172. Dalfiume, *Desegregation of the U.S. Armed Forces,* 201; Mershon and Schlossman, *Foxholes and Color Lines,* 158–217.

173. Nalty, *Strength for the Fight,* 256–262; Mershon and Schlossman, *Foxholes and Color Lines,* 74–81, 218–251.

174. Chang, *Friends and Enemies,* 80; Hunt, *Ideology and U.S. Foreign Policy,* 162.

175. Stueck, *The Korean War,* 127; Andrew J. Rotter, "In Retrospect: Harold R. Isaacs's *Scratches on Our Minds,*" *Reviews in American History* 24 (March 1996): 187.

176. Hastings, *The Korean War,* 69–70.

177. Ibid., 306–313.

178. "Smearing Negro GIs in Korea," editorial, *Crisis,* December 1950, 715; Clay Blair, *The Forgotten War: America in Korea, 1950–1953* (New York: Times Books, 1987), 683–684; Bowers, Hammond, and MacGarrigle, *Black Soldier,*

White Army, 186–188. For a summary of the controversy over the performance of African American soldiers in segregated units early in the war, see Mershon and Schlossman, *Foxholes and Color Lines,* 220–224.

179. Plummer, *Rising Wind,* 205.

180. Englehardt, *The End of Victory Culture,* 65; Plummer, *Rising Wind,* 209. Fears of Americans' having their minds removed and replaced by alien forces, whether from East Asia or outer space, were a common theme in science fiction literature and films of the period. See Peter Biskind, *Seeing Is Believing: How Hollywood Convinced Us to Stop Worrying and Love the Fifties* (New York: Pantheon, 1983).

181. Plummer, *Rising Wind,* 208–209.

182. On the war's effects on social reform efforts at home, see Alonzo Hamby, *Beyond the New Deal: Harry S. Truman and American Liberalism* (New York: Columbia University Press, 1973), 441–446.

183. South Africa's extensive supply of uranium ore makes the image of an expanding core—as in a nuclear reactor gone haywire—especially appropriate, in light of what might be seen as apartheid's cultural radioactivity.

3. The Last Hurrah of the Old Color Line

1. Louis Jefferson, *The John Foster Dulles Book of Humor* (New York: St. Martin's, 1986), 166–167 (emphasis in original).

2. Chester Bowles, "While the World Watches," *New York Times,* 17 January 1960, sec. 10, p. 6, clipping in Staff Files, Administrative Officer for Special Projects (Morrow), box 9, folder "Civil Rights Clippings and Data (2)," DDEL.

3. Robert F. Burk, *The Eisenhower Administration and Black Civil Rights* (Knoxville: University of Tennessee Press, 1984), 261–263; Stephen E. Ambrose, *Eisenhower,* vol. 2, *The President* (New York: Simon and Schuster, 1984), 18–19, 620.

4. Stephen E. Ambrose, *Eisenhower,* vol. 1, *Soldier, General of the Army, President-Elect, 1890–1952* (New York: Simon and Schuster, 1983), 47; Ambrose, *Eisenhower,* 2:125; Burk, *Eisenhower Administration,* 27; Herbert S. Parmet, *Eisenhower and the American Crusades* (New York: Macmillan, 1972), 438; *Universal Military Training: Hearings before the Committee on Armed Services, U.S. Senate,* 80th Cong., 2d sess., March and April 1948 (Washington: Government Printing Office, 1948), 995–997. For examples of Eisenhower's occasional stands against blatant discrimination, see Burk, *Eisenhower Administration,* 27; Dwight D. Eisenhower, *The White House Years: Waging Peace, 1956–1961* (Garden City, N.Y.: Doubleday, 1965), 148; Ambrose, *Eisenhower,* 1:34–35, 113–114, 370.

5. Chester J. Pach, Jr., and Elmo Richardson, *The Presidency of Dwight D. Eisenhower,* rev. ed. (Lawrence: University Press of Kansas, 1991), 34; Arthur Larson, *Eisenhower: The President Nobody Knew* (New York: Scribner's, 1968), 127; Burk, *Eisenhower Administration,* 192; Ambrose, *Eisenhower,* 1:437–438, 469, 2:75, 142, 385–387, 497–498 (quotation).

6. Ambrose, *Eisenhower,* 1:102, 274, 2:387; Roy Wilkins, *Standing Fast: The Autobiography of Roy Wilkins* (New York: Viking, 1982), 256.

7. Mark Stern, "Presidential Strategies and Civil Rights: Eisenhower, The Early Years, 1952–54," *Presidential Studies Quarterly* 19 (Fall 1989): 780; David Halberstam, *The Fifties* (New York: Ballantine, 1993), 685; O'Reilly, *Nixon's Piano,* 165–187.

8. Branch, *Parting the Waters,* 181; Stern, "Presidential Strategies," 781–782; Burk, *Eisenhower Administration,* 190–191, 255–257; Peter Grose, *Gentleman Spy: The Life of Allen Dulles* (Boston: Houghton Mifflin, 1994), 7; Lauren, *Power and Prejudice,* 226; O'Reilly, *"Racial Matters,"* 39–40.

9. E. Frederic Morrow, *Black Man in the White House: A Diary of the Eisenhower Years by the Administrative Officer for Special Projects, the White House, 1955–1961* (New York: Coward-McCann, 1963), 47–48 (Rabb), 276–277 (Persons); Branch, *Parting the Waters,* 234; Ambrose, *Eisenhower,* 2:305.

10. Morrow, *Black Man,* 17–18, 40, 82–84, 88, 118–119, 136–137, 164, 179, 209–211, 258–259; Burk, *Eisenhower Administration,* 77–85; Halberstam, *Fifties,* 424–427.

11. FBI, "The Nation of Islam," October 1960, WHO, OSANSA, FBI Series, box 7, DDEL; Ambrose, *Eisenhower,* 1:439; Burk, *Eisenhower Administration,* 90, 108, 121–127.

12. Shapiro, *White Violence and Black Response,* 410–411 (quotation); Dittmer, *Local People,* 83–84; Southeastern Office, American Friends Service Committee, Department of Racial and Cultural Relations, National Council of Churches in Christ in the United States of America, and Southern Regional Council, *Intimidation, Reprisal, and Violence in the South's Racial Crisis* (Atlanta, 1959); Arnold R. Hirsch, "Massive Resistance in the Urban North: Trumbull Park, Chicago, 1953–1966," *Journal of American History* 82 (September 1995): 522–550; John E. Fleming, "African-American Museums, History, and the American Ideal," *Journal of American History* 81 (1994): 1025.

13. Wilkins, *Standing Fast,* 222; Stern, "Presidential Strategies," 778; Pach and Richardson, *Presidency,* 148–149; Burk, *Eisenhower Administration,* 23, 162–165, 263.

14. *A Testament of Hope: The Essential Writings of Martin Luther King, Jr.,* ed. James M. Washington (San Francisco: Harper and Row, 1986), 85–86; Ambrose, *Eisenhower,* 2:309, 498; Stern, "Presidential Strategies," 774–777, 788–789; Pach and Richardson, *Presidency,* 62–63, 156–157; Gary W. Reichard, "Democrats,

Civil Rights, and Electoral Strategies in the 1950s," *Congress and the Presidency* 13 (Spring 1986): 59–60; Jeff Broadwater, *Adlai Stevenson and American Politics: The Odyssey of a Cold War Liberal* (New York: Twayne, 1994), 47, 116, 140; Burk, *Eisenhower Administration,* 15–20.

15. Steven F. Lawson, *Running for Freedom: Civil Rights and Black Politics in America since 1941* (Philadelphia: Temple University Press, 1991), 55–56; Burk, *Eisenhower Administration,* 16, 89, 109, 152–153, 204–207.

16. "Discrimination on the Run," editorial, *Washington Post,* 22 April 1955, clipping enclosed in Rabb to Eisenhower, 29 April 1955, AWF, Ann Whitman Diary, box 5, DDEL; Rowan, *Go South to Sorrow,* 224–228; Dwight D. Eisenhower, *The White House Years: Mandate for Change, 1953–1956* (Garden City, N.Y.: Doubleday, 1963), 236; Burk, *Eisenhower Administration,* 45–67; Stern, "Presidential Strategies," 783; Lawson, *Running for Freedom,* 51–52.

17. Burk, *Eisenhower Administration,* 49, 60–67, 70–77.

18. Nalty, *Strength for the Fight,* 267–72; Burk, *Eisenhower Administration,* 24–44.

19. *The Eisenhower Diaries,* ed. Robert H. Ferrell (New York: Norton, 1981), 246–247 (Byrnes); Eisenhower to McGill, 26 February 1959, AWF, Name Series, box 23, DDEL; Burk, *Eisenhower Administration,* 238 (publishers); Morrow, *Black Man,* 218–219.

20. Ambrose, *Eisenhower,* 2:337 (quotation); Halberstam, *Fifties,* 686; Morrow, *Black Man,* 299–300.

21. Eisenhower to McGill, 26 February 1959, AWF, Name Series, box 23, DDEL ("center line"); Ambrose, *Eisenhower,* 2:327 (Jones).

22. Burk, *Eisenhower Administration,* 150, 201; Pach and Richardson, *Presidency,* 143. The Court's announcement on 17 May 1954 was a great surprise. Most press reporters in Washington had spent that morning speculating about the consequences of another event freighted with racial implications: the Vietnamese defeat of the French at Dien Bien Phu just ten days before. Sitkoff, *The Struggle for Black Equality,* 21.

23. Dudziak, "Desegregation as a Cold War Imperative," passim, but esp. 61 (quotation), 113, n. 299.

24. "'Voice' Speaks in Thirty-four Languages to Flash Court Ruling to World," *New York Times,* 18 May 1954, 1; "Editorial Excerpts from the Nation's Press on Segregation Ruling," ibid., 19; Isaacs, *The New World of Negro Americans,* 332 (Frazier); Harold Cruse, *The Crisis of the Negro Intellectual* (New York: William Morrow, 1967), 240; Dudziak, "Desegregation as a Cold War Imperative," 113.

25. Russell Baker, "Eisenhower Hints Desire for 'Slower' Integration," *New York Times,* 28 August 1958, 1; Earl Warren, *The Memoirs of Earl Warren* (Garden City, N.Y.: Doubleday, 1977), 291; Branch, *Parting the Waters,* 113; Larson, *Eisenhower,* 124; Burk, *Eisenhower Administration,* 134, 144, 152; Lawson, *Running*

for Freedom, 51; Sitkoff, *Struggle for Black Equality,* 24; Wilkins, *Standing Fast,* 222. Eisenhower's retrospective claim in his memoirs that he agreed with the *Brown* decision in 1954 does not accord with either his behavior at the time or the accounts of others close to him, especially Larson. See Eisenhower, *Waging Peace,* 150.

26. Richard Wright, *The Color Curtain: A Report on the Bandung Conference* (Cleveland: World, 1956), 133–136 (quotation); Carlos P. Romulo, *The Meaning of Bandung* (Chapel Hill: University of North Carolina Press, 1956), 35, 97–99; Tinker, *Race, Conflict, and the International Order,* 103.

27. Romulo, *Meaning of Bandung,* 48, 69 (quotation); Lauren, *Power and Prejudice,* 211; *Bandung: Texts of Selected Speeches and Final Communique of the Asian-African Conference, Bandung, Indonesia, April 18–24, 1955* (New York: Far Eastern Reporter, 1955), 14–15 (Sukarno).

28. "National Security Implications of Future Developments Regarding Africa," enclosed in Boggs to NSC, 10 August 1960, WHO, OSANSA, NSC Series, Policy Papers Subseries, box 28, folder "NSC 6001," DDEL; Lauren, *Power and Prejudice,* 213–214; *Bandung: Texts,* 24–31; Wright, *Color Curtain,* 157; Segal, *The Race War,* 340–342; Chang, *Friends and Enemies,* 137; George McTurnan Kahin, *The Asian-African Conference: Bandung, Indonesia, April 1955* (Ithaca: Cornell University Press, 1956).

29. Editorial footnote, *FRUS, 1955–57,* 18:1–2; Chang, *Friends and Enemies,* 132; Gerig to Wilcox, 9 May 1957, *FRUS, 1955–1957,* 18:66–68 (quotation).

30. Rabb memo of Eisenhower meeting with Powell, 11 May 1955, AWF, Ann Whitman Diary, box 5, DDEL; Charles V. Hamilton, *Adam Clayton Powell, Jr.: The Political Biography of an American Dilemma* (New York: Atheneum, 1991), 237–248; Dulles to Aldrich, 9 January 1956, John Foster Dulles Papers, Subject Series, box 6, folder "Policy of Independence for Colonial Peoples," DDEL.

31. Von Eschen, *Race against Empire,* 168–173 (black American press); Frederickson, *Black Liberation,* 267 (King); Plummer, *Rising Wind,* 247–256; Isaacs, *The New World of Negro Americans,* 51; Branch, *Parting the Waters,* 140.

32. Brown to Eisenhower, 2 March 1956, and Trade Arts Theater to Eisenhower, 27 February 1956, Records of White House Telegraph Office—Incoming, box 132, DDEL; Lawson, *Running for Freedom,* 51; O'Reilly, *Nixon's Piano,* 171 (Eisenhower); Branch, *Parting the Waters,* 203; Wofford, *Of Kennedys and Kings,* 121 ("white Christmas").

33. Stephen J. Whitfield, *A Death in the Delta: The Story of Emmett Till* (New York: Free Press, 1988), passim; Dittmer, *Local People,* 53–54.

34. Whitfield, *Death in the Delta,* 34 ("that river"), 46–47, 67–68 (Faulkner), 78–79, 85 (Tobias); Burk, *Eisenhower Administration,* 207 ("jungle fury"); Rowan, *Go South to Sorrow,* 41, 50–51. The French socialist *Le Populaire* credited colo-

nialism and its underlying motivation, racism, with responsibility for the situation in the American South. While Western Europeans had overseas colonies, "the essential difference is that the U.S. has its colony in the very interior of the country, while a third kind of empire, the U.S.S.R., has it on the borders." Rowan, *Go South to Sorrow,* 52.

35. Harry McPherson, *A Political Education* (Boston: Little, Brown, 1972), 138; Lawson, *Running for Freedom,* 66–67; Wilkins, *Standing Fast,* 265; Shapiro, *White Violence and Black Response,* 458–459 (Williams); Timothy B. Tyson, *Radio Free Dixie: Robert Williams and the Roots of Black Power* (Chapel Hill: University of North Carolina Press, 1999); Englehardt, *The End of Victory Culture,* 110–111; O'Reilly, *"Racial Matters,"* 42; Rowan, *Go South to Sorrow,* 17.

36. Feaster to Byrnes, 20 August 1953, AWF, Administrative Series, box 28, folder "Nixon, Richard M. (5)," DDEL (newspaperwoman); Rowan, *Go South to Sorrow,* 25, 28, 84–85; Michael J. Klarman, "How *Brown* Changed Race Relations: The Backlash Thesis," *Journal of American History* 81 (June 1994): 81–82; Fite, *Richard Russell,* 332, 369–371; Egerton, *Speak Now against the Day,* 574 (Talmadge).

37. Fite, *Richard Russell,* 333–334; David R. Goldfield, *Black, White, and Southern: Race Relations and Southern Culture, 1940 to the Present* (Baton Rouge: Louisiana State University Press, 1990), 84–85; Ambrose, *Eisenhower,* 2:305–306; Rowan, *Pitiful and Proud,* 19; Rowan, *Go South to Sorrow,* 173–174.

38. Ambrose, *Eisenhower,* 2:306–307 (quotation); Burk, *Eisenhower Administration,* 161, 208–209, 218–220, 240, 249–250; Pach and Richardson, *Presidency,* 147; Jordan to Eisenhower, 22 January 1957, Staff Files, Administrative Officer for Special Projects (Morrow), box 11, DDEL. For examples of contemporary appeals for Eisenhower to assert more leadership on this issue, see Rowan, *Go South to Sorrow,* 244, and G. Mennen Williams, "A Plea to the President," *Reporter,* 18 February 1960, clipping in Staff Files, Administrative Officer for Special Projects (Morrow), box 9, folder "Civil Rights Clippings and Data (2)," DDEL.

39. Val Washington speech, 18 October 1955, Staff Files, Bryce N. Harlow, box 8, folder "Civil Rights," DDEL; Ambrose, *Eisenhower,* 2:327–328; Broadwater, *Adlai Stevenson,* 152–168; Burk, *Eisenhower Administration,* 165–169; Lawson, *Running for Freedom,* 49–55; Wilkins, *Standing Fast,* 231–242. At the Democratic convention in Chicago in August, Governor George Wallace of Alabama enjoyed pointing out the white violence against black families in the Trumbull Park public housing project in the South Deering neighborhood—"within shouting distance" of the convention—and advising liberal Northern Democrats to attend to their own problems rather than criticizing segregation in the South. Hirsch, "Massive Resistance," 543.

40. King to Nixon, 30 August 1957, William P. Rogers Papers, 1938–1962, box 50, DDEL; Robert Dallek, *Lone Star Rising: Lyndon Johnson and His Times, 1908–1960* (New York: Oxford University Press, 1991), 526; Ambrose, *Eisenhower,* 2:407, 412–413; Wilkins, *Standing Fast,* 251–252; Larson, *Eisenhower,* 129; Lawson, *Running for Freedom,* 56–65; Burk, *Eisenhower Administration,* 226–228; Pach and Richardson, *Presidency,* 146; O'Reilly, *Nixon's Piano,* 179–180; Fite, *Richard Russell,* 339–343.

41. "Suez," *Time,* 20 August 1956, 17; Ambrose, *Eisenhower,* 2:331 (quotation), 462–463.

42. Lodge to Eisenhower, 21 December 1956, Dulles Papers, General Correspondence and Memoranda Series, box 5, DDEL; "Suez," *Time,* 20 August 1956, 18 (Labour official); Ambrose, *Eisenhower,* 2:361, 371 (Eisenhower); Wm. Roger Louis, "Dulles, Suez, and the British," in *John Foster Dulles and the Diplomacy of the Cold War,* ed. Richard H. Immerman (Princeton: Princeton University Press, 1990), 150–153.

43. Benjamine Fine, "Troops on Guard at School," *New York Times,* 25 September 1957, 1; Sitkoff, *Struggle for Black Equality,* 31; Anthony Lewis, *Portrait of a Decade: The Second American Revolution* (New York: Random House, 1964), 51; Pach and Richardson, *Presidency,* 152–153; Wilkins, *Standing Fast,* 253–254.

44. Text of Eisenhower address, 24 September 1957, AWF, Administrative Series, box 23, DDEL; Halberstam, *Fifties,* 687; Ambrose, *Eisenhower,* 2:414; O'Reilly, *Nixon's Piano,* 183–184.

45. Nelson Mandela, "A New Menace in Africa," March 1958, in *Mandela, Tambo, and the African National Congress—The Struggle against Apartheid, 1948–1990: A Documentary Survey,* ed. Sheridan Johns and R. Hunt Davis (New York: Oxford University Press, 1991), 56–59; Foreign Broadcast Information Service, "World Radio and Press Reaction to Events in Little Rock, Arkansas," 27 September 1957, Harlow Files, box 11, folder "Integration," DDEL (French commentators); U.S. Information Agency, "Public Reactions to Little Rock in Major World Capitals," 29 October 1957, enclosed in Loomis to Dearborn, 22 November 1957, Staff Files, Administrative Officer for Special Projects (Morrow), box 11, DDEL; Caputa to Rogers, 9 October 1957, Rogers Papers, box 2, DDEL; *Little Rock, U.S.A.: Materials for Analysis,* ed. Wilson Record and Jane Cassels Record (San Francisco: Chandler, 1960), 175 (white captain); Mary L. Dudziak, "The Little Rock Crisis and Foreign Affairs: Race, Resistance, and the Image of American Democracy," *Southern California Law Review* 70 (September 1997): 1641–1716; Isaacs, *The New World of Negro Americans,* 10–12; Harold R. Isaacs, "World Affairs and U.S. Race Relations: A Note on Little Rock," *Public Opinion Quarterly* 22 (Fall 1958): 366; Branch, *Parting the Waters,* 223; Halberstam, *Fifties,* 678–679; McKay, *Africa,* 403–404.

46. Text of Eisenhower address, 24 September 1957, AWF, Administrative Series, box 23, DDEL; Foreign Broadcast Information Service, "World Radio and Press Reaction," 27 September 1957, Harlow Files, box 11, DDEL; Isaacs, "World Reaction," 366.

47. Memo of NSC meeting, 8 May 1958, AWF, NSC Series, box 10, DDEL; Morrow, *Black Man*, 212–213, 289; Noer, *Cold War and Black Liberation*, 34–35; Dulles to Lodge, 3 April 1956, with enclosures, Dulles Papers, General Correspondence and Memorabilia Series, box 4, DDEL; Sears to Dulles, 15 February 1956, *FRUS, 1955–57*, 18:37–38 (U.S. delegation to UN); Attwood, *The Reds and the Blacks*, 16 (chief of protocol); Rowan, *Pitiful and Proud*, 152 (Ramsey); Lauren, *Power and Prejudice*, 217.

48. Isaacs, *Scratches on Our Minds*, 63–64, 232–233 ("lunacy"), 236–237; Rotter, "In Retrospect: Harold R. Isaacs's *Scratches on Our Minds*," 187 ("jets"); Chang, *Friends and Enemies*, 170–174 (other quotations). American fears of the doubly subversive threat of nonwhite Communists were widespread; *Newsweek*, for example, spread a fictitious rumor in early 1953 that the tiny Guatemalan Communist Party had "hordes of armed Indians" invading plantations throughout that country. Piero Gleijeses, *Shattered Hope: The Guatemalan Revolution and the United States*, (Princeton: Princeton University Press, 1991), 232.

49. Chang, *Friends and Enemies*, 138, 173 (Eisenhower and Dulles); "The Kremlin," *Time*, 12 November 1956, 30–31.

50. DeConde, *Ethnicity, Race, and American Foreign Policy*, 142–143; "Liberating 'Enslaved Peoples,'" *Crisis*, April 1953, 228–229; Cole to Eisenhower, 22 February 1956, Records of White House Telegraph Office, Incoming, box 132, folder "Incoming Telegrams 'Alabama' Feb. 1956," DDEL; Nalty, *Strength for the Fight*, 274; Morrow, *Black Man*, 109–110, 116–117; Shapiro, *White Violence and Black Response*, 451 (King); McGregor memo, 28 December 1955, *FRUS, 1955–57*, 18:26; Rowan, *Pitiful and Proud*, 153–159; Harold R. Isaacs, "World Affairs and U.S. Race Relations: A Note on Little Rock," *Public Opinion Quarterly* 22 (Fall 1958): 364–365; Bowles, *Africa's Challenge*, 103–104; Ambrose, *Eisenhower*, 2:498; transcript of telephone conversation, Dulles and Lodge, 7 October 1953, Dulles Papers, Telephone Call Series, box 1, DDEL.

51. Shapiro, *White Violence and Black Response*, 416–417 (quotations), 458; Morrow, *Black Man*, 274; James Forman, *The Making of Black Revolutionaries* (1972; Washington: Open Hand, 1985), 173; Tyson, *Radio Free Dixie*, chaps. 4–5.

52. Randolph to Eisenhower, 17 June 1953, *FRUS, 1952–54*, 11:43–46; Burk, *Eisenhower Administration*, 159–160 (Powell); Man to Eisenhower, 24 February 1956, and Trade Arts Theater to Eisenhower, 27 February 1956 (other critics), Records of White House Telegraph Office, Incoming, box 132, DDEL; Rowan,

Pitiful and Proud, 142–164; McGregor memo, 28 December 1955, *FRUS, 1955–57*, 18:26–27.

53. Herman E. Talmadge, *You and Segregation* (Birmingham, Ala.: Vulcan Press, 1955), cited in Dudziak, "Desegregation as a Cold War Imperative," 117; Fite, *Richard Russell*, 345; Chapple to Eisenhower, 22 February 1956, Records of White House Telegraph Office—Incoming, box 132, DDEL; Richard M. Fried, *Nightmare in Red: The McCarthy Era in Perspective* (New York: Oxford University Press, 1990), 174–178.

54. Transcript of telephone conversation, Dulles with Hall, 6 May 1953, Dulles Papers, Telephone Call Series, box 1, DDEL; Burk, *Eisenhower Administration*, 112, 160, 166–167; Morrow, *Black Man*, 52.

55. Nixon, "Report to the President," 5 April 1957, AWF, Administrative Series, box 28, DDEL; O'Reilly, *Nixon's Piano*, 184; Isaacs, *The New World of Negro Americans*, 5–7 (quotation), 13; Isaacs, "World Affairs," 365; McKay, *Africa*, 402–403; Burk, *Eisenhower Administration*, 50–52, 69–70; Phaon Goldman, "The Significance of African Freedom for the Negro American," *Negro History Bulletin* 34 (October 1960): 6.

56. Paul Good, "Odyssey of a Man—and a Movement," in *Black Protest in the Sixties*, ed. August Meier, John H. Bracey, and Elliott M. Rudwick (New York: Markus Wiener, 1991), 265; Martin Staniland, *American Intellectuals and African Nationalists, 1955–1970* (New Haven: Yale University Press, 1991), 179–180 (Wilkins); *The Speeches of Malcolm X at Harvard*, ed. Archie Epps (New York: Morrow, 1968), 168–169; Isaacs, *The New World of Negro Americans*, 285–293; James Forman, "A Year of Resistance," in *Towards Revolution*, ed. John Gerassi, vol. 2, *The Americas* (London: Weidenfeld and Nicolson, 1971), 690; George M. Houser, "Freedom's Struggle Crosses Oceans and Mountains: Martin Luther King, Jr., and the Liberation Struggles in Africa and America," in *We Shall Overcome: Martin Luther King, Jr., and the Black Freedom Struggle*, ed. Peter J. Albert and Ronald Hoffman (New York: Pantheon, 1990), 170–171, 176–177, 184–185. For further discussion of this issue, see Plummer, *Rising Wind*, esp. 257–297; Von Eschen, *Race against Empire*; Adelaide Cromwell Hill and Martin Kilson, eds., *Apropos of Africa: Sentiments of Negro American Leaders on Africa from the 1800s to the 1950s* (London: Frank Cass, 1969).

57. Isaacs, *The New World of Negro Americans*, 152–154, 294, 298–299, 303–309 (quotation at 306), 314, 320–321; Attwood, *Reds and Blacks*, 46–47.

58. C. J. Bartlett, *"The Special Relationship": A Political History of Anglo-American Relations since 1945* (London: Longman, 1992), 84; Ambrose, *Eisenhower*, 2:333, 377–378, 587–588; Pach and Richardson, *Presidency*, 58, 180; Harry McPherson, *A Political Education* (Boston: Little, Brown, 1972), 136; Fite, *Richard B. Russell*,

346; Matthew Connelly, "Taking Off the Cold War Lens: Visions of North-South Conflict during the Algerian War for Independence," *American Historical Review* 105 (June 2000): 762–763.

59. Ambrose, *Eisenhower,* 1:496 (Eisenhower), 509; Thomas G. Paterson, *Contesting Castro: The United States and the Triumph of the Cuban Revolution* (New York: Oxford University Press, 1994), 235; Isaacs, *Scratches on Our Minds,* 39 ("all the world"); Richard H. Immerman, ed., *John Foster Dulles and the Diplomacy of the Cold War* (Princeton: Princeton University Press, 1990), 18–19, 280; Bartlett, *"Special Relationship,"* 82.

60. Hunt, *Ideology and U.S. Foreign Policy,* 164 (Jackson); Memo of NSC meeting, 14 April 1960, *FRUS, 1958–60,* 14:127 (Nixon); Paterson, *Contesting Castro,* 10 (Dulles), 257 (Herter and Eisenhower).

61. Marvin R. Zahniser and W. Michael Weis, "A Diplomatic Pearl Harbor? Richard Nixon's Goodwill Mission to Latin America in 1958," *Diplomatic History* 13 (Spring 1989): 163–190; Stephen G. Rabe, *Eisenhower and Latin America: The Foreign Policy of Anticommunism* (Chapel Hill: University of North Carolina Press, 1988), chap. 6; Pach and Richardson, *Presidency,* 188, 221; Ambrose, *Eisenhower,* 2:558 (quotation).

62. Chester Bowles, *Africa's Challenge to America* (Berkeley: University of California Press, 1956), vi-vii, 56–59; Sherman Adams, *Firsthand Report: The Story of the Eisenhower Administration* (New York: Harper and Brothers, 1961), 293; Chang, *Friends and Enemies,* 167; Ambrose, *Eisenhower,* 2:261–262; Robert J. McMahon, "Eisenhower and Third World Nationalism: A Critique of the Revisionists," *Political Science Quarterly* 101 (Fall 1986): passim.

63. Operations Coordinating Board, "Report on Africa South of the Sahara," 21 March 1958, WHO, OSANSA, NSC Series, Policy Papers Subseries, box 21, folder "NSC 5719/1," DDEL; USIA, "The Image of America," enclosed in Bundy to Coffey, 27 October 1958, WHO, OSANSA, OCB Series, Subject Subseries, box 3, folder "Image of America," DDEL.

64. John Foster Dulles, "The Cost of Peace," *Department of State Bulletin,* 18 June 1956, 999–1000; Murphy to Rockefeller, 19 August 1955, enclosing "Neutralism in the NEA Area," WHO, OSANSA, NSC Staff Papers, 1948–1961, box 2, folder "#9 Bandung (1)," DDEL (State Department); NSC 5719/1, "U.S. Policy toward Africa South of the Sahara," 23 August 1957, WHO, OSANSA, NSC Series, Policy Papers Subseries, box 21, DDEL.

65. John Gunther, *Inside Africa* (New York: Harper and Brothers, 1955), 4 (quotation), 587; National Intelligence Estimates 83, "Conditions and Trends in Tropical Africa," 22 December 1953, *FRUS, 1952–54,* 11:71–89; CIA, "Major U.S. Interests in Africa," 22 March 1954, *FRUS, 1952–54,* 11:107; Allen to Dulles, 12

August 1955, *FRUS, 1955–57,* 18:16; NSC 5719/1, 23 August 1957; Elsie May Bell Grosvenor, "Safari through Changing Africa," *National Geographic* 104 (August 1953): 145–198.

66. Dean to Dulles, 16 February 1953, *FRUS, 1952–54,* 2:1098; Office of Special Assistant to the Joint Chiefs of Staff for NSC Affairs, "Africa," 19 March 1954, ibid., 103; Harr to McCone, 14 October 1960, enclosing Johnson memo, WHO, OSANSA, OCB Series, Subject Subseries, box 1, folder "Africa," DDEL; Jonathan E. Helmreich, "U.S. Foreign Policy and the Belgian Congo in the 1950s," *Historian* 58 (Winter 1996): 316; Gunther, *Inside Africa,* 669–678.

67. Hadsel to Byroade, 3 August 1956, enclosing "U.S. Policy toward South Africa," *FRUS, 1955–57,* 18:788–790 (quotation); Wailes to State Department, 6 August 1956, ibid., 791; editorial footnote, ibid., 791; minutes of U.S. delegation to UN meeting, 25 October 1955, ibid., 778.

68. Sears to Dulles, 15 February 1956, *FRUS, 1955–57,* 18:31 ("police state"); Operations Coordinating Board, "Report on Africa South of the Sahara," 30 September 1959, WHO, OSANSA, NSC Series, Policy Papers Subseries, box 25, folder NSC 5818 ("retrogressive"); Crowe to State Department, 27 April 1959, *FRUS, 1958–60,* 14:732–733; Gallman to State Department, 2 March 1953, *FRUS, 1952–54,* 11:989–994.

69. Dulles to Gallman, 21 April 1953, *FRUS, 1952–54,* 11:995–996; Gallman to State Department, 2 March 1953, ibid., 989–992; Frederickson, *Black Liberation,* 266–267 (quotation). The USSR had officially outlawed racial discrimination soon after the Bolshevik Revolution.

70. Gunther, *Inside Africa,* 521.

71. David N. Gibbs, "Political Parties and International Relations: The United States and the Decolonization of Sub-Saharan Africa," *International History Review* 17 (May 1995): 314–315; Minter, *King Solomon's Mines Revisited,* 134–135 (quotation; emphasis added); Harold R. Isaacs, "Western Man and the African Crisis," *Saturday Review,* 2 May 1953, 10.

72. NSC 5719/1, 23 August 1957; Henry A. Byroade, "The World's Colonies and Ex-Colonies," *Department of State Bulletin,* 16 November 1953, 657–659; Crawford Young, "United States Policy Toward Africa: Silver Anniversary Reflections," *African Studies Review* 27 (September 1984): 2–4; Steven Metz, "American Attitudes toward Decolonization in Africa," *Political Science Quarterly* 99 (Fall 1984): 520–528; Staniland, *American Intellectuals,* 25; Helmreich, "U.S. Foreign Policy," 316–325.

73. Byroade address, 31 October 1953, *FRUS, 1952–54,* 11:54–65 (also printed as Byroade, "World's Colonies"); George V. Allen, "United States Foreign Policy in Africa," *Department of State Bulletin,* 30 April 1956, 716–718; Metz, "American Attitudes," 519–521; Waldemar A. Nielsen, *The Great Powers and Africa*

(New York: Praeger, 1969), 259–263; Vernon McKay, *Africa in World Politics* (New York: Harper and Row, 1963), 320–323.

74. Nixon, "Report to the President on Trip to Africa," 5 April 1957, AWF, Administrative Series, box 28, DDEL; Colin Legum, *Must We Lose Africa?* (London: W. H. Allen, 1954); Bowles, *Africa's Challenge*, 1–2; Gunther, *Inside Africa*, 3.

75. Bowles, *Africa's Challenge*, 49; Allen to Dulles, 12 August 1955, *FRUS, 1955–57*, 18:20; Byroade, "World's Colonies," 660; Gunther, *Inside Africa*, 14–15; Helmreich, "U.S. Foreign Policy," 315; Noer, *Cold War and Black Liberation*, 48; Nielsen, *Great Powers*, 274–277; Metz, "American Attitudes," 520–525; Cary Fraser, *Ambivalent Anti-Colonialism: The United States and the Genesis of West Indian Independence, 1940–1964* (Westport, Conn.: Greenwood, 1994), 28–29.

76. Feld memo, 17 August 1953, *FRUS, 1952–54*, 11:48; Anthony Low, "The End of the British Empire in Africa," in *Decolonization and African Independence: The Transfers of Power, 1960–1980*, ed. Prosser Gifford and Wm. Roger Louis (New Haven: Yale University Press, 1988), 43, 54–55; Gunther, *Inside Africa*, 358, 373; Staniland, *American Intellectuals*, 187–188; Stephen E. Ambrose, *Nixon: The Education of a Politician, 1913–1962* (New York: Simon and Schuster, 1987), 433.

77. Sears to Dulles, 15 February 1956, *FRUS, 1955–57*, 18:35–36 (quotations); Mallon to State Department, 19 January 1953, *FRUS, 1952–54* 11:30–31; Isaacs, "Western Man," 56; Bowles, *Africa's Challenge*, 47–48, 73, 81.

78. Operations Coordinating Board, "Report on Africa South of the Sahara (NSC 5719/1)," 21 March 1958, WHO, OSANSA, NSC Series, Policy Papers Subseries, box 21, DDEL; Jernegan to Lourie, 19 June 1953, *FRUS, 1952–54*, 11:325; Tibbetts to State Department, 30 July 1953, *FRUS, 1952–54*, 11:329; Byroade, "World's Colonies," 658; Steere to Utter, 22 October 1954, *FRUS, 1952–54*, 11:344–345; Prosser Gifford, "Misconceived Dominion: The Creation and Disintegration of Federation in British Central Africa," in *The Transfer of Power in Africa: Decolonization, 1940–1960*, ed. Gifford and Wm. Roger Louis (New Haven: Yale University Press, 1982), 396–400; Robert I. Rotberg, *The Rise of Nationalism in Central Africa: The Making of Malawi and Zambia, 1873–1964* (Cambridge, Mass.: Harvard University Press, 1965), 255–258, 292–310; Robert C. Good, *U.D.I.: The International Politics of the Rhodesian Rebellion* (Princeton: Princeton University Press, 1973), 30–32; Gunther, *Inside Africa*, 632–636 ("this country").

79. Ndabaningi Sithole, *African Nationalism* (Cape Town: Oxford University Press, 1959), 28, 36, 38 (quotation), 49; National Intelligence Estimates 76–59, "The Outlook in East, Central, and South Africa," 20 October 1959, *FRUS, 1958–60*, 14:63 (CIA); National Intelligence Estimates 72–56, "Conditions and Trends in Tropical Africa," 14 August 1956, *FRUS, 1955–57*, 18:45–46; Holmes to Dulles, 6 February 1958, *FRUS, 1958–60*, 14:8.

80. Memo of Nkrumah conversation with Nixon, 4 March 1957, *FRUS, 1955–57,* 18:376; Rotberg, *Rise of Nationalism,* 283; Branch, *Parting the Waters,* 214; Morrow, *Black Man,* 133.

81. Kwame Nkrumah, *The Autobiography of Kwame Nkrumah* (Edinburgh: Thomas Nelson, 1957), 24–34, 42–43 (quotation); Robin McKown, *Nkrumah: A Biography* (Garden City, N.Y.: Doubleday, 1973), 32–33; Adu Boahen, "Ghana since Independence," in *Decolonization and African Independence: The Transfers of Power, 1960–1980,* ed. Prosser Gifford and Wm. Roger Louis (New Haven: Yale University Press, 1988), 203–224. The other independent states on the continent in 1957 were South Africa (white-ruled), Ethiopia (not successfully colonized), Liberia (widely viewed as a pseudocolony of the United States), and Egypt, Morocco, Tunisia, Libya, and the Sudan (all predominantly Arab rather than black African).

82. Low, "End of British Empire," 55 (Macmillan), 57, 71; Rupert Emerson, *From Empire to Nation: The Rise to Self-Assertion of Asian and African Peoples* (Cambridge, Mass.: Harvard University Press, 1960), 5; McKay, *Africa,* 1, 11–12, 339.

83. Felix Belair, Jr., "United States Has Secret Sonic Weapon—Jazz," *New York Times,* 6 November 1955, 1; "Jazz Program on 'Voice' to Get World Booking," ibid., 2 April 1956, 17; "Jazz, Surgery Aid U.S. in Pakistan," ibid., 8 April 1956, 31; Bowles, *Africa's Challenge,* 127; McKay, *Africa,* 383; Penny M. Von Eschen, "'Satchmo Blows Up the World': Jazz, Race, and Empire during the Cold War," unpublished manuscript in author's possession; Dizzy Gillespie, *Dizzy: To Be or Not to Bop: The Autobiography of Dizzy Gillespie* (London: Quartet, 1982), 413–421 (quotations at 413, 421). I am grateful to Matt Abramovitz for bringing these newspaper citations to my attention. For examples of black speakers abroad, see Asa T. Spaulding statement, "Discrimination and the Negro in the United States," New Delhi, India, 26 November 1956, Staff Files, Administrative Officer for Special Projects (Morrow), box 11, folder "Inter-Racial Affairs," DDEL, and Rowan, *Pitiful and Proud,* passim.

84. Gary Giddins, *Satchmo* (New York: Doubleday, 1988), 159–163; Gillespie, *Dizzy,* 437–438; Lodge to Eisenhower, 28 March 1956, AWF, Dulles-Herter Series, box 6, folder "Dulles, John Foster-March 1956," DDEL. Lodge cited the case of his recent trip to Khartoum with his wife as an example: after a long state dinner ending at 1:00 A.M., they resisted weariness and agreed to the Sudanese foreign minister's suggestion to visit a nightclub. "It was an effort, but the Foreign Minister (who was coal black) danced with my wife and I heard later from our U.S. representative that the word went all over town that the Americans were not stuffy." The reaction of white American diplomats and officials who were not so "temperamentally inclined"—especially those from the South—can readily be imagined.

85. Memo of NSC meeting, 8 August 1958, AWF, NSC Series, box 10, DDEL (quotations); Nixon, "Report to the President," 5 April 1957, AWF, Administrative Series, box 28, DDEL; U.S. Congress, House of Representatives, Committee on Foreign Affairs, *Report of the Special Study Mission to Africa, South and East of the Sahara by Honorable Frances P. Bolton* (Washington: Government Printing Office, 1956), especially 12, 147; McKay, *Africa*, 342–346; Nielsen, *Great Powers*, 266–270. Some of Nixon's colleagues in the administration considered him the "father" of the new, friendlier American policy that was developing in response to African independence. See, for example, memo of NSC meeting, 23 August 1957, AWF, NSC Series, box 9, DDEL.

86. NSC 5818, "Statement of U.S. Policy toward Africa," 26 August 1958, *FRUS, 1958–60*, 14:25; CIA, Annex B to NSC 5818, 29 December 1958, enclosed in Smith to Lay, 20 January 1959, WHO, OSANSA, NSC Series, Policy Papers Subseries, box 25, DDEL; Hoover to Consulate General in Accra, 20 February 1956, *FRUS, 1955–57*, 18:362–364; John D. Hargreaves, *Decolonization in Africa* (London: Longman, 1988), 209–210; Young, "United States Policy," 1; McKay, *Africa*, 342. In a 1960 NSC meeting CIA chief Allen Dulles expressed his concern that the lower living standards of Chinese and even Eastern European technicians gave them an advantage in Africa over Americans accustomed to much greater material comforts. Dulles worried that the Eastern bloc workers' spirits might be higher in the poverty-stricken conditions of most of the continent, and that Africans would be impressed by this difference. See editorial note, *FRUS, 1958–60*, 14:708.

87. Lodge to State Department, 20 November 1957, *FRUS, 1955–57*, 18:834; Byroade to State Department, 7 November 1958, *FRUS, 1958–60*, 14:730–731; editorial footnotes and note, *FRUS, 1958–60*, 14:731, 739. For an example of the administration's awareness of the brutalities of apartheid for the majority of South Africans, see Walmsley to Herter, 9 July 1958, *FRUS, 1958–60*, 14:727–729.

88. Memo of NSC meeting, 9 May 1958, AWF, NSC Series, box 10, DDEL ("Spirit of 1776"); Jackson to Dulles, 9 February 1959, Dulles Papers, General Correspondence and Memoranda Series, box 2 ("juvenile delinquencies"); "National Security Implications of Future Developments Regarding Africa," enclosed in Boggs to NSC, 10 August 1960, WHO, OSANSA, NSC Series, Policy Paper Subseries, box 28, DDEL ("contagion"); Gibbs, "Political Parties," 314 ("serious and unnecessary"); memo of NSC meeting, 31 March 1960, AWF, NSC Series, box 12, DDEL (Nixon and Stans).

89. McGregor memo, 28 December 1955, *FRUS, 1955–57*, 18:26–27; memo of NSC meeting, 9 May 1958, AWF, NSC Series, box 10, DDEL; "The Kingdom of Silence: The Truth about Africa's Most Oppressed Colony," *Harper's*, May 1961,

31–32 ("have worked together"); Benjamin Welles, "Eisenhower Hails Portugal's Gains," *New York Times,* 21 May 1960, 3.

90. Ann Whitman diary, 10 October 1957, AWF, Ann Whitman Diary Series, box 9, DDEL; Eisenhower to Gallin-Douathe, 14 November 1960, AWF, International Series, box 6, folder "Central African Republic," DDEL; Dulles to U.S. Embassy in Ghana, 9 October 1957, *FRUS, 1955–57,* 18:378–379; John H. Morrow, *First American Ambassador to Guinea* (New Brunswick: Rutgers University Press, 1968), 101–102; Isaacs, *The New World of Negro Americans,* 16; Morrow, *Black Man,* 175; Rowan, *Pitiful and Proud,* 157; McKay, *Africa,* 402–403.

91. Goldman, "Significance of African Freedom," 6; Morris et al. to Eisenhower, 19 January 1960, WHCF, General File, box 44, folder "Civil Rights—Civil Liberties (14)," DDEL.

92. Crowe to State Department, 30 March 1960, *FRUS, 1958–60,* 14:747–748; memo of NSC meeting, 2 April 1960, AWF, NSC Series, box 12, DDEL; James Barber and John Barratt, *South Africa's Foreign Policy: The Search for Status and Security, 1945–1988* (Cambridge: Cambridge University Press, 1990), 70.

93. Editorial note, *FRUS, 1958–60,* 14:741 (State Department, Eisenhower, Herter); memo of conversation, 28 March 1960, ibid., 746 (Eisenhower to Macmillan); Crowe to State Department, 25 March 1960, ibid., 743–745.

94. Memo of NSC meeting, 24 March 1960, AWF, NSC Series, box 12, DDEL; NSC 6001, "Statement of U.S. Policy toward South, Central, and East Africa," 19 January 1960, *FRUS, 1958–60,* 14:86; National Intelligence Estimates 73–60, "The Outlook for the Union of South Africa," 19 July 1960, *FRUS, 1958–60,* 14:754–755; Hare to Williams, 19 August 1960, *FRUS, 1958–60,* 14:755–756; George M. Houser, *No One Can Stop the Rain: Glimpses of Africa's Liberation Struggle* (New York: Pilgrim Press, 1989), 88 (Mboya).

95. Burden to State Department, 10 July 1960, *FRUS, 1958–60,* 14:287; Minter, *King Solomon's Mines Revisited,* 147.

96. McIlvaine to State Department, 26 July 1960, *FRUS, 1958–60,* 14:356–357 ("an opportunist"); Tomlinson to State Department, 14 June 1960, ibid., 276 ("no better alternative"); Cumming to Herter, 25 July 1960, ibid., 356; memo of NSC meeting, 21 July 1960, ibid., 338–339 ("a Castro"); Timberlake to State Department, 16 September 1960, ibid., 491; Timberlake to State Department, 26 September 1960, ibid., 504; Richard D. Mahoney, *JFK: Ordeal in Africa* (New York: Oxford University Press, 1983), 40–50. A 26 August 1960 cable from Allen Dulles instructed the CIA station chief in Leopoldville that Lumumba's "removal must be an urgent and prime objective and . . . should be a high priority of our covert actions." Dulles cited "high quarters here" as the source of this "clearcut conclusion"—and the CIA director's only superior is the president. Editorial note, *FRUS, 1958–60,* 14:443. In a 19 September 1960

conversation with the British foreign secretary Lord Home that touched on the Congo, Eisenhower (according to notes of the meeting made by his son John) "expressed his wish that Lumumba would fall into a river of crocodiles; Lord Home said regretfully that we have lost many of the techniques of old-fashioned diplomacy." Despite his differences with the British over Suez, Eisenhower was perhaps more "old-fashioned" in this regard than his allies realized. Editorial note, *FRUS, 1958–60*, 14:495. For the best study to date of the U.S. role in the Congo, see David N. Gibbs, *The Political Economy of Third World Intervention: Mines, Money, and U.S. Policy in the Congo Crisis* (Chicago: University of Chicago Press, 1991).

97. Mahoney, *JFK*, 39 (Lumumba); *Time*, 16 May 1955, quoted in Staniland, *American Intellectuals*, 218; Crawford Young, "The Colonial State and Post-Colonial Africa," in *Decolonization and African Independence: The Transfers of Power, 1960–1980*, ed. Prosser Gifford and Wm. Roger Louis (New Haven: Yale University Press, 1988), 20–22.

98. Editorial footnotes, *FRUS, 1958–60*, 14:359 (Lodge), 366; Mahoney, *JFK*, 43.

99. Editorial footnote and notes, *FRUS, 1958–60*, 14:367 (Dillon), 528 (Hammarskjold), 354 (Bunche), 490 (Allen Dulles); memo of conversation, 11 October 1960, ibid., 522 (Wigny); Burden to State Department, 10 July 1960, ibid., 289.

100. Lewis, *Portrait of a Decade*, 162 (Perez); Mahoney, *JFK*, 38–39 (State Department).

101. Memo of NSC meeting, 18 August 1960, *FRUS, 1958–60*, 14:422–424 (Stans); Board of National Estimates to Allen Dulles, 22 August 1960, ibid., 439; Mahoney, *JFK*, 39.

102. Ilunga Kabongo, "The Catastrophe of Belgian Decolonization," in *Decolonization and African Independence: The Transfers of Power, 1960–1980*, ed. Prosser Gifford and Wm. Roger Louis (New Haven: Yale University Press, 1988), 388–389 (blackboard); Timberlake to State Department, 17 July 1960, *FRUS, 1958–60*, 14:318–319; Herter to Consul General in Leopoldville, 23 June 1960, ibid., 277–278; transcript of telephone conversations, Herter and Hammarskjold, and Eisenhower and Herter, ibid., 296–297; McIlvaine to State Department, 26 July 1960, ibid., 356–357; memo of conversation, 5 August 1960, ibid., 387. Allen Dulles also worried about the white South African response to the disorder in the Congo, especially the possibility of an alliance with Southern Rhodesia and the Portuguese colonies to create "a *cordon sanitaire*" across southern Africa for the containment of African self-rule. Memo of NSC meeting, 1 August 1960, *FRUS, 1958–60*, 14:376.

103. Paterson, *Confronting Castro*, 241 (Dulles); Hunt, *Ideology and U.S. Foreign Policy*, 166–167; Carlos Moore, *Castro, the Blacks, and Africa* (Los Angeles:

Center for Afro-American Studies, University of California, 1988), 59–63; Plummer, *Rising Wind*, 288–289.

104. Max Frankel, "Castro Stays in Hotel to Work Privately as Aides Take Up Fight before U.N.," *New York Times*, 22 September 1960, 14; Moore, *Castro*, 71–91; Plummer, *Rising Wind*, 289–297; Isaacs, *The New World of Negro Americans*, 337; Mary-Alice Waters, "Appendix: Fidel Castro's Arrival in Harlem," in Fidel Castro and Ernesto Che Guevara, *To Speak the Truth: Why Washington's 'Cold War' against Cuba Doesn't End* (New York: Pathfinder, 1992), 201–205; Van Gosse, "The African-American Press Greets the Cuban Revolution," in *Between Race and Empire: African-Americans and Cubans before the Cuban Revolution*, ed. Lisa Brock and Digna Castañeda Fuertes (Philadelphia: Temple University Press, 1998), 266–280.

105. In 1963 John Kennedy invited Eisenhower to endorse his legislative proposals for civil rights. The retired president declined, claiming that a "whole bunch of laws" were not the solution to the problem. Mark Stern, *Calculating Visions: Kennedy, Johnson, and Civil Rights* (New Brunswick: Rutgers University Press, 1992), 87. In 1966 Eisenhower passed along propaganda in support of apartheid to Clifford Roberts, the chairman of the Augusta National Golf Club, to which he attached a note: "Dear Clif, We hear so much misinformation about South Africa, I thought you might like to read this letter from the Ambassador of that nation." O'Reilly, *Nixon's Piano*, 187.

4. Revolutions in the American South and Southern Africa

1. Rusk to U.S. Embassy in South Africa, *FRUS, 1961–63*, 21:611.
2. ANC flyer, "Announcement of the Formation of Umkhonto we Sizwe," 16 December 1961, in *Mandela, Tambo, and the African National Congress—The Struggle against Apartheid, 1948–1990: A Documentary Survey*, ed. Sheridan Johns and R. Hunt Davis (New York: Oxford University Press, 1991), 138–139; CIA, "The Addis Ababa Conference and Its Aftermath," 11 July 1963, NSF, box 3, JFKL. Like the other Cold War, this one could also turn hot—as it did in Angola. On the support from the rest of Africa for southern African struggles against white rule, see Korry to Rusk, 25 May 1963, NSF, box 3, JFKL.
3. State Department paper, "Problems of Southern Africa," 25 March 1963, *FRUS, 1961–63*, 21:495; memorandum of conversation, 17 July 1963, ibid., 639.
4. Schlesinger to Robert Kennedy, 1 July 1963, ibid., 497; notes 1, 2, ibid., 612; Gerald E. Thomas, "The Black Revolt: The United States and Africa in the 1960s," in *The Diplomacy of the Crucial Decade: American Foreign Relations during the 1960s*, ed. Diane B. Kunz (New York: Columbia University Press, 1994), 324–325; Thomas J. Noer, "New Frontiers and Old Priorities in Africa," in *Kennedy's Quest for Victory: American Foreign Policy, 1961–1963*, ed. Thomas

G. Paterson (New York: Oxford University Press, 1989), 253–283; Lawson, *Running for Freedom,* 80.

5. James N. Giglio, *The Presidency of John F. Kennedy* (Lawrence: University of Kansas Press, 1991), 159–160; Gerald S. Strober and Deborah H. Strober, *"Let Us Begin Anew": An Oral History of the Kennedy Presidency* (New York: HarperCollins, 1993), 274; Richard Reeves, *President Kennedy: Profile of Power* (New York: Simon and Schuster, 1993), 62, 313–314.

6. J. Richard Snyder, "Introduction," in *John F. Kennedy: Person, Policy, Presidency,* ed. Snyder (Wilmington, Del.: Scholarly Resources, 1988), x; Branch, *Parting the Waters,* 687. Truman's significant support for civil rights as president and political ties to African American leaders did not preclude the private use of racial epithets indicative of more traditional white attitudes. Kennedy told White House aide Daniel Patrick Moynihan at one point, for example, "I can't get used to Harry Truman talking all the time about 'the niggers.'" Reeves, *President Kennedy,* 61–62.

7. Thomas C. Reeves, *A Question Of Character: A Life of John F. Kennedy* (New York: Free Press, 1991), 335–338; Herbert S. Parmet, *Jack: The Struggles of John F. Kennedy* (New York: Dial, 1980), 408–416; Giglio, *The Presidency of John F. Kennedy,* 160.

8. Brubeck to Bundy, 29 October 1963, NSF, boxes 158–161, folder "South Africa, General, 9/30/63–10/29/63," JFKL; Komer to Kennedy, 13 October 1962, *FRUS, 1961–63,* 21:102; Ronald J. Nurse, "Critic of Colonialism: JFK and Algerian Independence," *The Historian* 39 (February 1977): 307–326; Parmet, *Jack,* 399–416; Giglio, *The Presidency of John F. Kennedy,* 14, 221–223. After his much heralded 1957 Senate speech criticizing French colonialism in Algeria, for example, Kennedy subsequently muted his public comments on the subject, telling aides that he was "wary of being known as the Senator from Algeria." Parmet, *Jack,* 407.

9. Wofford, *Of Kennedys and Kings,* 133; Branch, *Parting the Waters,* 398, 808–809; Reeves, *President Kennedy,* 61–62.

10. Gary W. Reichard, "Democrats, Civil Rights, and Electoral Strategies in the 1950s," *Congress and the Presidency* 13 (Spring 1986): 59–81; Stern, *Calculating Visions,* 31–38; Burk, *Eisenhower Administration,* 258–260; Wofford, *Of Kennedys and Kings,* 25, 51–52, 61–63; Richard N. Goodwin, *Remembering America: A Voice from the Sixties* (Boston: Little, Brown, 1988), 136; Morrow, *Black Man,* 280–282.

11. Mahoney, *JFK,* 26–33, 52; Clifford M. Kuhn, "'There's a Footnote to History!': Martin Luther King's October 1960 Arrest and Its Aftermath," *Journal of American History* 84 (September 1997): 583–595; Lawson, *Running for Freedom,* 75–79; Wofford, *Of Kennedys and Kings,* 40–41; Parmet, *Jack,* 249–250.

12. Douglas Dales, "Candidates Urge End to Bias in U.S.," *New York Times,* 13 October 1960, 26; Wofford, *Of Kennedys and Kings,* 21.

13. Wofford, *Of Kennedys and Kings,* 67; Andrew Young, OHT, p. 23, LBJL; O'Reilly, *"Racial Matters,"* 102–103, 114, 126–131, 137, 150–151; Branch, *Parting the Waters,* 646, 692, 836, 902–906.

14. Reeves, *President Kennedy,* 270–271, 325, 466–468, 579; Victor S. Navasky, *Kennedy Justice* (New York: Atheneum, 1971), 244–248; Parmet, *JFK,* 251; Dittmer, *Local People,* 194–195; Giglio, *The Presidency of John F. Kennedy,* 37.

15. Randall Bennett Woods, *Fulbright: A Biography* (New York: Cambridge University Press, 1995), 252–260; J. William Fulbright, *The Price of Empire* (New York: Pantheon, 1989), 69; Richard N. Goodwin, *Remembering America: A Voice from the Sixties* (Boston: Little, Brown, 1988), 136; Wofford, *Of Kennedys and Kings,* 81–82; Steel, *Walter Lippman,* 523; Merle Miller, *Lyndon: An Oral Biography* (New York: G. P. Putnam's, 1980), 274–275.

16. Thomas J. Schoenbaum, *Waging Peace and War: Dean Rusk in the Truman, Kennedy, and Johnson Years* (New York: Simon and Schuster, 1988), 25, 242, 382–383 (quotations).

17. Isaacs, *The New World of Negro Americans,* 19 (quotation), 316.

18. Belk to Bundy, 17 July 1962, NSF, box 2, folder "Africa, General (6/62–7/62)," JFKL; Mahoney, *JFK,* 27.

19. Wofford to Kennedy, 17 July 1961, and Bowles to Kennedy, 21 August 1961, in President's Office Files, box 28, JFKL.

20. Noer, "New Frontiers and Old Priorities," 265 (Ball); Komer to Kennedy, 14 October 1963, *FRUS, 1961–63,* 21:11; Bundy to McNamara, 1 December 1962, *FRUS, 1961–63,* 21:153. It is not hard to imagine royal advisers in London in the 1790s referring to newly independent Americans as oversensitive. There is about the term, in this usage, a note of disdain for the dignity of the less powerful.

21. Bowles to Kennedy, 1 December 1962, President's Office Files, box 28, JFKL.

22. Belk to Bundy, 19 May 1961, NSF, box 2, JFKL.

23. Quoted in Branch, *Parting the Waters,* 477.

24. Elizabeth A. Cobbs, "Decolonization, the Cold War, and the Foreign Policy of the Peace Corps," *Diplomatic History* 20 (Winter 1996): 105. See also Elizabeth Cobbs Hoffman, *All You Need Is Love: The Peace Corps and the Spirit of the 1960s* (Cambridge, Mass.: Harvard University Press, 1998).

25. Wofford, *Of Kennedys and Kings,* 44–45.

26. Dean Rusk, "Building the Frontiers of Freedom," *Department of State Bulletin,* 17 June 1961, 948; Komer to Bundy, 14 October 1963, *FRUS, 1961–63,* 21:128; Gary R. Hess, "Commitment in the Age of Counterinsurgency: Kennedy's Vietnam Options and Decisions, 1961–1963," in *Shadow on the White House:*

Presidents and the Vietnam War, 1945–1975, ed. David L. Anderson (Lawrence: University of Kansas Press, 1993), 63; Mahoney, *JFK,* 188.

27. Wofford, *Of Kennedys and Kings,* 256.

28. "The White Redoubt," 28 June 1962, enclosed in Owen to McGhee et al., 6 July 1962, NSF, box 2, JFKL.

29. Chang, *Friends and Enemies,* 236; Marilyn B. Young, *The Vietnam Wars, 1945–1990* (New York: HarperCollins, 1991), 76.

30. Robert Buzzanco, *Masters of War: Military Dissent and Politics in the Vietnam Era* (New York: Cambridge University Press, 1996), 107.

31. Moore, *Castro, the Blacks, and Africa,* 93–94, 108–111. See also Thomas G. Paterson, *Confronting Castro,* and Paterson, "Fixation with Cuba: The Bay of Pigs, Missile Crisis, and Covert War against Castro," in *Kennedy's Quest for Victory: American Foreign Policy, 1961–1963,* ed. Paterson, (New York: Oxford University Press, 1989), 123–155.

32. Moore, *Castro, the Blacks, and Africa,* 129–135, 332; Mahoney, *JFK,* 31.

33. Komer to Kennedy, 15 July 1963, *FRUS, 1961–63,* 21:123; Komer to Bundy, 1 November 1963, ibid., 33.

34. Moore, *Castro, the Blacks, and Africa,* 112–114; Attwood, *The Reds and the Blacks,* 27.

35. Tyson, "Radio Free Dixie," 290–291; Moore, *Castro, the Blacks, and Africa,* 113, 121–122.

36. Plummer, *Rising Wind,* 305; Moore, *Castro, the Blacks, and Africa,* 112.

37. Moore, *Castro, the Blacks, and Africa,* 114.

38. Paterson, "Fixation with Cuba."

39. For recent summaries of this perspective, see Noer, "New Frontiers and Old Priorities," and David A. Dickson, "U.S. Foreign Policy toward Southern and Central Africa: The Kennedy and Johnson Years," *Presidential Studies Quarterly* 23 (Spring 1993): 301–315.

40. Rusk to Kennedy, 9 September 1961, enclosing report of Williams trip to Africa, NSF, box 2, JFKL; "Summary of Governor Williams' Trip to Africa," 25 September 1961," *FRUS, 1961–63,* 21:302–303; memorandum of conversation, 19 July 1963, *FRUS, 1961–63,* 21:420–423; Bowles to Kaysen, 13 November 1962, NSF, box 3, JFKL; Elizabeth Cobbs Hoffman, "Diplomatic History and the Meaning of Life: Toward a Global American History," *Diplomatic History* 21 (Fall 1997): 517; Mahoney, *JFK,* 20–33, 248; Henry F. Jackson, *From the Congo to Soweto: U.S. Foreign Policy toward Africa since 1960* (New York: William Morrow, 1982), 38–39; Roger Hilsman, *To Move a Nation: The Politics of Foreign Policy in the Administration of John F. Kennedy* (Garden City, N.Y.: Doubleday, 1967), 246.

41. Rusk to Williams, 25 February 1961, and Williams to Rusk, 26 February 1961,

in NSF, box 2, JFKL; note 5, *FRUS, 1961–63*, 21:347; Robert Kinloch Massie, *Loosing the Bonds: The United States and South Africa in the Apartheid Years* (New York: Doubleday, 1997), 115 (quotation); Cobbs, "Decolonization," 105; Richard J. Walton, *Cold War and Counterrevolution: The Foreign Policy of John F. Kennedy* (Baltimore: Viking, 1972), 206.

42. "The Strategic Importance of Africa," undated [c. May 1963], *FRUS, 1961–63*, 21:331; Dickson, "U.S. Foreign Policy," 304–305; Wofford, *Of Kennedys and Kings*, 380–381. Trade with Africa in 1960 constituted only 3–4 percent of total U.S. exports and imports, a point Europeanists liked to emphasize. Brinkley, *Dean Acheson*, 305. Richard Reeves's careful study of the daily course of Kennedy's presidency reveals how little attention the president and his most senior advisers usually paid to African issues. Reeves, *President Kennedy*, passim.

43. State Department, "Africa: Guidelines for Policy and Operations," March 1962, NSF, box 2, JFKL.

44. Williams to Rusk, 12 June 1963, NSF, box 3, JFKL; Rusk to certain African posts, 28 May 1963, *FRUS, 1961–63*, 21:332–334; Lauren, *Power and Prejudice*, 218–219.

45. Satterthwaite et al. to Williams, undated [September 1961], NSF, box 2, JFKL.

46. National Intelligence Estimates 60/70–2–61, "The Probable Interrelationships of the Independent African States," 31 August 1961, *FRUS, 1961–63*, 21:299–300.

47. Plummer, *Rising Wind*, 308.

48. Brubeck to Bundy, 29 October 1963, *FRUS, 1961–63*, 21:505–507.

49. Attwood, *The Reds and the Blacks*, 47; Wofford, *Of Kennedys and Kings*, 371.

50. Murrow to Bundy, 5 December 1962, *FRUS, 1961–63*, 21:325.

51. Freund to Rusk, 3 December 1962, and Rusk, Circular 1032, 4 December 1962, in NSF, box 3, folder "Africa, General (11/62–12/62)," JFKL; *Department of State Bulletin*, 24 December 1962, 961.

52. Geren to Rusk, 5 December 1962, NSF, box 155A, JFKL. Ellender repeated his attacks on the capacity of Africans for self-government in televised remarks on 16 June 1963, eliciting an angry response from several African ambassadors. Little to Bundy, 25 June 1963, NSF, box 3, JFKL.

53. "Problems of Southern Africa," 4 October 1962, enclosed in Rostow to members of Tuesday Planning Group, 18 October 1962, NSF, box 2, JFKL.

54. "The White Redoubt," 28 June 1962, enclosed in Owen to McGhee et al., 6 July 1962, NSF, box 2, JFKL. Roy Welensky was the prime minister of the Central African Federation, consisting until 1963 of Southern Rhodesia, Northern Rhodesia (Zambia), and Nyasaland (Malawi); Hendrik Verwoerd was the prime minister of South Africa; and Antonio Salazar was the president of Portugal.

55. Mandela, "Address to Conference of the Pan-African Freedom Movement of East and Central Africa," January 1962, Addis Ababa, in *Mandela, Tambo, and the African National Congress—The Struggle against Apartheid, 1948–1990: A Documentary Survey*, ed. Sheridan Johns and R. Hunt Davis (New York: Oxford University Press, 1991), 106–111.

56. Rusk to U.S. Consulate General in Salisbury, 30 November 1962, *FRUS, 1961–63*, 21:523; Ball to U.S. Embassy in United Kingdom, 7 January 1963, ibid., 524.

57. Williams to Rusk, 25 February 1963, ibid., 329.

58. "The White Redoubt," NSF, box 2, JFKL.

59. Stevenson to Rusk, 21 August 1963, NSF, boxes 158–161, JFKL.

60. "Problems of Southern Africa," NSF, box 2, JFKL.

61. "The White Redoubt," NSF, box 2, JFKL.

62. McNamara to Rusk, 11 July 1963, reprinted in *South Africa and the United States: The Declassified History*, ed. Kenneth Mokoena (New York: New Press, 1993), 62–63; State Department, "Guidelines for Policy and Operations: Republic of South Africa," May 1962, NSF, box 2, JFKL; memorandum of conversation, 21 August 1963, *FRUS, 1961–63*, 21:492–493; Williams to Rusk, 23 November 1963, *FRUS, 1961–63*, 21:339.

63. For examples, see memorandum of conversation, 7 November 1963, *FRUS, 1961–63*, 21:582–583; Rusk to Department of State, 6 October 1962, ibid., 621–622.

64. Brubeck to Bundy, 29 October 1963, NSF, boxes 158–161, JFKL.

65. A novel of unusual insights into this situation is Barbara Kingsolver, *The Poisonwood Bible* (New York: HarperFlamingo, 1998).

66. Noer, "New Frontiers and Old Priorities," 260–269.

67. Elbrick to Rusk, 11 December 1962, NSF, box 5, JFKL.

68. "Problems of Southern Africa," NSF, box 2, JFKL.

69. *New York Times*, 16 February 1961, 11; C. Eric Lincoln, "The Race Problem and International Relations," *New South* 21 (Fall 1966): 11; F. Chidozie Ogene, *Interest Groups and the Shaping of Foreign Policy: Four Case Studies of United States African Policy* (Lagos: Macmillan Nigeria, 1983), 32–33.

70. Gibbs, *Political Economy of Third World Intervention*, 122; Staniland, *American Intellectuals*, 236–243; Ogene, *Interest Groups and the Shaping of Foreign Policy*, 22–31; Noer, "New Frontiers and Old Priorities," 261–264. Katanga's American supporters sometimes slid over into explicit statements of racial prejudice about Congolese who did not agree with Tshombe. See, for example, Mahoney, *JFK*, 118.

71. Noer, "New Frontiers and Old Priorities," 266.

72. "Status Report of African Reactions to Civil Rights in the United States," 12

July 1963, NSF, box 295, JFKL; Minter, *King Solomon's Mines Revisited*, 146–150.

73. Williams to Rusk, 25 February 1963, *FRUS, 1961–63*, 21:328–329.

74. Mahoney, *JFK*, 136.

75. John Marcum, *The Angolan Revolution*, 2 vols. (Cambridge, Mass.: MIT Press, 1969–1978).

76. Stevenson to Kennedy, 26 April 1962, and Bowles to Kennedy, 4 June 1962, in NSF, boxes 154A–155, JFKL; Massie, *Loosing the Bonds*, 119; Noer, "New Frontiers and Old Priorities," 269–274; Mahoney, *JFK*, 188–189.

77. Bundy to Johnson, 28 July 1961, and Satterthwaite et al. to Williams, undated [September 1961], both in NSF, box 2, JFKL.

78. Hilsman to Bundy, 23 May 1961, *FRUS, 1961–63*, 21:543; Rusk to U.S. Embassy in Lisbon, 31 July 1961, NSF, box 5, JFKL; Rusk to U.S. Embassy in Congo, 5 October 1961, *FRUS, 1961–63*, 21:551–552; Mahoney, *JFK*, 190–204; Minter, *King Solomon's Mines Revisited*, 162; Noer, "New Frontiers and Old Priorities," 271.

79. Rusk to U.S. Embassy in Portugal, 23 April 1961, *FRUS, 1961–63*, 21:543 (quotation); Rusk to U.S. Embassy in Portugal, 11 August 1961, and Bundy to Kennedy, 31 August 1961, in NSF, box 5, JFKL; "The Kingdom of Silence"; Mahoney, *JFK*, 195.

80. Note 1, *FRUS, 1961–63*, 21:548.

81. Belk to Kennedy, 14 September 1961, ibid., 550.

82. Taft (for Williams) to Rusk, 16 August 1961, NSF, box 5, folder "Angola, General, 8/61–12/61," JFKL; report of Presidential Task Force on Portuguese Territories in Africa, 12 July 1961, NSF, box 5, JFKL; "The Kingdom of Silence," 30–31. One dissenting Portuguese colonial official had concluded that in Angola "only the dead are exempt from forced labor." Massie, *Loosing the Bonds*, 117.

83. Report of the Chairman of the Task Force on Portuguese Territories in Africa [Williams], 4 July 1961, enclosed in Belk to Bundy, 4 July 1961, NSF, box 5, JFKL. Ambassador William Attwood concisely explained the same point in a November 1961 speech to the Princeton Club of Washington: "The crisis in Berlin seems as remote to them [Africans] as Waterloo seemed remote to us 150 years ago." Attwood, *The Reds and the Blacks*, 60.

84. See, for example, Rusk to Johnson, 18 June 1961, *FRUS, 1961–63*, 21:544; Devine to Rusk, 19 February 1963, NSF, boxes 154A–155, JFKL.

85. Williams to Rusk, 23 October 1962, *FRUS, 1961–63*, 21:561.

86. Eldrick to Rusk, 11 July 1961, NSF, box 5, JFKL.

87. Rusk to U.S. Embassy in Lisbon, 23 August 1961, NSF, box 5, JFKL; CIA telegram, 31 January 1962, NSF, boxes 154A–155, JFKL; memorandum of conversa-

tion, 24 October 1962, *FRUS, 1961–63*, 21:562–563; Noer, *Cold War and Black Liberation*, 73; Brinkley, *Dean Acheson*, 307.

88. Quoted in Schoenbaum, *Waging Peace and War*, 376.

89. Belk to Dungan, 9 January 1962, NSF, box 2, JFKL; CIA, "Prospects for Nationalism and Revolt in Mozambique," 30 April 1963, NSF, box 3, JFKL. Tanganyika became Tanzania after merging with Zanzibar in 1964.

90. Hughes to Rusk, 5 November 1963, NSF, box 5, JFKL.

91. Stevenson to Kennedy, 26 April 1962, NSF, boxes 154A–155, JFKL; Noer, *Cold War and Black Liberation*, 75; Mahoney, *JFK*, 75.

92. Geren to Department of State, 25 October 1963, NSF, box 155A, JFKL.

93. Brinkley, *Dean Acheson*, 306–312, 389, n. 50 (quotation).

94. Bundy to Kennedy, 2 May 1962, NSF, boxes 154A–155, JFKL; note 1, *FRUS, 1961–63*, 21:562; Mahoney, *JFK*, 218–220. For samples of Kennedy's position, see Brubeck memorandum for the record, 15 July 1963, NSF, box 3, JFKL, and memorandums for the record, 18 July 1963 and 9 September 1963, *FRUS, 1961–63*, 21:573, 579.

95. Sorensen to Kennedy, 4 June 1962, NSF, boxes 154A–155, JFKL.

96. T. R. H. Davenport, *South Africa: A Modern History*, 3d ed. (Toronto: University of Toronto Press, 1987), 403–404; Joseph Lelyveld, *Move Your Shadow* (New York: Times Books, 1985), 327.

97. "The White Redoubt," NSF, box 2, JFKL.

98. Belk to Bundy, 11 July 1963, NSF, box 3, JFKL. On uranium in the earlier Cold War years, see Borstelmann, *Apartheid's Reluctant Uncle*.

99. Bundy to Gilpatric, 7 June 1961, *FRUS, 1961–63*, 21:595–597; Rusk to Kennedy, 16 March 1963, NSF, box 3, JFKL.

100. Belk to Bundy, 11 July 1963, NSF, box 3, JFKL; Hughes to Rusk, 12 August 1963, *FRUS, 1961–63*, 646–647. For further discussion of these issues, see Minter, *King Solomon's Mines Revisited*.

101. Bowles to Bundy, 21 September 1961, *FRUS, 1961–63*, 21:603–605; Stevenson to Rusk, 2 June 1961, ibid., 594. On the military importance of the tracking station, see Gilpatric to Bowles, 16 March 1961, ibid., 588–589; Brown to Deputy Assistant Secretary of Defense, 8 July 1963, NSF, box 3, JFKL.

102. See, for example, Rusk to U.S. Embassy in South Africa, 25 August 1961, *FRUS, 1961–63*, 21:598–600; Bowles to U.S. Embassy in South Africa, 28 September 1961, ibid., 605–606.

103. Memorandum of conversation, 20 November 1961, ibid., 615. See also Rusk to State Department, ibid., 620–622.

104. "Problems of Southern Africa," NSF, box 2, JFKL.

105. Allard Lowenstein, *Brutal Mandate: A Journey to South West Africa* (New York: Macmillan, 1962); William H. Chafe, *Never Stop Running: Allard Lowenstein*

and the Struggle to Save American Liberalism (New York: Basic Books, 1993), chap. 6, "The South Africa Years."

106. CIA, "The South-West Africa Issue in the United Nations," 19 April 1963, NSF, box 3, JFKL.

107. Satterthwaite to Rusk, 16 May 1963, NSF, box 3, JFKL.

108. Noer, "New Frontiers and Old Priorities," 277; Mahoney, *JFK,* 241–242; Christopher Coker, *The United States and South Africa, 1968–1985: Constructive Engagement and Its Critics* (Durham: Duke University Press, 1986), 27.

109. McKay, *Africa in World Politics,* 352; Noer, "New Frontiers and Old Priorities," 275.

110. Bundy to Kennedy, 13 July 1963, NSF, box 3, JFKL; Rusk to Harriman, 15 July 1963, *FRUS, 1961–63,* 21:635.

111. Rusk to U.S. Embassy in South Africa, 25 October 1961, *FRUS, 1961–63,* 21:611.

112. *Mandela, Tambo, and the African National Congress—The Struggle against Apartheid: A Documentary Survey,* ed. Sheridan Johns and R. Hunt Davis (New York: Oxford University Press, 1991), 89–90.

113. Satterthwaite to Rusk, 6 May 1963, enclosed in C. K. (Carl Kaysen) to Kennedy, 16 May 1963, NSF, box 3, JFKL.

114. Satterthwaite to Rusk, 18 December 1962, enclosing "Country Internal Defense Plan," NSF, box 3, JFKL.

115. Mandela, excerpt from courtroom statement, Rivonia trial, 20 April 1964, in Johns and Davis, *Mandela, Tambo, and the African National Congress,* 115–133. See also "The Communist Party in South Africa" [from *Africa Report,* March 1961], NSF, box 2, JFKL.

116. Massie, *Loosing the Bonds,* 129–132.

117. CIA, "Subversive Movements in South Africa," 10 May 1963, NSF, box 3, JFKL; Peter J. Schraeder, *United States Foreign Policy toward Africa: Incrementalism, Crisis, and Change* (New York: Cambridge University Press, 1994), 206; Minter, *King Solomon's Mines Revisited,* 192; Noer, *Cold War and Black Liberation,* 138–139. The supposedly coincidental timing of a visit to Durban by a U.S. naval task force on 31 May 1961, the day South Africa officially became an independent republic, was doubted by African observers. They tended instead to see U.S. marines demonstrating flame throwers and machine guns and U.S. helicopters flying over African neighborhoods as evidence of American support for the Pretoria government. Minter, *King Solomon's Mines Revisited,* 192.

118. Satterthwaite to State Department, 17 March 1962, *FRUS, 1961–63,* 618–619, esp. n. 4; Minter, *King Solomon's Mines Revisited,* 188.

119. CIA, "Subversive Movements in South Africa," NSF, box 3, JFKL. This docu-

ment has also been reprinted in Mokoena, *South Africa and the United States,* 188–194.

120. Joseph Albright and Marcia Kunstel, "CIA Tip Led to '62 Arrest of Mandela: Ex-Official Tells of US 'Coup' to Aid S. Africa," *Atlanta Journal and Constitution,* 10 June 1990, 1, 14.

121. Albright and Kunstel, "CIA Tip." Two days before the president's death, Robert Kennedy told Bundy that regarding black liberation fighters in South Africa, Southern Rhodesia, Angola, and Mozambique, "I gather that we really don't have much of a policy or we are just beginning to develop one." Such ignorance of the CIA's possible involvement in an area deemed peripheral seems perfectly plausible, despite Robert Kennedy's central role in administration policy on civil rights and on many international issues. Robert Kennedy to Bundy, 20 November 1963, NSF, Confidential File, box 76, LBJL.

122. Reeves, *President Kennedy,* 59–60.

123. Branch, *Parting the Waters,* 414 (quotation), 606; Chafe, "Kennedy and the Civil Rights Movement"; Strober, *"Let Us Begin Anew,"* 282; Reeves, *President Kennedy,* 60; Giglio, *The Presidency of John F. Kennedy,* 161; O'Reilly, *Nixon's Piano,* 201.

124. Stern, *Calculating Visions,* 52–53; Goodwin, *Remembering America,* 4–5; Giglio, *The Presidency of John F. Kennedy,* 162–164; Wilkins, *Standing Fast,* 281; Wofford, *Of Kennedys and Kings,* 141. Senior figures in the administration shared this attention to racial symbolism and its impact abroad, from Robert Kennedy initially recommending William Hastie for the first opening on the highest court ("It would mean so much overseas that we had a Negro on the Supreme Court") to Lyndon Johnson's reaction while watching the lift-off of the first American astronaut to orbit the Earth ("If John Glenn were only a Negro!"). Giglio, *The Presidency of John F. Kennedy,* 41; Miller, *Lyndon,* 278.

125. Sitkoff, *The Struggle for Black Equality,* 90–91, 95–105, 113–117, 128–131, 144–147; Chafe, "Kennedy and the Civil Rights Movement"; O'Reilly, *Nixon's Piano,* 236–237.

126. King to "My dear Fellow Clergymen," ("Letter from a Birmingham Jail"), 16 April 1963, copy in Burke Marshall Papers, box 8, JFKL (also reprinted in William H. Chafe and Harvard Sitkoff, eds., *A History of Our Time,* 3d ed. [New York: Oxford University Press, 1991]).

127. Wofford to Kennedy, 29 May 1961, Howard K. Smith, CBS weekly news analysis, 21 May 1961, and Wofford to Kennedy, 23 January 1962, in President's Office Files, box 67, JFKL.

128. Dittmer, *Local People,* 169; Stern, *Calculating Visions,* 44; O'Reilly, *"Racial Matters,"* 122.

129. "Negro Leaders Seek Halt in Freedom Ride Testing; Murrow Cites Reaction,"

New York Times, 25 June 1961, 1; O'Reilly, *Nixon's Piano,* 211–212 (quotation); Reeves, *President Kennedy,* 145; Wofford, *Of Kennedys and Kings,* 125, 155–156; Wilkins, *Standing Fast,* 280–283; Stern, *Calculating Visions,* 56–68; Branch, *Parting the Waters,* 472–479; Dittmer, *Local People,* 92–94.

130. Both antiracist organizers and some journalists were well aware of this ironic analogy of the South to the Soviet Union; see, for example, Dittmer, *Local People,* 90–91, and Dan T. Carter, *The Politics of Rage: George Wallace, the Origins of the New Conservatism, and the Transformation of American Politics* (New York: Simon and Schuster, 1995), 114–115.

131. Carter, *Politics of Rage,* 110–111; Reeves, *President Kennedy,* 356, 364.

132. Andrew Young, OHT, p. 5, LBJL; Dittmer, *Local People,* 138–142.

133. Branch, *Parting the Waters,* 665 (whites, president); Reeves, *President Kennedy,* 380 (attorney general). From the perspective of local civil rights workers, as John Dittmer has shown, the whole crisis at Oxford was a "sideshow"—created by a solo operator, Meredith—to the real agenda of the slow, steady organizing of black communities across the Magnolia State. See Dittmer, *Local People,* 138. The Kennedys, similarly, saw Meredith as a grandstander who was partly responsible for the crisis and the harm it caused to American interests abroad.

134. Chafe, "Kennedy and the Civil Rights Movement," 71; Sitkoff, *The Struggle for Black Equality,* 133–135; Glenn T. Askew, *But for Birmingham: The Local and National Movements in the Civil Rights Struggle* (Chapel Hill: University of North Carolina Press, 1997).

135. Korry to Rusk, 23 May 1963, and "Status Report of African Reactions to Civil Rights in the United States," 6 July 1963, enclosed in Read to Bundy, 8 July 1963, both in NSF, box 3, JFKL; "Reaction to Racial Tension in Birmingham, Alabama," 14 May 1963, and Wilson to Kennedy, 17 May 1963, both in President's Office Files, box 96, JFKL; Branch, *Parting the Waters,* 783–786.

136. Branch, *Parting the Waters,* 791.

137. Ibid., photograph no. 60, following p. 688.

138. O'Reilly, *Nixon's Piano,* 226; Stern, *Calculating Visions,* 86–87.

139. Memorandum of Kennedy-Wallace conversation, 18 May 1963, President's Office Files, box 96, JFKL.

140. Tom Wicker, "Kennedy Sees 'Moral Crisis' in U.S.," *New York Times,* 12 June 1963, 1, 20; Stern, *Calculating Visions,* 91–93 (Robert Kennedy); Reeves, *A Question of Character,* 354–356.

141. Tad Szulc, "Kennedy Asks Break in Cold War," *New York Times,* 11 June 1963, 1; Giglio, *The Presidency of John F. Kennedy,* 217.

142. Rusk to all American diplomatic and consular posts, 19 June 1963, and Ken-

nedy to all American diplomatic and consular posts, 19 June 1963, in NSF, box 3, JFKL.

143. "Status Report of African Reactions to Civil Rights in the United States," 6 July 1963, enclosed in Read to Bundy, 7 July 1963, NSF, box 3, JFKL; "Status Report of African Reactions to Civil Rights in the United States," 12 July 1963, and "Recent Worldwide Comment on the U.S. Racial Problem," 19 July 1963, in NSF, box 295, JFKL.

144. "Status Report of African Reactions to Civil Rights in the United States," 12 July 1963, and "Recent Worldwide Comment on the U.S. Racial Problem," 19 July 1963, in NSF, box 295, JFKL.

145. Stern, *Calculating Visions*, 89; Branch, *Parting the Waters*, 825. Ross Barnett's appearance at the subsequent trial of Evers's accused killer, Byron De La Beckwith, was no more encouraging. Medgar's wife, Myrlie, recalls Barnett walking into the courtroom, slapping Beckwith warmly on the shoulder, and sitting down next to him. Jurors had no trouble discerning the message from their governor that this man should be acquitted, and he was. Beckwith was re-tried and convicted thirty years later, when new evidence of official tampering with the trial came to light. Claudia Dreifus, "The Widow Gets Her Verdict," *New York Times Magazine*, 27 November 1994.

146. James Farmer, "The March on Washington: The Zenith of the Southern Movement," in *New Directions in Civil Rights Studies*, ed. Armstead L. Robinson and Patricia Sullivan (Charlottesville: University Press of Virginia, 1991), 31–32; Reeves, *President Kennedy*, 580; Giglio, *The Presidency of John F. Kennedy*, 185.

147. O'Reilly, *"Racial Matters,"* 120 (quotation); Dittmer, *Local People*, 128; Lawson, *Running for Freedom*, 83–86.

148. Giglio, *The Presidency of John F. Kennedy*, 169–170.

149. Reeves, *President Kennedy*, 431; Giglio, *The Presidency of John F. Kennedy*, 187.

150. Thomas J. Friedman, "Cold War without End," *New York Times Magazine*, 22 August 1993, 45 ("Why can"); O'Reilly, *Nixon's Piano*, 222 ("to persecute"); Carter, *Politics of Rage*, 161 ("the same people").

151. Lewis, *Portrait of a Decade*, 146.

152. Sitkoff, *The Struggle for Black Equality*, 139.

153. Carter, *Politics of Rage*, 120.

154. Dittmer, *Local People*, 255.

155. USIA, "Recent Worldwide Comment on the U.S. Racial Problem," 19 July 1963, NSF, box 295, JFKL; "Reaction to Racial Tension in Birmingham, Alabama," 14 May 1963, President's Office Files, box 96, JFKL; Harold R. Isaacs, "American Race Relations and the United States Image in World Affairs," *Journal of Human Relations* 10 (Winter–Spring 1962): 266–280.

156. Forman, *Making of Black Revolutionaries,* 177–178 (quotation); Wofford, *Of Kennedys and Kings,* 305.

157. Wallner to Rusk, 31 January 1961, NSF, box 2, JFKL; Romano, "No Diplomatic Immunity."

158. Wofford to Kennedy, 23 January 1962, President's Office Files, box 67, JFKL; Wofford, *Of Kennedys and Kings,* 149; Cabell Phillips, "U.S. Is Worried by Capital's Bias," *New York Times,* 26 May 1961, 26; "The Color Line in Diplomacy," *U.S. News and World Report,* 27 March 1961, 78–79; "When African Diplomats Come to Washington," *U.S. News and World Report,* 12 June 1961, 86–88.

159. Sanjuan to Chief of Protocol, 21 September 1961, enclosed in Battle to Dutton, 29 September 1961, NSF, box 2, JFKL.

160. Mahoney, *JFK,* 23. For Rusk's steep learning curve on this issue, see Schoenbaum, *Waging Peace and War,* 381–382.

161. Isaacs, *The New World of Negro Americans,* 17–18.

162. "Big Step Ahead on a High Road," *Life,* 8 December 1961, 32–39; "Department Urges Maryland to Pass Public Accommodations Bill," *Department of State Bulletin,* 2 October 1961, 551–552; "Troubled Route," *Time,* 13 October 1961; Timothy P. Maga, "Battling the 'Ugly American' at Home: The Special Protocol Service and the New Frontier, 1961–63," *Diplomacy and Statecraft* 3, 1 (1992): 126–142; Massie, *Loosing the Bonds,* 120–123; McKay, *Africa in World Politics,* 405–407; Wofford, *Of Kennedys and Kings,* 126–127.

163. Williams to Rusk, 23 November 1963, *FRUS, 1961–63,* 21:338.

164. Memorandum of conversation, 23 November 1962, ibid., 622–624.

165. Jon Nordheimer, "A Creek, Negro Run, Is the Source of Debate," *New York Times,* 3 November 1994, B6.

166. King, "'The Time For Freedom Has Come,'" *New York Times Magazine,* 10 September 1961, 25, 118–119.

167. King to "My dear Fellow Clergymen," 16 April 1963, in Chafe and Sitkoff, *A History of Our Time,* 187.

168. Horace Mann Bond, "Howe and Isaacs in the Bush: The Ram in the Thicket" [1961], reprinted in Hill and Kilson, *Apropos of Africa,* 278–288; Ferguson to State Department, 18 December 1953, *FRUS, 1952–54,* 11:70; McKay, *Africa,* 392–393.

169. James Reston, "'Copper Sun, Scarlet Sea, What Is Africa to Me?'" *New York Times,* 17 February 1961, 26; "Riot in Gallery Halts U.N. Debate," *New York Times,* 16 February 1961, 1, 10.

170. Wofford address to the National Civil Liberties Clearing House, 23 March 1961, President's Office Files, box 97, JFKL.

171. See Von Eschen, *Race against Empire;* Plummer, *Rising Wind;* Duberman, *Robeson;* Horne, *Black and Red.*

172. Massie, *Loosing the Bonds,* 129–132.

173. Houser, *No One Can Stop the Rain,* 266; Houser, "Freedom's Struggle Crosses Oceans and Mountains," 186–187.

174. Jonathan Zimmerman, "Beyond Double Consciousness: Black Peace Corps Volunteers in Africa, 1961–1971," *Journal of American History* 82 (December 1995): 1015–1018.

175. O'Reilly, *Nixon's Piano,* 225.

176. Branch, *Parting the Waters,* 705.

177. State Department, "Guidelines for Policy and Operations: Republic of South Africa," May 1962, NSF, box 2, JFKL.

178. Williams to Rusk, 12 June 1963, NSF, box 3, JFKL.

179. Rusk to Kennedy, 21 June 1963, NSF, boxes 158–161, JFKL (quotation); Massie, *Loosing the Bonds,* 144–146.

180. Isaacs, *The New World of Negro Americans,* 16.

181. See, for example, Dittmer, *Local People,* 211, and Peter G. Bourne, *Jimmy Carter: A Comprehensive Biography from Plains to Postpresidency* (New York: Simon and Schuster, 1997), 140.

5. The Perilous Path to Equality

1. Steven F. Lawson, "Mixing Moderation with Militancy," in *The Johnson Years,* vol. 3, *LBJ at Home and Abroad,* ed. Robert A. Divine (Lawrence: University Press of Kansas, 1993), 105; Bruce Miroff, "Presidential Leverage over Social Movements: The Johnson White House and Civil Rights," *Journal of Politics* 43, 1 (February 1981): 21–22. Julius Lester suggested black radicals' awareness of Johnson's desire to co-opt them when he referred to the president's extraordinary civil rights speech of 15 March 1965 with its declaration, "We shall overcome": "One began having nightmares of Lyndon Johnson standing before a joint session of Congress and closing an address with the words 'Black Power!'" Julius Lester, *Look Out, Whitey! Black Power's Gon' Get Your Mama!* (New York: Dial, 1968), 112–114.

2. Eric F. Goldman, *The Tragedy of Lyndon Johnson* (New York: Knopf, 1969), 352; Joseph A. Califano, Jr., *The Triumph and Tragedy of Lyndon Johnson: The White House Years* (New York: Simon and Schuster, 1991), 24, 196–203.

3. Clifford Alexander, OHT, 13–15, LBJL; Doris Kearns, *Lyndon Johnson and the American Dream* (New York: Harper and Row, 1976), x; Lawson, "Mixing Moderation with Militancy," 100.

4. Kearns, *Lyndon Johnson and the American Dream,* 340.

5. Ibid., 10–11.

6. Miller, *Lyndon,* 280–282 (quotation); Waldo Heinrichs, "Lyndon B. Johnson: Change and Continuity," in *Lyndon Johnson Confronts the World: American Foreign Policy, 1963–1968,* ed. Warren I. Cohen and Nancy Bernkopf Tucker (New York: Cambridge University Press, 1994), 25–26; Kearns, *Lyndon Johnson and the American Dream,* 194.

7. Philip Geyelin, *Lyndon B. Johnson and the World* (New York: Praeger, 1966), 15.

8. Miller, *Lyndon,* 455–456.

9. Ibid., 294; Heinrichs, "Lyndon B. Johnson," 25.

10. Lloyd C. Gardner, *Pay Any Price: Lyndon Johnson and the Wars for Vietnam* (Chicago: Ivan Dee, 1995).

11. Miller, *Lyndon,* 466.

12. Sandra C. Taylor, "Lyndon Johnson and the Vietnamese," in *Shadow on the White House: Presidents and the Vietnam War, 1945–1975,* ed. David L. Anderson (Lawrence: University of Kansas Press, 1993), 123. Johnson was not exaggerating: the speech Ky had just given in Honolulu had been prepared by the U.S. embassy in Saigon. Woods, *Fulbright,* 403–404.

13. Ronnie Dugger, *The Politician—The Life and Times of Lyndon Johnson: The Drive for Power, from the Frontier to Master of the Senate* (New York: Norton, 1982), 32.

14. Miller, *Lyndon,* 489–490; Bruce J. Schulman, *Lyndon B. Johnson and American Liberalism: A Brief Biography with Documents* (New York: St. Martin's, 1994), 142.

15. Wilkins, *Standing Fast,* 298; Roy Wilkins, OHT, 6, LBJL.

16. David M. Barrett, *Uncertain Warriors: Lyndon Johnson and His Vietnam Advisers* (Lawrence: University Press of Kansas, 1993), 31–32 (quotation); Geyelin, *Lyndon B. Johnson and the World,* 134; Terence Lyons, "Keeping Africa off the Agenda," in *Lyndon Johnson Confronts the World: American Foreign Policy, 1963–1968,* ed. Warren I. Cohen and Nancy Bernkopf Tucker (New York: Cambridge University Press, 1994), 245–248; Gerald E. Thomas, "The Black Revolt: The United States and Africa in the 1960s," in *The Diplomacy of the Crucial Decade: American Foreign Relations during the 1960s,* ed. Diane B. Kunz (New York: Columbia University Press, 1994), 326–327.

17. Roger Wilkins, *A Man's Life: An Autobiography* (New York: Simon and Schuster, 1982), 132; Louis Martin, OHT, 15, LBJL; Andrew Young, OHT, 19, LBJL; Ramsey Clark, OHT (III), 19–20, LBJL.

18. "Lyndon Johnson and the Civil Rights Revolution: A Panel Discussion," in *Lyndon Baines Johnson and the Uses of Power,* ed. Bernard J. Firestone and Robert C. Vogt (Westport, Conn.: Greenwood, 1988), 181 (quotation); Robert Dallek, *Lone Star Rising: Lyndon Johnson and His Times, 1908–1960* (New York:

Oxford University Press, 1991), 519–520; Robert Dallek, *Flawed Giant: Lyndon Johnson and His Times, 1961–1973* (New York: Oxford University Press, 1998), 111–121. Johnson liked to explain his conservative voting record as a congressman and senator from Texas as a matter of simple survival: "There's nothing in the world more useless than a dead liberal." Miller, *Lyndon,* 69.

19. Califano, *Triumph and Tragedy of Lyndon Johnson,* 56; Miller, *Lyndon,* 177.

20. Dallek, *Lone Star Rising,* 583–584.

21. Califano, *Triumph and Tragedy of Lyndon Johnson,* 177–178.

22. Ibid., 337 (quotation); Miller, *Lyndon,* 187–188.

23. Goodwin, *Remembering America,* 413.

24. O'Reilly, *"Racial Matters,"* 289.

25. Lawson, "Mixing Moderation with Militancy," 83–86, 96–99, 104–105.

26. Miller, *Lyndon,* 454–455.

27. Markman to Johnson, 1 February 1967, WHCF, Welfare, box 28, LBJL.

28. Robert Sherrill, *The Accidental President* (New York: Grossman, 1967), 191.

29. James Farmer, OHT (II), 27–28, LBJL; Lawson, "Mixing Moderation with Militancy," 97–100; Miller, *Lyndon,* 435; Wilkins, *A Man's Life,* 242.

30. Johnson, "The American Promise," remarks to a joint session of Congress, 15 March 1965, NSF, CF, box 77, LBJL; H. W. Brands, *The Wages of Globalism: Lyndon Johnson and the Limits of American Power* (New York: Oxford University Press, 1995), 4, 27–28.

31. Goodwin, *Remembering America,* 257–258 ("sneakers"); "Edison Dictaphone Recording: LBJ-Sorensen," 3 June 1963, George Reedy Office Files, box 1, LBJL.

32. Goldman, *Tragedy of Lyndon Johnson,* 245–248.

33. James Farmer, OHT (II), 19–20, 25, LBJL; Sitkoff, *The Struggle for Black Equality,* 172–73; Schulman, *Lyndon B. Johnson and American Liberalism,* 76.

34. Wilkins, *Standing Fast,* 243, 296, 300–301; Barrett, *Uncertain Warriors,* 45; Fite, *Richard B. Russell,* 404–405; Goodwin, *Remembering America,* 313–314.

35. Undated, unsigned memo, Marvin Watson Office Files, box 18, LBJL.

36. Geyelin, *Lyndon B. Johnson and the World,* 147 (quotation); Carl Rowan to Johnson, 21 July 1964, NSF, CF, box 76, LBJL; David J. Garrow, *Protest at Selma: Martin Luther King, Jr., and the Voting Rights Act of 1965* (New Haven: Yale University Press, 1978), 169; Piero Gleijeses, "'Flee! The White Giants Are Coming!': The United States, the Mercenaries, and the Congo, 1964–65," *Diplomatic History* 18, 2 (Spring 1994): 222, n. 81; C. Eric Lincoln, "The Race Problem and International Relations," *New South* 21, 4 (Fall 1966): 2–3.

37. Harold R. Isaacs, "Color in World Affairs," *Foreign Affairs* 47 (January 1969): 247–250; Lawson, "Mixing Moderation with Militancy," 104–106.

38. Lester, *Look Out, Whitey!* 22–23.

39. Carter, *Politics of Rage,* 222.

40. Thomas R. West and James W. Mooney, eds., *To Redeem a Nation: A History and Anthology of the Civil Rights Movement* (St. James, N.Y.: Brandywine, 1993), 219.

41. Haynes to Komer, 2 November 1965, NSF, CF, box 97, LBJL; transcript of telephone conversation, Ball and Greenfield, 10 November 1965, George Ball Papers, box 1, LBJL; Komer to Johnson, 16 June 1965, NSF, CF, box 76, LBJL; Anthony Lake, *The "Tar Baby" Option: American Policy toward Southern Rhodesia* (New York: Columbia University Press, 1976), 67–70; Lyons, "Keeping Africa off the Agenda," 246–251; Henry F. Graff, *The Tuesday Cabinet: Deliberation and Decision on Peace and War under Lyndon B. Johnson* (Englewood Cliffs, N.J.: Prentice-Hall, 1970), 75–76.

42. Staniland, *American Intellectuals and African Nationalists,* 99–139.

43. Ibid., 229–231.

44. Carl Rowan to Johnson, 21 July 1964, NSF, CF, box 76, LBJL; Rowan to Bundy, 29 September 1964, NSF, International Meetings and Travel File, boxes 33, 34, LBJL; Lincoln, "The Race Problem and International Relations," 12; Gleijeses, "'Flee! The White Giants Are Coming!'" 237.

45. "Angola Genocide Laid to Portugal," *New York Times,* 24 May 1964, clipping in NSF, CF, boxes 79, 80, LBJL; Attwood to Rusk, 1 June 1964, NSF, CF, box 76, LBJL.

46. Komer to Johnson, 16 June 1965, NSF, CF, box 76, LBJL; Noer, *Cold War and Black Liberation,* 117; Gleijeses, "'Flee! The White Giants Are Coming!'" 207–208, n. 5; Lauren, *Power and Prejudice,* 230–232.

47. Crawford Young, "Rebellion and the Congo," in *Protest and Power in Black Africa,* ed. Robert I. Rothberg and Ali A. Mazrui (New York: Oxford University Press, 1970), 969–1011, esp. 969–74, 1000–1007; Lyons, "Keeping Africa off the Agenda," 257.

48. CIA, "Short-Term Prospects for the Tshombe Government in the Congo," 5 August 1964, NSF, National Intelligence Estimates, boxes 8, 9, LBJL. From the agency's perspective, Tshombe evidently did not qualify as an "extremist."

49. Geyelin, *Lyndon B. Johnson and the World,* 38–39, 118–119; Gleijeses, "'Flee! The White Giants Are Coming!'" 208–209; Lyons, "Keeping Africa off the Agenda," 256; Mahoney, *JFK,* 289.

50. "How C.I.A. Put 'Instant Air Force' into Congo," *New York Times,* 26 April 1966, 1, 30; Mahoney, *JFK,* 230.

51. Transcript of telephone conversation, Lippmann and Ball, 25 August 1964, Ball Papers, box 2, LBJL.

52. Gleijeses, "'Flee! The White Giants Are Coming!'" 207–208, 216–217.

53. Mahoney, *JFK,* 230. Burke Marshall of the Justice Department remembered

the all-white Mississippi Highway Patrol in 1964 as "better than a lot of sheriffs" in the state but still "full of very, very low-grade types of police officers, to put it mildly." Marshall, OHT, 31, LBJL.

54. "Opinion: Englehard in Zambia" [editorial], *Africa Today* 11, 8 (October 1964): 3.

55. *Memos of the Special Assistant for National Security Affairs: McGeorge Bundy to President Johnson, 1963–1966* (Frederick, Md.: University Publications of America, 1985; microfilm, 4 reels), contains many of Bundy's key memos to Johnson and other administration officials. See Bundy to Johnson, 16 November 1964, reel 2, frame 47; Young, "Rebellion and the Congo," 974–975; Lake, *The "Tar Baby" Option*, 75–76.

56. Gleijeses, "'Flee! The White Giants Are Coming!'" 218, 231.

57. Segal, *The Race War*, 108–110 (quotation); Lyons, "Keeping Africa off the Agenda," 258–259.

58. Malcolm X, *The Speeches of Malcolm X at Harvard*, ed. Archie Epps (New York: Morrow, 1968), 165–168 (quotation); "Dr. King Advocates Congo Withdrawal," *New York Times*, 14 December 1964, 3; Houser, "Freedom's Struggle Crosses Oceans and Mountains," 184; Moore, *Castro, the Blacks, and Africa*, 188–192; Gleijeses, "'Flee! The White Giants Are Coming!'" 233, 236.

59. "The Congo Massacre," *Time*, 4 December 1964, 28–32; Lyons, "Keeping Africa off the Agenda," 260; Minter, *King Solomon's Mines Revisited*, 152–153; Gleijeses, "'Flee! The White Giants Are Coming!'" 230–233.

60. Bundy to Johnson, 19 January 1965, in Bundy, *Memos of the Special Assistant*, reel 2, frame 314.

61. Taylor Branch, *Pillar of Fire: America in the King Years, 1963–65* (New York: Simon and Schuster, 1998), 341–509.

62. The precise level of violence against blacks in the South in the 1960s eludes measurement, but tantalizing bits of evidence suggest that the visible part may have been merely the tip of a hidden iceberg. The FBI investigation of the murders of civil rights workers James Chaney, Michael Schwerner, and Andrew Goodman in Neshoba County, Mississippi, in the summer of 1964, for example, included the dredging of swamps, which recovered several black corpses and parts of many more. South American–style "disappearances" seem to have been more than a remote possibility. Other forms of violence abounded as well, of course, such as coerced sexual relations between white law officers and African American women. On both of these matters, see O'Reilly, *"Racial Matters,"* 173. On the analogy to war, see David Halberstam, *The Children* (New York: Random House, 1998), 7–8.

63. Richard H. King, *Civil Rights and the Idea of Freedom* (New York: Oxford University, 1992), 167–168, 174–175; Clayborne Carson, *In Struggle: SNCC and the*

Black Awakening of the 1960s (Cambridge, Mass.: Harvard University Press, 1981), 175.

64. Epps, *The Speeches of Malcolm X at Harvard,* 172; Herbert H. Haines, *Black Radicals and the Civil Rights Mainstream, 1954–1970* (Knoxville: University of Tennessee Press, 1988), 57.

65. James Farmer, OHT (II), 2–6, LBJL.

66. Sherrill, *The Accidental President,* 19 (quotation); Sitkoff, *The Struggle for Black Equality,* 168.

67. Wilkins, *Standing Fast,* 305; O'Reilly, *"Racial Matters,"* 185–190; David J. Garrow, *The FBI and Martin Luther King, Jr.: From "Solo" to Memphis* (New York: Norton, 1981), 118–119.

68. The increasingly different perspectives of the White House and the civil rights movement could also be seen in Johnson's use of prominent Republican and former CIA chief Allen Dulles as an emissary to Mississippi in the summer of 1964 to investigate Klan violence. The president and his advisers were clearly pleased to use Dulles's mission as an effective form of public pressure on the recalcitrant head of the FBI, J. Edgar Hoover, embarrassing the bureau into taking a more active role in heading off further Klan violence. Civil rights activists Aaron Henry, Bob Moses, Dave Dennis, and Lawrence Guyot, meeting briefly with Dulles in Jackson on 25 June, were less impressed with the prominent Republican. Pressed to explain his suggestion to them that "we want this mess cleaned up," Dulles continued: "Well, these civil rights demonstrations are causing this kind of friction, and we're just not gonna have it, even if we have to bring troops in here." Aghast, Henry muttered, "You talkin' to the wrong people." Burke Marshall, OHT, 30–31, LBJL; O'Reilly, *"Racial Matters,"* 167.

69. Memorandum of Conversation, 12 November 1964, NSF, CF, box 78, LBJL; Miller, *Lyndon,* 398.

70. Lawson, *Running for Freedom,* 112; O'Reilly, *"Racial Matters,"* 196. See also Garrow, *Protest at Selma.*

71. Wofford, *Of Kennedys and Kings,* 194.

72. Wilkins, *A Man's Life,* 150–153.

73. O'Reilly, *"Racial Matters,"* 197–198.

74. Wofford, *Of Kennedys and Kings,* 189, 193.

75. Carson, *In Struggle,* 159 (quotation); Wilkins, *Standing Fast,* 306.

76. Wofford, *Of Kennedys and Kings,* 178.

77. Bundy to Johnson, 8 March 1965, NSF, CF, box 76, LBJL.

78. "Selma Violence Likened to That in South Africa," *Washington Post,* 22 March 1965, clipping, NSF, CF, box 76, LBJL.

79. Johnson, "Outline of Informal Speech," undated [10 March 1965], NSF, CF, box 76, LBJL.

80. Carson, *In Struggle,* 160 (quotation); Lawson, *Running for Freedom,* 112–113.

81. Johnson, "The American Promise," 15 March 1965, NSF, CF, box 77, LBJL; Califano, *Triumph and Tragedy of Lyndon Johnson,* 56; Goodwin, *Remembering America,* 332. Even Johnson's most critical biographer has acknowledged the extraordinary power of this speech. Robert A. Caro, *The Years of Lyndon Johnson: Means of Ascent* (New York: Knopf, 1990), xix–xx.

82. Caro, *Means of Ascent,* xx–xxii; Wilkins, *Standing Fast,* 307, 311.

83. Bundy to Read, 10 May 1965, NSF, CF, box 77, LBJL.

84. Lyndon Baines Johnson, *The Vantage Point: Perspectives of the Presidency, 1963–1969* (New York: Holt, Rinehart, and Winston, 1971), 166; "An Interview with Ralph Ellison" [1967], Harry McPherson Office Files, box 55, LBJL.

85. Barrett, *Uncertain Warriors,* 38.

86. Larry Berman, "Coming to Grips with Lyndon Johnson's War," *Diplomatic History* 17 (Fall 1993): 523 (quotation), 525; Drinnon, *Facing West,* 447.

87. Califano, *Triumph and Tragedy of Lyndon Johnson,* 23.

88. Barrett, *Uncertain Warriors,* 32.

89. Bundy to Johnson, 12 August 1965, Bundy, *Memos of the Special Assistant,* reel 3, frame 442.

90. Young, *The Vietnam Wars,* 197–198 (emphasis in original).

91. David J. Garrow, *Bearing the Cross: Martin Luther King, Jr., and the Southern Christian Leadership Conference* (New York: Morrow, 1986), 429–430, 438, 453.

92. Carson, *In Struggle,* 183–184.

93. O'Reilly, *"Racial Matters,"* 191.

94. "Civil Rights: The Deacons," *Newsweek,* 2 August 1965, 28–29.

95. California Governor's Commission on the Los Angeles Riots, *Violence in the City—An End or a Beginning* (Los Angeles, 1965), 1–2.

96. Not simply an undirected "riot," the events in Watts constituted what CBS Radio called "an insurrection against all authority" that targeted almost exclusively white-owned properties. Gerald Horne, *Fire This Time: The Watts Uprising and the 1960s* (Charlottesville: University Press of Virginia, 1995), 36–39.

97. One of the most original comments came from former president Herbert Hoover, who implied, in what might be considered a vehicular theory of social control, that blacks had no reason to riot because "our 19 million Negroes probably own more automobiles than all the 220 million Russians and the 200 million African Negroes put together." O'Reilly, *"Racial Matters,"* 237.

98. Kearns, *Lyndon Johnson and the American Dream,* 304–305 (quotation); Wilkins, *Standing Fast,* 313.

99. Schulman, *Lyndon B. Johnson and American Liberalism,* 112.

100. Califano, *Triumph and Tragedy of Lyndon Johnson,* 59–63.

101. Ibid., 61–62; California Governor's Commission, *Violence in the City,* 22–23.

102. Takaki, *Strangers from a Different Shore*, 419–421; Reimers, *Still the Golden Door*, chap. 3.

103. Robert C. Good, *U.D.I.: The International Politics of the Rhodesian Rebellion* (Princeton: Princeton University Press, 1973), 16, 24, 39–45, 82. The Salisbury regime dropped "Southern" from the country's name, as there was no longer a Northern Rhodesia. Many of those opposed to the UDI continued to use "Southern Rhodesia" to emphasize the illegality of the declaration, but eventually the shorter "Rhodesia" became common usage, a practice followed here with no political intent.

104. Lake, *The "Tar Baby" Option*, 1, 42–44; Good, *U.D.I.*, 25–26, 53–65, 251–252; Minter, *King Solomon's Mines Revisited*, 201–208.

105. Good, *U.D.I.*, 20–23, 25, 73–75 (quotation); Young, "Rebellion and the Congo," 975; Lake, *The "Tar Baby" Option*, 53–56.

106. CIA, "African Response to the Rhodesian Rebellion," 3 January 1966, NSF, CF, box 97, LBJL.

107. Daisuke Kitagawa, "Race Tension Worldwide: Thoughts from Geneva," *Christian Century*, 5 October 1966, 1222.

108. Rusk to American Consul in Salisbury, 29 October 1965, NSF, CF, box 97, LBJL; Bundy to Secretary of Agriculture [undated], Bundy, *Memos of the Special Assistant*, reel 4, frame 127; Lake, *The "Tar Baby" Option*, 60–67, 80–103; Ogene, *Interest Groups and the Shaping of Foreign Policy*, 104–106; Minter, *King Solomon's Mines Revisited*, 209–213. Minter makes a strong case for the limited effectiveness of those sanctions when he points out the ready availability of both external capital and trading partners for the Rhodesian regime. The comparison he offers to the effectiveness of Western sanctions on Cuba is revealing.

109. Valenti to Bundy, 19 November 1965, NSF, CF, box 97, LBJL (emphasis in original); Lake, *The "Tar Baby" Option*, 3. Valenti did concede about white Rhodesians that "the best are admirable farmers and should be welcome in Australia or Nebraska or British Columbia," suggesting the enduring power of the idea of international white identity even among those Americans opposed to the Smith regime.

110. Brinkley, *Dean Acheson*, 318.

111. Bundy to Johnson, 6 October 1965, Bundy, *Memos of the Special Assistant*, reel 3, frame 845.

112. Notes of NSC meeting, 25 January 1967, NSF, NSC Meetings File, box 2, LBJL.

113. Garrow, *Bearing the Cross*, 453; Lake, *The "Tar Baby" Option*, 70–73.

114. Transcript of telephone conversation, Ball and Young, 15 December 1965, Ball Papers, box 6, LBJL.

115. Ogene, *Interest Groups and the Shaping of Foreign Policy,* 122–124.

116. Raymond Arsenault, "White on Chrome: Southern Congressmen and Rhodesia, 1962–1971," *Issue: A Quarterly Journal of Africanist Opinion* 2, 4 (Winter 1972): 47–51 (quotation), 55; Lake, *The "Tar Baby" Option,* 103–120.

117. Vernon McKay, "The Domino Theory of the Rhodesia Lobby," *Africa Report* 12, 6 (June 1967): 56–57.

118. Ogene, *Interest Groups and the Shaping of Foreign Policy,* 117–122.

119. McKay, "Domino Theory," 55.

120. Haynes to Komer, 29 March 1966, NSF, CF, box 78, LBJL; Lelyveld, *Move Your Shadow,* 229; Chafe, *Never Stop Running,* 270; Peter J. Schraeder, *United States Foreign Policy toward Africa: Incrementalism, Crisis, and Change* (New York: Cambridge University Press, 1994), 205–206.

121. Houser, "Freedom's Struggle Crosses Oceans and Mountains," 191; Miller, *Lyndon,* 436.

122. "Status Report on NSAM No. 295 of April 24, 1964—South Africa," 30 July 1964; this document may be found in the helpful collection *South Africa and the United States: The Declassified History,* ed. Kenneth Mokoena (New York: New Press, 1993), 74. Minter, *King Solomon's Mines Revisited,* 213.

123. CIA, "Sanctions and the South African Economy," 3 September 1965, NSF, CF, box 78, LBJL.

124. "Briefing for NSC Standing Group," 10 March 1964, NSF, CF, box 76, LBJL.

125. "Briefing for NSC Standing Group," 10 March 1964, NSF, CF, box 76, LBJL.

126. CIA, "South Africa on the Eve of Elections," 25 March 1966, NSF, CF, box 78, LBJL; CIA, "South Africa's New Foreign Policy Offensive," 19 May 1967, NSF, CF, box 78, LBJL.

127. CIA, "South Africa Builds an Airfield on the Caprivi Strip in South West Africa," 14 June 1965, NSF, CF, box 78, LBJL.

128. Benjamin H. Read to Bundy, 20 May 1964, NSF, CF, box 78, LBJL (emphasis added). See also "David" to Bundy, 2 April 1964, NSF, CF, box 78, LBJL.

129. "Briefing for NSC Standing Group," 10 March 1964, NSF, CF, box 76, LBJL; Coker, *The United States and South Africa,* 9; Houser, "Freedom's Struggle Crosses Oceans and Mountains," 179–180. For more on this parallel, see Frederickson, *Black Liberation.*

130. Memorandum of conversation, 17 June 1964, NSF, CF, box 78, LBJL.

131. Brubeck to Bundy, 5 December 1963, NSF, CF, box 78, LBJL.

132. Haynes to Komer, 31 August 1965, NSF, Ric Haynes Office Files, box 1, LBJL.

133. Satterthwaite to Rusk, 12 June 1964, NSF, CF, box 76, LBJL; memorandum of conversation, 30 July 1964, NSF, CF, box 78, LBJL; Haynes to Bundy, 12 March 1965, NSF, CF, box 78, LBJL.

134. "Briefing for NSC Standing Group," 10 March 1964, NSF, CF, box 76, LBJL

(quotation); Department of State to all AF Diplomatic and Consular Posts, 1 April 1964, NSF, CF, box 76, LBJL; Satterthwaite to Rusk, 14 May 1965, NSF, CF, box 78, LBJL. This debate can be followed, for example, in Satterthwaite to Rusk, 16 March 1964, NSF, CF, box 78, LBJL; Satterthwaite to Rusk, 15 April 1964, NSF, CF, box 76, LBJL; Bundy to Johnson, 21 June 1965, Bundy, *Memos of the Special Assistant,* reel 3, frame 102.

135. Johnson, "Remarks of the President . . . ," 26 May 1966, NSF, CF, box 76, LBJL; Moyers to Johnson, 26 May 1966, NSF, CF, box 76, LBJL; Lyons, "Keeping Africa off the Agenda," 261–266.

136. Notes of NSC meeting, 14 July 1966, NSF, NSC Meetings File, box 2, LBJL.

137. Student Nonviolent Coordinating Committee, "Statement on Vietnam," in *Vietnam and Black America: An Anthology of Protest and Resistance,* ed. Clyde Taylor (Garden City, N.Y.: Anchor, 1973), 258–259; Carson, *In Struggle,* 175–190; Paul Good, "Odyssey of a Man—and a Movement" [June 1967], in *Black Protest in the Sixties,* ed. August Meier, John Bracey, Jr., and Elliott Rudwick (New York: Markus Wiener, 1991), 275; James Forman, *High Tide of Black Resistance* (Seattle: Open Hand, 1994), 102–103; Gerald Gill, "From Maternal Pacifism to Revolutionary Solidarity: African-American Women's Opposition to the Vietnam War," in *Sights on the Sixties,* ed. Barbara L. Tischler (New Brunswick: Rutgers University Press, 1992), 182–183. SNCC was following in the footsteps of Malcolm X, one of the earliest critics of the American war in Vietnam, who had argued, "If it is right for America to draft us, and teach us how to be violent in defense of her, then it is right for you and me to do whatever is necessary to defend our own people right here in this country." Peter B. Levy, "Blacks and the Vietnam War," in *The Legacy: The Vietnam War in the American Imagination,* ed. D. Michael Shafer (Boston: Beacon, 1990), 217.

138. Carson, *In Struggle,* 180–182.

139. Good, "Odyssey of a Man," 272 (emphasis in original).

140. Segal, *The Race War,* 298; Garrow, *Bearing the Cross,* 458–459; Carson, *In Struggle,* 188–189.

141. O'Reilly, *"Racial Matters,"* 243 (quotation); Levy, "Blacks and the Vietnam War," 216; Adam Fairclough, "Martin Luther King, Jr., and the War in Vietnam," *Phylon* 45, 1 (March 1984): 21–26; James H. Cone, "Martin Luther King, Jr., and the Third World," *Journal of American History* 74 (September 1987): 463–464; Garrow, *Bearing the Cross,* 394, 444–445, 455, 469–470; Robert S. Browne, "The Civil Rights Movement and Vietnam," in West and Mooney, *To Redeem A Nation,* 153; Louis Martin, OHT, p. 31, LBJL; Lawson, "Mixing Moderation with Militancy," 88–89. Johnson deeply appreciated the continuing refusal of Whitney Young and Roy Wilkins to criticize his Viet-

nam policies. Young even traveled to South Vietnam twice in support of Johnson's efforts there. Nancy J. Weiss, *Whitney M. Young, Jr., and the Struggle for Civil Rights* (Princeton: Princeton University Press, 1989), 160–164; Steven F. Lawson, "Civil Rights," in *Exploring the Johnson Years,* ed. Robert A. Divine (Austin: University of Texas Press, 1981), 108.

142. Lester, *Look Out, Whitey!* xi; Epps, *The Speeches of Malcolm X,* 164.

143. Garrow, *Bearing the Cross,* 497.

144. King et al. to Carmichael, 3 August 1966, Watson Office Files, box 18, LBJL; SNCC Central Committee to Wilkins, 4 August 1966, Watson Office Files, box 18, LBJL; Lawson, "Mixing Moderation with Militancy," 91–92.

145. Lester, *Look Out, Whitey!* 117 (emphasis in original).

146. Stokely Carmichael, "Black Power and the Third World," in *Towards Revolution,* ed. John Gerassi, vol. 2, *The Americas* (London: Weidenfeld and Nicolson, 1971), 704–705.

147. O'Reilly, *"Racial Matters,"* 240–241; Carmichael, "Black Power and the Third World," 706; Eldridge Cleaver, *Soul on Ice* (New York: McGraw-Hill, 1968), 125; Lawrence Neal, "Black Power in the International Context," in *The Black Power Revolt: A Collection of Essays,* ed. Floyd B. Barbour (Boston: Porter Sargent, 1968), 136–142.

148. Carson, *In Struggle,* 201, 272, 276–278; King, *Civil Rights and the Idea of Freedom,* 183–184.

149. Malcolm X, *The Autobiography of Malcolm X* (New York: Grove, 1965), chapts. 17–19; Harold Cruse, *The Crisis of the Negro Intellectual* (New York: Morrow, 1967), 408–409; Houser, "Freedom's Struggle Crosses Oceans and Mountains," 180–182.

150. Carson, *In Struggle,* 265–286; Stokely Carmichael and Charles V. Hamilton, *Black Power: The Politics of Liberation in America* (New York: Random House, 1967), xi; Lester, *Look Out, Whitey!* 138; James Forman, "A Year of Resistance," in *Towards Revolution,* ed. John Gerassi, vol. 2, *The Americas* (London: Weidenfeld and Nicolson, 1971), 700–703.

151. Carmichael and Hamilton, *Black Power,* 6.

152. Carson, *In Struggle,* 198.

153. Forman, "A Year of Resistance," 699–700.

154. King, *Civil Rights and the Idea of Freedom,* 153–156, 172–200 (quotation at 182).

155. Cone, "Martin Luther King, Jr., and the Third World," 462.

156. Levy, "Blacks and the Vietnam War," 217.

157. Lawson, *Running for Freedom,* 129.

158. Goldman, *Tragedy of Lyndon B. Johnson,* 173–174, 315 (quotation). On Wallace, see Carter, *Politics of Rage.*

159. Garrow, *Bearing the Cross,* 532.

160. Fite, *Richard B. Russell,* 409; Engelhardt, *The End of Victory Culture,* 151. Southern senators also opposed Johnson's initial selection of Abe Fortas to replace Earl Warren, as Fortas was the first Jew nominated for chief justice of the Supreme Court. Miller, *Lyndon,* 483–485.

161. O'Reilly, *"Racial Matters,"* 240–241; Lawson, *Running for Freedom,* 129. White anger at black restiveness also showed up in the contrasting celebration of the hard-working, law-abiding "model minority" of Asian Americans, such as reported in the 26 December 1966 issue of *U.S. News and World Report.* See Gary Y. Okihiro, *Margins and Mainstreams: Asians in American History and Culture* (Seattle: University of Washington Press, 1994), 140.

162. Thomas E. Cronin to Johnson, 31 May 1967, WHCF, Welfare, box 29, LBJL.

163. Lawson, *Running for Freedom,* 133.

164. Miller, *Lyndon,* 467.

165. Garrow, *The FBI and Martin Luther King,* 106 (quotation), 151–166; O'Reilly, *"Racial Matters,"* 230.

166. Garrow, *The FBI and Martin Luther King,* 119–120 (quotation); O'Reilly, *"Racial Matters,"* 154–155, 179, 229, 235.

167. O'Reilly, *"Racial Matters,"* 261, 285; Levy, "Blacks and the Vietnam War," 224–225; Lawson, "Mixing Moderation with Militancy," 92–93.

168. Garrow, *The FBI and Martin Luther King,* 167–169; Califano, *Triumph and Tragedy of Lyndon Johnson,* 277; O'Reilly, *"Racial Matters,"* 229–230.

169. Levy, "Blacks and the Vietnam War," 217–218 (quotation); Thomas R. Peake, *Keeping the Dream Alive: A History of the Southern Christian Leadership Conference from King to the Nineteen-Eighties* (New York: Peter Lang, 1987), 212–217; Fairclough, "Martin Luther King, Jr., and the War in Vietnam," 26–29; Garrow, *Bearing the Cross,* 542–550.

170. Garrow, *Bearing the Cross,* 551–552.

171. Martin Luther King, Jr., "Beyond Vietnam," in *Vietnam and Black America: An Anthology of Protest and Resistance,* ed. Clyde Taylor (Garden City, N.Y.: Anchor, 1973), 81–84, 94; Garrow, *Bearing the Cross,* 553.

172. Roche to Johnson, 5 April 1967, WHCF, CF, Name File, box 147, LBJL; Califano, *Triumph and Tragedy of Lyndon Johnson,* 277; Lawson, "Mixing Moderation with Militancy," 90.

173. Cone, "Martin Luther King, Jr., and the Third World," 464 (quotation); O'Reilly, *"Racial Matters,"* 244–245.

174. Garrow, *Bearing the Cross,* 555. See also O'Reilly, *"Racial Matters,"* 243.

175. Garrow, *The FBI and Martin Luther King,* 180–181.

176. Garrow, *Bearing the Cross,* 555.

177. Levy, "Blacks and the Vietnam War," 217–219.

178. Allen J. Matusow, *The Unraveling of America: A History of Liberalism in the 1960s* (New York: Harper and Row, 1984), 363.

179. Wilkins, *A Man's Life,* 197.

180. Califano, *Triumph and Tragedy of Lyndon Johnson,* 217 (quotation); Cyrus Vance, "Final Report concerning the Detroit Riots," 23 July–2 August 1967, WHCF, CF, Oversize Attachments, box 183, LBJL.

181. J. Walter Yeagley to Clark, 4 August 1967, Ramsey Clark Papers, box 75, LBJL.

182. Carson, *In Struggle,* 256 (Brown); Fred Vinson to Clark, 7 August 1967, Clark Papers, box 75, LBJL (Carmichael).

183. Terry H. Anderson, *The Movement and The Sixties: Protest in America from Greensboro to Wounded Knee* (New York: Oxford University Press, 1995), 176–177. The Black Panthers were protesting a proposed law to ban loaded weapons within city limits.

184. "Carmichael Asks Revolution in U.S.," *New York Times,* 18 August 1967, 17. Johnson and his staff shared Goldwater's frustration and desire to lock Carmichael up (though not to kill him), but were reluctantly convinced by Clark that the administration lacked sufficient evidence to convict him of violating the Logan Act and its restrictions on contacts with a foreign government. Larry Temple to Johnson, 19 January 1968, WHCF, CF, Name File, box 144, LBJL; Califano, *Triumph and Tragedy of Lyndon Johnson,* 221–222.

185. Kearns, *Lyndon Johnson and the American Dream,* 308.

186. Walter LaFeber, "Johnson, Vietnam, and Tocqueville," in *Lyndon Johnson Confronts the World: American Foreign Policy, 1963–1968,* ed. Warren I. Cohen and Nancy Bernkopf Tucker (New York: Cambridge University Press, 1994), 32–33; Califano, *Triumph and Tragedy of Lyndon Johnson,* 212–216.

187. J. W. Fulbright, "The Great Society Is a Sick Society," *New York Times Magazine,* 20 August 1967, 30; Garrow, *Bearing the Cross,* 575. The NAACP and the National Urban League stuck by the president, despite comments like that of one black Detroit resident during the riot, who urged African Americans to stop being "house niggers and slaves like Whitney Young and Roy Wilkins—and to stand up and fight like Stokely Carmichael and Cassius Clay." Lawson, "Mixing Moderation with Militancy," 97. See also Roy Wilkins, "LBJ and the Negro," *New York Post,* 2 December 1967, clipping in Watson Office Files, box 18, LBJL.

188. Califano, *Triumph and Tragedy of Lyndon Johnson,* 199; Lawson, "Mixing Moderation with Militancy," 100; O'Reilly, *"Racial Matters,"* 245.

189. Good, *U.D.I.*, 233–239.

190. CIA, "Some Aspects of Subversion in Africa," 19 October 1967, NSF, CF, box 78, LBJL.

191. "Armed Forces," *Time,* 26 May 1967, 15–16; Levy, "Blacks and the Vietnam War," 214.

192. Wallace Terry, *Bloods: An Oral History of the Vietnam War by Black Veterans* (New York: Ballantine, 1984), 23, 239.

193. Charley Trujillo, *Soldados: Chicanos in Viet Nam* (San Jose, Calif.: Chusma House, 1990), 13.

194. Lea Ybarra, "Perceptions of Race and Class among Chicano Vietnam Veterans," *Vietnam Generation* 1, 2 (Spring 1989): 77; Stanley Goff, *Brothers: Black Soldiers in the Nam* (Novato, Calif.: Presidio Press, 1982), 160.

195. Thomas A. Johnson, "The U.S. Negro in Vietnam," *New York Times,* 29 April 1968, 1, 16.

196. "Fact Sheet: Negro Participation in the Armed Forces and in Vietnam," 18 January 1967, NSF, CF, box 191, LBJL; Harry McPherson to Nathaniel Patrick, 8 July 1966, McPherson Office Files, box 28, LBJL; D. Michael Shafer, "The Vietnam-Era Draft: Who Went, Who Didn't, and Why It Matters," in *The Legacy: The Vietnam War in the American Imagination,* ed. D. Michael Shafer (Boston: Beacon, 1990), 69–73; Levy, "Blacks and the Vietnam War," 211–214.

197. Terry, *Bloods,* 259–260.

198. James E. Westheider, *Fighting on Two Fronts: African Americans and the Vietnam War* (New York: New York University Press, 1997), 68–70; final report of the President's Committee on Equal Opportunity in the Armed Forces, November 1964, Lee C. White Office Files, box 3, LBJL.

199. Nalty, *Strength for the Fight,* 300–302, 305; Ronald H. Spector, *After Tet: The Bloodiest Year in Vietnam* (New York: Free Press, 1993), 246–247; Terry Whitmore, *Memphis Nam Sweden: The Autobiography of a Black American Exile* (Garden City, N.Y.: Doubleday, 1971), 101–103.

200. Spector, *After Tet,* 242–259 (quotation at 244–245); Westheider, *Fighting on Two Fronts,* 5, 66–93; Levy, "Blacks and the Vietnam War," 312; Nalty, *Strength for the Fight,* 308–309; Terry, *Bloods,* xiv, 12; Engelhardt, *The End of Victory Culture,* 248–249.

201. David Halberstam, *The Best and the Brightest* (1972; New York: Penguin, 1983), 560.

202. Taylor, "Lyndon Johnson and the Vietnamese," 120 (first quotation); Miller, *Lyndon,* 416 (second quotation).

203. Robert Jay Lifton, *Home from the War: Learning from Vietnam Veterans* (1973; Boston: Beacon, 1992), 41.

204. Neil Sheehan, *A Bright Shining Lie: John Paul Vann and America in Vietnam* (New York: Random House, 1988), 685.

205. Drinnon, *Facing West,* 457 (Kruch); Philip Caputo, *A Rumor of War* (New York: Ballantine, 1977), xix; Lifton, *Home from the War,* 59–65.

206. Terry, *Bloods,* 5, 90; Lifton, *Home from the War,* 42–45.

207. DeConde, *Ethnicity, Race, and American Foreign Policy,* 151 (quotation); Michael Bilton and Kevin Sim, *Four Hours in My Lai* (New York: Penguin, 1992), 21; Lifton, *Home from the War,* 45, 194–201; George C. Herring, "'Peoples Quite Apart': Americans, South Vietnamese, and the War in Vietnam," *Diplomatic History* 14 (Winter 1990): 11–13, 18. The National Liberation Front strategy of digging tunnel systems as a response to American bombing, bulldozers, and defoliants offered a particularly striking example of how Americans could associate their enemies with animals, literally burrowing into the ground. See Tom Mangold and John Penycate, *The Tunnels of Cu Chi* (New York: Random House, 1985).

208. Memorandum of meeting with the president, 17 February 1965, Meeting Notes File, box 1, LBJL.

209. Steel, *Walter Lippmann and the American Century,* 570.

210. Transcript of telephone conversation, Lippmann and Ball, 25 August 1964, Ball Papers, box 2, LBJL; Engelhardt, *The End of Victory Culture,* 199–200, 238–239.

211. Thomas A. Johnson, "Negro Expatriates Finding Wide Opportunity in Asia," *New York Times,* 30 April 1968, 18; Trujillo, *Soldados,* passim, esp. 24; Lifton, *Home from the War,* 204–205. For more on Latino soldiers, see *Aztlán and Viet Nam: Chicano and Chicana Experiences of the War,* ed. George Mariscal (Berkeley: University of California Press, 1999).

212. Goff, *Brothers,* 131–133. See also Young, *Vietnam Wars,* illustration following 242; Terry, *Bloods,* 137.

213. Tom Holm, *Strong Hearts, Wounded Souls: Native American Veterans of the Vietnam War* (Austin: University of Texas Press, 1996).

214. Levy, "Blacks and the Vietnam War," 214, 225–226; Lincoln, "The Race Problem and International Relations," 13.

215. Ellen Frey-Wouters and Robert S. Laufer, *Legacy of a War: The American Soldier in Vietnam* (Armonk, N.Y.: M. E. Sharpe, 1986), 135–136, 333–340.

216. Thomas A. Johnson, "Negro Veteran Is Confused and Bitter," *New York Times,* 29 July 1968, 14.

217. Holm, *Strong Hearts, Wounded Souls,* 148–149.

218. Levy, "Blacks and the Vietnam War," 220–224.

219. Neal, "Black Power in the International Context," 144; Gill, "From Maternal Pacifism to Revolutionary Solidarity," 191.

220. Lifton, *Home from the War,* 47 (quotation); Drinnon, *Facing West,* 368. John Wayne's 1968 film *The Green Berets* made explicit use of Indian war imagery. In some ways the movie was a reprise of his 1949 film *The Sands of Iwo Jima,* with Vietnamese enemies replacing Japanese ones but with the Special Forces outpost still called Dodge City, Americans still as the cowboys, and Vietnamese dividing up into good (South Vietnamese) and bad (National Liberation Front and North Vietnamese) Indians. Sounding much like Tonto, the South Vietnamese got to say things like "We kill all stinking Cong," while Wayne, as the Lone Ranger, responded, "Affirmative. I like the way you talk." Bruce Taylor, "The Vietnam War Movie," in *The Legacy: The Vietnam War in the American Imagination,* ed. D. Michael Shafer (Boston: Beacon, 1990), 188.

221. Bilton and Sim, *Four Hours,* 51–52.

222. *Report of the National Advisory Commission on Civil Disorders* (New York: Bantam, 1968), 1, 10; McPherson to Califano, 1 March 1968, McPherson Office Files, box 32, LBJL; Roy Wilkins, OHT, 20, LBJL; Wilkins, *Standing Fast,* 327–329; Goodwin, *Remembering America,* 514; Lawson, "Mixing Moderation with Militancy," 100–101.

223. Califano, *Triumph and Tragedy of Lyndon Johnson,* 287; Lawson, "Mixing Moderation with Militancy," 102–104.

224. "Race Riot at Long Binh," *Newsweek,* 30 September 1968, 35; Spector, *After Tet,* 242–243, 251–252, 254–256; Nalty, *Strength for the Fight,* 305–306.

225. Carl T. Rowan, *Dream Makers, Dream Breakers: The World of Justice Thurgood Marshall* (Boston: Little, Brown, 1993), 355.

226. Burke Marshall, OHT, 33–34, LBJL.

227. Rowan, *Dream Makers, Dream Breakers,* 356.

228. Wilkins, *A Man's Life,* 156; Lawson, "Mixing Moderation with Militancy," 106.

229. Donald Spivey, "Black Consciousness and Olympic Protest Movement, 1964–1980," in *Sport in America: New Historical Perspectives,* ed. Spivey (Westport, Conn.: Greenwood, 1985), 239–262.

6. The End of the Cold War and White Supremacy

1. The evidentiary trail grows fainter after 1968 as well, as many materials for more recent decades, especially on national security issues, remain classified.

2. O'Reilly, *Nixon's Piano,* 5–6, 329; Bruce H. Kalk, "Wormley's Hotel Revisited: Richard Nixon's Southern Strategy and the End of the Second Reconstruction," *North Carolina Historical Review* 71 (January 1994): 89.

3. Herbert S. Parmet, *Richard Nixon and His America* (Boston: Little, Brown, 1990), 60, 267–269; Joan Hoff, *Nixon Reconsidered* (New York: Basic, 1994),

77; Tom Wicker, *One of Us: Richard Nixon and the American Dream* (New York: Random House, 1991), 184, 238, 240.

4. Carter, *Politics of Rage*, 375.

5. William H. Chafe, *Civilities and Civil Rights: Greensboro, North Carolina, and the Black Struggle for Freedom* (New York: Oxford University Press, 1980), 192.

6. Ambrose, *Nixon*, 2:244.

7. O'Reilly, *Nixon's Piano*, 277–329, esp. 279; Goodwin, *Remembering America*, 106.

8. Thomas Byrne Edsall and Mary D. Edsall, *Chain Reaction: The Impact of Race, Rights, and Taxes on American Politics* (New York: Norton, 1992), 85.

9. Carter, *From George Wallace to Newt Gingrich*, 28–29 (quotation); Wicker, *One of Us*, 325–327.

10. Joseph A. Aistrup, *The Southern Strategy Revisited: Republican Top-Down Advancement in the South* (Lexington: University of Kentucky Press, 1996), 29 (quotation); Kalk, "Wormley's Hotel Revisited," 85–105; Mary C. Brennan, *Turning Right in the Sixties: The Conservative Capture of the GOP* (Chapel Hill: University of North Carolina Press, 1995), 125–126; Reg Murphy and Hal Gulliver, *The Southern Strategy* (New York: Charles Scribner's Sons, 1971). The perspective of those implementing the Southern strategy can be found in Harry S. Dent, *The Prodigal South Returns to Power* (New York: John Wiley, 1978).

11. Carter, *Politics of Rage*, 334–335, 358–360 (quotation), 369; Brennan, *Turning Right in the Sixties*, 132.

12. Sitkoff, *The Struggle for Black Equality*, 212; Carter, *Politics of Rage*, 368.

13. Jonathan Schell, *The Time of Illusion* (New York: Vintage, 1975), 58 (quotation); Carter, *Politics of Rage*, 328–331; Ambrose, *Nixon*, 2:163.

14. Edsall and Edsall, *Chain Reaction*, 10, 20.

15. Kalk, "Wormley's Hotel Revisited," 105.

16. Seymour Hersh, *The Price of Power: Kissinger in the Nixon White House* (New York: Summit, 1983), 110–111.

17. Seymour M. Hersh, "Nixon's Last Cover-Up: The Tapes He Wants the Archives to Suppress," *New Yorker*, 14 December 1992, 94.

18. Hugh Davis Graham, "Richard Nixon and Civil Rights: Explaining an Enigma," *Presidential Studies Quarterly* 26 (Winter 1996): 98.

19. O'Reilly, *Nixon's Piano*, 327.

20. Daniel P. Moynihan, "Report to the President," 19 March 1969, box 1, EX Human Rights 2, Equality, White House Central Files, NPM, NA II. Moynihan noted his own dissent from Jensen's views, even as he reported them: "I personally do not believe this is so, but the truth is that it is an open question."

21. Clifford Alexander, OHT, 1 November 1971, p. 34, LBJL.

22. Walter Isaacson, *Kissinger: A Biography* (New York: Simon and Schuster, 1992),

560–561. Nixon apparently enjoyed baiting Kissinger with diatribes against blacks and Jews, which Kissinger refused to respond to. When his aide Winston Lord asked him why he did not say something, Kissinger replied, "I have enough trouble fighting with him on the things that really matter [foreign relations]; his attitudes towards Jews and blacks are not my worry." Ibid., 148.

23. Ambrose, *Nixon,* 2:172.

24. Ibid., 124.

25. Graham, "Richard Nixon and Civil Rights," 98.

26. Hersh, *The Price of Power,* 110–111.

27. O'Reilly, *"Racial Matters,"* 294–297 (quotations), 326–327, 334.

28. Tinker, *Race, Conflict, and the International Order,* 122–123 (quotation); Lawson, *Running for Freedom,* 142; O'Reilly, *"Racial Matters,"* 295.

29. Moynihan, "Report to the President," 23.

30. Flora Lewis, "The Rumble at Camp Lejeune," *Atlantic Monthly,* January 1970, 35–41; Westheider, *Fighting on Two Fronts,* 94–130; Nalty, *Strength for the Fight,* 309–313.

31. Richard Hammer, *The Court-Martial of Lt. Calley* (New York: Coward, McCann, and Geoghegan, 1971), 46–47, 349.

32. Drinnon, *Facing West,* 452, 457 (quotation).

33. Bilton and Sim, *Four Hours,* 14.

34. Ibid., 20.

35. Ibid., 340.

36. Ibid., 315–316.

37. Ibid., 455.

38. Drinnon, *Facing West,* 444–445.

39. Hugh Davis Graham, *The Civil Rights Era: Origins and Development of National Policy, 1960–1972* (New York: Oxford University Press, 1990), 303.

40. Moynihan, "Report to the President," 1.

41. Robinson to Nixon, 22 January 1969, WHCF, Human Rights, box 1, NPM, NA II.

42. Wicker, *One of Us,* 502.

43. Ibid., 488 (quotation; emphasis in original); Sitkoff, *The Struggle for Black Equality,* 212–213; Edsall and Edsall, *Chain Reaction,* 83–84; Leon E. Panetta and Peter Gall, *Bring Us Together: The Nixon Team and the Civil Rights Retreat* (Philadelphia: Lippincott, 1971).

44. Ambrose, *Nixon,* 2:247–248 (quotation; emphasis in original), 363.

45. Roy Reed, "G.O.P., Aided by Agnew, Surges in the South," *New York Times,* 7 December 1969, clipping in Staff Member Office Files, Harry Dent, box 8, NPM, NA II.

46. Wicker, *One of Us,* 488–489 (quotation; emphasis in original); Edsall and

Edsall, *Chain Reaction,* 81; Schell, *The Time of Illusion,* 42; Ambrose, *Nixon,* 2:406–407.

47. Graham, "Richard Nixon and Civil Rights," 99.

48. Kalk, "Wormley's Hotel Revisited," 102.

49. Schell, *The Time of Illusion,* 81–82. Nixon's chief lobbyist on Capitol Hill, for example, admitted that many senators "think Carswell's a boob, a dummy. And what counter is there to that? He is." The president himself wrote "*My god!*" on a news summary of some of Carswell's segregationist activities. Kalk, "Wormley's Hotel Revisited," 103–104.

50. Wicker, *One of Us,* 492–493; Ambrose, *Nixon,* 2:187, 520–523; Edsall and Edsall, *Chain Reaction,* 76, 87ff.

51. Melvin Small, *The Presidency of Richard Nixon* (Lawrence: University Press of Kansas, 1999), 163.

52. Schell, *The Time of Illusion,* 219.

53. Anderson, *The Movement and the Sixties,* 397.

54. Seymour M. Hersh, "The Price of Power: Kissinger, Nixon, and Chile," *Atlantic Monthly,* December 1982, 35.

55. O'Reilly, *Nixon's Piano,* 479.

56. Kissinger to Haldeman, 6 June 1969, WHCF, CO, box 22, NPM, NA II. See also Butterfield to Nixon, 20 July 1971, WHCF, Country Files, Subject Categories, box 4, NPM, NA II.

57. John Marcum, "Southern Africa and United States Policy: A Consideration of Alternatives," in *Racial Influences on American Foreign Policy,* ed. George W. Shepherd, Jr. (New York: Basic Books, 1970), 202; Minter, *King Solomon's Mines Revisited,* 254 (quotation).

58. Minter, *King Solomon's Mines Revisited,* 276–277; Jackson, *From the Congo to Soweto,* 69–70.

59. Hersh, *The Price of Power,* 111.

60. DeConde, *Ethnicity, Race, and American Foreign Policy,* 178–179; Houser, "Freedom's Struggle Crosses Oceans and Mountains," 193.

61. Morris, *Uncertain Greatness,* 117 (quotations); Minter, *King Solomon's Mines Revisited,* 220–221.

62. *The Kissinger Study of Southern Africa: National Security Study Memorandum 39,* ed. Mohamed El-Khawas and Barry Cohen (Westport, Conn.: Lawrence Hill, 1976), 105.

63. Morris, *Uncertain Greatness,* 112.

64. Acheson to Kissinger, 30 April 1969, WHCF, CO, box 65, NPM, NA II ("the most stable"); Brinkley, *Dean Acheson,* 321 ("subjugation"); "Introduction," in *Kissinger Study of Southern Africa,* 31; Morris, *Uncertain Greatness,* 114–116.

65. The *Kissinger Study of Southern Africa,* 87. On this issue, see also Richard W.

Hull, *American Enterprise in South Africa: Historical Dimensions of Engagement and Disengagement* (New York: New York University Press, 1990), and Minter, *King Solomon's Mines Revisited.*

66. Jacobs to Nixon, 16 March 1970, WHCF, CO, box 63, NPM, NA II. See also Lake, *The "Tar Baby" Option*, chaps. 4–5; Raymond Arsenault, "White on Chrome: Southern Congressmen and Rhodesia, 1962–1971," *Issue: A Quarterly Journal of Africanist Opinion* 2 (Winter 1972): 51; Ogene, *Interest Groups and the Shaping of Foreign Policy*; Sensing to Dent, 24 October 1969, Tower to Nixon, 12 January 1970, and Wolfenberger to Nixon, 14 March 1970, all in WHCF, CO, box 63, NPM, NA II.

67. Abshire to Morgan, 4 June 1971, WHCF, CO, box 62, NPM, NA II.

68. Stokes et al. to Nixon, 13 December 1972, WHCF, CO, box 62, NPM, NA II.

69. McDonald et al. to Nixon, 24 July 1969, WHCF, CO, box 63, NPM, NA II.

70. Lake, *The "Tar Baby" Option*, chap. 6; Ogene, *Interest Groups and the Shaping of Foreign Policy*, 104–106; Good, *U.D.I.*, 324.

71. Schraeder, *United States Foreign Policy toward Africa*, 41; Lake, *The "Tar Baby" Option*, 216–226; Arsenault, "White on Chrome," 46–55.

72. Schaufele to Kissinger, 1 April 1976, document 39 in Mokoena, *South Africa and the United States*, 228–235; Larry Bowman, "US Policy towards Rhodesia," in *American Policy in Southern Africa: The Stakes and the Stance*, ed. René Lemarchand (Washington: University Press of America, 1978), 184–189; Thomas J. Noer, "International Credibility and Political Survival: The Ford Administration's Intervention in Angola," *Presidential Studies Quarterly* 23 (Fall 1993): 780–783.

73. Isaacson, *Kissinger*, 687.

74. Ibid., 688, 989–992.

75. Good, *U.D.I.*, 236–237.

76. Lauren, *Power and Prejudice*, 245–246.

77. Mahoney, *JFK*, 237–238 (quotation), 243; Horan to Roosevelt, 5 September 1973, WHCF, CO, box 4, NPM, NA II.

78. State Department, "U.S. Policy toward Angola," 16 December 1975, document 37 in Mokoena, *South Africa and the United States*, 219–225.

79. Noer, "International Credibility and Political Survival," 780 (quotation); *The Kissinger Transcripts: The Top-Secret Talks with Beijing and Moscow*, ed. William Burr (New York: New Press, 1999), xvii.

80. House Select Committee on Intelligence, selection from the Pike Report Relating to Angola, February 1976, document 38 in Mokoena, *South Africa and the United States*, 226–227; Noer, "International Credibility and Political Survival," 777–778; John Robert Greene, *The Presidency of Gerald R. Ford* (Lawrence: University Press of Kansas, 1995), 115.

81. Peter W. Rodman, *More Precious Than Peace: The Cold War and the Struggle for the Third World* (New York: Scribner's, 1994), 163–182. For a more critical view of U.S. involvement, see John Stockwell, *In Search of Enemies: A CIA Story* (New York: Norton, 1978).

82. State Department, "U.S. Policy toward Angola," 224 (quotation); William A. DePalo, Jr., "Cuban Internationalism: The Angola Experience, 1975–1988," *Parameters: US Army War College Quarterly* 23 (Autumn 1993): 61–74; Moore, *Castro, the Blacks, and Africa*, 328–329; Jackson, *From the Congo to Soweto*, 70; "Introduction," in Mokoena, *South Africa and the United States*, xxiii; Noer, "International Credibility and Political Survival," 776.

83. Rodman, *More Precious Than Peace*, 360 (quotation); James Barber and John Barratt, *South Africa's Foreign Policy: The Search for Status and Security, 1945–1988* (Cambridge: Cambridge University Press, 1990), 154.

84. Coker, *The United States and South Africa*, 14; Steven Metz, "Congress, the Antiapartheid Movement, and Nixon," *Diplomatic History* 12 (Spring 1988): 167; O'Reilly, *"Racial Matters,"* 337–338.

85. Lauren, *Power and Prejudice*, 238–259 (quotation at 238).

86. Barber and Barratt, *South Africa's Foreign Policy*, 204–205.

87. *The Kissinger Study of Southern Africa*, 87.

88. Morris, *Uncertain Greatness*, 131–132.

89. Wilkins, *A Man's Life*, 279–282; O'Reilly, *Nixon's Piano*, 7.

90. Morris, *Uncertain Greatness*, 119–120. Joan Hoff makes a similar point; see Hoff, *Nixon Reconsidered*, 247.

91. Carter, *From George Wallace to Newt Gingrich*, 35. Kissinger, a former Harvard political science professor, once wrote on a cable from the U.S. ambassador in Chile reporting on atrocities, "Cut out the political science lectures." "Kissinger and Pinochet," *Nation*, 29 March 1999, 5.

92. Chang, *Friends and Enemies*, 285 (quotations); C. L. Sulzberger, *The World and Richard Nixon* (New York: Prentice Hall, 1987), 199.

93. Krogh to Hoover, 27 June 1969, and Krogh to Nixon, 27 June 1969, WHSF, Staff Member Office Files, John Ehrlichman, box 15, NPM, NA II.

94. *The Kissinger Study of Southern Africa*, 134 (quotation); DeConde, *Ethnicity, Race, and American Foreign Policy*, 177–178.

95. Nixon to Nyerere, 4 May 1970, WHCF, CO, box 62, NPM, NA II.

96. Peter G. Bourne, *Jimmy Carter: A Comprehensive Biography from Plains to Postpresidency* (New York: Simon and Schuster, 1997), 385.

97. John Dumbrell, *The Carter Presidency: A Re-evaluation* (Manchester: Manchester University Press, 1993), 199; Gaddis Smith, *Morality, Reason, and Power: American Diplomacy in the Carter Years* (New York: Hill and Wang, 1986), 118–119; Michael McClintock, *Instruments of Statecraft: U.S. Guerrilla Warfare,*

Counterinsurgency, and Counterterrorism, 1940–1990 (New York: Pantheon, 1992), 313–314.

98. Bourne, *Jimmy Carter*, 23–29, 55.

99. Lawson, *Running for Freedom*, 190–191; Bourne, *Jimmy Carter*, 30, 98–99, 115 (quotation), 173.

100. Kenneth E. Morris, *Jimmy Carter: American Moralist* (Athens: University of Georgia Press, 1996), 151–154; Robert Scheer, *Thinking Tuna Fish, Talking Death: Essays on the Pornography of Power* (New York: Farrar, Straus, and Giroux, 1988), 233–239; Dumbrell, *The Carter Presidency*, 86–87; Bourne, *Jimmy Carter*, 71, 94–97, 377–378.

101. Murphy and Gulliver, *The Southern Strategy*, 186–187; Burton I. Kaufman, *The Presidency of James Earl Carter, Jr.* (Lawrence: University Press of Kansas, 1993), 8–9; Bourne, *Jimmy Carter*, 182, 190–197.

102. Lawson, *Running for Freedom*, 191; Dumbrell, *The Carter Presidency*, 87 (quotation); Bourne, *Jimmy Carter*, 212–213.

103. Carter graduation address to Cheyney State College, Cheyney, Pennsylvania, 20 May 1979, Louis Martin Files, box 19, JECL; Lawson, *Running for Freedom*, 192–193.

104. Dumbrell, *The Carter Presidency*, 89.

105. Kaufman, *The Presidency of James Earl Carter, Jr.*, 13–14.

106. Bourne, *Jimmy Carter*, 88, 302, 344.

107. Ibid., 342.

108. Ibid., 340.

109. Ibid., 328, 354.

110. Brzezinski to Costanza et al., 10 May 1978, DPS-Eizenstat, box 208, JECL; Kaufman, *The Presidency of James Earl Carter, Jr.*, 38.

111. McClintock, *Instruments of Statecraft*, 312.

112. Bourne, *Jimmy Carter*, 384.

113. Carter, "U.S. Interest and Ideals," address to the World Affairs Council, Philadelphia, 9 May 1980, Louis Martin Files, box 3, JECL.

114. Dumbrell, *The Carter Presidency*, 194.

115. Carter speech to Congressional Black Caucus legislative dinner, Washington, 30 September 1978, DPS—Civil Rights and Justice—Malson, box 2, JECL.

116. Fallows to Carter, 8 July 1977, Staff Offices—Speechwriter—Subject File, box 27, JECL.

117. Carter address at Cheyney State College.

118. Remarks of the president with Coretta King and Dr. Jonas E. Salk, 11 July 1977, DPS—Civil Rights and Justice—Malson, box 2, JECL; M. Glenn Abernathy, "The Carter Administration and Domestic Civil Rights," in *The Carter*

Years: The President and Policy Making, ed. Abernathy, Dilys M. Hill, and Phil Williams (New York: St. Martin's, 1984), 106–122; Kaufman, *The Presidency of James Earl Carter, Jr.,* 26–31; Dumbrell, *The Carter Presidency,* 94–98.

119. Joseph A. Califano, Jr., *Governing America: An Insider's Report from the White House and the Cabinet* (New York: Simon and Schuster, 1981), 231–243; Kaufman, *The Presidency of James Earl Carter, Jr.,* 68–70; Dumbrell, *The Carter Presidency,* 92–94.

120. Dees to Carter, 12 July 1977, DPS-Eizenstat, box 164, JECL; Walter LaFeber, *The American Age: United States Foreign Policy at Home and Abroad since 1750* (New York: Norton, 1989), 648–649.

121. Gutierrez to Eizenstat, 11 May 1978, and Aragon memo to Jordan et al., 25 May 1978, DPS-Eizenstat, box 165, JECL.

122. Eleanor Randolph, "Blacks Wary—Is Carter Their Man?" *Chicago Tribune,* 28 February 1977, clipping in DPS—Civil Rights and Justice—Malson, box 2, JECL.

123. Bruce Shapiro, "A House Divided: Racism at the State Department," *Nation,* 12 February 1996, 11–16.

124. Adamson to Powell and Eizenstat, 12 July 1978, DPS-Eizenstat, box 165, JECL; "NAACP Leaders Say Carter Is Sympathetic to Cause," *Washington Star,* 12 December 1978, clipping in Louis Martin Files, box 68, JECL; Martin to Carter, 4 September 1979, Martin Files, box 57; Kaufman, *The Presidency of James Earl Carter, Jr.,* 109–111, 136.

125. Califano, *Governing America,* 230–231; Sitkoff, *The Struggle for Black Equality,* 214–215.

126. Lawson, *Running for Freedom,* 201–202.

127. Jacqueline Trescott, "The Survival of Andy Young," *Washington Post,* 8 August 1979, E1 (quotation); Cyrus Vance, "U.S. Relations with Africa," address to the U.S. Jaycees, Atlantic City, N.J., 20 June 1978, *State Department Bulletin,* August 1978, 10–14; Jackson, *From the Congo to Soweto,* 157; Shapiro, "A House Divided," 12–13; Dumbrell, *The Carter Presidency,* 103–104; Carl Gardner, *Andrew Young: A Biography* (New York: Drake Publishers, 1978).

128. Trescott, "The Survival of Andy Young"; Pranay B. Gupte, "Young Bids Africa Build U.S. Ties," *New York Times,* 7 September 1979; Leon Dash, "African Respect for Young Aids Trade Mission," *Washington Post,* 11 September 1979; "Andrew Young's Coup for Pullman-Kellogg," *Business Week,* 1 October 1979.

129. "Young's Statements—and Reactions to Them," *Washington Post,* 16 August 1979, A7.

130. Califano, *Governing America,* 431; Dumbrell, *The Carter Presidency,* 102–104.

131. Wayne King, "North Carolina's Leaders Worried by Blemishes on the State's

Image," *New York Times,* 22 February 1978, 1; Lauren, *Power and Prejudice,* 247; Kaufman, *The Presidency of James Earl Carter, Jr.,* 105.

132. Bernard Gwertzman, "Young Resigns Post at U.N. in Furor over P.L.O. Talks," *New York Times,* 16 August 1979, 1; Joseph Kraft, "Andrew Young's Transgression," *Washington Post,* 16 August 1979.

133. George W. Ball and Douglas B. Ball, *The Passionate Attachment: America's Involvement with Israel, 1947 to the Present* (New York: Norton, 1992), 96.

134. Smith, *Morality, Reason, and Power,* 10.

135. Peake, *Keeping the Dream Alive,* 369–373; DeConde, *Ethnicity, Race, and American Foreign Policy,* 180.

136. Douglas Brinkley, *The Unfinished Presidency: Jimmy Carter's Journey beyond the White House* (New York: Penguin, 1998); Bourne, *Jimmy Carter,* 491.

137. Smith, *Morality, Reason, and Power,* 135.

138. Ibid., 134, 144 (quotation), 147.

139. Gardner, *Andrew Young,* 211.

140. Congressional Black Caucus to Carter, 31 May 1978, "Congressional Black Caucus Denounces Cold War in Africa," 6 June 1978, and Congressional Black Caucus to Carter, 28 June 1978," DPS—Civil Rights and Justice—Malson, box 2, JECL; Houser, "Freedom's Struggle Crosses Oceans and Mountains," 196; Minter, *King Solomon's Mines Revisited,* 280.

141. Hooks to Carter, 21 September 1978, Martin Files, box 68, JECL; Brzezinski to Carter, 2 August 1979, WHCF—Subject File, box CO-54, JECL.

142. "Rhetoric and Reality," editorial, *New York Times,* 20 March 1977; Kaufman, *The Presidency of James Earl Carter, Jr.,* 91–92.

143. Moore, *Castro, the Blacks, and Africa,* xiv–xv, 330–331, 372–373.

144. Zbigniew Brzezinski, ed., *Africa and the Communist World* (Stanford: Hoover Institution Publications, Stanford University, 1963).

145. Minter, *King Solomon's Mines Revisited,* 282.

146. Coker, *The United States and South Africa,* 148–149.

147. "An Overview of US Policy toward Africa," December 1977, DPS-Eizenstat, box 208, JECL; Lauren, *Power and Prejudice,* 261.

148. Peter Osnos and Caryle Murphy, "Vorster Derides Carter for 'Selective Morality' in Africa," *Washington Post,* 21 April 1978, A23. Andrew J. DeRoche makes this argument forcefully in his "Standing Firm for Principles: Jimmy Carter and Zimbabwe," *Diplomatic History* 23 (Fall 1999): 657–685.

149. Tarnoff to Brzezinski, 3 June 1978, WHCF—Subject File, box CO-54, JECL; Patricia Derian, "Human Rights in South Africa," 13 May 1980, Martin Files, box 3, JECL.

150. Barber and Barratt, *South Africa's Foreign Policy,* 275; Gardner, *Andrew Young,* 128–132.

151. Greeley to Carter, 19 July 1977, WHCF—Subject File, box CO-54, JECL.

152. Califano, *Governing America,* 228.

153. Macfarlane to Cary, 20 December 1977, WHCF—Subject File, box CO-54, JECL.

154. Vance to all African diplomatic posts, 1 November 1977, in Mokoena, *South Africa and the United States,* 129–133; Coker, *The United States and South Africa,* 148–149.

155. Van Buuren to Carter, 12 February 1980, WHCF—Subject File, box CO-51, JECL; Hoppenstein address, "Southern Africa—What Course the Future?" 25 March 1977, DPS-Eizenstat, box 208, JECL.

156. Barber and Barratt, *South Africa's Foreign Policy,* 275.

157. Carter, remarks upon signing Rhodesian chrome legislation, 18 March 1977, Speechwriters—Chronological File, box 3, JECL; Smith, *Morality, Reason, and Power,* 140.

158. DeRoche, "Standing Firm for Principles," 667.

159. Stephen Low, "The Zimbabwe Settlement, 1976–1979," in *International Mediation in Theory and Practice,* ed. Saadia Touval and I. William Zartman (Boulder, Colo.: Westview, 1985), 91–109.

160. Baker et al. to Sithole, 14 September 1978, Hodding Carter to Moose, 13 October 1978, and Hodding Carter to Vance, 27 October 1978, WHCF—Subject File, box CO-51, JECL; Watson to Jackson, 23 October 1978, WHCF—Subject File, box CO-54, JECL; Smith, *Morality, Reason, and Power,* 141; Kaufman, *The Presidency of James Earl Carter, Jr.,* 91.

161. Dole to Carter, 27 December 1977, Derwinski et al. to Carter, 19 January 1978, and Hatch et al. to Carter, 22 February 1978, WHCF—Subject File, box CO-50, JECL.

162. DeRoche, "Standing Firm for Principles," 677; Barber and Barratt, *South Africa's Foreign Policy,* 267; Kaufman, *The Presidency of James Earl Carter, Jr.,* 157. For earlier discussions of these points, see Mondale to Hatfield [April 1978], WHCF—Subject File, box CO-50, JECL; Martin to Carter, 29 May 1979, Staff Offices Counsel—Lipshutz, box 44, JECL.

163. Harriman to Brzezinski, 22 April 1980, WHCF—Subject File, box CO-51, JECL.

164. "An Overview of US Policy toward Africa," memorandum for president's meeting with black leaders' forum, 14 December 1977, DPS-Eizenstat, box 208, JECL; McClintock, *Instruments of Statecraft,* 314–315; Smith, *Morality, Reason, and Power,* chap. 9.

165. Dumbrell, *The Carter Presidency,* 193.

166. Bourne, *Jimmy Carter,* 369; Dumbrell, *The Carter Presidency,* 91.

167. Bourne, *Jimmy Carter,* 454.

168. Carter, "U.S. Interests and Ideals," address to the World Affairs Council, Philadelphia, 9 May 1980, Martin Files, box 3, JECL.

169. William E. Pemberton, *Exit with Honor: The Life and Presidency of Ronald Reagan* (Armonk: M. E. Sharpe, 1998), 138–139.

170. Bourne, *Jimmy Carter,* 463.

171. Edsall and Edsall, *Chain Reaction,* 10.

172. Sean Wilentz, "Color-Blinded," *New Yorker,* 2 October 1995, 91; Omi and Winant, *Racial Formation in the United States,* 132–134.

173. "Major Klan Group Gives Reagan Its Endorsement," *New York Times,* 1 August 1980, clipping in Martin Files, box 3, JECL. When Carter launched his own reelection campaign on 1 September in Tuscumbia, Alabama, with a speech appealing for racial harmony, hooded Klan members in the crowd held up "Reagan for President" signs. Bourne, *Jimmy Carter,* 463.

174. Adam Fairclough, "Martin Luther King, Jr., and the War in Vietnam," *Phylon* 45 (March 1984): 39.

175. Pemberton, *Exit with Honor,* 138.

176. Shapiro, "A House Divided," 13–14.

177. Lawson, *Running for Freedom,* 204–212; O'Reilly, *Nixon's Piano,* 355–378.

178. Schraeder, *United States Foreign Policy toward Africa,* 273, n. 73. One powerful ally of the administration's "far right hand" was Senator Jesse Helms, who on occasion found the White House too tough on racists. For nine months he blocked Chester Crocker's nomination as assistant secretary of state for African affairs because of Crocker's frank personal condemnation of apartheid, despite Crocker's design of the "constructive engagement" policy for drawing closer to the Pretoria regime. Coker, *The United States and South Africa,* 160.

179. Massie, *Loosing the Bonds,* 485.

180. Houser, "Freedom's Struggle Crosses Oceans and Mountains," 271, n. 44; Minter, *King Solomon's Mines Revisited,* 305–341.

181. Schraeder, *United States Foreign Policy toward Africa,* 34.

182. Massie, *Loosing the Bonds,* 616.

183. Moose statement to subcommittee on Africa of House Foreign Affairs Committee, 30 April 1980, Martin Files, box 3, JECL.

184. Lauren, *Power and Prejudice,* 262.

185. Ibid., 274.

186. Coker, *The United States and South Africa,* 227.

187. Massie, *Loosing the Bonds,* 506.

188. Rodman, *More Precious than Peace,* 364–365; Barber and Barratt, *South Africa's Foreign Policy,* 313–314.

189. Michael G. Schatzberg, *Mobutu or Chaos? The United States and Zaire, 1960–1990* (Lanham, Md.: University Press of America, 1991), 87.

190. Moore, *Castro, the Blacks, and Africa,* 348.

191. *New York Times* reporter Joseph Lelyveld won the Pulitzer Prize for his 1985 book *Move Your Shadow;* Hollywood's *Cry, Freedom* (1987) and *A World Apart* (1988) were widely viewed; and singer Paul Simon's album *Graceland* (1986), a collaboration with Ladysmith Black Mombazo, introduced millions of Americans to elements of black South African music.

192. Barber and Barratt, *South Africa's Foreign Policy,* 303–307; Coker, *The United States and South Africa,* 174–175; Lauren, *Power and Prejudice,* 282; Massie, *Loosing the Bonds,* 523–671; Donald R. Culverson, *Contesting Apartheid: U.S. Activism, 1960–87* (Boulder, Colo.: Westview, 1999).

193. Massie, *Loosing the Bonds,* 560–561.

194. Coker, *The United States and South Africa,* 199–200.

195. Massie, *Loosing the Bonds,* 587.

196. Ibid., 613.

197. O'Reilly, *Nixon's Piano,* 376.

198. Carter, *From George Wallace to Newt Gingrich,* xi–xiii.

199. O'Reilly, *Nixon's Piano,* 378–391.

200. Omi and Winant, *Racial Formation in the United States,* 146.

201. O'Reilly, *Nixon's Piano,* 391–399.

202. Henry Louis Gates, Jr., "Powell and the Political Elite," *New Yorker,* 25 September 1995, 68 (quotation); Richard L. Berke, "President Powell? Not Just If, But Why?" *New York Times,* 30 July 1995, sec. 4, p. 1; DeConde, *Ethnicity, Race, and American Foreign Policy,* 181.

203. Allister Sparks, *Tomorrow Is Another Country: The Inside Story of South Africa's Road to Change* (New York: Hill and Wang, 1995), 10.

204. LaFeber, *The American Age,* 752–760.

Epilogue

1. Kevin Sack, "In Little Rock, Clinton Warns of Racial Split," *New York Times,* 26 September 1997, 20; Katharine Q. Seelye, "Blacks Stand by the President in His Time of Need," ibid., 16 February 1998, 11 (Lewis); Steven A. Holmes, "On Civil Rights, Clinton Steers Bumpy Course between Right and Left," ibid., 20 October 1996, 16.

2. Elizabeth Olson, "U.N. Report Criticizes U.S. for 'Racist' Use of Death Penalty," ibid., 7 April 1998, 17.

3. William J. Broad, "Spies vs. Sweat: The Debate over China's Nuclear Arsenal," ibid., 7 September 1999, 1.

4. James Dao, "At Missouri Base, Clinton Hails Power of Weaponry and Diversity," ibid., 12 June 1999, 7.

5. Lauren, *Power and Prejudice,* 275.

6. "Rising Tide of Color" (editorial), *Crisis,* May 1960, 306–307.

7. Quoted in Sherry, *In the Shadow of War,* 208.

8. *New York Times,* 24 September 1998, A20.

9. Missouri qualifies most accurately as a border state, but its slaveholding traditions were strong, especially in Truman's own family of origin.

ARCHIVES AND MANUSCRIPT COLLECTIONS

James E. Carter, Jr., Library, Atlanta, Georgia
    Jimmy Carter Presidential Papers
    White House Central Files
    Staff Office Files
    Domestic Policy Staff
    White House Office of Counsel to the President
    Office of the Chief of Staff
    Presidential Speechwriters
    Special Assistant for Black Affairs (Louis Martin)
    Special Assistant for Hispanic Affairs (Esteban Torres)
    Assistant for Communications (Gerald Rafshoon)
    Records of the Cuban-Haitian Task Force

Dwight D. Eisenhower Library, Abilene, Kansas
    Dwight D. Eisenhower Presidential Papers
    Ann Whitman File
    Administration Series
    ACW Diary Series
    Cabinet Series
    Name Series
    Dulles-Herter Series
    NSC Series
    White House Central Files
    Confidential File
    Official File
    General File
    Alphabetical File

President's Personal File
White House Office, Telegraph Office Records
White House Office, Office of the Special Assistant for National Security Affairs
    Records
John Foster Dulles Papers
General Correspondence and Memoranda Series
Special Assistants Chronological Series
Telephone Conversation Series
White House Memoranda Series
Christian A. Herter Papers
Bryce Harlow Records
D. Jackson Papers
Frederic Morrow Records
Wilton E. Persons Records
William P. Rogers Papers
Alven C. Gillem, Jr., Papers

Lyndon B. Johnson Library, Austin, Texas
    Lyndon B. Johnson Presidential Papers
    National Security Files
    Office Files of the White House Aides
    Special Files
    White House Central Files
    George Ball Papers
    Warren Christopher Papers
    Ramsey Clark Papers
    John Roche Papers
    Oral Histories
    Ramsey Clark
    Clifford and Virginia Durr
    James O. Eastland
    Allen J. Ellender
    Charles Evers
    James Farmer
    John A. Hannah
    Aaron E. Henry
    Burke Marshall
    Thurgood Marshall
    Louis Martin
    Joseph Palmer, II

Philip Randolph
John P. Roche
Terry Sanford
John E. Stennis
Strom Thurmond
Lee White
Roy Wilkins
Mennen Williams
Andrew J. Young, Jr.

John F. Kennedy Library, Boston, Massachusetts
John F. Kennedy Presidential Papers
President's Office Files
Subject Series
Country Series
National Security Files
White House Central Files
White House Staff Files
McGeorge Bundy Files
Harris Wofford Files
Lee White Files
Wayne Fredericks Papers
Nicholas deB. Katzenbach Papers
Burke Marshall Papers

Richard M. Nixon Presidential Materials, National Archives, College Park,
    Maryland
    White House Special Files (from White House Central Files and Staff Member
    Office Files)
Charles W. Colson Files
John W. Dean Files
Harry Dent Files
John D. Ehrlichman Files
H. R. Haldeman Files
Egil M. Krogh Files
President's Office Files
President's Personal Files
White House Central Files—Subject Files
Country Files
Commission on Civil Rights Files

Foreign Affairs Files
Human Rights Files

Harry S. Truman Library, Independence, Missouri
Harry S. Truman Presidential Papers
President's Secretary's Files
Intelligence File
Political File
Subject File
White House Central Files
Confidential File
Official File
Dean Acheson Papers
Eleanor Bontecou Papers
Tom C. Clark Papers
Clark M. Clifford Files
Charles Fahy Papers
Philleo Nash Files
Philleo Nash Papers
Naval Aide to the President Office Files
David Niles Papers
U.S. President's Committee on Civil Rights Records
U.S. President's Committee on Equality of Treatment and Opportunity in the
    Armed Services Records
E. W. Kenworthy Oral History

INDEX